STUDENT'S
SOLUTIONS MANUAL

BEVERLY FUSFIELD

TRIGONOMETRY
ELEVENTH EDITION

Margaret L. Lial
American River College

John Hornsby
University of New Orleans

David I. Schneider
University of Maryland

Callie J. Daniels
St. Charles Community College

PEARSON

Boston Columbus Indianapolis New York San Francisco
Amsterdam Cape Town Dubai London Madrid Milan Munich Paris Montreal Toronto
Delhi Mexico City Sao Paulo Sydney Hong Kong Seoul Singapore Taipei Tokyo

ISBN-13: 978-0-13-431021-3
ISBN-10: 0-13-431021-7

www.pearsonhighered.com

PEARSON

CONTENTS

Appendices

Chapter 1

TRIGONOMETRIC FUNCTIONS

Section 1.1   Angles

1. One degree, written $1°$, represents $\frac{1}{360}$ of a complete rotation.

3. If the measure of an angle is $x°$, its supplement can be expressed as $\underline{180° - x°}$.

5. The measure of an angle that is its own supplement is $\underline{90°}$.

7. One second, written $1''$, is $\frac{1}{60}$ of a minute.

9. $55.25°$ written in degrees and minutes is $\underline{55°15'}$.

11. $30°$
   (a) $90° - 30° = 60°$
   (b) $180° - 30° = 150°$

13. $45°$
   (a) $90° - 45° = 45°$
   (b) $180° - 45° = 135°$

15. $54°$
   (a) $90° - 54° = 36°$
   (b) $180° - 54° = 126°$

17. $1°$
   (a) $90° - 1° = 89°$
   (b) $180° - 1° = 179°$

19. $14°20'$
   (a) $90° - 14°20' = 89°60' - 14°20' = 75°40'$
   (b) $180° - 14°20' = 179°60' - 14°20'$
   $\qquad = 165°40'$

21. $20°10'30''$
   (a) $90° - 20°10'30'' = 89°59'60'' - 20°10'30''$
   $\qquad = 69°49'30''$
   (b) $180° - 20°10'30''$
   $\qquad = 179°59'60'' - 20°10'30''$
   $\qquad = 159°49'30''$

23. The two angles form a straight angle.
   $7x + 11x = 180 \Rightarrow 18x = 180 \Rightarrow x = 10$
   The measures of the two angles are
   $(7x)° = \left[7(10)\right]° = 70°$ and
   $(11x)° = \left[11(10)\right]° = 110°$.

25. The two angles form a right angle.
   $4x + 2x = 90 \Rightarrow 6x = 90 \Rightarrow x = 15$
   The two angles have measures of
   $(4x)° = \left[4(15)\right]° = 60°$ and
   $(2x)° = \left[2(15)\right]° = 30°$.

27. The two angles form a straight angle.
   $(-4x) + (-14x) = 180 \Rightarrow -18x = 180 \Rightarrow$
   $x = -10$
   The measures of the two angles are
   $(-4x)° = \left[-4(-10)\right]° = 40°$ and
   $(-14x)° = \left[-14(-10)\right]° = 140°$.

29. The sum of the measures of two supplementary angles is $180°$.
   $(10x + 7) + (7x + 3) = 180$
   $17x + 10 = 180$
   $17x = 170 \Rightarrow x = 10$
   The measures of the two angles are
   $(10x + 7)° = \left[10(10) + 7\right]° = (100 + 7)° = 107°$
   and $(7x + 3)° = \left[7(10) + 3\right]° = (70 + 3)° = 73°$.

31. The sum of the measures of two complementary angles is $90°$.
   $(9x + 6) + 3x = 90 \Rightarrow 12x + 6 = 90 \Rightarrow$
   $12x = 84 \Rightarrow x = 7$
   The measures of the two angles are
   $(9x + 6)° = \left[9(7) + 6\right]° = (63 + 6)° = 69°$ and
   $(3x)° = \left[3(7)\right]° = 21°$.

33. $\dfrac{25 \text{ minutes}}{60 \text{ minutes}} = \dfrac{x}{360°}$
   $$x = \frac{25}{60}(360) = 25(6) = 150°$$

35. At 15 minutes after the hour, the minute hand is $\frac{1}{4}$ the way around, so the hour hand is $\frac{1}{4}$ of the way between the 3 and 4. Thus, the hour hand is located 16.25 minutes past 12. The minute hand is 15 minutes after the 12. The smaller angle formed by the hands of the clock can be found by solving the proportion
$$\frac{(16.25-15)\text{ minutes}}{60\text{ minutes}} = \frac{x}{360°}.$$

$$\frac{(16.25-15)\text{ minutes}}{60\text{ minutes}} = \frac{x}{360°} \Rightarrow \frac{1.25}{60} = \frac{x}{360} \Rightarrow$$

$$x = \frac{1.25}{60}(360) = 1.25(6) = 7.5° = 7°30'$$

37. At 20 minutes after the hour, the minute hand is $\frac{1}{3}$ the way around, so the hour hand is $\frac{1}{3}$ of the way between the 8 and 9. Thus, the hour hand is located $41\frac{2}{3}$ minutes past 12. The minute hand is 20 minutes after the 12. The smaller angle formed by the hands of the clock can be found by solving the proportion
$$\frac{\left(41\frac{2}{3}-20\right)\text{ minutes}}{60\text{ minutes}} = \frac{x}{360°}.$$

$$\frac{\left(41\frac{2}{3}-20\right)\text{ minutes}}{60\text{ minutes}} = \frac{x}{360°} \Rightarrow \frac{21\frac{2}{3}}{60} = \frac{x}{360} \Rightarrow$$

$$x = \frac{21\frac{2}{3}}{60}(360) = \left(21\frac{2}{3}\right)(6) = 130°$$

39.
$$\begin{array}{r} 62°\ 18' \\ +21°\ 41' \\ \hline 83°\ 59' \end{array}$$

41. $97°42' + 81°37' = 178°79' = 179°19'$

43. $47°\ 29' - 71°18' = -\left(71°18' - 47°\ 29'\right)$
$$= -\left(70°\ 78' - 47°\ 29'\right)$$
$$= -23°\ 49'$$

45. $90° - 51°\ 28' = 89°\ 60' - 51°\ 28' = 38°32'$

47. $180° - 119°\ 26' = 179°\ 60' - 119°\ 26' = 60°\ 34'$

49. $90° - 72°\ 58'\ 11'' = 89°\ 59'\ 60'' - 72°\ 58'\ 11''$
$$= 17°01'\ 49''$$

51. $26°20' + 18°17' - 14°10' = 44°37' - 14°10'$
$$= 30°27'$$

53. $35°30' = 35° + \frac{30}{60}° = 35° + 0.5° = 35.5°$

55. $112°15' = 112° + \frac{15}{60}° = 112° + 0.25° = 112.25°$

57. $-60°12' = -\left(60° + \frac{12}{60}°\right) = -\left(60° + 0.2°\right)$
$$= -60.2°$$

59. $20°54'36'' = 20° + \frac{54}{60}° + \frac{36}{3600}°$
$$= 20° + 0.900° + 0.01° = 20.91°$$

61. $91°35'\ 54'' = 91° + \frac{35}{60}° + \frac{54}{3600}°$
$$\approx 91° + 0.5833° + 0.0150°$$
$$\approx 91.598°$$

63. $274°18'\ 59'' = 274° + \frac{18}{60}° + \frac{59}{3600}°$
$$\approx 274° + 0.3000° + 0.0164°$$
$$\approx 274.316°$$

65. $39.25° = 39° + 0.25° = 39° + 0.25(60')$
$$= 39° + 15' + 0'' = 39°15'00''$$

67. $126.76° = 126° + 0.76° = 126° + 0.76(60')$
$$= 126° + 45.6' = 126° + 45' + 0.6'$$
$$= 126° + 45' + 0.6(60'')$$
$$= 126° + 45' + 36'' = 126°45'36''$$

69. $-18.515° = -\left(18° + 0.515°\right)$
$$= -\left(18° + 0.515(60')\right)$$
$$= -\left(18° + 30.9'\right) = -\left(18° + 30' + 0.9'\right)$$
$$= -\left(18° + 30' + 0.9(60'')\right)$$
$$= -\left(18° + 30' + 54''\right) = -18°30'54''$$

71. $31.4296° = 31° + 0.4296° = 31° + 0.4296(60')$
$$= 31° + 25.776' = 31° + 25' + 0.776'$$
$$= 31° + 25' + 0.776(60'')$$
$$= 31°25'\ 46.56'' \approx 31°\ 25'\ 47''$$

73. $89.9004° = 89° + 0.9004° = 89° + 0.9004(60')$
$$= 89° + 54.024' = 89° + 54' + 0.024'$$
$$= 89° + 54' + 0.024(60'')$$
$$= 89°54'1.44'' \approx 89°54'01''$$

75. $178.5994° = 178° + 0.5994°$
$$= 178° + 0.5994(60')$$
$$= 178° + 35.964'$$
$$= 178° + 35' + 0.964'$$
$$= 178° + 35' + 0.964(60'')$$
$$= 178°35'\ 57.84'' \approx 178°35'\ 58''$$

77. $32°$ is coterminal with $360° + 32° = 392°$.

79. $26°30'$ is coterminal with $360° + 26°30' = 386°30'$.

81. $-40°$ is coterminal with $360° + (-40°) = 320°$

83. $-125°30'$ is coterminal with $360° + (-125°30') = 359°60' - 125°30' = 234°30'$.

85. 361° is coterminal with 361° − 360° = 1°.

87. −361° is coterminal with
−361° + 2(360°) = 359°.

89. 539° is coterminal with 539° − 360° = 179°.

91. 850° is coterminal with
850° − 2(360°) = 850° − 720° = 130°.

93. 5280° is coterminal with
5280° − 14 · 360° = 5280° − 5040° = 240°.

95. −5280° is coterminal with
−5280° + 15 · 360° = −5280° + 5400° = 120°.

In exercises 97−99, answers may vary.

97. 90° is coterminal with
$$90° + 360° = 450°$$
$$90° + 2(360°) = 810°$$
$$90° − 360° = −270°$$
$$90° − 2(360°) = −630°$$

99. 0° is coterminal with
$$0° + 360° = 360°$$
$$0° + 2(360°) = 720°$$
$$0° − 360° = −360°$$
$$0° − 2(360°) = −720°$$

101. 30°
A coterminal angle can be obtained by adding
an integer multiple of 360°.
$$30° + n · 360°$$

103. 135°
A coterminal angle can be obtained by adding
an integer multiple of 360°.
$$135° + n · 360°$$

105. −90°
A coterminal angle can be obtained by adding
an integer multiple of 360°.
$$−90° + n · 360°$$

107. 0°
A coterminal angle can be obtained by adding
integer multiple of 360°.
$$0° + n · 360° = n · 360°$$

109. The answers to Exercises 107 and 108 give the
same set of angles because 0° is coterminal
with 360°.

For Exercises 111−121, other answers are possible.

111.

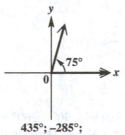

**435°; −285°;
quadrant I**

75° is coterminal with 75° + 360° = 435°
and 75° − 360° = −285°. These angles are in
quadrant I.

113.

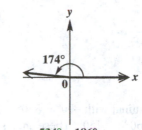

**534°; −186°;
quadrant II**

174° is coterminal with 174° + 360° = 534°
and 174° − 360° = −186°. These angles are
in quadrant II.

115.

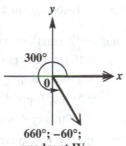

**660°; −60°;
quadrant IV**

300° is coterminal with 300° + 360° = 660°
and 300° − 360° = −60°. These angles are in
quadrant IV.

117.

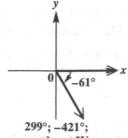

**299°; −421°;
quadrant IV**

−61° is coterminal with −61° + 360° = 299°
and −61° − 360° = −421°. These angles are in
quadrant IV.

119.

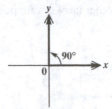

450°; –270°;
no quadrant

90° is coterminal with $90° + 360° = 450°$
and $90° - 360° = -270°$. These angles are
not in a quadrant.

121.

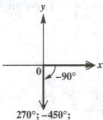

270°; –450°;
no quadrant

–90° is coterminal with $-90° + 360° = 270°$
and $-90° - 360° = -450°$. These angles are
not in a quadrant.

123. 45 revolutions per min $= \frac{45}{60}$ revolution per sec

$= \frac{3}{4}$ revolution per sec

A turntable will make $\frac{3}{4}$ revolution in 1 sec.

125. 600 rotations per min $= \frac{600}{60}$ rotations per sec

$= 10$ rotations per sec

$= 5$ rotations per $\frac{1}{2}$ sec

$= 5(360°)$ per $\frac{1}{2}$ sec

$= 1800°$ per $\frac{1}{2}$ sec

A point on the edge of the tire will move
1800° in $\frac{1}{2}$ sec.

127. 75° per min $= 75°(60)$ per hr $= 4500°$ per hr

$= \frac{4500°}{360°}$ rotations per hr

$= 12.5$ rotations per hr

The pulley makes 12.5 rotations in 1 hour.

129. Earth rotates 360° in 24 hr. 360° is equal to
$360(60') = 21,600'$.

$$\frac{24 \, \text{hr}}{21,600'} = \frac{x}{1'}$$

$$x = \frac{24}{21,600} \text{hr} = \frac{24}{21,600}(60 \, \text{min})$$

$$= \frac{1}{15} \text{min} = \frac{1}{15}(60 \, \text{sec}) = 4 \, \text{sec}$$

It should take the motor 4 sec to rotate the
telescope through an angle of 1 min.

Section 1.2   Angle Relationships and
Similar Triangles

1. The sum of the measures of the angles of any
triangle is <u>180°</u>.

3. An equilateral triangle has <u>three</u> equal sides.

5. In the figure, there are two parallel lines and a
transversal, so the measures of angles 1, 2, 5
and 6 are all the same. Also, the measures of
the angle marked 131° and angles 3, 4, and 7
are the same. Angle 1 is supplementary to the
angle marked 131°, so the measure of angle 1
is 49°, as are the measures of angles 2, 5, and
6.

7. Corresponding angles are A and P, B and Q, C
and R. Corresponding sides are AC and PR,
BC and QR, AB and PQ.

9. Corresponding angles are A and C, E and D,
ABE and CBD. Corresponding sides are EB
and DB, AB and CB, AE and CD

11. The two indicated angles are vertical angles,
so their measures are equal.
$5x - 129 = 2x - 21 \Rightarrow 3x = 108 \Rightarrow x = 36$
$5(36) - 129 = 51$ and $2(36) - 21 = 51$, so both
angles measure 51°.

13. The three angles are the interior angles of a
triangle, so the sum of their measures is 180°.
$x + (x + 20) + (210 - 3x) = 180 \Rightarrow$
$230 - x = 180 \Rightarrow x = 50$
$50 + 20 = 70$ and $210 - 3(50) = 60$, so the
three angles measure 50°, 60°, and 70°.

15. The three angles are the interior angles of a
triangle, so the sum of their measures is 180°.
$$(2x - 120) + \left(\tfrac{1}{2}x + 15\right) + (x - 30) = 180$$
$$\tfrac{7}{2}x - 135 = 180$$
$$\tfrac{7}{2}x = 315$$
$$x = 90$$
$2(90) - 120 = 60, \tfrac{1}{2}(90) + 15 = 60,$ and
$90 - 30 = 60$, so the three angles each measure
60°.

17. In a triangle, the measure of an exterior angle
equals the sum of the measures of the nonadjacent interior angles. Thus,
$$(6x + 3) + (4x - 3) = 9x + 12$$
$$10x = 9x + 12 \Rightarrow x = 12$$
$6(12) + 3 = 75, 4(12) - 3 = 45,$ and
$9(12) + 12 = 120$, so the three angles measure
45°, 75°, and 120°.

19. The two angles are alternate interior angles, so their measures are equal.
    $2x - 5 = x + 22 \Rightarrow x = 27$
    $2(27) - 5 = 49$ and $27 + 22 = 49$, so both angles measure 49°.

21. The two angles are interior angles on the same side of the transversal, so the sum of their measures is 180°.
    $(x + 1) + (4x - 56) = 180 \Rightarrow 5x - 55 = 180 \Rightarrow$
    $5x = 235 \Rightarrow x = 47$
    $47 + 1 = 48$ and $4(47) - 56 = 132$, so the angles measure 48° and 132°.

23. Let x = the measure of the third angle. Then
    $37 + 52 + x = 180 \Rightarrow x = 91$
    The third angle of the triangle measures 91°.

25. Let x = the measure of the third angle. Then
    $147°12' + 30°19' + x = 180°$
    $177°31' + x = 180°$
    $177°31' + x = 179°60' \Rightarrow x = 2°29'$
    The third angle of the triangle measures $2°29'$.

27. Let x = the measure of the third angle. Then
    $74.2° + 80.4° + x = 180° \Rightarrow x = 25.4°$
    The third angle of the triangle measures 25.4°.

29. Let x = the measure of the third angle. Then
    $51°20'14'' + 106°10'12'' + x = 180°$
    $157°30'26'' + x = 180°$
    $157°30'26'' + x = 179°59'60''$
    $x = 22°29'34''$
    The third angle of the triangle measures $22°29'34''$.

31. No, a triangle cannot have angles of measures 85° and 100°. The sum of the measures of these two angles is 85° + 100° = 185°, which exceeds 180°.

33. The triangle has a right angle, but each side has a different measure. The triangle is a right triangle and a scalene triangle.

35. The triangle has three acute angles and three equal sides, so it is acute and equilateral.

37. The triangle has a right angle and three unequal sides, so it is right and scalene.

39. The triangle has a right angle and two equal sides, so it is right and isosceles.

41. The triangle has one obtuse angle and three unequal sides, so it is obtuse and scalene.

43. The triangle has three acute angles and two equal sides, so it is acute and isosceles.

45. Angles 1, 2, and 3 form a straight angle on line m and, therefore, sum to 180°. It follows that the sum of the measures of the angles of triangle PQR is 180°, because the angles marked 1 are alternate interior angles whose measures are equal, as are the angles marked 2.

47. Angle Q corresponds to angle A, so the measure of angle Q is 42°. Angles A, B, and C are interior angles of a triangle, so the sum of their measures is 180°.
    $$m\angle A + m\angle B + m\angle C = 180°$$
    $$42° + m\angle B + 90° = 180°$$
    $$132° + m\angle B = 180°$$
    $$m\angle B = 48°$$
    Angle R corresponds to angle B, so the measure of angle R is 48°.

49. Angle B corresponds to angle K, so the measure of angle B is 106°. Angles A, B, and C are interior angles of a triangle, so the sum of their measures is 180°.
    $$m\angle A + m\angle B + m\angle C = 180°$$
    $$m\angle A + 106° + 30° = 180°$$
    $$m\angle A + 136° = 180°$$
    $$m\angle A = 44°$$
    Angle M corresponds to angle A, so the measure of angle M is 44°.

51. Angles X, Y, and Z are interior angles of a triangle, so the sum of their measures is 180°.
    $$m\angle X + m\angle Y + m\angle Z = 180°$$
    $$m\angle X + 90° + 38° = 180°$$
    $$m\angle X + 128° = 180°$$
    $$m\angle X = 52°$$
    Angle M corresponds to angle X, so the measure of angle M is 52°.

In Exercises 53−57, corresponding sides of similar triangles are proportional. Other proportions are possible in solving these exercises.

53. $\dfrac{25}{10} = \dfrac{a}{8} \Rightarrow 8(25) = 10a \Rightarrow 200 = 10a \Rightarrow 20 = a$
    $\dfrac{25}{10} = \dfrac{b}{6} \Rightarrow 6(25) = 10b \Rightarrow 150 = 10b \Rightarrow 15 = b$

55. $\dfrac{6}{12} = \dfrac{a}{12} \Rightarrow a = 6$
    $\dfrac{6}{12} = \dfrac{b}{15} \Rightarrow \dfrac{1}{2} = \dfrac{b}{15} \Rightarrow b = \dfrac{15}{2} = 7\dfrac{1}{2}$

57. $\dfrac{4}{6} = \dfrac{x}{9} \Rightarrow 6x = 36 \Rightarrow x = 6$

59. Let x = the height of the tree.
The triangle formed by the tree and its shadow is similar to the triangle formed by the stick and its shadow.

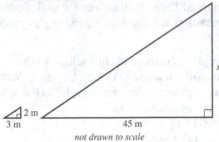

not drawn to scale

$\dfrac{x}{2} = \dfrac{45}{3} \Rightarrow \dfrac{x}{2} = \dfrac{15}{1} \Rightarrow x = 30$
The tree is 30 m high.

61. Let x = the middle side of the actual triangle (in meters); y = the longest side of the actual triangle (in meters).
The triangles in the photograph and the piece of land are similar. The shortest side on the land corresponds to the shortest side on the photograph.
$\dfrac{400 \text{ m}}{4 \text{ cm}} = \dfrac{x}{5 \text{ cm}} \Rightarrow \dfrac{100}{1} = \dfrac{x}{5} \Rightarrow x = 500$ and
$\dfrac{400 \text{ m}}{4 \text{ cm}} = \dfrac{y}{7 \text{ cm}} \Rightarrow \dfrac{100}{1} = \dfrac{y}{7} \Rightarrow y = 700$
The other two sides are 500 m and 700 m long.

63. Let x = the height of the building.
The triangle formed by the house and its shadow is similar to the triangle formed by the building and its shadow.

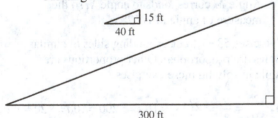

$\dfrac{15}{40} = \dfrac{x}{300} \Rightarrow 40x = 4500 \Rightarrow x = 112.5$
The building is 112.5 ft tall.

65. $\dfrac{x}{50} = \dfrac{100+120}{100} \Rightarrow \dfrac{x}{50} = \dfrac{220}{100} \Rightarrow \dfrac{x}{50} = \dfrac{11}{5} \Rightarrow$
$5x = 550 \Rightarrow x = 110$

67. $\dfrac{c}{100} = \dfrac{10+90}{90} \Rightarrow \dfrac{c}{100} = \dfrac{100}{90} \Rightarrow \dfrac{c}{100} = \dfrac{10}{9} \Rightarrow$
$9c = 1000 \Rightarrow c = \dfrac{1000}{9} \approx 111.1$

69. (a) Let D_s = the distance from the Earth to the sun; d_s = the diameter of the sun, D_m = the distance from the Earth to the moon; and d_m = the diameter of the moon.

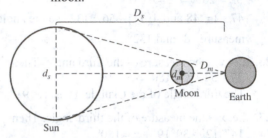

$\dfrac{D_s}{D_m} = \dfrac{d_s}{d_m}$
$\dfrac{94,500,000}{D_m} = \dfrac{865,000}{2159}$
$865,000 D_m = 94,500,000 \cdot 2159$
$D_m = \dfrac{94,500,000 \cdot 2159}{865,000}$
$\approx 236,000 \text{ mi}$

(b) No, a total solar eclipse cannot occur every time. The moon must be less than 236,000 miles away from Earth for an eclipse to occur, and sometimes it is farther than this.

71. (a) Let D_s = the distance from Mars to the sun; d_s = the diameter of the sun, D_m = the distance from Mars to Phobos; and d_m = the diameter of Phobos.
$\dfrac{D_s}{D_m} = \dfrac{d_s}{d_m}$
$\dfrac{142,000,000}{D_m} = \dfrac{865,000}{17.4}$
$865,000 D_m = 142,000,000 \cdot 17.4$
$D_m = \dfrac{142,000,000 \cdot 17.4}{865,000}$
$\approx 2900 \text{ mi}$

(b) No, Phobos does not come close enough to the surface of Mars.

73. (a) The thumb covers about 2 arc degrees or about 120 arc minutes. This is $\frac{31}{120}$ or approximately $\frac{1}{4}$ of the thumb would cover the moon.

(b) $20° + 10° = 30°$
The stars are 30 arc degrees apart.

Chapter 1 Quiz
(Sections 1.1−1.2)

1. 19°

(a) complement: $90° − 19° = 71°$

(b) supplement: $180° − 19° = 161°$

3. The two angles form a right angle.
$(5x − 1) + 2x = 90 \Rightarrow 7x − 1 = 90 \Rightarrow$
$7x = 91 \Rightarrow x = 13$
The measures of the two angles are
$(5x − 1)° = [5(13) − 1] = 64°$ and
$2x = 2(13) = 26°$.

5. The two marked angles are supplements, so their sum is 180°.
$(−14x + 18) + (−6x + 2) = 180$
$-20x + 20 = 180$
$-20x = 160 \Rightarrow x = −8$
The measures of the angles are
$−14(−8) + 18 = 130°$ and $−6(−8) + 2 = 50°$.

7. (a) 410° is coterminal with
$410° − 360° = 50°$.

(b) −60° is coterminal with
$−60° + 360° = 300°$.

(c) 890° is coterminal with
$890° − 2(360°) = 890° − 720° = 170°$.

(d) 57° is coterminal with $57° + 360° = 417°$.

9.

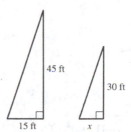

Using similar triangles, we have
$\frac{45}{15} = \frac{30}{x} \Rightarrow \frac{3}{1} = \frac{30}{x} \Rightarrow 3x = 30 \Rightarrow x = 10$.
The pole's shadow is 10 ft.

Section 1.3   Trigonometric Functions

1. The Pythagorean theorem for right triangles states that the sum of the squares of the lengths of the legs is equal to the square of the <u>hypotenuse</u>.

3. For any nonquadrantal angle θ, $\sin\theta$ and $\csc\theta$ will have the <u>same</u> sign.

5. If the terminal side of an angle θ lies in quadrant III, then the value of $\tan\theta$ and $\cot\theta$ are <u>positive</u>, and all other trigonometric function values are <u>negative</u>.

7. $r = \sqrt{(−3)^2 + (−3)^2} = \sqrt{18} = 3\sqrt{2}$

9. $\cos\theta = \frac{x}{r} = \frac{−3}{3\sqrt{2}} = −\frac{1}{\sqrt{2}} = −\frac{\sqrt{2}}{2}$

11.

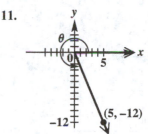

$x = 5, y = −12,$ and
$r = \sqrt{x^2 + y^2} = \sqrt{(−12)^2 + 5^2}$
$= \sqrt{144 + 25} = \sqrt{169} = 13$

$\sin\theta = \frac{y}{r} = \frac{−12}{13} = −\frac{12}{13}$

$\cos\theta = \frac{x}{r} = \frac{5}{13}$

$\tan\theta = \frac{y}{x} = \frac{−12}{5} = −\frac{12}{5}$

$\cot\theta = \frac{x}{y} = \frac{5}{−12} = −\frac{5}{12}$

$\sec\theta = \frac{r}{x} = \frac{13}{5};\ \csc\theta = \frac{r}{y} = \frac{13}{−12} = −\frac{13}{12}$

13.

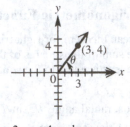

$x = 3$, $y = 4$ and

$$r = \sqrt{x^2 + y^2} = \sqrt{3^2 + 4^2} = \sqrt{25} = 5$$

$$\sin\theta = \frac{y}{r} = \frac{4}{5}; \quad \cos\theta = \frac{x}{r} = \frac{3}{5}$$

$$\tan\theta = \frac{y}{x} = \frac{4}{3}; \quad \cot\theta = \frac{x}{y} = \frac{3}{4}$$

$$\sec\theta = \frac{r}{x} = \frac{5}{3}; \quad \csc\theta = \frac{r}{y} = \frac{5}{4}$$

15.

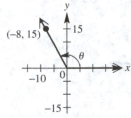

$x = -8$, $y = 15$, and

$$r = \sqrt{x^2 + y^2} = \sqrt{(-8)^2 + 15^2}$$
$$= \sqrt{64 + 225} = \sqrt{289} = 17$$

$$\sin\theta = \frac{y}{r} = \frac{15}{17}$$

$$\cos\theta = \frac{x}{r} = \frac{-8}{17} = -\frac{8}{17}$$

$$\tan\theta = \frac{y}{x} = \frac{15}{-8} = -\frac{15}{8}$$

$$\cot\theta = \frac{x}{y} = \frac{-8}{15} = -\frac{8}{15}$$

$$\sec\theta = \frac{r}{x} = \frac{17}{-8} = -\frac{17}{8}$$

$$\csc\theta = \frac{r}{y} = \frac{17}{15}$$

17.

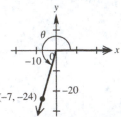

$x = -7$, $y = -24$, and

$$r = \sqrt{x^2 + y^2} = \sqrt{(-7)^2 + (-24)^2}$$
$$= \sqrt{49 + 576} = \sqrt{625} = 25$$

$$\sin\theta = \frac{y}{r} = \frac{-24}{25} = -\frac{24}{25}$$

$$\cos\theta = \frac{x}{r} = \frac{-7}{25} = -\frac{7}{25}$$

$$\tan\theta = \frac{y}{x} = \frac{-24}{-7} = \frac{24}{7}$$

$$\cot\theta = \frac{x}{y} = \frac{-7}{-24} = \frac{7}{24}$$

$$\sec\theta = \frac{r}{x} = \frac{25}{-7} = -\frac{25}{7}$$

$$\csc\theta = \frac{r}{y} = \frac{25}{-24} = -\frac{25}{24}$$

19.

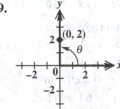

$x = 0$, $y = 2$, and

$$r = \sqrt{x^2 + y^2} = \sqrt{0^2 + 2^2} = \sqrt{0 + 4} = \sqrt{4} = 2$$

$$\sin\theta = \frac{y}{r} = \frac{2}{2} = 1$$

$$\cos\theta = \frac{x}{r} = \frac{0}{2} = 0$$

$$\tan\theta = \frac{y}{x} = \frac{2}{0} \text{ undefined}$$

$$\cot\theta = \frac{x}{y} = \frac{0}{2} = 0$$

$$\sec\theta = \frac{r}{x} = \frac{2}{0} \text{ undefined}$$

$$\csc\theta = \frac{r}{y} = \frac{2}{2} = 1$$

21.

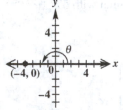

$x = 4$, $y = 0$, and

$$r = \sqrt{x^2 + y^2} = \sqrt{(-4)^2 + 0^2}$$
$$= \sqrt{16 + 0} = \sqrt{16} = 4$$

$$\sin\theta = \frac{y}{r} = \frac{0}{4} = 0$$

$$\cos\theta = \frac{x}{r} = \frac{-4}{4} = -1$$

(continued on next page)

(continued)

$$\tan \theta = \frac{y}{x} = \frac{0}{-4} = 0$$

$$\cot \theta = \frac{x}{y} = \frac{-4}{0} \text{ undefined}$$

$$\sec \theta = \frac{r}{x} = \frac{4}{-4} = -1$$

$$\csc \theta = \frac{r}{y} = \frac{4}{0} \text{ undefined}$$

23.

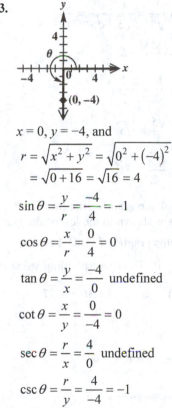

$x = 0, y = -4,$ and

$$r = \sqrt{x^2 + y^2} = \sqrt{0^2 + (-4)^2}$$
$$= \sqrt{0 + 16} = \sqrt{16} = 4$$

$$\sin \theta = \frac{y}{r} = \frac{-4}{4} = -1$$

$$\cos \theta = \frac{x}{r} = \frac{0}{4} = 0$$

$$\tan \theta = \frac{y}{x} = \frac{-4}{0} \text{ undefined}$$

$$\cot \theta = \frac{x}{y} = \frac{0}{-4} = 0$$

$$\sec \theta = \frac{r}{x} = \frac{4}{0} \text{ undefined}$$

$$\csc \theta = \frac{r}{y} = \frac{4}{-4} = -1$$

25.

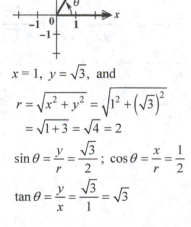

$x = 1, y = \sqrt{3},$ and

$$r = \sqrt{x^2 + y^2} = \sqrt{1^2 + \left(\sqrt{3}\right)^2}$$
$$= \sqrt{1 + 3} = \sqrt{4} = 2$$

$$\sin \theta = \frac{y}{r} = \frac{\sqrt{3}}{2}; \ \cos \theta = \frac{x}{r} = \frac{1}{2}$$

$$\tan \theta = \frac{y}{x} = \frac{\sqrt{3}}{1} = \sqrt{3}$$

$$\cot \theta = \frac{x}{y} = \frac{1}{\sqrt{3}} = \frac{1}{\sqrt{3}} \cdot \frac{\sqrt{3}}{\sqrt{3}} = \frac{\sqrt{3}}{3}$$

$$\sec \theta = \frac{r}{x} = \frac{2}{1} = 2$$

$$\csc \theta = \frac{r}{y} = \frac{2}{\sqrt{3}} = \frac{2}{\sqrt{3}} \cdot \frac{\sqrt{3}}{\sqrt{3}} = \frac{2\sqrt{3}}{3}$$

27.

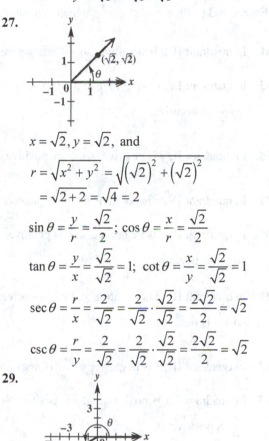

$x = \sqrt{2}, y = \sqrt{2},$ and

$$r = \sqrt{x^2 + y^2} = \sqrt{\left(\sqrt{2}\right)^2 + \left(\sqrt{2}\right)^2}$$
$$= \sqrt{2 + 2} = \sqrt{4} = 2$$

$$\sin \theta = \frac{y}{r} = \frac{\sqrt{2}}{2}; \ \cos \theta = \frac{x}{r} = \frac{\sqrt{2}}{2}$$

$$\tan \theta = \frac{y}{x} = \frac{\sqrt{2}}{\sqrt{2}} = 1; \ \cot \theta = \frac{x}{y} = \frac{\sqrt{2}}{\sqrt{2}} = 1$$

$$\sec \theta = \frac{r}{x} = \frac{2}{\sqrt{2}} = \frac{2}{\sqrt{2}} \cdot \frac{\sqrt{2}}{\sqrt{2}} = \frac{2\sqrt{2}}{2} = \sqrt{2}$$

$$\csc \theta = \frac{r}{y} = \frac{2}{\sqrt{2}} = \frac{2}{\sqrt{2}} \cdot \frac{\sqrt{2}}{\sqrt{2}} = \frac{2\sqrt{2}}{2} = \sqrt{2}$$

29.

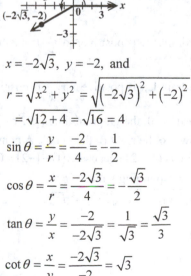

$x = -2\sqrt{3}, y = -2,$ and

$$r = \sqrt{x^2 + y^2} = \sqrt{\left(-2\sqrt{3}\right)^2 + (-2)^2}$$
$$= \sqrt{12 + 4} = \sqrt{16} = 4$$

$$\sin \theta = \frac{y}{r} = \frac{-2}{4} = -\frac{1}{2}$$

$$\cos \theta = \frac{x}{r} = \frac{-2\sqrt{3}}{4} = -\frac{\sqrt{3}}{2}$$

$$\tan \theta = \frac{y}{x} = \frac{-2}{-2\sqrt{3}} = \frac{1}{\sqrt{3}} = \frac{\sqrt{3}}{3}$$

$$\cot \theta = \frac{x}{y} = \frac{-2\sqrt{3}}{-2} = \sqrt{3}$$

(continued on next page)

(*continued*)

$$\sec\theta = \frac{r}{x} = \frac{4}{-2\sqrt{3}} = -\frac{2}{\sqrt{3}} = -\frac{2\sqrt{3}}{3}$$

$$\csc\theta = \frac{r}{y} = \frac{4}{-2} = -2$$

In Exercises 31–49, $r = \sqrt{x^2 + y^2}$, which is positive.

31. In quadrant II, x is negative, so $\dfrac{x}{r}$ is negative.

33. In quadrant IV, x is positive and y is negative, so $\dfrac{y}{x}$ is negative.

35. In quadrant II, y is positive, so $\dfrac{y}{r}$ is positive.

37. In quadrant IV, x is positive, so $\dfrac{x}{r}$ is positive.

39. In quadrant II, x is negative and y is positive, so $\dfrac{x}{y}$ is negative.

41. In quadrant III, x is negative and y is negative, so $\dfrac{y}{x}$ is positive.

43. In quadrant III, x is negative, so $\dfrac{r}{x}$ is negative.

45. In quadrant I, x is positive and y is positive, so $\dfrac{x}{y}$ is positive.

47. In quadrant I, y is positive, so $\dfrac{y}{r}$ is positive.

49. In quadrant I, x is positive, so $\dfrac{r}{x}$ is positive.

51. Because $x \geq 0$, the graph of the line $2x + y = 0$ is shown to the right of the y-axis. A point on this line is $(1, -2)$ because $2(1) + (-2) = 0$. The corresponding value of r is
$$r = \sqrt{1^2 + (-2)^2} = \sqrt{1+4} = \sqrt{5}.$$

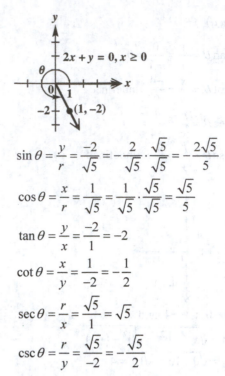

$$\sin\theta = \frac{y}{r} = \frac{-2}{\sqrt{5}} = -\frac{2}{\sqrt{5}} \cdot \frac{\sqrt{5}}{\sqrt{5}} = -\frac{2\sqrt{5}}{5}$$

$$\cos\theta = \frac{x}{r} = \frac{1}{\sqrt{5}} = \frac{1}{\sqrt{5}} \cdot \frac{\sqrt{5}}{\sqrt{5}} = \frac{\sqrt{5}}{5}$$

$$\tan\theta = \frac{y}{x} = \frac{-2}{1} = -2$$

$$\cot\theta = \frac{x}{y} = \frac{1}{-2} = -\frac{1}{2}$$

$$\sec\theta = \frac{r}{x} = \frac{\sqrt{5}}{1} = \sqrt{5}$$

$$\csc\theta = \frac{r}{y} = \frac{\sqrt{5}}{-2} = -\frac{\sqrt{5}}{2}$$

53. Because $x \leq 0$, the graph of the line $-6x - y = 0$ is shown to the left of the y-axis. A point on this graph is $(-1, 6)$ because $-6(-1) - 6 = 0$. The corresponding value of r is $r = \sqrt{(-1)^2 + 6^2} = \sqrt{1 + 36} = \sqrt{37}$.

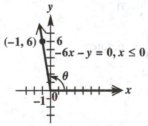

$$\sin\theta = \frac{y}{r} = \frac{6}{\sqrt{37}} = \frac{6}{\sqrt{37}} \cdot \frac{\sqrt{37}}{\sqrt{37}} = \frac{6\sqrt{37}}{37}$$

$$\cos\theta = \frac{x}{r} = \frac{-1}{\sqrt{37}} = -\frac{1}{\sqrt{37}} \cdot \frac{\sqrt{37}}{\sqrt{37}} = -\frac{\sqrt{37}}{37}$$

$$\tan\theta = \frac{y}{x} = \frac{6}{-1} = -6; \quad \cot\theta = \frac{x}{y} = \frac{-1}{6} = -\frac{1}{6}$$

$$\sec\theta = \frac{r}{x} = \frac{\sqrt{37}}{-1} = -\sqrt{37}; \quad \csc\theta = \frac{r}{y} = \frac{\sqrt{37}}{6}$$

55. Because $x \le 0$, the graph of the line $-4x + 7y = 0$ is shown to the left of the y-axis. A point on this line is $(-7, -4)$ because $-4(-7) + 7(-4) = 0$. The corresponding value of r is $r = \sqrt{(-7)^2 + (-4)^2} = \sqrt{49 + 16} = \sqrt{65}$.

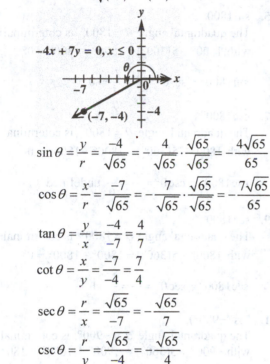

$$\sin\theta = \frac{y}{r} = \frac{-4}{\sqrt{65}} = -\frac{4}{\sqrt{65}} \cdot \frac{\sqrt{65}}{\sqrt{65}} = -\frac{4\sqrt{65}}{65}$$

$$\cos\theta = \frac{x}{r} = \frac{-7}{\sqrt{65}} = -\frac{7}{\sqrt{65}} \cdot \frac{\sqrt{65}}{\sqrt{65}} = -\frac{7\sqrt{65}}{65}$$

$$\tan\theta = \frac{y}{x} = \frac{-4}{-7} = \frac{4}{7}$$

$$\cot\theta = \frac{x}{y} = \frac{-7}{-4} = \frac{7}{4}$$

$$\sec\theta = \frac{r}{x} = \frac{\sqrt{65}}{-7} = -\frac{\sqrt{65}}{7}$$

$$\csc\theta = \frac{r}{y} = \frac{\sqrt{65}}{-4} = -\frac{\sqrt{65}}{4}$$

57. Because $x \ge 0$, the graph of the line $x + y = 0$ is shown to the right of the y-axis. A point on this line is $(2, -2)$ because $2 + (-2) = 0$. The corresponding value of r is

$$r = \sqrt{2^2 + (-2)^2} = \sqrt{4 + 4} = \sqrt{8} = 2\sqrt{2}.$$

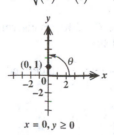

$$\sin\theta = \frac{y}{r} = \frac{-2}{2\sqrt{2}} = -\frac{1}{\sqrt{2}} \cdot \frac{\sqrt{2}}{\sqrt{2}} = -\frac{\sqrt{2}}{2}$$

$$\cos\theta = \frac{x}{r} = \frac{2}{2\sqrt{2}} = \frac{1}{\sqrt{2}} \cdot \frac{\sqrt{2}}{\sqrt{2}} = \frac{\sqrt{2}}{2}$$

$$\tan\theta = \frac{y}{x} = \frac{-2}{2} = -1; \quad \cot\theta = \frac{x}{y} = \frac{2}{-2} = -1$$

$$\sec\theta = \frac{r}{x} = \frac{2\sqrt{2}}{2} = \sqrt{2}$$

$$\csc\theta = \frac{r}{y} = \frac{2\sqrt{2}}{-2} = -\sqrt{2}$$

59. Because $x \le 0$, the graph of the line $-\sqrt{3}x + y = 0$ is shown to the left of the y-axis. A point on this line is $\left(-1, -\sqrt{3}\right)$ because $-\sqrt{3}(-1) - \sqrt{3} = \sqrt{3} - \sqrt{3} = 0$. The corresponding value of r is

$$r = \sqrt{(-1)^2 + \left(-\sqrt{3}\right)^2} = \sqrt{1 + 3} = \sqrt{4} = 2.$$

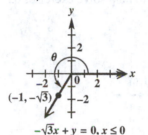

$$\sin\theta = \frac{y}{r} = \frac{-\sqrt{3}}{2} = -\frac{\sqrt{3}}{2}$$

$$\cos\theta = \frac{x}{r} = \frac{-1}{2} = -\frac{1}{2}$$

$$\tan\theta = \frac{y}{x} = \frac{-\sqrt{3}}{-1} = \sqrt{3}$$

$$\cot\theta = \frac{x}{y} = \frac{-1}{-\sqrt{3}} = \frac{1}{\sqrt{3}} \cdot \frac{\sqrt{3}}{\sqrt{3}} = \frac{\sqrt{3}}{3}$$

$$\sec\theta = \frac{r}{x} = \frac{2}{-1} = -2$$

$$\csc\theta = \frac{r}{y} = \frac{2}{-\sqrt{3}} = -\frac{2}{\sqrt{3}} \cdot \frac{\sqrt{3}}{\sqrt{3}} = -\frac{2\sqrt{3}}{3}$$

61. Because $x = 0$ and $y \ge 0$, the graph is the positive portion of the y-axis. A point on this line is $(0, 1)$. The corresponding value of r is

$$r = \sqrt{(0)^2 + (1)^2} = \sqrt{1} = 1.$$

(continued on next page)

(*continued*)

$$\sin\theta = \frac{y}{r} = \frac{1}{1} = 1; \ \cos\theta = \frac{x}{r} = \frac{0}{1} = 0$$

$$\tan\theta = \frac{y}{x} = \frac{1}{0}, \text{ undefined}$$

$$\cot\theta = \frac{x}{y} = \frac{0}{1} = 0$$

$$\sec\theta = \frac{r}{x} = \frac{1}{0}, \text{ undefined}$$

$$\csc\theta = \frac{r}{y} = \frac{1}{1} = 1$$

Use the figure below to help solve exercises 63–97.

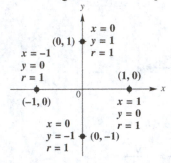

63. $\cos 90°$

$$\cos 90° = \frac{x}{r} = \frac{0}{1} = 0$$

65. $\tan 180°$

$$\tan 180° = \frac{y}{x} = \frac{0}{-1} = 0$$

67. $\sec 180°$

$$\sec 180° = \frac{r}{x} = \frac{1}{-1} = -1$$

69. $\sin\left(-270°\right)$

The quadrantal angle $\theta = -270°$ is coterminal with $-270° + 360° = 90°$.

$$\sin\left(-270°\right) = \sin 90 = \frac{y}{r} = \frac{1}{1} = 1$$

71. $\cot 540°$

The quadrantal angle $\theta = 540°$ is coterminal with $540° - 360° = 180°$.

$$\cot 540° = \cot 180° = \frac{x}{y} = \frac{-1}{0} \text{ undefined}$$

73. $\csc\left(-450°\right)$

The quadrantal angle $\theta = -450°$ is coterminal with $720° - 450° = 270°$.

$$\csc\left(-450°\right) = \csc 270° = \frac{r}{y} = \frac{1}{-1} = -1$$

75. $\sin 1800°$

The quadrantal angle $\theta = 1800°$ is coterminal with $1800° - 5\left(360°\right) = 1800° - 1800° = 0°$

$$\sin 1800° = \sin 0° = \frac{y}{r} = \frac{0}{1} = 0$$

77. $\csc 1800°$

The quadrantal angle $\theta = 1800°$ is coterminal with $1800° - 5\left(360°\right) = 1800° - 1800° = 0°$

$$\csc 1800° = \csc 0° = \frac{r}{y} = \frac{1}{0} \text{ undefined}$$

79. $\sec 1800°$

The quadrantal angle $\theta = 1800°$ is coterminal with $1800° - 5\left(360°\right) = 1800° - 1800° = 0°$

$$\sec 1800° = \sec 0° = \frac{r}{x} = \frac{1}{1} = 1$$

81. $\cos\left(-900°\right)$

The quadrantal angle $\theta = -900°$ is coterminal with $-900° + 3(360°) = -900° + 1080° = 180°$.

$$\cos\left(-900°\right) = \cos 180° = \frac{x}{r} = \frac{-1}{1} = -1$$

83. $\tan\left(-900°\right)$

The quadrantal angle $\theta = -900°$ is coterminal with $-900° + 3(360°) = -900° + 1080° = 180°$.

$$\tan\left(-900°\right) = \tan 180° = \frac{y}{x} = \frac{0}{-1} = 0$$

85. $\cos 90° + 3\sin 270°$

$$\cos 90° = \frac{x}{r} = \frac{0}{1} = 0$$

$$\sin 270° = \frac{y}{r} = \frac{-1}{1} = -1$$

$$\cos 90° + 3\sin 270° = 0 + 3\left(-1\right) = -3$$

87. $3\sec 180° - 5\tan 360°$

$$\sec 180° = \frac{r}{x} = \frac{1}{-1} = -1 \text{ and}$$

$$\tan 360° = \tan 0° = \frac{y}{x} = \frac{0}{1} = 0$$

$$3\sec 180° - 5\tan 360° = 3\left(-1\right) - 5\left(0\right)$$
$$= -3 - 0 = -3$$

89. $\tan 360° + 4\sin 180° + 5\cos^2 180°$

$$\tan 360° = \tan 0° = \frac{y}{x} = \frac{0}{1} = 0,$$

$$\sin 180° = \frac{y}{r} = \frac{0}{1} = 0, \text{ and}$$

$$\cos 180° = \frac{x}{r} = \frac{-1}{1} = -1$$

$$\tan 360° + 4\sin 180° + 5\cos^2 180°$$
$$= 0 + 4(0) + 5(-1)^2 = 0 + 0 + 5(1) = 5$$

91. $\sin^2 180° + \cos^2 180°$

$$\sin 180° = \frac{y}{r} = \frac{0}{1} = 0 \text{ and}$$

$$\cos 180° = \frac{x}{r} = \frac{-1}{1} = -1$$

$$\sin^2 180° + \cos^2 180° = 0^2 + (-1)^2 = 0 + 1 = 1$$

93. $\sec^2 180° - 3\sin^2 360° + \cos 180°$

$$\sec 180° = \frac{r}{x} = \frac{1}{-1} = -1,$$

$$\sin 360° = \sin 0° = \frac{y}{r} = \frac{0}{1} = 0 \text{ and}$$

$$\cos 180° = \frac{x}{r} = \frac{-1}{1} = -1$$

$$\sec^2 180° - 3\sin^2 360° + \cos 180°$$
$$= (-1)^2 - 3(0)^2 + (-1) = 1 - 1 = 0$$

95. $-2\sin^4 0° + 3\tan^2 0°$

$$\sin 0° = \frac{y}{r} = \frac{0}{1} = 0; \tan 0° = \frac{y}{x} = \frac{0}{1} = 0$$

$$-2\sin^4 0° + 3\tan^2 0° = -2(0)^4 + 3(0)^2 = 0$$

97. $\sin^2(-90°) + \cos^2(-90°)$

$$\sin(-90°) = \sin(-90° + 360°)$$

$$= \sin 270° = \frac{y}{r} = \frac{-1}{1} = -1$$

$$\cos(-90°) = \cos(-90° + 360°)$$

$$= \cos 270° = \frac{x}{r} = \frac{0}{1} = 0$$

$$\sin^2(-90°) + \cos^2(-90°) = (-1)^2 + 0^2 = 1$$

99. $\cos\left[(2n+1)\cdot 90°\right]$

This angle is a quadrantal angle whose terminal side lies on either the positive part of the y-axis or the negative part of the y-axis.

Any point on these terminal sides would have the form $(0, k)$, where k is any real number, $k \neq 0$.

$$\cos\left[(2n+1)\cdot 90°\right] = \frac{x}{r} = \frac{0}{\sqrt{0^2 + k^2}}$$

$$= \frac{0}{\sqrt{k^2}} = \frac{0}{|k|} = 0$$

101. $\tan\left[n\cdot 180°\right]$

The angle is a quadrantal angle whose terminal side lies on either the positive part of the x-axis or the negative part of the x-axis. Any point on these terminal sides would have the form $(k, 0)$, where k is any real number, $k \neq 0$.

$$\tan\left[n\cdot 180°\right] = \frac{y}{x} = \frac{0}{k} = 0$$

103. $\tan\left[(2n+1)\cdot 90°\right]$

This angle is a quadrantal angle whose terminal side lies on either the positive part of the y-axis or the negative part of the y-axis. Any point on these terminal sides would have the form $(0, k)$, where k is any real number, $k \neq 0$.

$$\tan\left[(2n+1)\cdot 90°\right] = \frac{y}{x} = \frac{k}{0} \text{ undefined}$$

105. $\cot\left[(2n+1)\cdot 90°\right]$

This angle is a quadrantal angle whose terminal side lies on either the positive part of the y-axis or the negative part of the y-axis. Any point on these terminal sides would have the form $(0, k)$, where k is any real number, $k \neq 0$.

$$\cot\left[(2n+1)\cdot 90°\right] = \frac{x}{y} = \frac{0}{k} = 0$$

107. $\sec\left[(2n+1)\cdot 90°\right]$

This angle is a quadrantal angle whose terminal side lies on either the positive part of the y-axis or the negative part of the y-axis. Any point on these terminal sides would have the form $(0, k)$, where k is any real number, $k \neq 0$.

$$\sec\left[(2n+1)\cdot 90°\right] = \frac{r}{x} = \frac{\sqrt{0^2 + k^2}}{0} \text{ undefined}$$

109. Using a calculator, $\sin 15° = 0.258819045$ and $\cos 75° = 0.258819045$. We can conjecture that the sines and cosines of complementary angles are equal. Trying another pair of complementary angles, we obtain $\sin 30° = \cos 60° = 0.5$. Therefore, our conjecture appears to be true.

111. Using a calculator, $\sin 10° = 0.173648178$ and $\sin(-10°) = -0.173648178$. We can conjecture that the sines of an angle and its negative are opposites of each other. Using a circle, an angle θ having the point (x, y) on its terminal side has a corresponding angle $-\theta$ with a point $(x, -y)$ on its terminal side. From the definition of sine, $\sin(-\theta) = \dfrac{-y}{r} = -\dfrac{y}{r}$ and $\sin\theta = \dfrac{y}{r}$. The sines are negatives of each other.

In Exercises 113–117, make sure your calculator is in the modes indicated in the instructions.

113. Use the TRACE feature to move around the circle in quadrant I to find $T = 20$. We see that $\cos 20° \approx 0.940$, and $\sin 20° \approx 0.342$.

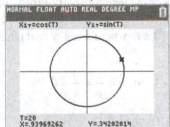

115. Use the TRACE feature to move around the circle in quadrant I. We see that $\sin 35° \approx 0.574$, so $T = 35°$.

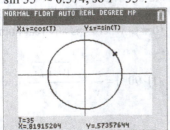

117. As T increases from $0°$ to $90°$, the cosine decreases and the sine increases.

Section 1.4 Using the Definitions of the Trigonometric Functions

1. Given $\cos\theta = \dfrac{1}{\sec\theta}$, two equivalent forms of this identity are $\sec\theta = \dfrac{1}{\cos\theta}$ and $\cos\theta \cdot \sec\theta = 1$.

3. For an angle θ measuring $105°$, the trigonometric functions $\underline{\sin\theta}$ and $\underline{\csc\theta}$ are positive, and the remaining trigonometric functions are negative.

5. $\sin\theta = \dfrac{1}{2}$ is possible because the range of $\sin\theta$ is $[-1, 1]$. Furthermore, when $\sin\theta = \dfrac{1}{2}$, $\csc\theta = \dfrac{1}{\frac{1}{2}} = 2$. Thus, $\sin\theta = \dfrac{1}{2}$ and $\csc\theta = 2$ is possible.

7. $\sin\theta > 0$, $\csc\theta < 0$ is impossible because $\csc\theta = \dfrac{1}{\sin\theta}$, so both functions must have the same sign.

9. $\cot\theta = -1.5$ is possible because the range of $\cot\theta$ is $(-\infty, \infty)$.

11. $\sec\theta = \dfrac{1}{\cos\theta} = \dfrac{1}{\frac{2}{3}} = \dfrac{3}{2}$

13. $\csc\theta = \dfrac{1}{\sin\theta} = \dfrac{1}{-\frac{3}{7}} = -\dfrac{7}{3}$

15. $\cot\theta = \dfrac{1}{\tan\theta} = \dfrac{1}{5}$

17. $\cos\theta = \dfrac{1}{\sec\theta} = \dfrac{1}{-\frac{5}{2}} = -\dfrac{2}{5}$

19. $\sin\theta = \dfrac{1}{\csc\theta} = \dfrac{1}{\frac{\sqrt{8}}{2}} = \dfrac{2}{\sqrt{8}} = \dfrac{2}{2\sqrt{2}} \cdot \dfrac{\sqrt{2}}{\sqrt{2}} = \dfrac{\sqrt{2}}{2}$

21. $\tan\theta = \dfrac{1}{\cot\theta} = \dfrac{1}{-2.5} = -0.4$

23. $\sin\theta = \dfrac{1}{\csc\theta} = \dfrac{1}{1.25} = 0.8$

25. The value of $\cos\theta$ cannot exceed 1.

27. A 74° angle in standard position lies in quadrant I, so all its trigonometric functions are positive.

29. A 218° angle in standard position lies in quadrant III, so its sine, cosine, secant, and cosecant are negative, while its tangent and cotangent are positive.

31. A 178° angle in standard position lies in quadrant II, so its sine and cosecant are positive, while its cosine, secant, tangent and cotangent are negative.

33. A −80° angle in standard position lies in quadrant IV, so its cosine and secant are positive, while its sine, cosecant, tangent and cotangent are negative.

35. An 855° angle in standard position is coterminal with a 135° angle, and thus, lies in quadrant II. So its sine and cosecant are positive, while its cosine, secant, tangent and cotangent are negative.

37. A −345° angle in standard position lies in quadrant I, so all its trigonometric functions are positive.

39. $\sin\theta > 0$, so $\csc\theta$ is also greater than 0. The functions are greater than 0 (positive) in quadrants I and II.

41. $\cos\theta > 0$ in quadrants I and IV, while $\sin\theta > 0$ in quadrants I and II. Both conditions are met only in quadrant I.

43. $\tan\theta < 0$ in quadrants II and IV, while $\cos\theta < 0$ in quadrants II and III. Both conditions are met only in quadrant II.

45. $\sec\theta > 0$ in quadrants I and IV, while $\csc\theta > 0$ in quadrants I and II. Both conditions are met only in quadrant I.

47. $\sec\theta < 0$ in quadrants II and III, while $\csc\theta < 0$ in quadrants III and IV. Both conditions are met only in quadrant III.

49. $\sin\theta < 0$, so $\csc\theta$ is also less than 0. The functions are less than 0 (negative) in quadrants III and IV.

51. The answers to exercises 41 and 45 are the same because functions in exercise 45 are the reciprocals of the functions in exercise 41.

53. Impossible because the range of $\sin\theta$ is $[-1, 1]$.

55. Possible because the range of $\cos\theta$ is $[-1, 1]$.

57. Possible because the range of $\tan\theta$ is $(-\infty, \infty)$.

59. Impossible because the range of $\sec\theta$ is $(-\infty, -1] \cup [1, \infty)$.

61. Possible because the range of $\csc\theta$ is $(-\infty, -1] \cup [1, \infty)$.

63. Possible because the range of $\cot\theta$ is $(-\infty, \infty)$.

65. If $\sin\theta = \dfrac{3}{5}$, then $y = 3$ and $r = 5$. So
$$r^2 = x^2 + y^2 \Rightarrow 5^2 = x^2 + 3^2 \Rightarrow 25 = x^2 + 9 \Rightarrow$$
$$16 = x^2 \Rightarrow \pm 4 = x. \ \theta \text{ is in quadrant II, so}$$
$x = -4$. Therefore, $\cos\theta = -\dfrac{4}{5}$.

Alternatively, use the identity
$$\sin^2\theta + \cos^2\theta = 1: \left(\frac{3}{5}\right)^2 + \cos^2\theta = 1 \Rightarrow$$
$$\frac{9}{25} + \cos^2\theta = 1 \Rightarrow \cos^2\theta = \frac{16}{25} \Rightarrow \cos\theta = \pm\frac{4}{5}$$
θ is in quadrant II, so $\cos\theta$ is negative.

Thus, $\cos\theta = -\dfrac{4}{5}$.

67. If $\cot\theta = -\dfrac{1}{2}$ and θ is in quadrant IV, then $x = 1$ and $y = -2$. So
$$r^2 = x^2 + y^2 \Rightarrow r^2 = 1^2 + (-2)^2 \Rightarrow$$
$$r^2 = 1 + 4 \Rightarrow r^2 = 5 \Rightarrow r = \sqrt{5}$$

Therefore, $\csc\theta = \dfrac{r}{y} = -\dfrac{\sqrt{5}}{2}$. Alternatively,

use the identity $1 + \cot^2\theta = \csc^2\theta$:
$$1 + \left(-\frac{1}{2}\right)^2 = \csc^2\theta \Rightarrow 1 + \frac{1}{4} = \csc^2\theta \Rightarrow$$
$$\frac{5}{4} = \csc^2\theta \Rightarrow \pm\frac{\sqrt{5}}{2} = \csc\theta. \ \theta \text{ is in quadrant}$$
IV, so $\csc\theta$ is negative. Thus, $\csc\theta = -\dfrac{\sqrt{5}}{2}$

69. If $\sin\theta = \dfrac{1}{2}$, then $y = 1$ and $r = 2$. So

$r^2 = x^2 + y^2 \Rightarrow 2^2 = x^2 + 1^2 \Rightarrow 4 = x^2 + 1 \Rightarrow$
$3 = x^2 \Rightarrow \pm\sqrt{3} = x$

θ is in quadrant II, so $x = -\sqrt{3}$. Therefore,

$\tan\theta = -\dfrac{1}{\sqrt{3}} = -\dfrac{1}{\sqrt{3}} \cdot \dfrac{\sqrt{3}}{\sqrt{3}} = -\dfrac{\sqrt{3}}{3}$.

Alternatively, use the identity
$\sin^2\theta + \cos^2\theta = 1$ to obtain

$\left(\dfrac{1}{2}\right)^2 + \cos^2\theta = 1 \Rightarrow \cos^2\theta = \dfrac{3}{4} \Rightarrow$

$\cos\theta = \pm\dfrac{\sqrt{3}}{2}$.

θ is in quadrant II, so $\cos\theta = -\dfrac{\sqrt{3}}{2}$. Then,

$\tan\theta = \dfrac{\sin\theta}{\cos\theta} = \dfrac{\frac{1}{2}}{-\frac{\sqrt{3}}{2}} = -\dfrac{1}{\sqrt{3}}$

$= -\dfrac{1}{\sqrt{3}} \cdot \dfrac{\sqrt{3}}{\sqrt{3}} = -\dfrac{\sqrt{3}}{3}$

71. Using the identity $1 + \cot^2\theta = \csc^2\theta$ gives
$1 + \cot^2\theta = (-1.45)^2 \Rightarrow \cot^2\theta = 1.1025 \Rightarrow$
$\cot\theta = \pm 1.05$
Because θ is in quadrant III, $\cot\theta = 1.05$.

For Exercises 73–77, remember that r is always positive.

73. $\tan\theta = -\dfrac{15}{8} = \dfrac{15}{-8}$, with θ in quadrant II

$\tan\theta = \dfrac{y}{x}$ and θ is in quadrant II, so let
$y = 15$ and $x = -8$.

$x^2 + y^2 = r^2 \Rightarrow (-8)^2 + 15^2 = r^2 \Rightarrow$
$64 + 225 = r^2 \Rightarrow 289 = r^2 \Rightarrow r = 17$

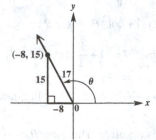

$\sin\theta = \dfrac{y}{r} = \dfrac{15}{17}$; $\cos\theta = \dfrac{x}{r} = \dfrac{-8}{17} = -\dfrac{8}{17}$

$\tan\theta = \dfrac{y}{x} = \dfrac{15}{-8} = -\dfrac{15}{8}$

$\cot\theta = \dfrac{x}{y} = \dfrac{-8}{15} = -\dfrac{8}{15}$

$\sec\theta = \dfrac{r}{x} = \dfrac{17}{-8} = -\dfrac{17}{8}$; $\csc\theta = \dfrac{r}{y} = \dfrac{17}{15}$

75. $\sin\theta = \dfrac{\sqrt{5}}{7}$, with θ in quadrant I

$\sin\theta = \dfrac{y}{r}$ and θ in quadrant I, so let
$y = \sqrt{5}, r = 7$.

$x^2 + y^2 = r^2 \Rightarrow x^2 + \left(\sqrt{5}\right)^2 = 7^2 \Rightarrow$
$x^2 + 5 = 49 \Rightarrow x^2 = 44 \Rightarrow x = \pm\sqrt{44} \Rightarrow$
$x = \pm 2\sqrt{11}$

θ is in quadrant I, so $x = 2\sqrt{11}$.

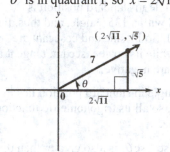

Drawing not to scale

$\sin\theta = \dfrac{y}{r} = \dfrac{\sqrt{5}}{7}$

$\cos\theta = \dfrac{x}{r} = \dfrac{2\sqrt{11}}{7}$

$\tan\theta = \dfrac{y}{x} = \dfrac{\sqrt{5}}{2\sqrt{11}} = \dfrac{\sqrt{5}}{2\sqrt{11}} \cdot \dfrac{\sqrt{11}}{\sqrt{11}} = \dfrac{\sqrt{55}}{22}$

$\cot\theta = \dfrac{x}{y} = \dfrac{2\sqrt{11}}{\sqrt{5}} = \dfrac{2\sqrt{11}}{\sqrt{5}} \cdot \dfrac{\sqrt{5}}{\sqrt{5}} = \dfrac{2\sqrt{55}}{5}$

$\sec\theta = \dfrac{r}{x} = \dfrac{7}{2\sqrt{11}} = \dfrac{7}{2\sqrt{11}} \cdot \dfrac{\sqrt{11}}{\sqrt{11}} = \dfrac{7\sqrt{11}}{22}$

$\csc\theta = \dfrac{r}{y} = \dfrac{7}{\sqrt{5}} = \dfrac{7}{\sqrt{5}} \cdot \dfrac{\sqrt{5}}{\sqrt{5}} = \dfrac{7\sqrt{5}}{5}$

77. $\cot\theta = \dfrac{\sqrt{3}}{8}$, with θ in quadrant I

$\cot\theta = \dfrac{x}{y}$ and θ is in quadrant I, so let
$x = \sqrt{3}$ and $y = 8$.

$x^2 + y^2 = r^2 \Rightarrow \left(\sqrt{3}\right)^2 + (8)^2 = r^2 \Rightarrow$
$3 + 64 = r^2 \Rightarrow 67 = r^2 \Rightarrow \sqrt{67} = r$

(continued on next page)

(continued)

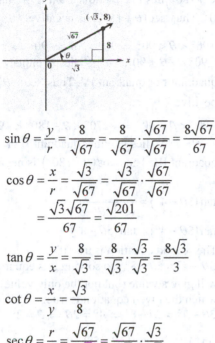

$$\sin\theta = \frac{y}{r} = \frac{8}{\sqrt{67}} = \frac{8}{\sqrt{67}} \cdot \frac{\sqrt{67}}{\sqrt{67}} = \frac{8\sqrt{67}}{67}$$

$$\cos\theta = \frac{x}{r} = \frac{\sqrt{3}}{\sqrt{67}} = \frac{\sqrt{3}}{\sqrt{67}} \cdot \frac{\sqrt{67}}{\sqrt{67}}$$
$$= \frac{\sqrt{3}\sqrt{67}}{67} = \frac{\sqrt{201}}{67}$$

$$\tan\theta = \frac{y}{x} = \frac{8}{\sqrt{3}} = \frac{8}{\sqrt{3}} \cdot \frac{\sqrt{3}}{\sqrt{3}} = \frac{8\sqrt{3}}{3}$$

$$\cot\theta = \frac{x}{y} = \frac{\sqrt{3}}{8}$$

$$\sec\theta = \frac{r}{x} = \frac{\sqrt{67}}{\sqrt{3}} = \frac{\sqrt{67}}{\sqrt{3}} \cdot \frac{\sqrt{3}}{\sqrt{3}}$$
$$= \frac{\sqrt{67}\sqrt{3}}{3} = \frac{\sqrt{201}}{3}$$

$$\csc\theta = \frac{r}{y} = \frac{\sqrt{67}}{8}$$

79. $\sin\theta = \dfrac{\sqrt{2}}{6}$, given that $\cos\theta < 0$

$\sin\theta$ is positive and $\cos\theta$ is negative when θ is in quadrant II.

$\sin\theta = \dfrac{y}{r}$ and θ in quadrant II, so let

$y = \sqrt{2}, r = 6.$

$$x^2 + y^2 = r^2 \Rightarrow x^2 + \left(\sqrt{2}\right)^2 = 6^2 \Rightarrow$$
$$x^2 + 2 = 36 \Rightarrow x^2 = 34 \Rightarrow$$
$$x = \pm\sqrt{34}$$

θ is in quadrant II, so $x = -\sqrt{34}$.

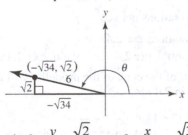

$$\sin\theta = \frac{y}{r} = \frac{\sqrt{2}}{6}; \; \cos\theta = \frac{x}{r} = -\frac{\sqrt{34}}{6}$$

$$\tan\theta = \frac{y}{x} = -\frac{\sqrt{2}}{\sqrt{34}} = -\frac{\sqrt{2}}{\sqrt{34}} \cdot \frac{\sqrt{34}}{\sqrt{34}} = -\frac{\sqrt{68}}{34}$$
$$= -\frac{2\sqrt{17}}{34} = -\frac{\sqrt{17}}{17}$$

$$\cot\theta = \frac{x}{y} = -\frac{\sqrt{34}}{\sqrt{2}} = -\frac{\sqrt{34}}{\sqrt{2}} \cdot \frac{\sqrt{2}}{\sqrt{2}} = -\frac{\sqrt{68}}{2}$$
$$= -\frac{2\sqrt{17}}{2} = -\sqrt{17}$$

$$\sec\theta = \frac{r}{x} = -\frac{6}{\sqrt{34}} = -\frac{6}{\sqrt{34}} \cdot \frac{\sqrt{34}}{\sqrt{34}} = -\frac{6\sqrt{34}}{34}$$
$$= -\frac{3\sqrt{34}}{17}$$

$$\csc\theta = \frac{r}{y} = \frac{6}{\sqrt{2}} = \frac{6}{\sqrt{2}} \cdot \frac{\sqrt{2}}{\sqrt{2}} = \frac{6\sqrt{2}}{2} = 3\sqrt{2}$$

81. $\sec\theta = -4$, given that $\sin\theta > 0$

$\sec\theta$ is negative and $\sin\theta$ is positive when θ is in quadrant II.

$\sec\theta = \dfrac{r}{x}$ and θ in quadrant II, so let

$x = -1, r = 4.$

$$x^2 + y^2 = r^2 \Rightarrow (-1)^2 + y^2 = 4^2 \Rightarrow$$
$$1 + y^2 = 16 \Rightarrow y^2 = 15 \Rightarrow y = \pm\sqrt{15}$$

θ is in quadrant II, so $y = \sqrt{15}$.

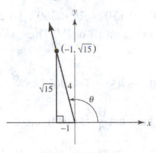

$$\sin\theta = \frac{y}{r} = \frac{\sqrt{15}}{4}; \; \cos\theta = \frac{x}{r} = -\frac{1}{4}$$

$$\tan\theta = \frac{y}{x} = -\frac{\sqrt{15}}{1} = -\sqrt{15}$$

$$\cot\theta = \frac{x}{y} = -\frac{1}{\sqrt{15}} = -\frac{1}{\sqrt{15}} \cdot \frac{\sqrt{15}}{\sqrt{15}} = -\frac{\sqrt{15}}{15}$$

$$\sec\theta = \frac{r}{x} = -4$$

$$\csc\theta = \frac{r}{y} = \frac{4}{\sqrt{15}} = \frac{4}{\sqrt{15}} \cdot \frac{\sqrt{15}}{\sqrt{15}} = \frac{4\sqrt{15}}{15}$$

83. $\sin\theta = 1$

$\sin^2\theta + \cos^2\theta = 1^2 \Rightarrow 1^2 + \cos^2\theta = 1 \Rightarrow$
$\cos^2\theta = 0 \Rightarrow \cos\theta = 0$

$\tan\theta = \dfrac{\sin\theta}{\cos\theta} = \dfrac{1}{0} \Rightarrow \tan\theta$ is undefined

$\cot\theta = \dfrac{\cos\theta}{\sin\theta} = \dfrac{0}{1} = 0$

$\sec\theta = \dfrac{1}{\cos\theta} = \dfrac{1}{0} \Rightarrow \sec\theta$ is undefined

$\csc\theta = \dfrac{1}{\sin\theta} = \dfrac{1}{1} = 1$

85. $x^2 + y^2 = r^2 \Rightarrow \dfrac{x^2 + y^2}{y^2} = \dfrac{r^2}{y^2} \Rightarrow$

$\dfrac{x^2}{y^2} + \dfrac{y^2}{y^2} = \dfrac{r^2}{y^2} \Rightarrow \left(\dfrac{x}{y}\right)^2 + 1 = \left(\dfrac{r}{y}\right)^2 \Rightarrow$

$1 + \left(\dfrac{x}{y}\right)^2 = \left(\dfrac{r}{y}\right)^2 \Rightarrow$

Because $\cot\theta = \dfrac{x}{y}$ and $\csc\theta = \dfrac{r}{y}$, we have

$1 + (\cot\theta)^2 = (\csc\theta)^2$ or $1 + \cot^2\theta = \csc^2\theta$.

87. The statement is false. For example,
$\sin 180° + \cos 180° = 0 + (-1) = -1 \neq 1.$

89. $90° < \theta < 180° \Rightarrow 180° < 2\theta < 360°$, so 2θ lies in either quadrant III or IV. Thus, $\sin 2\theta$ is negative.

91. $90° < \theta < 180° \Rightarrow 45° < \dfrac{\theta}{2} < 90°$, so $\dfrac{\theta}{2}$ lies in

quadrant I. Thus, $\tan\dfrac{\theta}{2}$ is positive.

93. $90° < \theta < 180° \Rightarrow 270° < \theta + 180° < 360°$, so $\theta + 180°$ lies in quadrant IV. Thus, $\cot(\theta + 180°)$ is negative.

95. $90° < \theta < 180° \Rightarrow -90° > -\theta > -180° \Rightarrow$ $-180° < \theta < -90°$, so $-\theta$ lies in quadrant III ($-180°$ is coterminal with $180°$, and $-90°$ is coterminal with $270°$.) Thus, $\cos(-\theta)$ is negative.

97. $-90° < \theta < 90° \Rightarrow -45° < \dfrac{\theta}{2} < 45°$, so $\dfrac{\theta}{2}$ lies

in either quadrant I or quadrant IV. Thus $\cos\dfrac{\theta}{2}$ is positive.

99. $-90° < \theta < 90° \Rightarrow 90° < \theta + 180° < 270°$, so $\theta + 180°$ lies in either quadrant II or quadrant III. Thus $\sec(\theta + 180°)$ is negative.

101. $-90° < \theta < 90° \Rightarrow 90° > -\theta > -90° \Rightarrow$ $-90° < -\theta < 90°$, so $-\theta$ lies in either quadrant I or quadrant IV. Thus $\sec(-\theta)$ is positive.

103. $-90° < \theta < 90° \Rightarrow -270° < \theta - 180° < -90°$, so $\theta - 180°$ lies in either quadrant II or quadrant III. Thus $\cos(\theta - 180°)$ is negative.

105. $\tan(3\theta - 4°) = \dfrac{1}{\cot(5\theta - 8°)} \Rightarrow$
$\tan(3\theta - 4°) = \tan(5\theta - 8°)$
The second equation is true if $3\theta - 4° = 5\theta - 8°$, so solving this equation will give a value (but not the only value) for which the given equation is true.
$3\theta - 4° = 5\theta - 8° \Rightarrow 4° = 2\theta \Rightarrow \theta = 2°$

107. $\sin(4\theta + 2°)\csc(3\theta + 5°) = 1 \Rightarrow$

$\sin(4\theta + 2°) = \dfrac{1}{\csc(3\theta + 5°)} \Rightarrow$
$\sin(4\theta + 2°) = \sin(3\theta + 5°)$

The third equation is true if $4\theta + 2° = 3\theta + 5°$, so solving this equation will give a value (but not the only value) for which the given equation is true.
$4\theta + 2° = 3\theta + 5° \Rightarrow \theta = 3°$

109. In quadrant II, the cosine is negative and the sine is positive.

Chapter 1   Review Exercises

1. The complement of $35°$ is $90° - 35° = 55°$.
The supplement of $35°$ is $180° - 35° = 145°$.

3. $-174°$. is coterminal with
$-174° + 360° = 186°$

5. 650 rotations per min $= \dfrac{650}{60}$ rotations per sec

$= \dfrac{65}{6}$ rotations per sec

$= 26$ rotations per 2.4 sec

$= 26(360°)$ per 2.4 sec $= 9360°$ per 2.4 sec

A point of the edge of the propeller will move $9360°$ in 2.4 sec.

7. $119°8'3'' = 119° + \dfrac{8}{60}° + \dfrac{3}{3600}°$
$\approx 119° + 0.1333° + 0.0008°$
$\approx 119.134°$

9. $275.1005 = 275° + 0.1005(60') = 275° + 6.03'$
$= 275°06' + 0.03'$
$= 275°06' + 0.03(60'')$
$= 275°6'1.8'' \approx 275°06'02''$

11. The three angles are the interior angles of a triangle, so the sum of their measures is 180°.
$4x + (5x + 5) + (4x - 20) = 180$
$13x - 15 = 180$
$13x = 195 \Rightarrow x = 15$
$4(15) = 60$, $5(15) + 5 = 80$, and
$4(15) - 20 = 40$, so the measures of the angles are 40°, 60°, and 80°.

13. The two indicated angles are alternate exterior angles, so their measures are equal.
$4x + 5 = 6x - 45 \Rightarrow 50 = 2x \Rightarrow 25 = x$
$4(25) + 5 = 105$

The measure of each indicated angle is 105°.

15. Assuming PQ and BA are parallel, $\triangle PCQ$ is similar to $\triangle ACB$ because the measure of $\angle PCQ$ is equal to the measure of $\angle ACB$ (they are vertical angles). The corresponding sides of similar triangles are proportional, so $\dfrac{PQ}{BA} = \dfrac{PC}{AC}$. Thus, we have $\dfrac{PQ}{BA} = \dfrac{PC}{AC} \Rightarrow$
$\dfrac{1.25 \text{ mm}}{BA} = \dfrac{150 \text{ mm}}{30 \text{ km}} \Rightarrow BA = \dfrac{1.25 \cdot 30}{150} \Rightarrow$
$BA = 0.25$ km

17. Angle R corresponds to angle P, so the measure of angle R is 82°. Angle M corresponds to angle S, so the measure of angle M is 86°. Angle N corresponds to angle Q, so the measure of angle N is 12°. Note: $12° + 82° + 86° = 180°$.

19. The large triangle is equilateral, so the smaller triangle is also equilateral, and $p = q = 7$.

21. $\dfrac{k}{6} = \dfrac{12 + 9}{9} \Rightarrow k = \dfrac{6 \cdot 21}{9} = 14$

23. Let x = the shadow of the 30-ft tree.
$\dfrac{20}{8} = \dfrac{30}{x} \Rightarrow 20x = 240 \Rightarrow x = 12$ ft

25. $x = 1$, $y = -\sqrt{3}$ and
$r = \sqrt{x^2 + y^2} = \sqrt{1^2 + (-\sqrt{3})^2}$
$= \sqrt{1 + 3} = \sqrt{4} = 2$

$\sin\theta = \dfrac{y}{r} = \dfrac{-\sqrt{3}}{2} = -\dfrac{\sqrt{3}}{2}$

$\cos\theta = \dfrac{x}{r} = \dfrac{1}{2}$

$\tan\theta = \dfrac{y}{x} = \dfrac{-\sqrt{3}}{1} = -\sqrt{3}$

$\cot\theta = \dfrac{x}{y} = \dfrac{1}{-\sqrt{3}} = -\dfrac{1}{\sqrt{3}} \cdot \dfrac{\sqrt{3}}{\sqrt{3}} = -\dfrac{\sqrt{3}}{3}$

$\sec\theta = \dfrac{r}{x} = \dfrac{2}{1} = 2$

$\csc\theta = \dfrac{r}{y} = \dfrac{2}{-\sqrt{3}} = -\dfrac{2}{\sqrt{3}} \cdot \dfrac{\sqrt{3}}{\sqrt{3}} = -\dfrac{2\sqrt{3}}{3}$

27. $x = 3$, $y = -4$ and $r = \sqrt{3^2 + (-4)^2} = \sqrt{25} = 5$

$\sin\theta = \dfrac{y}{r} = \dfrac{-4}{5} = -\dfrac{4}{5}$; $\cos\theta = \dfrac{x}{r} = \dfrac{3}{5}$

$\tan\theta = \dfrac{y}{x} = \dfrac{-4}{3} = -\dfrac{4}{3}$

$\cot\theta = \dfrac{x}{y} = \dfrac{3}{-4} = -\dfrac{3}{4}$

$\sec\theta = \dfrac{r}{x} = \dfrac{5}{3}$; $\csc\theta = \dfrac{r}{y} = \dfrac{5}{-4} = -\dfrac{5}{4}$

29. $x = -8$, $y = 15$, and
$r = \sqrt{(-8)^2 + 15^2} = \sqrt{289} = 17$

$\sin\theta = \dfrac{y}{r} = \dfrac{15}{17}$; $\cos\theta = \dfrac{x}{r} = \dfrac{-8}{17} = -\dfrac{8}{17}$

$\tan\theta = \dfrac{y}{x} = \dfrac{15}{-8} = -\dfrac{15}{8}$

$\cot\theta = \dfrac{x}{y} = \dfrac{-8}{15} = -\dfrac{8}{15}$

$\sec\theta = \dfrac{r}{x} = \dfrac{17}{-8} = -\dfrac{17}{8}$

$\csc\theta = \dfrac{r}{y} = \dfrac{17}{15}$

31. $x = 6\sqrt{3}$, $y = -6$, and
$r = \sqrt{(6\sqrt{3})^2 + (-6)^2} = \sqrt{108 + 36} = \sqrt{144} = 12$

$\sin\theta = \dfrac{y}{r} = \dfrac{-6}{12} = -\dfrac{1}{2}$

$\cos\theta = \dfrac{x}{r} = \dfrac{6\sqrt{3}}{12} = \dfrac{\sqrt{3}}{2}$

$\tan\theta = \dfrac{y}{x} = \dfrac{-6}{6\sqrt{3}} = -\dfrac{1}{\sqrt{3}} \cdot \dfrac{\sqrt{3}}{\sqrt{3}} = -\dfrac{\sqrt{3}}{3}$

(*continued on next page*)

(*continued*)

$$\cot\theta = \frac{x}{y} = \frac{6\sqrt{3}}{-6} = -\sqrt{3}$$

$$\sec\theta = \frac{r}{x} = \frac{12}{6\sqrt{3}} = \frac{2}{\sqrt{3}} \cdot \frac{\sqrt{3}}{\sqrt{3}} = \frac{2\sqrt{3}}{3}$$

$$\csc\theta = \frac{r}{y} = \frac{12}{-6} = -2$$

33. The terminal side of the angle is defined by
$5x - 3y = 0, x \geq 0,$ so a point on this terminal
side is $(3, 5)$.

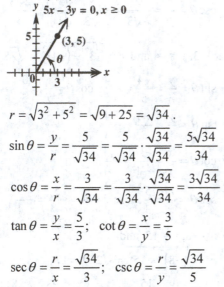

$$r = \sqrt{3^2 + 5^2} = \sqrt{9 + 25} = \sqrt{34}.$$

$$\sin\theta = \frac{y}{r} = \frac{5}{\sqrt{34}} = \frac{5}{\sqrt{34}} \cdot \frac{\sqrt{34}}{\sqrt{34}} = \frac{5\sqrt{34}}{34}$$

$$\cos\theta = \frac{x}{r} = \frac{3}{\sqrt{34}} = \frac{3}{\sqrt{34}} \cdot \frac{\sqrt{34}}{\sqrt{34}} = \frac{3\sqrt{34}}{34}$$

$$\tan\theta = \frac{y}{x} = \frac{5}{3}; \quad \cot\theta = \frac{x}{y} = \frac{3}{5}$$

$$\sec\theta = \frac{r}{x} = \frac{\sqrt{34}}{3}; \quad \csc\theta = \frac{r}{y} = \frac{\sqrt{34}}{5}$$

35. The terminal side of the angle is defined by
$12x + 5y = 0, x \geq 0,$ so a point on this terminal
side is $(5, -12)$.

$$r = \sqrt{(-12)^2 + 5^2} = \sqrt{144 + 25} = \sqrt{169} = 13.$$

$$\sin\theta = \frac{y}{r} = \frac{-12}{13} = -\frac{12}{13}; \quad \cos\theta = \frac{x}{r} = \frac{5}{13}$$

$$\tan\theta = \frac{y}{x} = \frac{-12}{5} = -\frac{12}{5}$$

$$\cot\theta = \frac{x}{y} = \frac{5}{-12} = -\frac{5}{12}$$

$$\sec\theta = \frac{r}{x} = \frac{13}{5}; \quad \csc\theta = \frac{r}{y} = \frac{13}{-12} = -\frac{13}{12}$$

37. $\sin(-90)° = -1$
$\cos(-90)° = 0$
$\tan(-90)°$ is undefined
$\cot(-90)° = 0$
$\sec(-90)°$ is undefined
$\csc(-90)° = -1$

39. $\cos\theta = -\dfrac{5}{8}$ and θ is in quadrant III

$\cos\theta = \dfrac{x}{r} = \dfrac{-5}{8}$ and θ in quadrant III, so let
$x = -5, r = 8.$
$x^2 + y^2 = r^2 \Rightarrow (-5)^2 + y^2 = 8^2 \Rightarrow$
$25 + y^2 = 64 \Rightarrow y^2 = 39 \Rightarrow y = \pm\sqrt{39}$
θ is in quadrant III, so $y = -\sqrt{39}.$

$$\sin\theta = \frac{y}{r} = \frac{-\sqrt{39}}{8} = -\frac{\sqrt{39}}{8}$$

$$\cos\theta = \frac{x}{r} = \frac{-5}{8} = -\frac{5}{8}$$

$$\tan\theta = \frac{y}{x} = \frac{-\sqrt{39}}{-5} = \frac{\sqrt{39}}{5}$$

$$\cot\theta = \frac{x}{y} = \frac{-5}{-\sqrt{39}} = \frac{5}{\sqrt{39}} \cdot \frac{\sqrt{39}}{\sqrt{39}} = \frac{5\sqrt{39}}{39}$$

$$\sec\theta = \frac{r}{x} = \frac{8}{-5} = -\frac{8}{5}$$

$$\csc\theta = \frac{r}{y} = \frac{8}{-\sqrt{39}} = -\frac{8}{\sqrt{39}}$$

$$= -\frac{8}{\sqrt{39}} \cdot \frac{\sqrt{39}}{\sqrt{39}} = -\frac{8\sqrt{39}}{39}$$

41. $\sec\theta = -\sqrt{5}$ and θ is in quadrant II
θ in quadrant II $\Rightarrow x < 0, y > 0$, so

$$\sec\theta = -\sqrt{5} = \frac{r}{x} = \frac{\sqrt{5}}{-1} \Rightarrow x = -1, r = \sqrt{5}$$

$$x^2 + y^2 = r^2 \Rightarrow (-1)^2 + y^2 = \left(\sqrt{5}\right)^2 \Rightarrow$$
$$1 + y^2 = 5 \Rightarrow y^2 = 4 \Rightarrow y = 2$$

$$\sin\theta = \frac{y}{r} = \frac{2}{\sqrt{5}} = \frac{2}{\sqrt{5}} \cdot \frac{\sqrt{5}}{\sqrt{5}} = \frac{2\sqrt{5}}{5}$$

$$\cos\theta = \frac{x}{r} = \frac{-1}{\sqrt{5}} = -\frac{1}{\sqrt{5}} \cdot \frac{\sqrt{5}}{\sqrt{5}} = -\frac{\sqrt{5}}{5}$$

$$\tan\theta = \frac{y}{x} = \frac{2}{-1} = -2; \quad \cot\theta = \frac{x}{y} = \frac{-1}{2} = -\frac{1}{2}$$

$$\sec\theta = \frac{r}{x} = \frac{\sqrt{5}}{-1} = -\sqrt{5}; \quad \csc\theta = \frac{r}{y} = \frac{\sqrt{5}}{2}$$

43. $\sec\theta = \dfrac{5}{4}$ and θ is in quadrant IV

θ in quadrant IV $\Rightarrow x > 0, y < 0$, so

$\sec\theta = \dfrac{r}{x} = \dfrac{5}{4} \Rightarrow x = 4, r = 5$

$x^2 + y^2 = r^2 \Rightarrow (4)^2 + y^2 = 5^2 \Rightarrow$

$16 + y^2 = 25 \Rightarrow y^2 = 9 \Rightarrow y = -3$

$\sin\theta = \dfrac{y}{r} = \dfrac{-3}{5} = -\dfrac{3}{5}$; $\cos\theta = \dfrac{x}{r} = \dfrac{4}{5}$

$\tan\theta = \dfrac{y}{x} = \dfrac{-3}{4} = -\dfrac{3}{4}$

$\cot\theta = \dfrac{x}{y} = \dfrac{4}{-3} = -\dfrac{4}{3}$

$\sec\theta = \dfrac{r}{x} = \dfrac{5}{4}$

$\csc\theta = \dfrac{r}{y} = \dfrac{5}{-3} = -\dfrac{5}{3}$

45. **(a)** Impossible because the range of $\sec\theta$ is $(-\infty, -1] \cup [1, \infty)$.

(b) Possible because the range of $\tan\theta$ is $(-\infty, \infty)$.

(c) Impossible because the range of $\cos\theta$ is $[-1, 1]$.

47. The triangles are similar, so

$\dfrac{20}{30} = \dfrac{x}{100 - x} \Rightarrow 20(100 - x) = 30x \Rightarrow$

$2000 - 20x = 30x \Rightarrow 2000 = 50x \Rightarrow 40 = x$

The lifeguard will enter the water 40 yards east of his original position.

49. Let x = the depth of the crater Autolycus. Then

$\dfrac{x}{11,000} = \dfrac{1.3}{1.5} \Rightarrow 1.5x = 1.3(11,000) \Rightarrow$

$1.5x = 14,300 \Rightarrow x \approx 9500$

Autoclycus is about 9500 feet deep.

Chapter 1 Test

1. $67°$

complement: $90° - 67° = 23°$

supplement: $180° - 67° = 113°$

2. The two angles are supplements, so their sum is $180°$.

$(7x + 19) + (2x - 1) = 180 \Rightarrow 9x + 18 = 180 \Rightarrow$

$\qquad\qquad 9x = 162 \Rightarrow x = 18$

$7(18) + 19 = 145; 2(18) - 1 = 35$

The measures of the angles are $145°$ and $35°$.

3. The two angles are complements, so their sum is $90°$.

$(-8x + 30) + (-3x + 5) = 90$

$\qquad\qquad -11x + 35 = 90$

$\qquad\qquad -11x = 55 \Rightarrow x = -5$

$-8(-5) + 30 = 70; -3(-5) + 5 = 20$

The angles measure $20°$ and $70°$.

4. The angles are vertical angles, so their measures are equal.

$4x - 30 = 5x - 70 \Rightarrow 40 = x$

$4(40) - 30 = 130; 5(40) - 70 = 130$

The angles each measure $130°$.

5. The angles are alternate interior angles, so their measures are equal.

$8x + 14 = 10x - 10 \Rightarrow 24 = 2x \Rightarrow 12 = x$

$8(12) + 14 = 110; 10(12) - 10 = 110$

The angles each measure $110°$.

6. The angles are interior angles of a triangle, so the sum of the three angles is $180°$.

$(2x + 18) + (20x + 10) + (32 - 2x) = 180$

$\qquad\qquad 20x + 60 = 180$

$\qquad\qquad 20x = 120$

$\qquad\qquad x = 6$

$2(6) + 18 = 30; 20(6) + 10 = 130; 32 - 2(6) = 20$

The three angles measure $30°$, $130°$, and $20°$.

7. Two of the angles are interior angles, but one is an exterior angle. From geometry, we know that the measure of an exterior angle is equal to the sum of the two nonadjacent interior angles.

$12x + 40 = 12x + 8x$

$\qquad 40 = 8x$

$\qquad 5 = x$

$12(5) + 40 = 100; 12(5) = 60; 8(5) = 40$

The three angles measure $100°$, $60°$, and $40°$.

8. $74°18'36'' = 74° + \dfrac{18}{60}° + \dfrac{36}{3600}°$

$\qquad\qquad \approx 74° + 0.3° + 0.01° = 74.31°$

9. $45.2025° = 45° + 0.2025°$

$\qquad\qquad = 45° + 0.2025(60')$

$\qquad\qquad = 45° + 12.15'$

$\qquad\qquad = 45° + 12' + 0.15'$

$\qquad\qquad = 45° + 12' + 0.15(60'')$

$\qquad\qquad = 45° + 12' + 09'' = 45°12'09''$

10. **(a)** $390°$ is coterminal with $390° - 360° = 30°$.

(b) $-80°$ is coterminal with $-80° + 360° = 280°$.

(c) 810° is coterminal with
$810° - 2(360°) = 810° - 720° = 90°$.

11. $\dfrac{450(360°)}{1 \text{ min}} = \dfrac{450(360°)}{60 \text{ sec}} = \dfrac{450(6°)}{\text{sec}}$
$\qquad = 2700°/\text{sec}$

A point on the tire rotates 2700° in one second.

12. $\dfrac{8}{30} = \dfrac{x}{40} \Rightarrow \dfrac{8 \cdot 40}{30} = x \Rightarrow x = 10\frac{2}{3}$

The shadow of the 40-ft pole is $10\frac{2}{3}$ ft, or 10 ft, 8 in.

13. $\dfrac{10}{25} = \dfrac{x}{20} \Rightarrow x = \dfrac{10 \cdot 20}{25} = 8$

$\dfrac{10}{25} = \dfrac{y}{15} \Rightarrow y = \dfrac{10 \cdot 15}{25} = 6$

14.

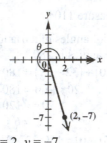

$x = 2, y = -7$

$r = \sqrt{x^2 + y^2} = \sqrt{2^2 + (-7)^2} = \sqrt{4 + 49} = \sqrt{53}$

$\sin\theta = \dfrac{y}{r} = \dfrac{-7}{\sqrt{53}} = -\dfrac{7}{\sqrt{53}} \cdot \dfrac{\sqrt{53}}{\sqrt{53}} = -\dfrac{7\sqrt{53}}{53}$

$\cos\theta = \dfrac{x}{r} = \dfrac{2}{\sqrt{53}} = \dfrac{2}{\sqrt{53}} \cdot \dfrac{\sqrt{53}}{\sqrt{53}} = \dfrac{2\sqrt{53}}{53}$

$\tan\theta = \dfrac{y}{x} = \dfrac{-7}{2} = -\dfrac{7}{2}$; $\quad \cot\theta = \dfrac{x}{y} = \dfrac{2}{-7} = -\dfrac{2}{7}$

$\sec\theta = \dfrac{r}{x} = \dfrac{\sqrt{53}}{2}$; $\quad \csc\theta = \dfrac{r}{y} = \dfrac{\sqrt{53}}{-7} = -\dfrac{\sqrt{53}}{7}$

15.

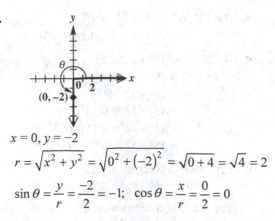

$x = 0, y = -2$

$r = \sqrt{x^2 + y^2} = \sqrt{0^2 + (-2)^2} = \sqrt{0 + 4} = \sqrt{4} = 2$

$\sin\theta = \dfrac{y}{r} = \dfrac{-2}{2} = -1$; $\quad \cos\theta = \dfrac{x}{r} = \dfrac{0}{2} = 0$

$\tan\theta = \dfrac{y}{x} = \dfrac{-2}{0}$ undefined

$\cot\theta = \dfrac{x}{y} = \dfrac{0}{-2} = 0$

$\sec\theta = \dfrac{r}{x} = \dfrac{2}{0}$ undefined

$\csc\theta = \dfrac{r}{y} = \dfrac{2}{-2} = -1$

16. Because $x \le 0$, the graph of the line $3x - 4y = 0$ is shown to the left of the y-axis.

A point on this graph is $(-4, -3)$. The corresponding value of r is

$r = \sqrt{(-4)^2 + (-3)^2} = \sqrt{16 + 9} = \sqrt{25} = 5$.

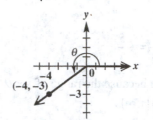

$$3x - 4y = 0, x \le 0$$
$x = -4, y = -3$

$\sin\theta = \dfrac{y}{r} = -\dfrac{3}{5}$; $\quad \cos\theta = \dfrac{x}{r} = -\dfrac{4}{5}$

$\tan\theta = \dfrac{y}{x} = \dfrac{3}{4}$; $\quad \cot\theta = \dfrac{x}{y} = \dfrac{4}{3}$

$\sec\theta = \dfrac{r}{x} = -\dfrac{5}{4}$; $\quad \csc\theta = \dfrac{r}{y} = -\dfrac{5}{3}$

17.

θ	90°	−360°	630°
$\sin\theta$	1	0	−1
$\cos\theta$	0	1	0
$\tan\theta$	undefined	0	undefined
$\cot\theta$	0	undefined	0
$\sec\theta$	undefined	1	undefined
$\csc\theta$	1	undefined	−1

18. If the terminal side of a quadrantal angle lies on the terminal side of a quadrantal angle lies on the negative part of the x-axis, any point on the terminal side would have the form $(k, 0)$, where k is any real number < 0.

$$\sin\theta = \frac{y}{r} = \frac{0}{r} = 0; \quad \cos\theta = \frac{x}{r} = \frac{k}{r}$$

$$\tan\theta = \frac{y}{x} = \frac{0}{k} = 0; \quad \cot\theta = \frac{x}{y} = \frac{k}{0} \text{ undefined}$$

$$\sec\theta = \frac{r}{x} = \frac{r}{k}; \quad \csc\theta = \frac{r}{y} = \frac{r}{0} \text{ undefined}$$

Thus, the cotangent and the cosecant are undefined.

19. (a) $\cos\theta > 0$ in quadrants I and IV, while $\tan\theta > 0$ in quadrants I and III. So, both conditions are met only in quadrant I.

(b) $\sin\theta < 0$ in quadrants III and IV. $\csc\theta$ is the reciprocal of $\sin\theta < 0$, so $\csc\theta < 0$ also in quadrants III and IV. Thus, both conditions are met in quadrants III and IV.

(c) $\cot\theta > 0$ in quadrants I and III, while $\cos < 0$ in quadrants II and III. Both conditions are met only in quadrant III.

20. (a) Impossible because the range of $\sin\theta$ is $[-1, 1]$.

(b) Possible because the range of $\sec\theta$ is $(-\infty, -1] \cup [1, \infty)$.

(c) Possible because the range of $\tan\theta$ is $(-\infty, \infty)$.

21. $\cos\theta = -\dfrac{7}{12} \Rightarrow \sec\theta = -\dfrac{12}{7}$

22. $\sin\theta = \dfrac{3}{7}$ with θ in quadrant II

θ in quadrant II $\Rightarrow x < 0, y > 0$

$$\sin\theta = \frac{y}{r} = \frac{3}{7} \Rightarrow y = 3, r = 7$$

$$r^2 = x^2 + y^2 \Rightarrow 7^2 = x^2 + 3^2 \Rightarrow 49 = x^2 + 9 \Rightarrow$$
$$x^2 = 40 \Rightarrow x = -\sqrt{40} = -2\sqrt{10}$$

$$\cos\theta = \frac{x}{r} = \frac{-2\sqrt{10}}{7} = -\frac{2\sqrt{10}}{7}$$

$$\tan\theta = \frac{y}{x} = \frac{3}{-2\sqrt{10}} = -\frac{3}{2\sqrt{10}} \cdot \frac{\sqrt{10}}{\sqrt{10}} = -\frac{3\sqrt{10}}{20}$$

$$\cot\theta = \frac{x}{y} = \frac{-2\sqrt{10}}{3}$$

$$\sec\theta = \frac{r}{x} = \frac{7}{-2\sqrt{10}} = -\frac{7}{2\sqrt{10}} \cdot \frac{\sqrt{10}}{\sqrt{10}} = -\frac{7\sqrt{10}}{20}$$

$$\csc\theta = \frac{r}{y} = \frac{7}{3}$$

Chapter 2

Acute Angles and Right Triangles

Section 2.1 Trigonometric Functions of Acute Angles

For Exercises 1–6, refer to the Function Values of Special Angles chart on page 52 of the text.

1. C; $\sin 30° = \dfrac{1}{2}$

2. H; $\cos 45° = \dfrac{\sqrt{2}}{2}$

3. B; $\tan 45° = 1$

5. E; $\csc 60° = \dfrac{1}{\sin 60°} = \dfrac{1}{\frac{\sqrt{3}}{2}} = \dfrac{2}{\sqrt{3}}$

$= \dfrac{2}{\sqrt{3}} \cdot \dfrac{\sqrt{3}}{\sqrt{3}} = \dfrac{2\sqrt{3}}{3}$

7. $\sin A = \dfrac{\text{side opposite}}{\text{hypotenuse}} = \dfrac{21}{29}$

$\cos A = \dfrac{\text{side adjacent}}{\text{hypotenuse}} = \dfrac{20}{29}$

$\tan A = \dfrac{\text{side opposite}}{\text{side adjacent}} = \dfrac{21}{20}$

9. $\sin A = \dfrac{\text{side opposite}}{\text{hypotenuse}} = \dfrac{n}{p}$

$\cos A = \dfrac{\text{side adjacent}}{\text{hypotenuse}} = \dfrac{m}{p}$

$\tan A = \dfrac{\text{side opposite}}{\text{side adjacent}} = \dfrac{n}{m}$

11. $a = 5, b = 12$
$c^2 = a^2 + b^2 \Rightarrow c^2 = 5^2 + 12^2 \Rightarrow c^2 = 169 \Rightarrow$
$c = 13$

$\sin B = \dfrac{\text{side opposite}}{\text{hypotenuse}} = \dfrac{b}{c} = \dfrac{12}{13}$

$\cos B = \dfrac{\text{side adjacent}}{\text{hypotenuse}} = \dfrac{a}{c} = \dfrac{5}{13}$

$\tan B = \dfrac{\text{side opposite}}{\text{side adjacent}} = \dfrac{b}{a} = \dfrac{12}{5}$

$\cot B = \dfrac{\text{side adjacent}}{\text{side opposite}} = \dfrac{a}{b} = \dfrac{5}{12}$

$\sec B = \dfrac{\text{hypotenuse}}{\text{side adjacent}} = \dfrac{c}{a} = \dfrac{13}{5}$

$\csc B = \dfrac{\text{hypotenuse}}{\text{side opposite}} = \dfrac{c}{b} = \dfrac{13}{12}$

13. $a = 6, c = 7$
$c^2 = a^2 + b^2 \Rightarrow 7^2 = 6^2 + b^2 \Rightarrow$
$49 = 36 + b^2 \Rightarrow 13 = b^2 \Rightarrow \sqrt{13} = b$

$\sin B = \dfrac{\text{side opposite}}{\text{hypotenuse}} = \dfrac{b}{c} = \dfrac{\sqrt{13}}{7}$

$\cos B = \dfrac{\text{side adjacent}}{\text{hypotenuse}} = \dfrac{a}{c} = \dfrac{6}{7}$

$\tan B = \dfrac{\text{side opposite}}{\text{side adjacent}} = \dfrac{b}{a} = \dfrac{\sqrt{13}}{6}$

$\cot B = \dfrac{\text{side adjacent}}{\text{side opposite}} = \dfrac{a}{b} = \dfrac{6}{\sqrt{13}}$

$= \dfrac{6}{\sqrt{13}} \cdot \dfrac{\sqrt{13}}{\sqrt{13}} = \dfrac{6\sqrt{13}}{13}$

$\sec B = \dfrac{\text{hypotenuse}}{\text{side adjacent}} = \dfrac{c}{a} = \dfrac{7}{6}$

$\csc B = \dfrac{\text{hypotenuse}}{\text{side opposite}} = \dfrac{c}{b} = \dfrac{7}{\sqrt{13}}$

$= \dfrac{7}{\sqrt{13}} \cdot \dfrac{\sqrt{13}}{\sqrt{13}} = \dfrac{7\sqrt{13}}{13}$

15. $a = 3, c = 10$
$c^2 = a^2 + b^2 \Rightarrow 10^2 = 3^2 + b^2 \Rightarrow$
$100 = 9 + b^2 \Rightarrow 91 = b^2 \Rightarrow \sqrt{91} = b$

$\sin B = \dfrac{\text{side opposite}}{\text{hypotenuse}} = \dfrac{b}{c} = \dfrac{\sqrt{91}}{10}$

$\cos B = \dfrac{\text{side adjacent}}{\text{hypotenuse}} = \dfrac{a}{c} = \dfrac{3}{10}$

$\tan B = \dfrac{\text{side opposite}}{\text{side adjacent}} = \dfrac{b}{a} = \dfrac{\sqrt{91}}{3}$

$\cot B = \dfrac{\text{side adjacent}}{\text{side opposite}} = \dfrac{a}{b} = \dfrac{3}{\sqrt{91}}$

$= \dfrac{3}{\sqrt{91}} \cdot \dfrac{\sqrt{91}}{\sqrt{91}} = \dfrac{3\sqrt{91}}{91}$

(*continued on next page*)

(continued)

$$\sec B = \frac{\text{hypotenuse}}{\text{side adjacent}} = \frac{c}{a} = \frac{10}{3}$$

$$\csc B = \frac{\text{hypotenuse}}{\text{side opposite}} = \frac{c}{b} = \frac{10}{\sqrt{91}}$$

$$= \frac{10}{\sqrt{91}} \cdot \frac{\sqrt{91}}{\sqrt{91}} = \frac{10\sqrt{91}}{91}$$

17. $a = 1,\ c = 2$

$$c^2 = a^2 + b^2 \Rightarrow 2^2 = 1^2 + b^2 \Rightarrow$$
$$4 = 1 + b^2 \Rightarrow 3 = b^2 \Rightarrow \sqrt{3} = b$$

$$\sin B = \frac{\text{side opposite}}{\text{hypotenuse}} = \frac{b}{c} = \frac{\sqrt{3}}{2}$$

$$\cos B = \frac{\text{side adjacent}}{\text{hypotenuse}} = \frac{a}{c} = \frac{1}{2}$$

$$\tan B = \frac{\text{side opposite}}{\text{side adjacent}} = \frac{b}{a} = \frac{\sqrt{3}}{1} = \sqrt{3}$$

$$\cot B = \frac{\text{side adjacent}}{\text{side opposite}} = \frac{a}{b} = \frac{1}{\sqrt{3}}$$

$$= \frac{1}{\sqrt{3}} \cdot \frac{\sqrt{3}}{\sqrt{3}} = \frac{\sqrt{3}}{3}$$

$$\sec B = \frac{\text{hypotenuse}}{\text{side adjacent}} = \frac{c}{a} = \frac{2}{1} = 2$$

$$\csc B = \frac{\text{hypotenuse}}{\text{side opposite}} = \frac{c}{b} = \frac{2}{\sqrt{3}}$$

$$= \frac{2}{\sqrt{3}} \cdot \frac{\sqrt{3}}{\sqrt{3}} = \frac{2\sqrt{3}}{3}$$

19. $b = 2,\ c = 5$

$$c^2 = a^2 + b^2 \Rightarrow 5^2 = a^2 + 2^2 \Rightarrow$$
$$25 = a^2 + 4 \Rightarrow 21 = a^2 \Rightarrow \sqrt{21} = a$$

$$\sin B = \frac{\text{side opposite}}{\text{hypotenuse}} = \frac{b}{c} = \frac{2}{5}$$

$$\cos B = \frac{\text{side adjacent}}{\text{hypotenuse}} = \frac{a}{c} = \frac{\sqrt{21}}{5}$$

$$\tan B = \frac{\text{side opposite}}{\text{side adjacent}} = \frac{b}{a} = \frac{2}{\sqrt{21}}$$

$$= \frac{2}{\sqrt{21}} \cdot \frac{\sqrt{21}}{\sqrt{21}} = \frac{2\sqrt{21}}{21}$$

$$\cot B = \frac{\text{side adjacent}}{\text{side opposite}} = \frac{a}{b} = \frac{\sqrt{21}}{2}$$

$$\sec B = \frac{\text{hypotenuse}}{\text{side adjacent}} = \frac{c}{a} = \frac{5}{\sqrt{21}}$$

$$= \frac{5}{\sqrt{21}} \cdot \frac{\sqrt{21}}{\sqrt{21}} = \frac{5\sqrt{21}}{21}$$

$$\csc B = \frac{\text{hypotenuse}}{\text{side opposite}} = \frac{c}{b} = \frac{5}{2}$$

21. $\cos 30° = \sin(90° - 30°) = \sin 60°$

23. $\csc 60° = \sec(90° - 60°) = \sec 30°$

25. $\sec 39° = \csc(90° - 39°) = \csc 51°$

27. $\sin 38.7° = \cos(90° - 38.7°) = \cos 51.3°$

29. $\sec(\theta + 15°) = \csc\left[90° - (\theta + 15°)\right]$
$$= \csc(75° - \theta)$$

For exercises 31–39, if the functions in the equations are cofunctions, then the equations are true if the sum of the angles is 90°.

31. $\tan \alpha = \cot(\alpha + 10°)$
$$\alpha + (\alpha + 10°) = 90°$$
$$2\alpha + 10° = 90°$$
$$2\alpha = 80° \Rightarrow \alpha = 40°$$

33. $\sin(2\theta + 10°) = \cos(3\theta - 20°)$
$$(2\theta + 10°) + (3\theta - 20°) = 90°$$
$$5\theta - 10° = 90°$$
$$5\theta = 100° \Rightarrow \theta = 20°$$

35. $\tan(3\beta + 4°) = \cot(5\beta - 10°)$
$$(3\beta + 4°) + (5\beta - 10°) = 90°$$
$$8\beta - 6° = 90°$$
$$8\beta = 96° \Rightarrow \beta = 12°$$

37. $\sin(\theta - 20°) = \cos(2\theta + 5°)$
$$(\theta - 20°) + (2\theta + 5°) = 90°$$
$$3\theta - 15° = 90°$$
$$3\theta = 105° \Rightarrow \theta = 35°$$

39. $\sec(3\beta + 10°) = \csc(\beta + 8°)$
$$(3\beta + 10°) + (\beta + 8°) = 90°$$
$$4\beta + 18° = 90°$$
$$4\beta = 72° \Rightarrow \beta = 18°$$

41. $\sin 50° > \sin 40°$
In the interval from 0° to 90°, as the angle increases, so does the sine of the angle, so $\sin 50° > \sin 40°$ is true.

43. $\sin 46° < \cos 46°$.
Using the cofunction identity, $\cos 46° = \sin(90° - 46°) = \sin 44°$. In the interval from 0° to 90°, as the angle increases, so does the sine of the angle, so $\sin 46° < \sin 44° \Rightarrow \sin 46° < \cos 46°$ is false.

45. $\tan 41° < \cot 41°$
Using the cofunction identity,
$\cot 41° = \tan\left(90° - 41°\right) = \tan 49°$. In the interval from 0° to 90°, as the angle increases, the tangent of the angle increases, so $\tan 41° < \tan 49° \Rightarrow \tan 41° < \cot 41°$ is true.

47. $\sec 60° > \sec 30°$
In the interval from 0° to 90°, as the angle increases, the cosine of the angle decreases, so the secant of the angle increases. Thus, $\sec 60° > \sec 30°$ is true.

Use the following figures for exercises 49–63.

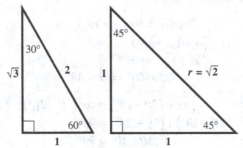

49. $\tan 30° = \dfrac{\text{side opposite}}{\text{side adjacent}} = \dfrac{1}{\sqrt{3}}$
$= \dfrac{1}{\sqrt{3}} \cdot \dfrac{\sqrt{3}}{\sqrt{3}} = \dfrac{\sqrt{3}}{3}$

51. $\sin 30° = \dfrac{\text{side opposite}}{\text{hypotenuse}} = \dfrac{1}{2}$

53. $\sec 30° = \dfrac{\text{hypotenuse}}{\text{side adjacent}} = \dfrac{2}{\sqrt{3}}$
$= \dfrac{2}{\sqrt{3}} \cdot \dfrac{\sqrt{3}}{\sqrt{3}} = \dfrac{2\sqrt{3}}{3}$

55. $\csc 45° = \dfrac{\text{hypotenuse}}{\text{side opposite}} = \dfrac{\sqrt{2}}{1} = \sqrt{2}$

57. $\cos 45° = \dfrac{\text{side adjacent}}{\text{hypotenuse}} = \dfrac{1}{\sqrt{2}}$
$= \dfrac{1}{\sqrt{2}} \cdot \dfrac{\sqrt{2}}{\sqrt{2}} = \dfrac{\sqrt{2}}{2}$

59. $\tan 45° = \dfrac{\text{side opposite}}{\text{side adjacent}} = \dfrac{1}{1} = 1$

61. $\sin 60° = \dfrac{\text{side opposite}}{\text{hypotenuse}} = \dfrac{\sqrt{3}}{2}$

63. $\tan 60° = \dfrac{\text{side opposite}}{\text{side adjacent}} = \dfrac{\sqrt{3}}{1} = \sqrt{3}$

65. Because $\sin 60° = \dfrac{\sqrt{3}}{2}$ and 60° is between 0° and 90°, A = 60°.

67.

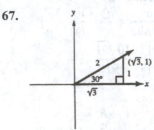

The line passes through (0, 0) and $\left(\sqrt{3},\ 1\right)$. The slope is change in y over the change in x.

Thus, $m = \dfrac{1}{\sqrt{3}} = \dfrac{1}{\sqrt{3}} \cdot \dfrac{\sqrt{3}}{\sqrt{3}} = \dfrac{\sqrt{3}}{3}$ and the equation of the line is $y = \dfrac{\sqrt{3}}{3}x$.

69. One point on the line $y = \sqrt{3}x$ is the origin $(0,0)$. Let (x, y) be any other point on this line. Then, by the definition of slope,
$m = \dfrac{y-0}{x-0} = \dfrac{y}{x} = \sqrt{3}$, but also, by the definition of tangent, $\tan \theta = \sqrt{3}$. Because $\tan 60° = \sqrt{3}$, the line $y = \sqrt{3}x$ makes a 60° angle with the positive x-axis (See exercise 68).

71. (a) Each of the angles of the equilateral triangle has measure $\frac{1}{3}(180°) = 60°$.

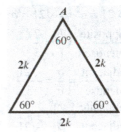

(b) The perpendicular bisects the opposite side so the length of each side opposite each 30° angle is k.

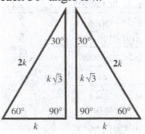

(c) Let x = the length of the perpendicular. Then apply the Pythagorean theorem.

$$x^2 + k^2 = (2k)^2 \Rightarrow x^2 + k^2 = 4k^2 \Rightarrow$$
$$x^2 = 3k^2 \Rightarrow x = \sqrt{3}k$$

The length of the perpendicular is $\sqrt{3}k$.

(d) In a 30°-60° right triangle, the hypotenuse is always $\underline{2}$ times as long as the shorter leg, and the longer leg has a length that is $\underline{\sqrt{3}}$ times as long as that of the shorter leg. Also, the shorter leg is opposite the $\underline{30°}$ angle, and the longer leg is opposite the $\underline{60°}$ angle.

73. Apply the relationships between the lengths of the sides of a $30° - 60°$ right triangle first to the triangle on the left to find the values of y and x, and then to the triangle on the right to find the values of z and w. In the $30° - 60°$ right triangle, the side opposite the 30° angle is $\frac{1}{2}$ the length of the hypotenuse. The longer leg is $\sqrt{3}$ times the shorter leg.

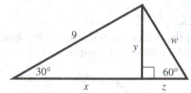

Thus, we have

$$y = \frac{1}{2}(9) = \frac{9}{2} \text{ and } x = y\sqrt{3} = \frac{9\sqrt{3}}{2}$$

$$y = z\sqrt{3}, \text{ so } z = \frac{y}{\sqrt{3}} = \frac{\frac{9}{2}}{\sqrt{3}} = \frac{9\sqrt{3}}{6} = \frac{3\sqrt{3}}{2},$$

and $w = 2z$, so $w = 2\left(\frac{3\sqrt{3}}{2}\right) = 3\sqrt{3}$

75. Apply the relationships between the lengths of the sides of a $45° - 45°$ right triangle to the triangle on the left to find the values of p and r. In the $45° - 45°$ right triangle, the sides opposite the 45° angles measure the same. The hypotenuse is $\sqrt{2}$ times the measure of a leg.

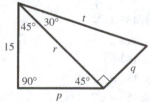

Thus, we have $p = 15$ and $r = p\sqrt{2} = 15\sqrt{2}$
Apply the relationships between the lengths of the sides of a $30° - 60°$ right triangle next to the triangle on the right to find the values of q and t. In the $30° - 60°$ right triangle, the side opposite the 60° angle is $\sqrt{3}$ times as long as the side opposite to the 30° angle. The length of the hypotenuse is 2 times as long as the shorter leg (opposite the 30° angle).
Thus, we have $r = q\sqrt{3} \Rightarrow$

$$q = \frac{r}{\sqrt{3}} = \frac{15\sqrt{2}}{\sqrt{3}} = \frac{15\sqrt{2}}{\sqrt{3}} \cdot \frac{\sqrt{3}}{\sqrt{3}} = 5\sqrt{6} \text{ and}$$

$$t = 2q = 2(5\sqrt{6}) = 10\sqrt{6}$$

77. Because $A = \frac{1}{2}bh$, we have

$$A = \frac{1}{2} \cdot s \cdot s = \frac{1}{2}s^2 \text{ or } A = \frac{s^2}{2}.$$

79. Graph the equations in the window $[-1.5, 1.5] \times [-1.2, 1.2]$. The point of intersection is $(0.70710678, 0.70710678)$. This corresponds to the point $\left(\frac{\sqrt{2}}{2}, \frac{\sqrt{2}}{2}\right)$.

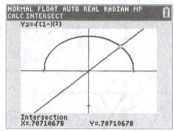

These coordinates are the sine and cosine of 45°.

81.

83. The legs of the right triangle provide the coordinates of P, $\left(2\sqrt{2}, 2\sqrt{2}\right)$.

Section 2.2 Trigonometric Functions of Non-Acute Angles

1. The value of sin 240° is <u>negative</u> because 240° is in quadrant <u>III</u>. The reference angle is <u>60°</u>, and the exact value of sin 240° is $-\frac{\sqrt{3}}{2}$.

3. The value of tan (−150°) is <u>positive</u> because −150° is in quadrant <u>III</u>. The reference angle is <u>30°</u> and the exact value of tan (−150°) is $\frac{\sqrt{3}}{3}$.

5. C; $180° - 98° = 82°$
    (98° is in quadrant II)

7. A; $-135° + 360° = 225°$ and
    $225° - 180° = 45°$
    (225° is in quadrant III)

9. D; $750° - 2 \cdot 360° = 30°$
    (30° is in quadrant I)

	θ	$\sin\theta$	$\cos\theta$	$\tan\theta$	$\cot\theta$	$\sec\theta$	$\csc\theta$
11.	30°	$\frac{1}{2}$	$\frac{\sqrt{3}}{2}$	$\frac{\sqrt{3}}{3}$	$\sqrt{3}$	$\frac{2\sqrt{3}}{3}$	2
13.	60°	$\frac{\sqrt{3}}{2}$	$\frac{1}{2}$	$\sqrt{3}$	$\frac{\sqrt{3}}{3}$	2	$\frac{2\sqrt{3}}{3}$
15.	135°	$\frac{\sqrt{2}}{2}$	$-\frac{\sqrt{2}}{2}$	$\tan 135° = -\tan 45° = -1$	$\cot 135° = -\cot 45° = -1$	$-\sqrt{2}$	$\sqrt{2}$
17.	210°	$-\frac{1}{2}$	$\cos 210° = -\cos 30° = -\frac{\sqrt{3}}{2}$	$\frac{\sqrt{3}}{3}$	$\sqrt{3}$	$\sec 210° = -\sec 30° = -\frac{2\sqrt{3}}{3}$	-2

19. To find the reference angle for 300°, sketch this angle in standard position.

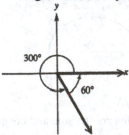

The reference angle is $360° - 300° = 60°$. Because 300° lies in quadrant IV, the sine, tangent, cotangent, and cosecant are negative.

$\sin 300° = -\sin 60° = -\frac{\sqrt{3}}{2}$

$\cos 300° = \cos 60° = \frac{1}{2}$

$\tan 300° = -\tan 60° = -\sqrt{3}$

$\cot 300° = -\cot 60° = -\frac{\sqrt{3}}{3}$

$\sec 300° = \sec 60° = 2$

$\csc 300° = -\csc 60° = -\frac{2\sqrt{3}}{3}$

21. To find the reference angle for 405°, sketch this angle in standard position.

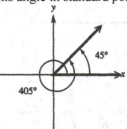

The reference angle for 405° is $405° - 360° = 45°$. Because 405° lies in quadrant I, the values of all of its trigonometric functions will be positive, so these values will be identical to the trigonometric function values for 45° See the Function Values of Special Angles table on page 52.)

$\sin 405° = \sin 45° = \frac{\sqrt{2}}{2}$

$\cos 405° = \cos 45° = \frac{\sqrt{2}}{2}$

$\tan 405° = \tan 45° = 1$

$\cot 405° = \cot 45° = 1$

$\sec 405° = \sec 45° = \sqrt{2}$

$\csc 405° = \csc 45° = \sqrt{2}$

23. To find the reference angle for 480°, sketch this angle in standard position.

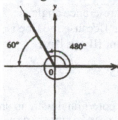

480° is coterminal with 480° − 360° = 120°. The reference angle is 180° − 120° = 60°. Because 480° lies in quadrant II, the cosine, tangent, cotangent, and secant are negative.

$$\sin\left(480°\right) = \sin 60° = \frac{\sqrt{3}}{2}$$

$$\cos\left(480°\right) = -\cos 60° = -\frac{1}{2}$$

$$\tan\left(480°\right) = -\tan 60° = -\sqrt{3}$$

$$\cot\left(480°\right) = -\cot 60° = -\frac{\sqrt{3}}{3}$$

$$\sec\left(80°\right) = -\sec 60° = -2$$

$$\csc\left(480°\right) = \csc 60° = \frac{2\sqrt{3}}{3}$$

25. To find the reference angle for 570° sketch this angle in standard position.

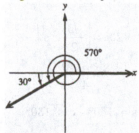

570° is coterminal with 570° − 360° = 210°. The reference angle is 210° − 180° = 30°. Because 570° lies in quadrant III, the sine, cosine, secant, and cosecant are negative.

$$\sin 570° = -\sin 30° = -\frac{1}{2}$$

$$\cos 570° = -\cos 30° = -\frac{\sqrt{3}}{2}$$

$$\tan 570° = \tan 30° = \frac{\sqrt{3}}{3}$$

$$\cot 570° = \cot 30° = \sqrt{3}$$

$$\sec 570° = -\sec 30° = -\frac{2\sqrt{3}}{3}$$

$$\csc 570° = -\csc 30° = -2$$

27. 1305° is coterminal with
1305° − 3 · 360° = 1305° = 1080° = 225°. The reference angle is 225° − 180° = 45°. Because 1305° lies in quadrant III, the sine, cosine, and secant and cosecant are negative.

$$\sin 1305° = -\sin 45° = -\frac{\sqrt{2}}{2}$$

$$\cos 1305° = -\cos 45° = -\frac{\sqrt{2}}{2}$$

$$\tan 1305° = \tan 45° = 1$$

$$\cot 1305° = \cot 45° = 1$$

$$\sec 1305° = -\sec 45° = -\sqrt{2}$$

$$\csc 1305° = -\csc 45° = -\sqrt{2}$$

29. To find the reference angle for −300°, sketch this angle in standard position.

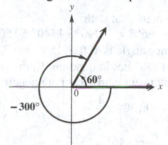

The reference angle for −300° is −300° + 360° = 60°. Because −300° lies in quadrant I, the values of all of its trigonometric functions will be positive, so these values will be identical to the trigonometric function values for 60°. See the Function Values of Special Angles table on page 52.)

$$\sin\left(-300°\right) = \sin 60° = \frac{\sqrt{3}}{2}$$

$$\cos\left(-300°\right) = \cos 60° = \frac{1}{2}$$

$$\tan\left(-300°\right) = \tan 60° = \sqrt{3}$$

$$\cot\left(-300°\right) = \cot 60° = \frac{\sqrt{3}}{3}$$

$$\sec\left(-300°\right) = \sec 60° = 2$$

$$\csc\left(-300°\right) = \csc 60° = \frac{2\sqrt{3}}{3}$$

31. $-510°$ is coterminal with
$-510° + 2 \cdot 360° = -510° + 720° = 210°$. The reference angle is $210° - 180° = 30°$. Because $-510°$ lies in quadrant III, the sine, cosine, and secant and cosecant are negative.

$$\sin(-510°) = -\sin 30° = -\frac{1}{2}$$

$$\cos(-510°) = -\cos 30° = -\frac{\sqrt{3}}{2}$$

$$\tan(-510°) = \tan 30° = \frac{\sqrt{3}}{3}$$

$$\cot(-510°) = \cot 30° = \sqrt{3}$$

$$\sec(-510°) = -\sec 30° = -\frac{2\sqrt{3}}{3}$$

$$\csc(-510°) = -\csc 30° = -2$$

33. $-1290°$ is coterminal with
$-1290° + 4 \cdot 360° = -1290° + 1440° = 150°$. The reference angle is $180° - 150° = 30°$. Because $-1290°$ lies in quadrant II, the cosine, tangent, cotangent, and secant are negative.

$$\sin 2670° = \sin 30° = \frac{1}{2}$$

$$\cos 2670° = -\cos 30° = -\frac{\sqrt{3}}{2}$$

$$\tan 2670° = -\tan 30° = -\frac{\sqrt{3}}{3}$$

$$\cot 2670° = -\cot 30° = -\sqrt{3}$$

$$\sec 2670° = -\sec 30° = -\frac{2\sqrt{3}}{3}$$

$$\csc 2670° = \csc 30° = 2$$

35. $-1860°$ is coterminal with
$-1860° + 6 \cdot 360° = -1860° + 2160° = 300°$. The reference angle is $360° - 300° = 60°$. Because $-1860°$ lies in quadrant IV, the sine, tangent, cotangent, and cosecant are negative.

$$\sin(-1860°) = -\sin 60° = -\frac{\sqrt{3}}{2}$$

$$\cos(-1860°) = \cos 60° = \frac{1}{2}$$

$$\tan(-1860°) = -\tan 60° = -\sqrt{3}$$

$$\cot(-1860°) = -\cot 60° = -\frac{\sqrt{3}}{3}$$

$$\sec(-1860°) = \sec 60° = 2$$

$$\csc(-1860°) = -\csc 60° = -\frac{2\sqrt{3}}{3}$$

37. Because $1305°$ is coterminal with an angle of $1305° - 3 \cdot 360° = 1305° - 1080° = 225°$, it lies in quadrant III. Its reference angle is $225° - 180° = 45°$. Because the sine is negative in quadrant III, we have

$$\sin 1305° = -\sin 45° = -\frac{\sqrt{2}}{2}.$$

39. Because $-510°$ is coterminal with an angle of $-510° + 2 \cdot 360° = -510° + 720° = 210°$, it lies in quadrant III. Its reference angle is $210° - 180° = 30°$. The cosine is negative in quadrant III, so

$$\cos(-510°) = -\cos 30° = -\frac{\sqrt{3}}{2}.$$

41. Because $-855°$ is coterminal with
$-855° + 3 \cdot 360° = -855° + 1080° = 225°$, it lies in quadrant III. Its reference angle is $225° - 180° = 45°$. The cosecant is negative in quadrant III, so

$$\csc(-855°) = -\csc 45° = -\sqrt{2}.$$

43. Because $3015°$ is coterminal with
$3015° - 8 \cdot 360° = 3015° - 2880° = 135°$, it lies in quadrant II. Its reference angle is $180° - 135° = 45°$. The tangent is negative in quadrant II, so $\tan 3015° = -\tan 45° = -1$.

45. $\sin^2 120° + \cos^2 120° = \left(\frac{\sqrt{3}}{2}\right)^2 + \left(\frac{1}{2}\right)^2$

$$= \frac{3}{4} + \frac{1}{4} = 1$$

47. $2\tan^2 120° + 3\sin^2 150° - \cos^2 180°$

$$= 2\left(-\sqrt{3}\right)^2 + 3\left(\frac{1}{2}\right)^2 - (-1)^2$$

$$= 2(3) + 3\left(\frac{1}{4}\right) - 1 = \frac{23}{4}$$

49. $\sin^2 225° - \cos^2 270° + \tan^2 60°$

$$= \left(-\frac{\sqrt{2}}{2}\right)^2 + 0^2 + \left(\sqrt{3}\right)^2 = \frac{2}{4} + 3 = \frac{7}{2}$$

51. $\cos^2 60° + \sec^2 150° - \csc^2 210°$

$$= \left(\frac{1}{2}\right)^2 + \left(-\frac{2\sqrt{3}}{3}\right)^2 - (-2)^2$$

$$= \frac{1}{4} + \frac{4}{3} - 4 = -\frac{29}{12}$$

53. $\cos(30° + 60°) \overset{?}{=} \cos 30° + \cos 60°$

Evaluate each side to determine whether this statement is true or false.

$\cos(30° + 60°) = \cos 90° = 0$ and

$\cos 30° + \cos 60° = \dfrac{\sqrt{3}}{2} + \dfrac{1}{2} = \dfrac{\sqrt{3}+1}{2}$

$0 \neq \dfrac{\sqrt{3}+1}{2}$, so the statement is false.

55. $\cos 60° \overset{?}{=} 2\cos 30°$

Evaluate each side to determine whether this statement is true or false.

$\cos 60° = \dfrac{1}{2}$ and $2\cos 30° = 2\left(\dfrac{\sqrt{3}}{2}\right) = \sqrt{3}$

Because $\dfrac{1}{2} \neq \sqrt{3}$, the statement is false.

57. $\sin^2 45° + \cos^2 45° \overset{?}{=} 1$

$\left(\dfrac{\sqrt{2}}{2}\right)^2 + \left(\dfrac{\sqrt{2}}{2}\right)^2 = \dfrac{2}{4} + \dfrac{2}{4} = 1$

Because $1 = 1$, the statement is true.

59. $\cos(2 \cdot 45)° \overset{?}{=} 2\cos 45°$

Evaluate each side to determine whether this statement is true or false.

$\cos(2 \cdot 45)° = \cos 90° = 0$ and

$2\cos 45° = 2\left(\dfrac{\sqrt{2}}{2}\right) = \sqrt{2}$

Because $0 \neq \sqrt{2}$, the statement is false.

61. $\sin \theta = \dfrac{1}{2}$

Because $\sin \theta$ is positive, θ must lie in quadrants I or II. Because one angle, namely 30°, lies in quadrant I, that angle is also the reference angle, θ'. The angle in quadrant II will be $180° - \theta' = 180° - 30° = 150°$.

63. $\tan \theta = -\sqrt{3}$

Because $\tan \theta$ is negative, θ must lie in quadrants II or IV. The absolute value of $\tan \theta$ is $\sqrt{3}$, so the reference angle, θ' must be 60°. The quadrant II angle θ equals $180° - \theta' = 180° - 60° = 120°$, and the quadrant IV angle θ equals $360° - \theta' = 360° - 60° = 300°$.

65. $\cos \theta = \dfrac{\sqrt{2}}{2}$

Because $\cos \theta$ is positive, θ must lie in quadrants I or IV. One angle, namely 45°, lies in quadrant I, so that angle is also the reference angle, θ'. The angle in quadrant IV will be $360° - \theta' = 360° - 45° = 315°$.

67. $\csc \theta = -2$

Because $\csc \theta$ is negative, θ must lie in quadrants III or IV. The absolute value of $\csc \theta$ is 2, so the reference angle, θ', is 30°. The angle in quadrant III will be $180° + \theta' = 180° + 30° = 210°$, and the quadrant IV angle is $360° - \theta' = 360° - 30° = 330°$.

69. $\tan \theta = \dfrac{\sqrt{3}}{3}$

Because $\tan \theta$ is positive, θ must lie in quadrants I or III. One angle, namely 30°, lies in quadrant I, so that angle is also the reference angle, θ'. The angle in quadrant III will be $180° + \theta' = 180° + 30° = 210°$.

71. $\csc \theta = -\sqrt{2}$

Because $\csc \theta$ is negative, θ must lie in quadrants III or IV. The absolute value of $\csc \theta$ is $\sqrt{2}$, so the reference angle, θ' must be 45°. The quadrant III angle θ equals $180° + \theta' = 180° + 45° = 225°$. and the quadrant IV angle θ equals $360° - \theta' = 360° - 45° = 315°$.

73. 150° is in quadrant II, so the reference angle is $180° - 150° = 30°$.

$\cos 30° = \dfrac{x}{r} \Rightarrow x = r\cos 30° = 6 \cdot \dfrac{\sqrt{3}}{2} = 3\sqrt{3}$

and $\sin 30° = \dfrac{y}{r} \Rightarrow y = r\sin 30° = 6 \cdot \dfrac{1}{2} = 3$

Because 150° is in quadrant II, the x- coordinate will be negative. The coordinates of P are $\left(-3\sqrt{3}, 3\right)$.

75. For every angle θ, $\sin^2 \theta + \cos^2 \theta = 1$. Because $(-0.8)^2 + (0.6)^2 = 0.64 + 0.36 = 1$, there is an angle θ for which $\cos \theta = 0.6$ and $\sin \theta = -0.8$. Because $\cos \theta > 0$ and $\sin \theta < 0$, θ lies in quadrant IV.

77. If θ is in the interval $(90°, 180°)$, then

$$90° < \theta < 180° \Rightarrow 45° < \frac{\theta}{2} < 90°. \text{ Thus } \frac{\theta}{2}$$

lies in quadrant I, and $\cos\frac{\theta}{2}$ is positive.

79. If θ is in the interval $(90°, 180°)$, then

$90° < \theta < 180° \Rightarrow 270° < \theta + 180° < 360°$.
Thus $\theta + 180°$ lies in quadrant IV, and
$\sec(\theta + 180°)$ is positive.

81. If θ is in the interval $(90°, 180°)$, then

$90° < \theta < 180° \Rightarrow -90° > -\theta > -180° \Rightarrow$
$-180° < -\theta < -90°$
Because $180°$ is coterminal with
$-180° + 360° = 180°$ and $-90°$ is coterminal
with $-90° + 360° = 270°$, $-\theta$ lies in quadrant
III, and $\sin(-\theta)$ is negative.

83. When an integer multiple of $360°$ is added to
θ, the resulting angle is coterminal with θ.
The sine values of coterminal angles are equal.

85. The reference angle for $115°$ is
$180° - 115° = 65°$. Because $115°$ is in quadrant
II the cosine is negative. Sin θ decreases on
the interval $(90°, 180°)$ from 1 to 0. Therefore,
$\sin 115°$ is closest to 0.9.

87. When $\theta = 45°$, $\sin\theta = \cos\theta = \frac{\sqrt{2}}{2}$. Sine and
cosine are both positive in quadrant I and both
negative in quadrant III. Because
$\theta + 180° = 45° + 180° = 225°$, $45°$ is the
quadrant I angle, and $225°$ is the quadrant III
angle.

Section 2.3 Approximations of Trigonometric Function Values

For Exercises 1–10, be sure your calculator is in
degree mode.

1.	J	**3.**	E
5.	D	**7.**	H
9.	G		

For Exercises 11–40, be sure your calculator is in
degree mode. If your calculator accepts angles in
degrees, minutes, and seconds, it is not necessary to
change angles to decimal degrees. Keystroke
sequences may vary on the type and/or model of
calculator being used. Screens shown will be from a
TI-84 Plus C calculator. To obtain the degree (°) and
(′) symbols, go to the ANGLE menu (2nd APPS).

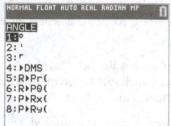

For Exercises 11–39, we include TI-84 screens only
for those exercises involving cotangent, secant, and
cosecant.

11. $\sin 38°42' \approx 0.625243$

13. $\sec 13°15' \approx 1.027349$

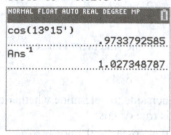

15. $\cot 183°48' \approx 15.055723$

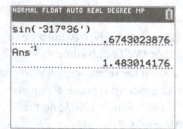

17. $\sin(-312°12') \approx 0.740805$

19. $\csc(-317°36') \approx 1.483014$

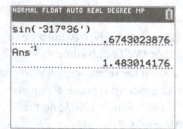

21. $\dfrac{1}{\cot 23.4°} = \tan 23.4° \approx 0.432739$

23. $\dfrac{\cos 77°}{\sin 77°} = \cot 77° \approx 0.230868$

25. $\cot(90° - 4.72°) = \tan 4.72° \approx 0.082566$

27. $\dfrac{1}{\csc(90° - 51°)} = \cos 51° \approx 0.629320$

29. $\tan \theta = 1.4739716$
   $\theta = \tan^{-1}(1.4739716) \approx 55.845496°$

31. $\sin \theta = 0.27843196$
   $\theta = \sin^{-1}(0.27843196) \approx 16.166641°$

33. $\cot \theta = 1.2575516$
   $\theta = \cot^{-1}(1.2575516) = \tan^{-1}\left(\dfrac{1}{1.2575516}\right)$
   $\approx 38.491580°$

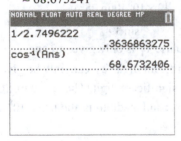

35. $\sec \theta = 2.7496222$
   $\theta = \sec^{-1}(2.7496222) = \cos^{-1}\left(\dfrac{1}{2.7496222}\right)$
   $\approx 68.673241°$

37. $\cos \theta = 0.70058013$
   $\theta = \cos^{-1}(0.70058013) \approx 45.526434°$

39. $\csc \theta = 4.7216543 \Rightarrow$
   $\theta = \csc^{-1}(4.7216543) = \sin^{-1}\left(\dfrac{1}{4.7216543}\right)$
   $\approx 12.227282°$

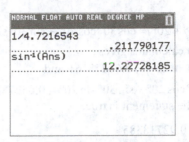

41. A common mistake is to have the calculator in radian mode, when it should be in degree mode (and vice verse).

43. Find $\tan^{-1}(1.482560969)$.
   $A = 56°$.

45. $\sin 35° \cos 55° + \cos 35° \sin 55° = 1$

47. $\sin^2 36° + \cos^2 36° = 1$

49. $\cos 75°29' \cos 14°31' - \sin 75°29' \sin 14°31' = 0$

51. $\sin 10° + \sin 10° \overset{?}{=} \sin 20°$
   Using a calculator gives
   $\sin 10° + \sin 10° \approx 0.34729636$ and
   $\sin 20° \approx 0.34202014$. Thus, the statement is false.

53. $\sin 50° \overset{?}{=} 2\sin 25° \cos 25°$
   Using a calculator gives $\sin 50° \approx 0.76604444$ and $2\sin 25° \cos 25° \approx 0.76604444$. Thus, the statement is true.

55. $\cos 40° \overset{?}{=} 1 - 2\sin^2 80°$
   Using a calculator gives
   $\cos 40° \approx 0.76604444$ and
   $1 - 2\sin^2 80° \approx -0.93969262$. Thus, the statement is false.

57. $\sin 39°48' + \cos 39°48' \overset{?}{=} 1$
   Using a calculator gives
   $\sin 39°48' + \cos 39°48' \approx 1.40839322 \ne 1$. Thus, the statement is false.

59. $1 + \cot^2 42.5° \overset{?}{=} \csc^2 42.5°$
   Using a calculator gives
   $1 + \cot^2 42.5° \approx 2.1909542$ and
   $\csc^2 42.5° \approx 2.1909542$. Thus, the statement is true.

61. $\cos(30° + 20°) \overset{?}{=} \cos 30° \cos 20° - \sin 30° \sin 20°$

Using a calculator gives

$\cos(30° + 20°) \approx 0.64278761$ and

$\cos 30° \cos 20° - \sin 30° \sin 20° \approx 0.64278761$.

Thus, the statement is true.

63. $\sin \theta = 0.92718385$

$\sin \theta$ is positive in quadrants I and II.

$\sin^{-1}(0.92718385) = 68°$

The angle in quadrant II with the same sine is $180° - 68° = 112°$.

65. $\cos \theta = 0.71933980$

$\cos \theta$ is positive in quadrants I and IV.

$\cos^{-1}(0.71933980) = 44°$

The angle in quadrant IV with the same cosine is $360° - 44° = 316°$.

67. $\tan \theta = 1.2348971$

$\tan \theta$ is positive in quadrants I and III.

$\tan^{-1}(1.2348971) = 51°$

The angle in quadrant III with the same tangent is $180° + 51° = 231°$.

69. $F = W \sin \theta$

$F = 2100 \sin 1.8° \approx 65.96 \approx 70$ lb

71. $F = W \sin \theta$

$-130 = 2600 \sin \theta \Rightarrow \dfrac{-130}{2600} = \sin \theta \Rightarrow$

$-0.05 = \sin \theta \Rightarrow \theta = \sin^{-1}(-0.05) \approx -2.9°$

73. $F = W \sin \theta$

$120 = W \sin(2.7°) \Rightarrow \dfrac{120}{\sin(2.7°)} = W \Rightarrow$

$W \approx 2547 \approx 2500$ lb

75. $F = W \sin \theta$

$F = 2200 \sin 2° \approx 76.77889275$ lb

$F = 2000 \sin 2.2° \approx 76.77561818$ lb

The 2200-lb car on a 2° uphill grade has the greater grade resistance.

77. 45 mph = 66 ft/sec, $V = 66$, $\theta = 3°$, $g = 32.2$,

$f = 0.14$

$R = \dfrac{V^2}{g(f + \tan \theta)} = \dfrac{66^2}{32.2(0.14 + \tan 3°)}$

≈ 703 ft

79. Intuitively, increasing θ would make it easier to negotiate the curve at a higher speed much like is done at a race track. Mathematically, a larger value of θ (acute) will lead to a larger value for $\tan \theta$. If $\tan \theta$ increases, then the ratio determining R will *decrease*. Thus, the radius can be smaller and the curve sharper if θ is increased.

$R = \dfrac{V^2}{g(f + \tan \theta)} = \dfrac{66^2}{32.2(0.14 + \tan 4°)}$

≈ 644 ft

$R = \dfrac{V^2}{g(f + \tan \theta)} \approx \dfrac{102.67^2}{32.2(0.14 + \tan 4°)}$

≈ 1559 ft

As predicted, both values are less.

81. **(a)** $\theta_1 = 46°$, $\theta_2 = 31°$, $c_1 = 3 \times 10^8$ m per sec

$\dfrac{c_1}{c_2} = \dfrac{\sin \theta_1}{\sin \theta_2} \Rightarrow c_2 = \dfrac{c_1 \sin \theta_2}{\sin \theta_1} \Rightarrow$

$c_2 = \dfrac{(3 \times 10^8)(\sin 31°)}{\sin 46°} \approx 2 \times 10^8$

Because c_1 is only given to one significant digit, c_2 can only be given to one significant digit. The speed of light in the second medium is about 2×10^8 m per sec.

(b) $\theta_1 = 39°$, $\theta_2 = 28°$, $c_1 = 3 \times 10^8$ m per sec

$\dfrac{c_1}{c_2} = \dfrac{\sin \theta_1}{\sin \theta_2} \Rightarrow c_2 = \dfrac{c_1 \sin \theta_2}{\sin \theta_1} \Rightarrow$

$c_2 = \dfrac{(3 \times 10^8)(\sin 28°)}{\sin 39°} \approx 2 \times 10^8$

Because c_1 is only given to one significant digit, c_2 can only be given to one significant digit. The speed of light in the second medium is about 2×10^8 m per sec.

83. $\theta_1 = 90°$, $c_1 = 3 \times 10^8$ m per sec, and

$c_2 = 2.254 \times 10^8$

$\dfrac{c_1}{c_2} = \dfrac{\sin \theta_1}{\sin \theta_2} \Rightarrow \sin \theta_2 = \dfrac{c_2 \sin \theta_1}{c_1}$

$\sin \theta_2 = \dfrac{(2.254 \times 10^8)(\sin 90°)}{3 \times 10^8}$

$= \dfrac{2.254 \times 10^8 (1)}{3 \times 10^8} = \dfrac{2.254}{3} \Rightarrow$

$\theta_2 = \sin^{-1}\left(\dfrac{2.254}{3}\right) \approx 48.7°$

85.
$$V_1 = 55 \text{ mph} = 55 \text{ mph} \cdot \frac{1 \text{ hr}}{3600 \text{ sec}} \cdot \frac{5280 \text{ ft}}{1 \text{ mi}}$$
$$= 80\frac{2}{3} \text{ ft per sec} \approx 80.67 \text{ ft per sec},$$
$$V_2 = 30 \text{ mph} = 30 \text{ mph} \cdot \frac{1 \text{ hr}}{3600 \text{ sec}} \cdot \frac{5280 \text{ ft}}{1 \text{ mi}}$$
$$= 44 \text{ ft per sec}$$
$$\theta = 3.5°, \ K_1 = 0.4, \ K_2 = 0.02.$$
$$D = \frac{1.05\left(V_1^2 - V_2^2\right)}{64.4\left(K_1 + K_2 + \sin\theta\right)}$$
$$= \frac{1.05\left(80.67^2 - 44^2\right)}{64.4\left(0.4 + 0.02 + \sin 3.5°\right)} \approx 155 \text{ ft}$$

87. Negative values of θ require greater distances for slowing down than positive values.

89. For Auto A, calculate $70 \cdot \cos 10° \approx 68.94$.

Auto A's reading is approximately 69 mph. For Auto B, calculate $70 \cdot \cos 20° \approx 65.78$. Auto B's reading is approximately 66 mph.

91. $h = 1.9 \text{ ft}, \ \alpha = 0.9°, \ \theta_1 = -3°, \ \theta_2 = 4°,$
$S = 336 \text{ ft}:$
$$L = \frac{[4 - (-3)]336^2}{200(1.9 + 336\tan .9°)} = 550 \text{ ft}$$

Chapter 2 Quiz
(Sections 2.1–2.3)

1.
$$\sin A = \frac{\text{side opposite}}{\text{hypotenuse}} = \frac{24}{40} = \frac{3}{5}$$
$$\cos A = \frac{\text{side adjacent}}{\text{hypotenuse}} = \frac{32}{40} = \frac{4}{5}$$
$$\tan A = \frac{\text{side opposite}}{\text{side adjacent}} = \frac{24}{32} = \frac{3}{4}$$
$$\cot A = \frac{\text{side adjacent}}{\text{side opposite}} = \frac{32}{24} = \frac{4}{3}$$
$$\sec A = \frac{\text{hypotenuse}}{\text{side adjacent}} = \frac{40}{32} = \frac{5}{4}$$
$$\csc A = \frac{\text{hypotenuse}}{\text{side opposite}} = \frac{40}{24} = \frac{5}{3}$$

3.
$$\sin 30° = \frac{w}{36} \Rightarrow w = 36\sin 30° = 36 \cdot \frac{1}{2} = 18$$
$$\cos 30° = \frac{x}{36} \Rightarrow x = 36\cos 30° = 36 \cdot \frac{\sqrt{3}}{2} = 18\sqrt{3}$$
$$\tan 45° = \frac{w}{y} \Rightarrow 1 = \frac{18}{y} \Rightarrow y = 18$$
$$\sin 45° = \frac{w}{z} \Rightarrow \frac{\sqrt{2}}{2} = \frac{18}{z} \Rightarrow z = \frac{36}{\sqrt{2}} = 18\sqrt{2}$$

5. $180° - 135° = 45°$, so the reference angle is 45°. The original angle (135°) lies in quadrant II, so the sine and cosecant are positive, while the remaining trigonometric functions are negative.
$$\sin 135° = \frac{\sqrt{2}}{2}; \ \cos 135° = -\frac{\sqrt{2}}{2}$$
$$\tan 135° = -1; \ \cot 135° = -1$$
$$\sec 135° = -\sqrt{2}; \ \csc 135° = \sqrt{2}$$

7. $1020°$ is coterminal with $1020° - 720° = 300°$. This lies in quadrant IV, so the reference angle is $360° - 300° = 60°$. In quadrant IV, the cosine and secant are positive, while the remaining trigonometric functions are negative.
$$\sin 1020° = -\sin 60° = -\frac{\sqrt{3}}{2}$$
$$\cos 1020° = \cos 60° = \frac{1}{2}$$
$$\tan 1020° = -\tan 60° = -\sqrt{3}$$
$$\cot 1020° = -\cot 60° = -\frac{\sqrt{3}}{3}$$
$$\sec 1020° = \sec 60° = 2$$
$$\csc 1020° = -\csc 60° = -\frac{2\sqrt{3}}{3}$$

9. $\sec\theta = -\sqrt{2}$
Because $\sec\theta$ is negative, θ must lie in quadrants II or III. The absolute value of $\sec\theta$ is $\sqrt{2}$, so the reference angle, θ' must be 45°. The quadrant II angle θ equals $180° - \theta' = 180° - 45° = 135°$, and the quadrant III angle θ equals $180° + \theta' = 180° + 45° = 225°$.

11. $\sec(-212°12') \approx -1.181763$

13. $\csc\theta = 2.3861147 \Rightarrow \theta \approx 24.777233°$

15. The statement is true. Using the cofunction identity, $\tan(90° - 35°) = \cot 35°$.

Section 2.4   Solutions and Applications of Right Triangles

1. B   **3.** A   **5.** C

7. 23.825 to 23.835

9. 8958.5 to 8959.5

11. If h is the actual height of a building and the height is measure as 58.6 ft, then
$$|h - 58.6| \leq \underline{0.05}.$$

13. $A = 36°20'$, $c = 964$ m

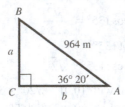

$$A + B = 90° \Rightarrow B = 90° - A \Rightarrow$$
$$B = 90° - 36°20'$$
$$= 89°60' - 36°20' = 53°40'$$
$$\sin A = \frac{a}{c} \Rightarrow \sin 36°20' = \frac{a}{964} \Rightarrow$$
$$a = 964 \sin 36°20' \approx 571 \text{m} \text{ (rounded to}$$
three significant digits)
$$\cos A = \frac{b}{c} \Rightarrow \cos 36°20' = \frac{b}{964} \Rightarrow$$
$$b = 964 \cos 36°20' \approx 777 \text{ m} \text{ (rounded to}$$
three significant digits)

15. $N = 51.2°$, $m = 124$ m

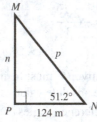

$$M + N = 90° \Rightarrow M = 90° - N \Rightarrow$$
$$M = 90° - 51.2° = 38.8°$$
$$\tan N = \frac{n}{m} \Rightarrow \tan 51.2° = \frac{n}{124} \Rightarrow$$
$$n = 124 \tan 51.2° \approx 154 \text{ m} \text{ (rounded to}$$
three significant digits)
$$\cos N = \frac{m}{p} \Rightarrow \cos 51.2° = \frac{124}{p} \Rightarrow$$
$$p = \frac{124}{\cos 51.2°} \approx 198 \text{ m} \text{ (rounded to three}$$
significant digits)

17. $B = 42.0892°$, $b = 56.851$

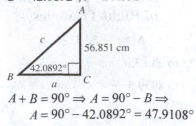

$$A + B = 90° \Rightarrow A = 90° - B \Rightarrow$$
$$A = 90° - 42.0892° = 47.9108°$$

$$\sin B = \frac{b}{c} \Rightarrow \sin 42.0892° = \frac{56.851}{c} \Rightarrow$$
$$c = \frac{56.851}{\sin 42.0892°} \approx 84.816 \text{ cm}$$
(rounded to five significant digits)
$$\tan B = \frac{b}{a} \Rightarrow \tan 42.0892° = \frac{56.851}{a} \Rightarrow$$
$$a = \frac{56.851}{\tan 42.0892°} \approx 62.942 \text{ cm}$$
(rounded to five significant digits)

19. $a = 12.5, b = 16.2$

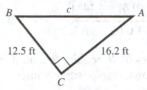

Using the Pythagorean theorem, we have
$$a^2 + b^2 = c^2 \Rightarrow 12.5^2 + 16.2^2 = c^2 \Rightarrow$$
$$418.69 = c^2 \Rightarrow c \approx 20.5 \text{ ft (rounded to three}$$
significant digits)
$$\tan A = \frac{a}{b} \Rightarrow \tan A = \frac{12.5}{16.2} \Rightarrow$$
$$A = \tan^{-1} \frac{12.5}{16.2} \approx 37.6540°$$
$$\approx 37° + (0.6540 \cdot 60)' \approx 37°39' \approx 37°40'$$
(rounded to three significant digits)
$$\tan B = \frac{b}{a} \Rightarrow \tan B = \frac{16.2}{12.5} \Rightarrow$$
$$B = \tan^{-1} \frac{16.2}{12.5} \approx 52.3460°$$
$$\approx 52° + (0.3460 \cdot 60)' \approx 52°21'$$
$$\approx 52°20' \text{ (rounded to three significant}$$
digits)

21. No. You need to have at least one side to solve the triangle. There are infinitely many similar right triangles satisfying the given conditions.

23. Answers will vary. If you know one acute angle, the other acute angle may be found by subtracting the given acute angle from 90°. If you know one of the sides, then choose two of the trigonometric ratios involving sine, cosine or tangent that involve the known side in order to find the two unknown sides.

25. $A = 28.0°$, $c = 17.4$ ft

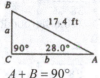

$A + B = 90°$
$B = 90° - A$
$B = 90° - 28.0° = 62.0°$

$\sin A = \dfrac{a}{c} \Rightarrow \sin 28.0° = \dfrac{a}{17.4} \Rightarrow$
$a = 17.4 \sin 28.0° \approx 8.17$ ft   (rounded to three significant digits)

$\cos A = \dfrac{b}{c} \Rightarrow \cos 28.00° = \dfrac{b}{17.4} \Rightarrow$
$b = 17.4 \cos 28.00° \approx 15.4$ ft   (rounded to three significant digits)

27. Solve the right triangle with $B = 73.0°$, $b = 128$ in. and $C = 90°$

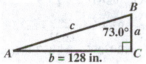

$A + B = 90° \Rightarrow A = 90° - B \Rightarrow$
$A = 90° - 73.0° = 17.0°$

$\tan B° = \dfrac{b}{a} \Rightarrow \tan 73.0° = \dfrac{128}{a} \Rightarrow$
$a = \dfrac{128}{\tan 73.0°} \Rightarrow a = 39.1$ in (rounded to three significant digits)

$\sin B° = \dfrac{b}{c} \Rightarrow \sin 73.0° = \dfrac{128}{c} \Rightarrow$
$c = \dfrac{128}{\sin 73.0°} \Rightarrow c = 134$ in (rounded to three significant digits)

29. $A = 61.0°$, $b = 39.2$ cm
$A + B = 90° \Rightarrow B = 90° - A \Rightarrow$
$B = 90° - 61.0° = 29.0°$

$\tan A = \dfrac{a}{b} \Rightarrow \tan 61.0° = \dfrac{a}{39.2} \Rightarrow$
$a = 39.2 \tan 61.0 \approx 70.7$ cm
(rounded to three significant digits)

$\cos A = \dfrac{b}{c} \Rightarrow \cos 61.0° = \dfrac{39.2}{c} \Rightarrow$
$c = \dfrac{39.2}{\cos 61.0°} \approx 80.9$ cm
(rounded to three significant digits)

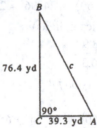

31. $a = 13$ m, $c = 22$m

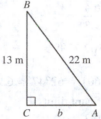

$c^2 = a^2 + b^2 \Rightarrow 22^2 = 13^2 + b^2 \Rightarrow$
$484 = 169 + b^2 \Rightarrow 315 = b^2 \Rightarrow b \approx 18$ m
(rounded to two significant digits)

We will determine the measurements of both A and B by using the sides of the right triangle. In practice, once you find one of the measurements, subtract it from $90°$ to find the other.

$\sin A = \dfrac{a}{c} \Rightarrow \sin A = \dfrac{13}{22} \Rightarrow$
$A \approx \sin^{-1}\left(\dfrac{13}{22}\right) \approx 36.2215° \approx 36°$ (rounded to two significant digits)

$\cos B = \dfrac{b}{c} \Rightarrow \cos B = \dfrac{13}{22} \Rightarrow$
$B \approx \cos^{-1}\left(\dfrac{13}{22}\right) \approx 53.7784° \approx 54°$

(rounded to two significant digits)

33. $a = 76.4$ yd, $b = 39.3$ yd

$c^2 = a^2 + b^2 \Rightarrow c = \sqrt{a^2 + b^2}$
$= \sqrt{(76.4)^2 + (39.3)^2} = \sqrt{5836.96 + 1544.49}$
$= \sqrt{7381.45} \approx 85.9$ yd (rounded to three significant digits)

We will determine the measurements of both A and B by using the sides of the right triangle. In practice, once you find one of the measurements, subtract it from $90°$ to find the other.

(*continued on next page*)

(*continued*)

$$\tan A = \frac{a}{b} \Rightarrow \tan A = \frac{76.4}{39.3} \Rightarrow$$

$$A \approx \tan^{-1}\left(\frac{76.4}{39.3}\right) \approx 62.7788°$$

$$\approx 62° + \left(0.7788 \cdot 60\right)' \approx 62°47' \approx 62°50'$$

(rounded to three significant digits)

$$\tan B = \frac{b}{a} \Rightarrow \tan B = \frac{39.3}{76.4} \Rightarrow$$

$$B \approx \tan^{-1}\left(\frac{39.3}{76.4}\right) \approx 27.2212°$$

$$\approx 27° + \left(0.2212 \cdot 60\right)' \approx 27°13' \approx 27°10'$$

(rounded to three significant digits)

35. $a = 18.9$ cm, $c = 46.3$ cm

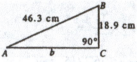

$$c^2 = a^2 + b^2 \Rightarrow 46.3^2 = 18.9^2 + b^2 \Rightarrow$$

$$2143.69 = 357.21 + b^2 \Rightarrow 1786.48 = b^2 \Rightarrow$$

$$b \approx 42.3 \text{ cm (rounded to three}$$

significant digits)

$$\sin A = \frac{a}{c} \Rightarrow \sin A = \frac{18.9}{46.3} \Rightarrow$$

$$A \approx \sin^{-1}\left(\frac{18.9}{46.3}\right) \approx 24.09227°$$

$$\approx 24° + \left(0.09227 \cdot 60\right)' \approx 24°06' \approx 24°10'$$

(rounded to three significant digits)

$$\cos B = \frac{a}{c} \Rightarrow \cos B = \frac{18.9}{46.3} \Rightarrow$$

$$B \approx \cos^{-1}\left(\frac{18.9}{46.3}\right) \approx 65.9077°$$

$$\approx 65° + \left(0.9077 \cdot 60\right)' \approx 65°54' \approx 65°50'$$

(rounded to three significant digits)

37. $A = 53°24'$, $c = 387.1$ ft

$$A + B = 90°$$
$$B = 90° - A$$
$$B = 90° - 53°24'$$
$$= 89°60' - 53°24'$$
$$= 36°36'$$

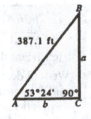

$$\sin A = \frac{a}{c} \Rightarrow \sin 53°24' = \frac{a}{387.1} \Rightarrow$$
$$a = 387.1 \sin 53°24' \approx 310.8 \text{ ft (rounded}$$
to four significant digits)

$$\cos A = \frac{b}{c} \Rightarrow \cos 53°24' = \frac{b}{387.1} \Rightarrow$$
$$b = 387.1 \cos 53°24' \approx 230.8 \text{ ft (rounded}$$
to four significant digits)

39. $B = 39°09'$, $c = 0.6231$ m

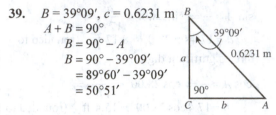

$$A + B = 90°$$
$$B = 90° - A$$
$$B = 90° - 39°09'$$
$$= 89°60' - 39°09'$$
$$= 50°51'$$

$$\sin B = \frac{b}{c} \Rightarrow \sin 39°09' = \frac{b}{0.6231} \Rightarrow$$
$$b = 0.6231 \sin 39°09' \approx 0.3934 \text{ m (rounded}$$
to four significant digits)

$$\cos B = \frac{a}{c} \Rightarrow \cos 39°09' = \frac{a}{0.6231} \Rightarrow$$
$$a = 0.6231 \cos 39°09' \approx 0.4832 \text{ m (rounded}$$
to four significant digits)

41. If B is a point above point A as shown in the figure, the angle of elevation from A to B is the acute angle formed by the horizontal line through A and the line of sight from A to B.

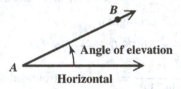

43. Angles DAB and ABC are alternate interior angle formed by the transversal AB intersecting parallel lines AD and BC. Thus they have the same measure.

45. $$\sin 43°50' = \frac{d}{13.5}$$
$$d = 13.5 \sin 43°50' \approx 9.3496000$$
The ladder goes up the wall 9.35 m. (rounded to three significant digits)

47. Let x represent the horizontal distance between the two buildings and y represent the height of the portion of the building across the street that is higher than the window.

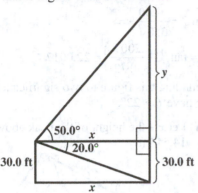

$$\tan 20.0° = \frac{30.0}{x} \Rightarrow x = \frac{30.3}{\tan 20.0°} \approx 82.4$$

$$\tan 50.0° = \frac{y}{x} \Rightarrow$$

$$y = x \tan 50.0° = \left(\frac{30.0}{\tan 20.0°}\right) \tan 50.0°$$

$$\text{height} = y + 30.0$$

$$= \left(\frac{30.0}{\tan 20.0°}\right) \tan 50.0° + 30.0$$

$$\approx 128.2295$$

The height of the building across the street is about 128 ft. (rounded to three significant digits)

49. The altitude of an isosceles triangle bisects the base as well as the angle opposite the base. The two right triangles formed have interior angles which have the same measure. The lengths of the corresponding sides also have the same measure. The altitude bisects the base, so each leg (base) of the right triangles is $\frac{42.36}{2} = 21.18$ in.

Let x = the length of each of the two equal sides of the isosceles triangle.

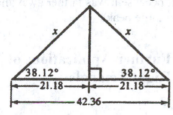

$$\cos 38.12° = \frac{21.18}{x} \Rightarrow x \cos 38.12° = 21.18 \Rightarrow$$

$$x = \frac{21.18}{\cos 38.12°} \approx 26.921918$$

The length of each of the two equal sides of the triangle is 26.92 in. (rounded to four significant digits)

51. Let h represent the height of the tower. In triangle ABC we have

$$\tan 34.6° = \frac{h}{40.6}$$

$$h = 40.6 \tan 34.6° \approx 28.0081$$

The height of the tower is 28.0 m. (rounded to three significant digits)

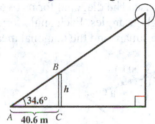

53. Let x = the length of the shadow.

$$\tan 23.4° = \frac{5.75}{x}$$

$$x \tan 23.4° = 5.75$$

$$x = \frac{5.75}{\tan 23.4°} \approx 13.2875$$

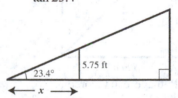

The length of the shadow is 13.3 ft. (rounded to three significant digits)

55. Let θ = the angle of depression.

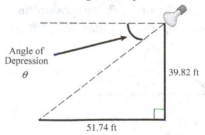

$$\tan \theta = \frac{39.82}{51.74} \Rightarrow \theta = \tan^{-1}\left(\frac{39.82}{51.74}\right)$$

$$\theta \approx 37.58° \approx 37°35'$$

57. Let $\theta =$ the angle of elevation of the sun.

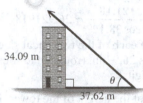

$$\tan \theta = \frac{34.09}{37.62} \Rightarrow \theta = \tan^{-1}\left(\frac{34.09}{37.62}\right)$$
$$\theta \approx 42.18°$$

59. In order to find the angle of elevation, θ, we need to first find the length of the diagonal of the square base. The diagonal forms two isosceles right triangles. Each angle formed by a side of the square and the diagonal measures 45°.

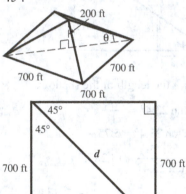

By the Pythagorean theorem,
$$700^2 + 700^2 = d^2 \Rightarrow 2 \cdot 700^2 = d^2 \Rightarrow$$
$$d = \sqrt{2 \cdot 700^2} \Rightarrow d = 700\sqrt{2}$$
Thus, length of the diagonal is $700\sqrt{2}$ ft. To to find the angle, θ, we consider the following isosceles triangle.

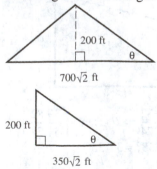

The height of the pyramid bisects the base of this triangle and forms two right triangles. We can use one of these triangles to find the angle of elevation, θ.
$$\tan \theta = \frac{200}{350\sqrt{2}} \Rightarrow$$
$$\theta \approx \tan^{-1}\left(\frac{200}{350\sqrt{2}}\right) \approx 22.0017$$
Rounding this figure to two significant digits, we have $\theta \approx 22°$.

61. (a) Let $x =$ the height of the peak above 14,545 ft.

The diagonal of the right triangle formed is in miles, so we must first convert this measurement to feet. Because there are 5280 ft in one mile, we have the length of the diagonal is $27.0134(5280) =$ $142,630.752$. Find the value of x by solving $\sin 5.82° = \dfrac{x}{142,630.752}$.
$$x = 142,630.752 \sin 5.82°$$
$$\approx 14,463.2674$$
Thus, the value of x rounded to five significant digits is 14,463 ft. Thus, the total height is about $14,545 + 14,463 = 29,008 \approx 29,000$ ft.

(b) The curvature of the earth would make the peak appear shorter than it actually is. Initially the surveyors did not think Mt. Everest was the tallest peak in the Himalayas. It did not look like the tallest peak because it was farther away than the other large peaks.

Section 2.5 Further Applications of Right Triangles

1. C   **3.** A

5. B   **7.** F

9. I

11. $(-4, 0)$

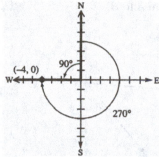

The bearing of the airplane measured in a clockwise direction from due north is 270°. The bearing can also be expressed as N 90° W, or S 90° W.

13. $(0, 4)$

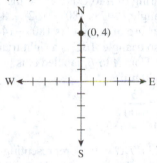

The bearing of the airplane measured in a clockwise direction from due north is 0°. The bearing can also be expressed as N 0° E or N 0° W.

15. $(-5, 5)$

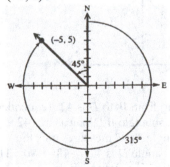

The bearing of the airplane measured in a clockwise direction from due north is 315°. The bearing can also be expressed as N 45° W.

17. $(2, -2)$

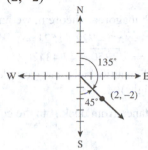

The bearing of the airplane measured in a clockwise direction from due north is 135°. The bearing can also be expressed as S 45° E.

19. Let x = the distance the plane is from its starting point. In the figure, the measure of angle ACB is $38° + (180° - 128° = 38° + 52° = 90°$. Therefore, triangle ACB is a right triangle.

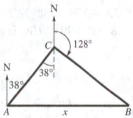

Because $d = rt$, the distance traveled in 1.5 hr is $(1.5 \text{ hr})(110 \text{ mph}) = 165$ mi. The distance traveled in 1.3 hr is $(1.3 \text{ hr})(110 \text{ mph}) = 143$ mi. Using the Pythagorean theorem, we have
$$x^2 = 165^2 + 143^2 \Rightarrow x^2 = 27,225 + 20,449 \Rightarrow$$
$$x^2 = 47,674 \Rightarrow x \approx 218.3438$$
The plane is 220 mi from its starting point. (rounded to two significant digits)

21. Let x = distance the ships are apart. In the figure, the measure of angle CAB is $130° - 90° = 40$ Therefore, triangle CAB is a right triangle. Because $d = rt$, the distance traveled by the first ship in 1.5 hr is $(1.5 \text{ hr})(18 \text{ knots}) = 27$ nautical mi and the second ship is $(1.5\text{hr})(26 \text{ knots}) = 39$ nautical mi.

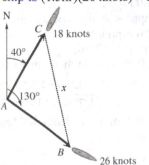

(*continued on next page*)

(*continued*)

Applying the Pythagorean theorem, we have
$$x^2 = 27^2 + 39^2 \Rightarrow x^2 = 729 + 1521 \Rightarrow$$
$$x^2 = 2250 \Rightarrow x = \sqrt{2250} \approx 47.4342$$
The ships are 47 nautical mi apart (rounded to 2 significant digits).

23. Let b = the distance from dock A to the coral reef C.

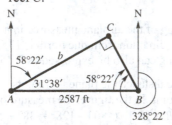

In the figure, the measure of angle CAB is $90° - 58°22' = 31°38'$, and the measure of angle CBA is $328°22' - 270° = 58°22'$. Because $31°38' + 58°22' = 90°$, ABC is a right triangle.
$$\cos A = \frac{b}{2587}$$
$$\cos 31°38' = \frac{b}{2587} \Rightarrow b = 2587 \cos 31°38'$$
$$b \approx 2203 \text{ ft}$$

25. Let x = distance between the two ships.

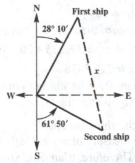

The angle between the bearings of the ships is $180° - (28°10' + 61°50') = 90°$. The triangle formed is a right triangle. The distance traveled at 24.0 mph is
(4 hr) (24.0 mph) = 96 mi. The distance traveled at 28.0 mph is
(4 hr)(28.0 mph) = 112 mi.
Applying the Pythagorean theorem we have
$$x^2 = 96^2 + 112^2 \Rightarrow x^2 = 9216 + 12,544 \Rightarrow$$
$$x^2 = 21,760 \Rightarrow x = \sqrt{21,760} \approx 147.5127$$
The ships are 148 mi apart. (rounded to three significant digits)

27. Let b = the distance from A to C and let c = the distance from A to B.

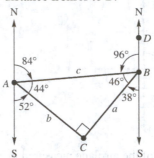

Because the bearing from A to B is N 84° E, the measure of angle ABD is $180° - 84° = 96°$. The bearing from B to C is 38°, so the measure of angle $ABC = 180° - (96° + 38°) = 46°$. The bearing of A to C is 52°, so the measure of angle BAC is $180° - (52° + 84°) = 44°$. The measure of angle C is $180° - (44° + 46°) = 90°$, so triangle ABC is a right triangle. The distance from A to B, labeled c, is $2.4(250) = 600$ miles.
$$\sin 46° = \frac{b}{c} = \frac{b}{600}$$
$$b = 600 \sin 46° \approx 430 \text{ mi}$$

29. Draw triangle WDG with W representing Winston-Salem, D representing Danville, and G representing Goldsboro. Name any point X on the line due south from D.

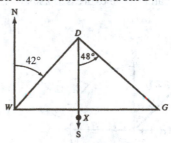

The bearing from W to D is 42° (equivalent to N 42° E), so angle WDX measures 42°. Because angle XDG measures 48°, the measure of angle D is $42° + 48° = 90°$. Thus, triangle WDG is a right triangle.
Using $d = rt$ and the Pythagorean theorem, we have
$$WG = \sqrt{(WD)^2 + (DG)^2}$$
$$= \sqrt{[65(1.1)]^2 + [65(1.8)]^2}$$
$$WG = \sqrt{71.5^2 + 117^2} = \sqrt{5112.25 + 13,689}$$
$$= \sqrt{18,801.25} \approx 137$$

The distance from Winston-Salem to Goldsboro is approximately 140 mi. (rounded to two significant digits)

31. Let x = the distance from the closer point on the ground to the base of height h of the pyramid.

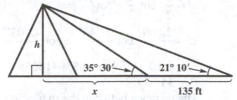

In the larger right triangle, we have

$$\tan 21°10' = \frac{h}{135 + x} \Rightarrow h = (135 + x)\tan 21°10'$$

In the smaller right triangle, we have

$$\tan 35°30' = \frac{h}{x} \Rightarrow h = x\tan 35°30'.$$

Substitute for h in this equation, and solve for x to obtain the following.

$$(135 + x)\tan 21°10' = x\tan 35°30'$$
$$135\tan 21°10' + x\tan 21°10' = x\tan 35°30'$$
$$135\tan 21°10' = x\tan 35°30' - x\tan 21°10'$$
$$135\tan 21°10' = x(\tan 35°30' - \tan 21°10')$$
$$\frac{135\tan 21°10'}{\tan 35°30' - \tan 21°10'} = x$$

Substitute for x in the equation for the smaller triangle.

$$h = \frac{135\tan 21°10'}{\tan 35°30' - \tan 21°10'}\tan 35°30'$$
$$\approx 114.3427$$

The height of the pyramid is 114 ft. (rounded to three significant digits)

33. Let x = the height of the antenna; h = the height of the house.

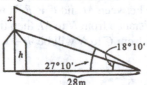

In the smaller right triangle, we have

$$\tan 18°10' = \frac{h}{28} \Rightarrow h = 28\tan 18°10'.$$

In the larger right triangle, we have

$$\tan 27°10' = \frac{x + h}{28} \Rightarrow x + h = 28\tan 27°10' \Rightarrow$$
$$x = 28\tan 27°10' - h$$
$$x = 28\tan 27°10' - 28\tan 18°10'$$
$$\approx 5.1816$$

The height of the antenna is 5.18 m. (rounded to three significant digits)

35. Algebraic solution:
Let x = the side adjacent to 49.2° in the smaller triangle.

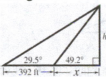

In the larger right triangle, we have

$$\tan 29.5° = \frac{h}{392 + x} \Rightarrow h = (392 + x)\tan 29.5°.$$

In the smaller right triangle, we have

$$\tan 49.2° = \frac{h}{x} \Rightarrow h = x\tan 49.2°.$$

Substituting, we have

$$x\tan 49.2° = (392 + x)\tan 29.5°$$
$$x\tan 49.2° = 392\tan 29.5° + x\tan 29.5°$$
$$x\tan 49.2° - x\tan 29.5° = 392\tan 29.5°$$
$$x(\tan 49.2° - \tan 29.5°) = 392\tan 29.5°$$
$$x = \frac{392\tan 29.5°}{\tan 49.2° - \tan 29.5°}$$

Now substitute this expression for x in the equation for the smaller triangle to obtain
$h = x\tan 49.2°$

$$h = \frac{392\tan 29.5°}{\tan 49.2° - \tan 29.5°} \cdot \tan 49.2°$$
$$\approx 433.4762 \approx 433 \text{ ft (rounded to three}$$

significant digits.

Graphing calculator solution:

Graph $y_1 = (\tan 29.5°)x$ and

$y_2 = (\tan 29.5°)(x - 392)$ in the window $[0, 1000] \times [0, 500]$. Then find the intersection.

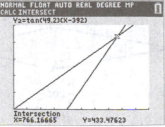

The height of the triangle is 433 ft (rounded to three significant digits.

37. Let x = the minimum distance that a plant needing full sun can be placed from the fence.

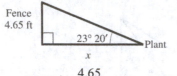

$$\tan 23°20' = \frac{4.65}{x} \Rightarrow x \tan 23°20' = 4.65 \Rightarrow$$

$$x = \frac{4.65}{\tan 23°20'} \approx 10.7799$$

The minimum distance is 10.8 ft. (rounded to three significant digits)

39. Let h = the minimum height above the surface of Earth so a pilot at A can see an object on the horizon at C.

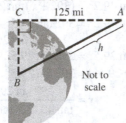

Using the Pythagorean theorem, we have

$$\left(4.00 \times 10^3 + h\right)^2 = \left(4.00 \times 10^3\right)^2 + 125^2$$

$$\left(4000 + h\right)^2 = 4000^2 + 125^2$$

$$\left(4000 + h\right)^2 = 16,000,000 + 15,625$$

$$\left(4000 + h\right)^2 = 16,015,625$$

$$4000 + h = \sqrt{16,015,625} \Rightarrow$$

$$h = \sqrt{16,015,625} - 4000$$

$$\approx 4001.9526 - 4000 = 1.9526$$

The minimum height above the surface of Earth would be 1.95 mi. (rounded to 3 significant digits)

41. (a)

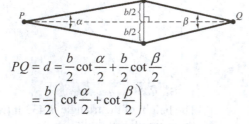

$$PQ = d = \frac{b}{2}\cot\frac{\alpha}{2} + \frac{b}{2}\cot\frac{\beta}{2}$$

$$= \frac{b}{2}\left(\cot\frac{\alpha}{2} + \cot\frac{\beta}{2}\right)$$

(b) Using the result of part (a), let $\alpha = 37'48''$, $\beta = 42'03''$, and $b = 2.000$

$$d = \frac{b}{2}\left(\cot\frac{\alpha}{2} + \cot\frac{\beta}{2}\right) \Rightarrow$$

$$d = \frac{2.000}{2}\left(\cot\frac{37'48''}{2} + \cot\frac{42'03''}{2}\right)$$

$$= \cot 0.315° + \cot 0.3504166667°$$

$$\approx 345.3951$$

The distance between the two points P and Q is about 345.4 cm.

43. (a) If $\theta = 37°$, then $\dfrac{\theta}{2} = \dfrac{37°}{2} = 18.5°$.

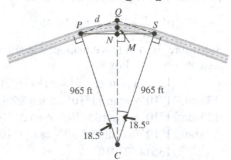

To find the distance between P and Q, d, we first note that angle QPC is a right angle. Hence, triangle QPC is a right triangle and we can solve

$$\tan 18.5° = \frac{d}{965}$$

$$d = 965 \tan 18.5° \approx 322.8845$$

The distance between P and Q, is 320 ft. (rounded to two significant digits)

(b) We are dealing with a circle, so the distance between M and C is R. If we let x be the distance from N to M, then the distance from C to N will be $R - x$.

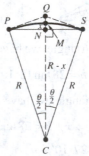

Triangle CNP is a right triangle, so we can set up the following equation.

$$\cos\frac{\theta}{2} = \frac{R-x}{R} \Rightarrow R\cos\frac{\theta}{2} = R - x \Rightarrow$$

$$x = R - R\cos\frac{\theta}{2} \Rightarrow x = R\left(1 - \cos\frac{\theta}{2}\right)$$

44. (a) $\theta \approx \dfrac{57.3S}{R} = \dfrac{57.3(336)}{600} = 32.088°$

$d = R\left(1 - \cos\dfrac{\theta}{2}\right)$

$= 600\left(1 - \cos 16.044°\right) \approx 23.3702 \text{ ft}$

The distance is 23 ft. (rounded to two significant digits)

(b) $\theta \approx \dfrac{57.3S}{R} = \dfrac{57.3(485)}{600} = 46.3175°$

$d = R\left(1 - \cos\dfrac{\theta}{2}\right)$

$= 600\left(1 - \cos 23.15875°\right) \approx 48.3488$

The distance is 48 ft. (rounded to two significant digits)

(c) The faster the speed, the more land needs to be cleared on the inside of the curve.

45. $y = \tan\theta(x - a) \Rightarrow y = \tan 35°(x - 25)$

46. $y = \tan\theta(x - a) \Rightarrow y = \tan 15°(x - 5)$

47.

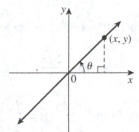

The line that bisects quadrants I and III passes through the origin, so $a = 0$. In addition, $\theta = 45°$ because the line bisects the angle formed by the axes. $\tan 45° = 1$, so an equation of the line is $y = \tan 45°(x)$ or $y = x$.

49. All points whose bearing from the origin is $240°$ lie in quadrant III.

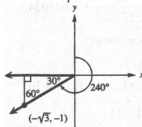

The reference angle, θ', is 30°. For any point, (x, y) on the ray $\dfrac{x}{r} = -\cos\theta'$ and

$\dfrac{y}{r} = -\sin\theta'$, where r is the distance from the point to the origin. Let $r = 2$, so

$\dfrac{x}{r} = -\cos\theta'$

$x = -r\cos\theta' = -2\cos 30° = -2 \cdot \dfrac{\sqrt{3}}{2} = -\sqrt{3}$

$\dfrac{y}{r} = -\sin\theta'$

$y = -r\sin\theta' = -2\sin 30° = -2 \cdot \dfrac{1}{2} = -1$

Thus, a point on the ray is $\left(-\sqrt{3}, -1\right)$. The ray contains the origin, the equation is of the form $y = mx$. Substituting the point $\left(-\sqrt{3}, -1\right)$, we have $-1 = m\left(-\sqrt{3}\right) \Rightarrow$

$m = \dfrac{-1}{-\sqrt{3}} = \dfrac{1}{\sqrt{3}} \cdot \dfrac{\sqrt{3}}{\sqrt{3}} = \dfrac{\sqrt{3}}{3}$. Thus, an equation of

the ray is $y = \dfrac{\sqrt{3}}{3}x, x \leq 0$ (because the ray lies in quadrant III).

Chapter 2   Review Exercises

1. $\sin A = \dfrac{\text{side opposite}}{\text{hypotenuse}} = \dfrac{60}{61}$

$\cos A = \dfrac{\text{side adjacent}}{\text{hypotenuse}} = \dfrac{11}{61}$

$\tan A = \dfrac{\text{side opposite}}{\text{side adjacent}} = \dfrac{60}{11}$

$\cot A = \dfrac{\text{side adjacent}}{\text{side opposite}} = \dfrac{11}{60}$

$\sec A = \dfrac{\text{hypotenuse}}{\text{side adjacent}} = \dfrac{61}{11}$

$\csc A = \dfrac{\text{hypotenuse}}{\text{side opposite}} = \dfrac{61}{60}$

3. $\sin 4\beta = \cos 5\beta$

Sine and cosine are cofunctions, so the sum of the angles is 90°.

$4\beta + 5\beta = 90° \Rightarrow 9\beta = 90° \Rightarrow \beta = 10°$

5. $\tan(5x + 11°) = \cot(6x + 2°)$

Tangent and cotangent are cofunctions, so the sum of the angles is 90°.

$(5x + 11°) + (6x + 2°) = 90°$

$11x + 13° = 90°$

$11x = 77° \Rightarrow x = 7°$

7. $\sin 46° < \sin 58°$

In the interval from $0°$ to $90°$, as the angle increases, so does the sine of the angle, so $\sin 46° < \sin 58°$ is true.

9. $\tan 60° \geq \cot 40°$

Using the cofunction identity,
$\cot 40° = \tan(90° - 40°) = \tan 50°$. In quadrant I, the tangent function is increasing. Thus $\cot 40° = \tan 50° < \tan 60°$, and the statement is true.

11.

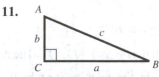

$\cos A = \frac{b}{c}$ and $\sin B = \frac{b}{c}$, so $\cos A = \sin B$.

This is an example of equality of cofunctions of complementary angles.

13. $1020°$ is coterminal with
$1020° - 2 \cdot 360° = 300°$. The reference angle is $360° - 300° = 60°$. Because $1020°$ lies in quadrant IV, the sine, tangent, cotangent, and cosecant are negative.

$$\sin 1020° = -\sin 60° = -\frac{\sqrt{3}}{2}$$
$$\cos 1020° = \cos 60° = \frac{1}{2}$$
$$\tan 1020° = -\tan 60° = -\sqrt{3}$$
$$\cot 1020° = -\cot 60° = -\frac{\sqrt{3}}{3}$$
$$\sec 1020° = \sec 60° = 2$$
$$\csc 1020° = -\csc 60° = -\frac{2\sqrt{3}}{3}$$

15. $-1470°$ is coterminal with
$-1470° + 5 \cdot 360° = 330°$. This angle lies in quadrant IV. The reference angle is $360° - 330° = 30°$. Because $-1470°$ is in quadrant IV, the sine, tangent, cotangent, and cosecant are negative.

$$\sin(-1470°) = -\sin 30° = -\frac{1}{2}$$
$$\cos(-1470°) = \cos 30° = \frac{\sqrt{3}}{2}$$
$$\tan(-1470°) = -\tan 30° = -\frac{\sqrt{3}}{3}$$
$$\cot(-1470°) = -\cot 30° = -\sqrt{3}$$
$$\sec(-1470°) = \sec 30° = \frac{2\sqrt{3}}{3}$$
$$\csc(-1470°) = -\csc 30° = -2$$

17. $\cos \theta = -\frac{1}{2}$

Because $\cos \theta$ is negative, θ must lie in quadrants II or III. The absolute value of $\cos \theta$ is $\frac{1}{2}$, so the reference angle, θ' must be $60°$. The quadrant II angle θ equals $180° - \theta' = 180° - 60° = 120°$, and the quadrant III angle θ equals $180° + \theta' = 180° + 60° = 240°$.

19. $\sec \theta = -\frac{2\sqrt{3}}{3}$

Because $\sec \theta$ is negative, θ must lie in quadrants II or III. The absolute value of $\sec \theta$ is $\frac{2\sqrt{3}}{3}$, so the reference angle, θ' must be $30°$. The quadrant II angle θ equals $180° - \theta' = 180° - 30° = 150°$, and the quadrant III angle θ equals $180° + \theta' = 180° + 30° = 210°$.

21. $\tan^2 120° - 2\cot 240° = \left(-\sqrt{3}\right)^2 - 2\left(\frac{\sqrt{3}}{3}\right)$
$$= 3 - \frac{2\sqrt{3}}{3}$$

23. $\sec^2 300° - 2\cos^2 150° = 2^2 - 2\left(-\frac{\sqrt{3}}{2}\right)^2$
$$= 4 - \frac{3}{2} = \frac{5}{2}$$

For the exercises in this section, be sure your calculator is in degree mode.

25. $\sec 222°30' = \dfrac{1}{\cos 222°30'} \approx -1.356342$

27. $\csc 78°21' = \dfrac{1}{\sin 78°21'} \approx 1.021034$

29. $\tan 11.7689° \approx 0.208344$

31. $\sin \theta = 0.8254121$
$\theta = \sin^{-1}(0.8254121) \approx 55.673870°$

33. $\cos \theta = 0.97540415$
$\theta = \cos^{-1}(0.97540415) \approx 12.733938°$

35. $\tan \theta = 1.9633124$
$\theta = \tan^{-1}(1.9633124) \approx 63.008286°$

37. $\sin\theta = 0.73135370$

$\theta = \sin^{-1}(0.73135370) \approx 47°$

The value of $\sin\theta$ is positive in quadrants I and II, so the two angles in $[0°, 360°)$ are $47°$ and $180° - 47° = 133°$.

39. $\sin 50° + \sin 40° \overset{?}{=} \sin 90°$

Using a calculator gives

$\sin 50° + \sin 40° = 1.408832053$ while $\sin 90° = 1$. Thus, the statement is false.

41. $\sin 240° \overset{?}{=} 2\sin 120° \cos 120°$

$\sin 240° = -\sin 60° = -\dfrac{\sqrt{3}}{2}$ and

$2\sin 120° \cos 120° = 2\sin 60°(-\cos 60°)$

$\qquad = 2\left(\dfrac{\sqrt{3}}{2}\right)\left(\dfrac{1}{2}\right) = -\dfrac{\sqrt{3}}{2}$

Thus, the statement is true.

43. No, this will result in an angle having tangent equal to 25. The function $\tan^{-1}$ is not the reciprocal of the tangent (the cotangent), but is the *inverse tangent* function. To find $\cot 25°$, the student must find the reciprocal of $\tan 25°$.

$\cot 25° = \dfrac{1}{\tan 25°} \ne \tan^{-1} 25°.$

45. $A = 58°\,30',\ c = 748$

$A + B = 90° \Rightarrow B = 90° - A \Rightarrow$

$\quad B = 90° - 58°30' = 89°60' - 58°30'$

$\qquad = 31°30'$

$\sin A = \dfrac{a}{c} \Rightarrow \sin 58°30' = \dfrac{a}{748} \Rightarrow$

$\quad a = 748\sin 58°30' \approx 638$ (rounded to three significant digits)

$\cos A = \dfrac{b}{c} \Rightarrow \cos 58°30' = \dfrac{b}{748} \Rightarrow$

$\quad b = 748\cos 58°30' \approx 391$ (rounded to three significant digits)

47. $A = 39.72°,\ b = 38.97$ m

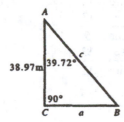

49. Let x = the height of the tree.

$\tan 70° = \dfrac{x}{50} \Rightarrow x = 50\tan 70° \approx 137$ ft

51. Let x = height of the tower.

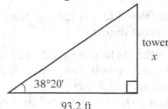

$\tan 38°20' = \dfrac{x}{93.2}$

$x = 93.2\tan 38°20'$

$x \approx 73.6930$

The height of the tower is 73.7 ft. (rounded to three significant digits)

53. Let x = length of the diagonal

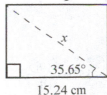

$\cos 35.65° = \dfrac{15.24}{x}$

$x = \dfrac{15.24}{\cos 35.65°} \approx 18.7548$

The length of the diagonal is 18.75 cm (rounded to three significant digits).

$A + B = 90° \Rightarrow B = 90° - A \Rightarrow$

$\quad B = 90° - 39.72° = 50.28°$

$\tan A = \dfrac{a}{b} \Rightarrow \tan 39.72° = \dfrac{a}{38.97} \Rightarrow$

$\quad a = 38.97\tan 39.72° \approx 32.38$ m (rounded to four significant digits)

$\cos A = \dfrac{b}{c} \Rightarrow \cos 39.72° = \dfrac{38.97}{c} \Rightarrow$

$c\cos 39.72° = 38.97 \Rightarrow$

$\quad c = \dfrac{38.97}{\cos 39.72°} \approx 50.66$ m

(rounded to five significant digits)

55. Draw triangle ABC and extend the north-south lines to a point X south of A and S to a point Y, north of C.

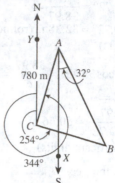

Angle $ACB = 344° - 254° = 90°$, so ABC is a right triangle.

Angle $BAX = 32°$ because it is an alternate interior angle to $32°$.

Angle $YCA = 360° - 344° = 16°$

Angle $XAC = 16°$ because it is an alternate interior angle to angle YCA.

Angle $BAC = 32° + 16° = 48°$.

In triangle ABC,

$$\cos A = \frac{AC}{AB} \Rightarrow \cos 48° = \frac{780}{AB} \Rightarrow$$

$$AB \cos 48° = 780 \Rightarrow AB = \frac{780}{\cos 48°} \approx 1165.6917$$

The distance from A to B is 1200 m. (rounded to two significant digits)

57. Suppose A is the car heading south at 55 mph, B is the car heading west, and point C is the intersection from which they start. After two hours, using $d = rt$, $AC = 55(2) = 110$. Angle ACB is a right angle, so triangle ACB is a right triangle. The bearing of A from B is 324°, so angle $CAB = 360° - 324° = 36°$.

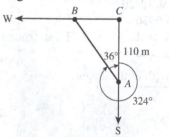

$$\cos \angle CAB = \frac{AC}{AB} \Rightarrow \cos 36° = \frac{110}{AB} \Rightarrow$$

$$AB = \frac{110}{\cos 36°} \approx 135.9675$$

The distance from A to B is about 140 mi (rounded to two significant digits).

59. Answers will vary. Sample answer: Find the value of x.

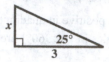

61. $h = R\left(\dfrac{1}{\cos\left(\frac{180T}{P}\right)} - 1\right)$

(a) Let $R = 3955$ mi, $T = 25$ min, $P = 140$ min.

$$h = R\left(\frac{1}{\cos\left(\frac{180T}{P}\right)} - 1\right)$$

$$h = 3955\left(\frac{1}{\cos\left(\frac{180\cdot25}{140}\right)} - 1\right) \approx 715.9424$$

The height of the satellite is approximately 716 mi.

(b) Let $R = 3955$ mi, $T = 30$ min, $P = 140$ min.

$$h = R\left(\frac{1}{\cos\left(\frac{180T}{P}\right)} - 1\right)$$

$$h = 3955\left(\frac{1}{\cos\left(\frac{180\cdot30}{140}\right)} - 1\right) \approx 1103.6349$$

The height of the satellite is approximately 1104 mi.

Chapter 2   Chapter Test

1. $\sin A = \dfrac{\text{side opposite}}{\text{hypotenuse}} = \dfrac{12}{13}$

$\cos A = \dfrac{\text{side adjacent}}{\text{hypotenuse}} = \dfrac{5}{13}$

$\tan A = \dfrac{\text{side opposite}}{\text{side adjacent}} = \dfrac{12}{5}$

$\cot A = \dfrac{\text{side adjacent}}{\text{side opposite}} = \dfrac{5}{12}$

$\sec A = \dfrac{\text{hypotenuse}}{\text{side adjacent}} = \dfrac{13}{5}$

$\csc A = \dfrac{\text{hypotenuse}}{\text{side opposite}} = \dfrac{13}{12}$

2. Apply the relationships between the lengths of the sides of a $30° - 60°$ right triangle first to the triangle on the right to find the values of y and w. In the $30° - 60°$ right triangle, the side opposite the $60°$ angle is $\sqrt{3}$ times as long as the side opposite to the $30°$ angle. The length of the hypotenuse is 2 times as long as the shorter leg (opposite the $30°$ angle).

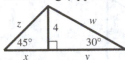

Thus, we have $y = 4\sqrt{3}$ and $w = 2(4) = 8$.

Apply the relationships between the lengths of the sides of a $45° - 45°$ right triangle next to the triangle on the left to find the values of x and z. In the $45° - 45°$ right triangle, the sides opposite the $45°$ angles measure the same. The hypotenuse is $\sqrt{2}$ times the measure of a leg. Thus, we have $x = 4$ and $z = 4\sqrt{2}$

3. $\sin(\theta + 15°) = \cos(2\theta + 30°)$

   Sine and cosine are cofunctions, so the sum of the angles is 90°. So,
   $$(\theta + 15°) + (2\theta + 30°) = 90°$$
   $$3\theta + 45° = 90°$$
   $$3\theta = 45° \Rightarrow \theta = 15°$$

4. (a) $\sin 24° < \sin 48°$

   In the interval from 0° to 90°, as the angle increases, so does the sine of the angle, so $\sin 24° < \sin 48°$ is true.

   (b) $\cos 24° < \cos 48°$

   In the interval from 0° to 90°, as the angle increases, so the cosine of the angle decreases, so $\cos 24° < \cos 48°$ is false.

   (c) $\cos(60° + 30°)$
   $$= \cos 60° \cos 30° - \sin 60° \sin 30°$$
   $$\cos(60° + 30°) = \cos 90° = 0$$
   $$\cos 60° \cos 30° - \sin 60° \sin 30°$$
   $$= \frac{1}{2}\left(\frac{\sqrt{3}}{2}\right) - \frac{\sqrt{3}}{2}\left(\frac{1}{2}\right) = 0$$

   Thus, the statement is true.

5. A 240° angle lies in quadrant III, so the reference angle is $240° - 180° = 60°$. Because 240° is in quadrant III, the sine, cosine, secant, and cosecant are negative.

   $$\sin 240° = -\sin 60° = -\frac{\sqrt{3}}{2}$$
   $$\cos 240° = -\cos 60° = -\frac{1}{2}$$
   $$\tan 240° = \tan 60° = \sqrt{3}$$
   $$\cot 240° = \cot 60° = \frac{1}{\sqrt{3}} = \frac{\sqrt{3}}{3}$$
   $$\sec 240° = -\sec 60° = -2$$
   $$\csc 240° = -\csc 60° = -\frac{2}{\sqrt{3}} = -\frac{2\sqrt{3}}{3}$$

6. $-135°$ is coterminal with $-135° + 360° = 225°$. This angle lies in quadrant III. The reference angle is $225° - 180° = 45°$. Because $-135°$ is in quadrant III, the sine, cosine, secant, and cosecant are negative.

   $$\sin(-135°) = -\sin 45° = -\frac{\sqrt{2}}{2}$$
   $$\cos(-35°) = -\cos 45° = -\frac{\sqrt{2}}{2}$$
   $$\tan(-135°) = \tan 45° = 1$$
   $$\cot(-135°) = \cot 45° = 1$$
   $$\sec(-135°) = -\sec 45° = -\sqrt{2}$$
   $$\csc(-135°) = -\csc 45° = -\sqrt{2}$$

7. 990° is coterminal with $990° - 2 \cdot 360° = 270°$, which is the reference angle.
   $$\sin 990° = \sin 270° = -1$$
   $$\cos 990° = \cos 270° = 0$$
   $$\tan 990° = \tan 270° \text{ undefined}$$
   $$\cot 990° = \cot 270° = 0$$
   $$\sec 990° = \sec 270° \text{ undefined}$$
   $$\csc 990° = \csc 270° = -1$$

8. $\cos \theta = -\frac{\sqrt{2}}{2}$

   Because $\cos \theta$ is negative, θ must lie in quadrant II or quadrant III. The absolute value of $\cos \theta$ is $\frac{\sqrt{2}}{2}$, so $\theta' = 45°$. The quadrant II angle θ equals $180° - \theta' = 180° - 45° = 135°$, and the quadrant III angle θ equals $180° + \theta' = 180° + 45° = 225°$.

9. $\csc\theta = -\dfrac{2\sqrt{3}}{3}$

   Because $\csc\theta$ is negative, θ must lie in quadrant III or quadrant IV. The absolute value of $\csc\theta$ is $\dfrac{2\sqrt{3}}{3}$, so $\theta' = 60°$. The quadrant III angle θ equals $180° + \theta' = 180° + 60° = 240°$, and the quadrant IV angle θ equals $360° - \theta' = 360° - 60° = 300°$.

10. $\tan\theta = 1 \Rightarrow \theta = 45°$ or $\theta = 225°$

11. $\tan\theta = 1.6778490$

    $\cot\theta = \dfrac{1}{\tan\theta} = (\tan\theta)^{-1}$, so we can use division or the inverse key (multiplicative inverse).

    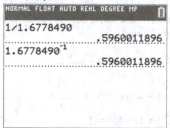

12. **(a)** $\sin 78°21' \approx 0.979399$

    (b) $\tan 117.689° \approx -1.905608$

    (c) $\sec 58.9041° = \dfrac{1}{\cos 58.9041°} \approx 1.936213$

13. $\sin\theta = 0.27843196$

    $\theta = \sin^{-1}(0.27843196) \approx 16.166641°$

14. $A = 58°30', a = 748$

    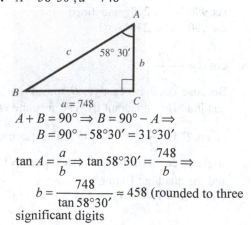

    $A + B = 90° \Rightarrow B = 90° - A \Rightarrow$
    $B = 90° - 58°30' = 31°30'$

    $\tan A = \dfrac{a}{b} \Rightarrow \tan 58°30' = \dfrac{748}{b} \Rightarrow$

    $b = \dfrac{748}{\tan 58°30'} \approx 458$ (rounded to three significant digits

    $\sin A = \dfrac{a}{c} \Rightarrow \sin 58°30' = \dfrac{748}{c} \Rightarrow$

    $c = \dfrac{748}{\sin 58°30'} \approx 877$ (rounded to three significant digits

15. Let θ = the measure of the angle that the guy wire makes with the ground.

    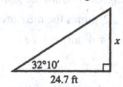

    $\sin\theta = \dfrac{71.3}{77.4}$

    $\theta = \sin^{-1}\left(\dfrac{71.3}{77.4}\right) \approx 67.1° \approx 67°10'$

16. Let x = the height of the flagpole.

    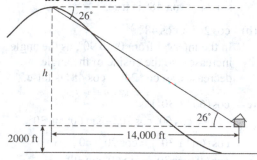

    $\tan 32°10' = \dfrac{x}{24.7}$

    $x = 24.7\tan 32°10' \approx 15.5344$

    The flagpole is approximately 15.5 ft high. (rounded to three significant digits)

17. Let h = the height of the top of mountain above the cabin. Then $2000 + h$ = the height of the mountain.

    $\tan 26° = \dfrac{h}{14,000} \Rightarrow h \approx 6800$ (rounded to two significant digits). Thus, the height of the mountain is about $6800 + 2000 = 8800$ ft.

18. Let x = distance the ships are apart.
In the figure, the measure of angle CAB is $122° - 32° = 90°$. Therefore, triangle CAB is a right triangle.

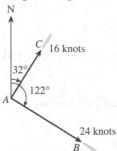

Because $d = rt$, the distance traveled by the first ship in 2.5 hr is
(2.5 hr)(16 knots) = 40 nautical mi and the second ship is
(2.5hr)(24 knots) = 60 nautical mi.
Applying the Pythagorean theorem, we have
$x^2 = 40^2 + 60^2 \Rightarrow x^2 = 1600 + 3600 \Rightarrow$
$x^2 = 5200 \Rightarrow x = \sqrt{5200} \approx 72.111$
The ships are 72 nautical mi apart (rounded to 2 significant digits).

19. Draw triangle ACB and extend north-south lines from points A and C. Angle ACD is 62° (alternate interior angles of parallel lines cut by a transversal have the same measure), so Angle ACB is $62° + 28° = 90°$.

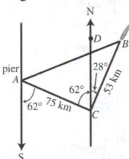

Angle ACB is a right angle, so use the Pythagorean theorem to find the distance from A to B.
$(AB)^2 = 75^2 + 53^2 \Rightarrow (AB)^2 = 5625 + 2809 \Rightarrow$
$(AB)^2 = 8434 \Rightarrow AB = \sqrt{8434} \approx 91.8368$
It is 92 km from the pier to the boat, rounded to two significant digits.

20. Let x = the side adjacent to 52.5° in the smaller triangle.

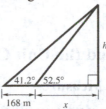

In the larger triangle, we have
$$\tan 41.2° = \frac{h}{168 + x} \Rightarrow h = (168 + x)\tan 41.2°.$$
In the smaller triangle, we have
$$\tan 52.5° = \frac{h}{x} \Rightarrow h = x \tan 52.5°.$$
Substitute for h in this equation to solve for x.
$$(168 + x)\tan 41.2° = x \tan 52.5°$$
$$168 \tan 41.2° + x \tan 41.2° = x \tan 52.5°$$
$$168 \tan 41.2° = x \tan 52.5° - x \tan 41.2°$$
$$168 \tan 41.2° = x(\tan 52.5° - \tan 41.2°)$$
$$\frac{168 \tan 41.2°}{\tan 52.5° - \tan 41.2°} = x$$
Substituting for x in the equation for the smaller triangle gives
$$h = x \tan 52.5°$$
$$h = \frac{168 \tan 41.2° \tan 52.5°}{\tan 52.5° - \tan 41.2°} \approx 448.0432$$
The height of the triangle is approximately 448 m (rounded to three significant digits).

Chapter 3

Radian Measure and the Unit Circle

Section 3.1   Radian Measure

1. An angle with its vertex at the center of a circle that intercepts an arc on the circle equal in length to the <u>radius</u> of the circle has measure 1 radian.

3. Multiply a degree measure by $\frac{\pi}{180}$ radian and simplify to convert to radians.

5. θ is in quadrant I, so $0 < \theta < \frac{\pi}{2}$. Because $\frac{\pi}{2} \approx 1.57$, 1 is the only integer value in the interval. Thus, the radian measure of θ is 1 radian.

7. θ is in quadrant II, so $\frac{\pi}{2} < \theta < \pi$. Because $\frac{\pi}{2} \approx 1.57$ and $\pi \approx 3.14$, 2 and 3 are the only integers in the interval. θ is closer to π, so the radian measure of θ is 3 radians.

9. θ is an angle in quadrant III drawn in a clockwise direction, so $-\pi < \theta < -\frac{\pi}{2}$. Also $-\pi \approx -3.14$ and $-\frac{\pi}{2} \approx -1.57$, so -2 and -3 are the only integers in the interval. θ is closer to $-\pi$, so the radian measure of θ is -3 radians.

11. $60° = 60\left(\frac{\pi}{180} \text{ radian}\right) = \frac{\pi}{3}$ radians

13. $90° = 90\left(\frac{\pi}{180} \text{ radian}\right) = \frac{\pi}{2}$ radians

15. $150° = 150\left(\frac{\pi}{180} \text{ radian}\right) = \frac{5\pi}{6}$ radians

17. $-300° = -300\left(\frac{\pi}{180} \text{ radian}\right) = -\frac{5\pi}{3}$ radians

19. $450° = 450\left(\frac{\pi}{180} \text{ radian}\right) = \frac{5\pi}{2}$ radians

21. $1800° = 1800°\left(\frac{\pi}{180°} \text{ radian}\right) = 10\pi$ radians

23. $0° = 0°\left(\frac{\pi}{180°} \text{ radian}\right) = 0$ radians

25. $-900° = -900°\left(\frac{\pi}{180°} \text{ radian}\right)$
$\qquad = -5\pi$ radians

27. Radian measure provides a method for measuring angles in which the central angle, θ, of a circle is the ratio of the intercepted arc, s, to the radius of the circle, r. That is, $\theta = \frac{s}{r}$.

29. $\frac{\pi}{3} = \frac{\pi}{3}\left(\frac{180°}{\pi}\right) = 60°$

31. $\frac{7\pi}{4} = \frac{7\pi}{4}\left(\frac{180°}{\pi}\right) = 315°$

33. $\frac{11\pi}{6} = \frac{11\pi}{6}\left(\frac{180°}{\pi}\right) = 330°$

35. $-\frac{\pi}{6} = -\frac{\pi}{6}\left(\frac{180°}{\pi}\right) = -30°$

37. $\frac{7\pi}{10} = \frac{7\pi}{10}\left(\frac{180°}{\pi}\right) = 126°$

39. $-\frac{4\pi}{15} = -\frac{4\pi}{15}\left(\frac{180°}{\pi}\right) = -48°$

41. $\frac{17\pi}{20} = \frac{17\pi}{20}\left(\frac{180°}{\pi}\right) = 153°$

43. $-5\pi = -5\pi\left(\frac{180°}{\pi}\right) = -900°$

45. $39° = 39\left(\frac{\pi}{180} \text{ radian}\right) \approx 0.681$ radian

47. $42.5° = 42.5\left(\frac{\pi}{180} \text{ radian}\right) \approx 0.742$ radian

49. $139°10' = \left(139 + \frac{10}{60}\right)°$

$\approx 139.1666667\left(\frac{\pi}{180} \text{ radian}\right)$

$\approx 2.429 \text{ radians}$

51. $64.29° = 64.29\left(\frac{\pi}{180} \text{ radian}\right) \approx 1.122 \text{ radians}$

53. $56°25' = \left(56 + \frac{25}{60}\right)°$

$\approx 56.41666667\left(\frac{\pi}{180} \text{ radian}\right)$

$\approx 0.985 \text{ radian}$

55. $-47.69° = -47.69\left(\frac{\pi}{180} \text{ radian}\right)$

$\approx -0.832 \text{ radian}$

57. $2 \text{ radians} = 2\left(\frac{180°}{\pi}\right) \approx 114.591559°$

$= 114° + (0.591559 \cdot 60)'$

$\approx 114°35'$

59. $1.74 \text{ radians} = 1.74\left(\frac{180°}{\pi}\right) \approx 99.69465635°$

$= 99° + (0.69465635 \cdot 60)'$

$\approx 99°42'$

61. $0.3417 \text{ radian} = .3417\left(\frac{180°}{\pi}\right) \approx 19.57796786°$

$= 19° + (0.57796786 \cdot 60)'$

$= 19°35'$

63. $-5.01095 \text{ radian} = -5.01095\left(\frac{180°}{\pi}\right)$

$\approx -287.1062864°$

$= -\left[287° + (0.1062864 \cdot 60)'\right]$

$\approx -287°06'$

65. Without the degree symbol on the 30, it is assumed that 30 is measured in radians. Thus, the approximate value of sin 30 is -0.98803, not $\frac{1}{2}$.

67. $\sin\frac{\pi}{3} = \sin\left(\frac{\pi}{3} \cdot \frac{180°}{\pi}\right) = \sin 60° = \frac{\sqrt{3}}{2}$

69. $\tan\frac{\pi}{4} = \tan\left(\frac{\pi}{4} \cdot \frac{180°}{\pi}\right) = \tan 45° = 1$

71. $\sec\frac{\pi}{6} = \sec\left(\frac{\pi}{6} \cdot \frac{180°}{\pi}\right) = \sec 30° = \frac{2\sqrt{3}}{3}$

73. $\sin\frac{\pi}{2} = \sin\left(\frac{\pi}{2} \cdot \frac{180°}{\pi}\right) = \sin 90° = 1$

75. $\tan\frac{5\pi}{3} = \tan\left(\frac{5\pi}{3} \cdot \frac{180°}{\pi}\right) = \tan 300°$

$= -\tan 60° = -\sqrt{3}$

77. $\sin\left(\frac{5\pi}{6}\right) = \sin\left(\frac{5\pi}{6} \cdot \frac{180°}{\pi}\right) = \sin 150°$

$= \sin 30° = \frac{1}{2}$

79. $\cos 3\pi = \cos\left(3\pi \cdot \frac{180°}{\pi}\right) = \cos 540°$

$= \cos(540° - 360°) = \cos 180° = -1$

81. $\sin\left(-\frac{8\pi}{3}\right) = \sin\left(-\frac{8\pi}{3} \cdot \frac{180°}{\pi}\right)$

$= \sin(-480°)$

$= \sin(-480° + 2 \cdot 360°)$

$= \sin 240° = -\sin 60° = -\frac{\sqrt{3}}{2}$

83. $\sin\left(-\frac{7\pi}{6}\right) = \sin\left(-\frac{7\pi}{6} \cdot \frac{180°}{\pi}\right)$

$= \sin(-210°)$

$= \sin(-210° + 360°)$

$= \sin 150° = \sin 30° = \frac{1}{2}$

85. $\tan\left(-\frac{14\pi}{3}\right) = \tan\left(-\frac{14\pi}{3} \cdot \frac{180°}{\pi}\right)$

$= \tan(-840°)$

$= \tan(-840° + 3 \cdot 360) = \tan 240°$

$= \tan 60° = \sqrt{3}$

87. Begin the calculation with the blank next to 30°, and then proceed counterclockwise from there.

$30° = 30\left(\frac{\pi}{180} \text{ radian}\right) = \frac{\pi}{6} \text{ radian}$

$\frac{\pi}{4} \text{ radians} = \frac{\pi}{4}\left(\frac{180°}{\pi}\right) = 45°$

$60° = 60\left(\frac{\pi}{180} \text{ radian}\right) = \frac{\pi}{3} \text{ radians}$

$\frac{2\pi}{3} \text{ radians} = \frac{2\pi}{3}\left(\frac{180°}{\pi}\right) = 120°$

(*continued on next page*)

(*continued*)

$$\frac{3\pi}{4} \text{ radians} = \frac{3\pi}{4}\left(\frac{180°}{\pi}\right) = 135°$$

$$150° = 150\left(\frac{\pi}{180} \text{ radian}\right) = \frac{5\pi}{6} \text{ radians}$$

$$180° = 180\left(\frac{\pi}{180} \text{ radian}\right) = \pi \text{ radians}$$

$$210° = 210\left(\frac{\pi}{180} \text{ radian}\right) = \frac{7\pi}{6} \text{ radians}$$

$$225° = 225\left(\frac{\pi}{180} \text{ radian}\right) = \frac{5\pi}{4} \text{ radians}$$

$$\frac{4\pi}{3} \text{ radians} = \frac{4\pi}{3}\left(\frac{180°}{\pi}\right) = 240°$$

$$\frac{5\pi}{3} \text{ radians} = \frac{5\pi}{3}\left(\frac{180°}{\pi}\right) = 300°$$

$$315° = 315\left(\frac{\pi}{180} \text{ radian}\right) = \frac{7\pi}{4} \text{ radians}$$

$$330° = 330\left(\frac{\pi}{180} \text{ radian}\right) = \frac{11\pi}{6} \text{ radians}$$

89. Answers may vary. Sample answer:
$$\pi + 2\pi = 3\pi$$
$$\pi - 2\pi = -\pi$$
Two angles that are coterminal with an angle of π radians are 3π and $-\pi$.

91. (a) In 24 hours, the hour hand will rotate twice around the clock. One complete rotation is 2π radians, the two rotations will measure $2 \cdot 2\pi = 4\pi$ radians.

(b) In 4 hours, the hour hand will rotate $\frac{4}{12} = \frac{1}{3}$ of the way around the clock, which is $\frac{1}{3} \cdot 2\pi = \frac{2\pi}{3}$ radians.

93. In each rotation around Earth, the space vehicle would rotate 2π radians.

(a) In 2.5 orbits, the space vehicle travels $2.5 \cdot 2\pi = 5\pi$ radians.

(b) In $\frac{4}{3}$ of an orbit, the space vehicle travels $\frac{4}{3} \cdot 2\pi = \frac{8\pi}{3}$ radians.

95. In each revolution, the horse revolves 2π radians. In 12 revolutions, the horse revolves $12 \cdot 2\pi = 24\pi$ radians.

Section 3.2 Applications of Radian Measure

1. $r = 4,\ \theta = \frac{\pi}{2}$
$$s = r\theta = 4\left(\frac{\pi}{2}\right) = 2\pi$$

3. $s = 6\pi,\ \theta = \frac{3\pi}{4}$
$$s = r\theta \Rightarrow r = \frac{s}{\theta} = \frac{6\pi}{\frac{3\pi}{4}} = 6\pi \cdot \frac{4}{3\pi} = 8$$

5. $r = 3,\ s = 3$
$$s = r\theta \Rightarrow \theta = \frac{s}{r} = \frac{3}{3} = 1$$

7. $r = 6,\ s = 2\pi$
$$s = r\theta \Rightarrow 2\pi = 6\theta \Rightarrow \theta = \frac{2\pi}{6} = \frac{\pi}{3}$$
$$A = \frac{1}{2}r^2\theta \Rightarrow$$
$$A = \frac{1}{2}(6)^2\left(\frac{\pi}{3}\right) = \frac{1}{2}(36)\left(\frac{\pi}{3}\right) = 6\pi$$

9. $A = 3$ sq units, $r = 2$
$$A = \frac{1}{2}r^2\theta \Rightarrow 3 = \frac{1}{2}(2)^2\,\theta \Rightarrow 3 = \frac{1}{2}(4)\theta \Rightarrow$$
$$3 = 2\theta \Rightarrow \theta = \frac{3}{2} = 1.5 \text{ radians}$$

11. $A = 6\pi$ sq units, $r = 6$
$$A = \frac{1}{2}r^2\theta \Rightarrow 6\pi = \frac{1}{2}(6)^2\,\theta \Rightarrow$$
$$6\pi = \frac{1}{2}(36)\theta \Rightarrow 6\pi = 18\theta \Rightarrow$$
$$\theta = \frac{6\pi}{18} = \frac{\pi}{3} \text{ radians}$$
$$\frac{\pi}{3} \text{ radians} = \frac{\pi}{3} \cdot \frac{180}{\pi} = 60°$$
The measure of the central angle is $60°$.

13. $r = 12.3$ cm, $\theta = \frac{2\pi}{3}$ radians
$$s = r\theta = 12.3\left(\frac{2\pi}{3}\right) = 8.2\pi \approx 25.8 \text{ cm}$$

15. $r = 1.38$ ft, $\theta = \frac{5\pi}{6}$ radians
$$s = r\theta = 1.38\left(\frac{5\pi}{6}\right)$$
$$= 1.15\pi \text{ ft} \approx 3.61 \text{ ft (rounded to three significant digits)}$$

17. $r = 4.82$ m, $\theta = 60°$
Converting θ to radians, we have
$$\theta = 60° = 60\left(\frac{\pi}{180} \text{ radian}\right) = \frac{\pi}{3} \text{ radians}.$$
Thus, the arc is
$$s = r\theta = 4.82\left(\frac{\pi}{3}\right) = \frac{4.82\pi}{3} \approx 5.05 \text{ m}.$$
(rounded to three significant digits)

19. $r = 15.1$ in., $\theta = 210°$
Converting θ to radians, we have
$$\theta = 210° = 210\left(\frac{\pi}{180} \text{ radian}\right) = \frac{7\pi}{6} \text{ radians}.$$
Thus, the arc is
$$s = r\theta = 15.1\left(\frac{7\pi}{6}\right) = \frac{105.7\pi}{6} \approx 55.3 \text{ in}.$$
(rounded to three significant digits)

21. The formula for arc length is $s = r\theta$.
Substituting $2r$ for r we obtain
$s = (2r)\theta = 2(r\theta)$. The length of the arc is
doubled.

For Exercises 23−27, note that 6400 has two significant digits and the angles are given to the nearest degree, so we can have only two significant digits in the answers.

23. 9° N, 40° N
$$\theta = 40° - 9° = 31° = 31\left(\frac{\pi}{180} \text{ radian}\right)$$
$$= \frac{31\pi}{180} \text{ radian}$$
$$s = r\theta = 6400\left(\frac{31\pi}{180}\right) \approx 3500 \text{ km}$$

25. 41° N, 12° S
12° S = −12° N
$$\theta = 41° - (-12°) = 53° = 53\left(\frac{\pi}{180} \text{ radian}\right)$$
$$= \frac{53\pi}{180} \text{ radian}$$
$$s = r\theta = 6400\left(\frac{53\pi}{180}\right) \approx 5900 \text{ km}$$

27. $r = 6400$ km, $s = 1200$ km
$$s = r\theta \Rightarrow 1200 = 6400\theta \Rightarrow \theta = \frac{1200}{6400} = \frac{3}{16}$$
Converting $\frac{3}{16}$ radian to degrees, we have
$$\theta = \frac{3}{16}\left(\frac{180°}{\pi}\right) \approx 11°. \text{ The north-south}$$
distance between the two cities is 11°.
Let x = the latitude of Madison.
$x - 33° = 11° \Rightarrow x = 44°$ N
The latitude of Madison is 44° N.

29. The arc length on the smaller gear is
$$s = r\theta = 3.7\left(300 \cdot \frac{\pi}{180}\right) = 3.7\left(\frac{5\pi}{3}\right)$$
$$= \frac{18.5\pi}{3} \text{ cm}$$
An arc with this length on the larger gear corresponds to an angle measure θ where
$$s = r\theta \Rightarrow \frac{18.5\pi}{3} = 7.1\theta \Rightarrow \frac{18.5\pi}{21.3} = \theta \Rightarrow$$
$$\theta = \frac{18.5\pi}{21.3} \cdot \frac{180}{\pi} \approx 156°$$
The larger gear will rotate through approximately 156°.

31. A rotation of $\theta = 60.0\left(\frac{\pi}{180} \text{ radian}\right) = \frac{\pi}{3}$
radians on the smaller wheel moves through an arc length of
$$s = r\theta = 5.23\left(\frac{\pi}{3}\right) = \frac{5.23\pi}{3} \text{ cm} \text{ (holding on}$$
to more digits for the intermediate steps)
Both wheels move together, so the larger
wheel also moves $\frac{5.23\pi}{3}$ cm, which rotates it
through an angle θ, where
$$\frac{5.23\pi}{3} = 8.16\theta$$
$$\theta = \frac{5.23\pi}{24.48} \text{ radian} = \frac{5.23\pi}{24.48}\left(\frac{180°}{\pi}\right) \approx 38.5°$$
The larger wheel rotates through 38.5°.

33. The arc length s represents the distance traveled by a point on the rim of a wheel. The two wheels rotate together, so s will be the same for both wheels. For the smaller wheel,

$$\theta = 80° = 80\left(\frac{\pi}{180}\right) = \frac{4\pi}{9} \text{ radians and}$$

$$s = r\theta = 11.7\left(\frac{4\pi}{9}\right) = 5.2\pi \text{ cm.}$$

For the larger wheel,

$$\theta = 50° = 50\left(\frac{\pi}{180} \text{ radian}\right) = \frac{5\pi}{18} \text{ radian.}$$

Thus, we can solve

$$s = r\theta \Rightarrow 5.2\pi = r\left(\frac{5\pi}{18}\right) \Rightarrow$$

$$r = 5.2\pi \cdot \frac{18}{5\pi} = 18.72$$

The radius of the larger wheel is 18.7 cm. (rounded to 3 significant digits)

35. (a) The number of inches lifted is the arc length in a circle with $r = 9.27$ in. and $\theta = 71°50'$.

$$71°50' = \left(71 + \frac{50}{60}\right)\left(\frac{\pi}{180°}\right)$$

$$s = r\theta \Rightarrow$$

$$s = 9.27\left(71 + \frac{50}{60}\right)\left(\frac{\pi}{180°}\right) \approx 11.6221$$

The weight will rise 11.6 in. (rounded to three significant digits)

(b) When the weight is raised 6 in., we have $s = r\theta \Rightarrow 6 = 9.27\theta \Rightarrow$

$$\theta = \frac{6}{9.27} \text{ radian} = \frac{6}{9.27}\left(\frac{180°}{\pi}\right)$$

$$\approx 37.0846° = 37° + 0.0846(60')$$

$$\approx 37°05'$$

The pulley must be rotated through 37°05'.

37. A rotation of

$$\theta = 180\left(\frac{\pi}{180} \text{ radian}\right) = \pi \text{ radians. The chain}$$

moves a distance equal to half the arc length of the larger gear. So, for the large gear and pedal, $s = r\theta \Rightarrow 4.72\pi$. Thus, the chain moves 4.72π in. The small gear rotates through an angle as follows.

$$\theta = \frac{s}{r} \Rightarrow \theta = \frac{4.72\pi}{1.38} \approx 3.42\pi$$

θ for the wheel and θ for the small gear are the same, or 3.42π. So, for the wheel, we have

$$s = r\theta \Rightarrow r = 13.6(3.42\pi) \approx 146.12$$

The bicycle will move 146 in. (rounded to three significant digits)

39. Because $\dfrac{30}{60} = \dfrac{1}{2}$ rotation, we have

$$\theta = \frac{1}{2}(2\pi) = \pi. \text{ Thus, } s = r\theta \Rightarrow s = 3\pi \text{ in.}$$

41. Because $\theta = 4.5(2\pi) = 9\pi$, we have

$$s = r\theta \Rightarrow s = 3(9\pi) = 27\pi \text{ in.}$$

43. Let $t =$ the length of the train. t is approximately the arc length subtended by 3° 20'. First convert $\theta = 3°20'$ to radians.

$$\theta = 3°20' = \left(3 + \frac{20}{60}\right)° = 3\frac{1}{3}°$$

$$= \left(3\frac{1}{3}\right)\left(\frac{\pi}{180} \text{ radian}\right) = \left(\frac{10}{3}\right)\left(\frac{\pi}{180} \text{ radian}\right)$$

$$= \frac{\pi}{54} \text{ radian}$$

The length of the train is

$$t = r\theta \Rightarrow t = 3.5\left(\frac{\pi}{54}\right) \approx 0.20 \text{ km long.}$$

(rounded to two significant digits)

45. Let $r =$ the distance of the boat. The height of the mast, 32.0 ft, is approximately the arc length subtended by 2° 11'. First convert $\theta = 2°11'$ to radians.

$$\theta = 2°11' = \left(2 + \frac{11}{60}\right)° = 2\frac{11}{60}°$$

$$= \left(2\frac{11}{60}\right)\left(\frac{\pi}{180} \text{ radian}\right) = \left(\frac{131}{60}\right)\left(\frac{\pi}{180} \text{ radian}\right)$$

$$= \frac{131\pi}{10800} \text{ radian}$$

We must now find the radius, r.

$$s = r\theta \Rightarrow r = \frac{s}{\theta} \Rightarrow$$

$$r = \frac{32}{\frac{131\pi}{10800}} = 32 \cdot \frac{10800}{131\pi} \approx 839.7549$$

The boat is about 840 ft away. (rounded to two significant digits)

In Exercises 47–54, we will be rounding to the nearest tenth.

47. $r = 29.2$ m, $\theta = \dfrac{5\pi}{6}$ radians 0.517°.

The area of the sector is 1116.1 m². (1120 m² rounded to three significant digits)

49. $r = 30.0$ ft, $\theta = \dfrac{\pi}{2}$ radians

$A = \dfrac{1}{2}r^2\theta \Rightarrow$

$A = \dfrac{1}{2}(30.0)^2\left(\dfrac{\pi}{2}\right) = \dfrac{1}{2}(900)\left(\dfrac{\pi}{2}\right) = 225\pi$

≈ 706.8583

The area of the sector is 706.9 ft². (707 ft² rounded to three significant digits)

51. $r = 12.7$ cm, $\theta = 81°$

The formula $A = \dfrac{1}{2}r^2\theta$ requires that θ be measured in radians. Converting 81° to radians, we have

$\theta = 81\left(\dfrac{\pi}{180}\text{ radian}\right) = \dfrac{9\pi}{20}$ radians.

$A = \dfrac{1}{2}(12.7)^2\left(\dfrac{9\pi}{20}\right) = \dfrac{1}{2}(161.29)\left(\dfrac{9\pi}{20}\right)$

≈ 114.0092

The area of the sector is 114.0 cm². (114 cm² rounded to three significant digits)

53. $r = 40.0$ mi, $\theta = 135°$

The formula $A = \dfrac{1}{2}r^2\theta$ requires that θ be measured in radians. Converting 135° to radians, we have

$\theta = 135\left(\dfrac{\pi}{180}\text{ radian}\right) = \dfrac{3\pi}{4}$ radians.

$A = \dfrac{1}{2}(40.0)^2\left(\dfrac{3\pi}{4}\right) = \dfrac{1}{2}(1600)\left(\dfrac{3\pi}{4}\right)$

$= 600\pi \approx 1884.9556$

The area of the sector is 1885.0 mi². (1880 mi² rounded to three significant digits)

55. $A = 16$ in.², $r = 3.0$ in.

$A = \dfrac{1}{2}r^2\theta \Rightarrow 16 = \dfrac{1}{2}(3)^2\theta \Rightarrow 16 = \dfrac{9}{2}\theta \Rightarrow$

$\theta = 16 \cdot \dfrac{2}{9} = \dfrac{32}{9} \approx 3.6$ radians

(rounded to two significant digits)

57. The formula $A = \dfrac{1}{2}r^2\theta$ requires that θ be measured in radians. Converting 40.0° to radians, we have

$\theta = 40.0\left(\dfrac{\pi}{180}\text{ radian}\right) = \dfrac{2\pi}{9}$ radians.

$A = \dfrac{1}{2}r^2\theta = \dfrac{1}{2}(152)^2\left(\dfrac{2\pi}{9}\right)$

$= \dfrac{1}{2}(23,104)\left(\dfrac{2\pi}{9}\right) = \dfrac{23,104\pi}{9}$

≈ 8060 yd²

(rounded to three significant digits)

59. $A = 50$ in.², $r = 5$ in.

First find θ.

$A = \dfrac{1}{2}r^2\theta \Rightarrow 50 = \dfrac{1}{2}(5)^2\theta \Rightarrow 50 = \dfrac{25}{2}\theta \Rightarrow$

$\theta = 4$ radians

To find the arc length, apply the formula $s = r\theta$.

$s = 5 \cdot 4 = 20$ in.

61. (a) The central angle in degrees measures

$\dfrac{360°}{27} = 13\frac{1}{3}°$. Converting to radians, we have the following.

$13\frac{1}{3}° = \left(13\frac{1}{3}\right)\left(\dfrac{\pi}{180}\text{ radian}\right)$

$= \left(\dfrac{40}{3}\right)\left(\dfrac{\pi}{180}\text{ radian}\right) = \dfrac{2\pi}{27}$ radian

(b) $C = 2\pi r$, and $r = 76$ ft, so

$C = 2\pi(76) = 152\pi \approx 477.5221$. The circumference is about 478 ft.

(c) $r = 76$ ft and $\theta = \dfrac{2\pi}{27}$, so

$s = r\theta = 76\left(\dfrac{2\pi}{27}\right) = \dfrac{152\pi}{27} \approx 17.6860$.

Thus, the length of the arc is 17.7 ft. Note: If this measurement is approximated to be $\dfrac{160}{9}$, then the approximated value would be 17.8 ft, rounded to three significant digits.

(d) Area of sector with $r = 76$ ft and $\theta = \dfrac{2\pi}{27}$ is as follows.

$A = \dfrac{1}{2}r^2\theta \Rightarrow$

$A = \dfrac{1}{2}(76^2)\dfrac{2\pi}{27} = \dfrac{1}{2}(5776)\dfrac{2\pi}{27} = \dfrac{5776\pi}{27}$

$\approx 672.0681 \approx 672$ ft²

63. (a)

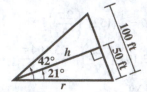

The triangle formed by the central angle and the chord is isosceles. Therefore, the bisector of the central angle is also the perpendicular bisector of the chord.

$$\sin 21° = \frac{50}{r} \Rightarrow r = \frac{50}{\sin 21°} \approx 140 \text{ ft}$$

(b) $r = \dfrac{50}{\sin 21°}; \ \theta = 42°$

Converting θ to radians, we have

$42\left(\dfrac{\pi}{180} \text{ radians}\right) = \dfrac{7\pi}{30}$ radians. Solving

for the arc length, we have
$s = r\theta \Rightarrow$

$$s = \frac{50}{\sin 21°} \cdot \frac{7\pi}{30} = \frac{35\pi}{3\sin 21°} \approx 102 \text{ ft}$$

(c)

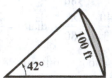

The area of the portion of the circle can be found by subtracting the area of the triangle from the area of the sector. From the figure in part (a), we have

$$\tan 21° = \frac{50}{h} \quad \text{so} \quad h = \frac{50}{\tan 21°}.$$

$\mathcal{A}_{\text{sector}} = \dfrac{1}{2}r^2\theta \Rightarrow$

$$\mathcal{A}_{\text{sector}} = \frac{1}{2}\left(\frac{50}{\sin 21°}\right)^2\left(\frac{7\pi}{30}\right) \approx 7135 \text{ ft}^2$$

and

$\mathcal{A}_{\text{triangle}} = \dfrac{1}{2}bh \Rightarrow$

$$\mathcal{A}_{\text{triangle}} = \frac{1}{2}(100)\left(\frac{50}{\tan 21°}\right) \approx 6513 \text{ ft}^2$$

The area bounded by the arc and the chord is $7135 - 6513 = 622 \text{ ft}^2$.

65. Use the Pythagorean theorem to find the hypotenuse of the triangle, which is also the radius of the sector of the circle.

$r^2 = 30^2 + 40^2 \Rightarrow r^2 = 900 + 1600 \Rightarrow$
$r^2 = 2500 \Rightarrow r = 50$

The total area of the lot is the sum of the areas of the triangle and the sector.
Converting $\theta = 60°$ to radians, we have

$$60\left(\frac{\pi}{180} \text{ radian}\right) = \frac{\pi}{3} \text{ radians.}$$

$$\mathcal{A}_{\text{triangle}} = \frac{1}{2}bh = \frac{1}{2}(30)(40) = 600 \text{ yd}^2$$

$$\mathcal{A}_{\text{sector}} = \frac{1}{2}r^2\theta = \frac{1}{2}(50)^2\left(\frac{\pi}{3}\right)$$

$$= \frac{1}{2}(2500)\left(\frac{\pi}{3}\right) = \frac{1250\pi}{3} \text{ yd}^2$$

Total area

$$\mathcal{A}_{\text{triangle}} + \mathcal{A}_{\text{sector}} = 600 + \frac{1250\pi}{3} \approx 1908.9969$$

or 1900 yd^2, rounded to two significant digits.

67. Converting $\theta = 7°12' = \left(7 + \frac{12}{60}\right)° = 7.2°$ to radians, we have

$$7.2\left(\frac{\pi}{180} \text{ radian}\right) = \frac{7.2\pi}{180} = \frac{\pi}{25} \text{ radian.}$$

Solving for the radius with the arc length formula, we have

$$s = r\theta \Rightarrow 496 = r \cdot \frac{\pi}{25} \Rightarrow$$

$$r = 496 \cdot \frac{25}{\pi} = \frac{12,400}{\pi} \approx 3947.0426$$

Thus, the radius is approximately 3950 mi. (rounded to three significant digits) Using this approximate radius, we can find the circumference of the Earth. Because $C = 2\pi r$, we have $C \approx 2\pi(3950) \approx 24,800$. Thus, the approximate circumference is 24,800 mi. (rounded to three significant digits)

69. If we let $r' = 2r$, then

$$\mathcal{A}_{\text{sector}} = \frac{1}{2}(r')^2\theta = \frac{1}{2}(2r)^2\theta \quad \text{Thus, the}$$

$$= \frac{1}{2}(4r^2)\theta = 4\left(\frac{1}{2}r^2\theta\right)$$

area, $\dfrac{1}{2}r^2\theta$, is quadrupled.

71. The base area is $\mathcal{A}_{\text{sector}} = \dfrac{1}{2}r^2\theta$. Thus, the

volume is $V = \dfrac{1}{2}r^2\theta h$ or $V = \dfrac{r^2\theta h}{2}$, where

θ is in radians.

73. L is the arc length, so $L = r\theta$. Thus, $r = \dfrac{L}{\theta}$.

75. $d = r - h \Rightarrow d = r - r\cos\dfrac{\theta}{2} \Rightarrow$

$$d = r\left(1 - \cos\dfrac{\theta}{2}\right)$$

Section 3.3   The Unit Circle and Circular Functions

1.

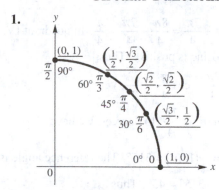

3. $\sin\dfrac{\pi}{4} = \dfrac{\sqrt{2}}{2}$ **5.** $\tan\dfrac{\pi}{4} = 1$

7. $\sin\theta = y = \dfrac{\sqrt{2}}{2}$; $\cos\theta = x = \dfrac{\sqrt{2}}{2}$

$\tan\theta = \dfrac{y}{x} = \dfrac{\frac{\sqrt{2}}{2}}{\frac{\sqrt{2}}{2}} = 1$; $\cot\theta = \dfrac{x}{y} = \dfrac{\frac{\sqrt{2}}{2}}{\frac{\sqrt{2}}{2}} = 1$

$\sec\theta = \dfrac{1}{x} = \dfrac{1}{\frac{\sqrt{2}}{2}} = \dfrac{2}{\sqrt{2}} = \dfrac{2}{\sqrt{2}} \cdot \dfrac{\sqrt{2}}{\sqrt{2}} = \sqrt{2}$

$\csc\theta = \dfrac{1}{y} = \dfrac{1}{\frac{\sqrt{2}}{2}} = \dfrac{2}{\sqrt{2}} = \dfrac{2}{\sqrt{2}} \cdot \dfrac{\sqrt{2}}{\sqrt{2}} = \sqrt{2}$

9. $\sin\theta = y = -\dfrac{12}{13}$; $\cos\theta = x = \dfrac{5}{13}$

$\tan\theta = \dfrac{y}{x} = \dfrac{-\frac{12}{13}}{\frac{5}{13}} = -\dfrac{12}{13}\left(\dfrac{13}{5}\right) = -\dfrac{12}{5}$

$\cot\theta = \dfrac{x}{y} = \dfrac{\frac{5}{13}}{-\frac{12}{13}} = \dfrac{5}{13}\left(-\dfrac{13}{12}\right) = -\dfrac{5}{12}$

$\sec\theta = \dfrac{1}{x} = \dfrac{1}{\frac{5}{13}} = \dfrac{13}{5}$; $\csc\theta = \dfrac{1}{y} = \dfrac{1}{-\frac{12}{13}} = -\dfrac{13}{12}$

11. An angle of $s = \dfrac{\pi}{2}$ radians intersects the unit circle at the point $(0,1)$.

    (a) $\sin s = y = 1$

    (b) $\cos s = x = 0$

    (c) $\tan s = \dfrac{y}{x} = \dfrac{1}{0}$; undefined

13. An angle of $s = 2\pi$ radians intersects the unit circle at the point $(1,0)$.

    (a) $\sin s = y = 0$

    (b) $\cos s = x = 1$

    (c) $\tan s = \dfrac{y}{x} = \dfrac{0}{1} = 0$

15. An angle of $s = -\pi$ radians intersects the unit circle at the point $(-1,0)$.

    (a) $\sin s = y = 0$

    (b) $\cos s = x = -1$

    (c) $\tan s = \dfrac{y}{x} = \dfrac{0}{-1} = 0$

For Exercises 17–31, use the following copy of Figure 13 on page 117 of the text.

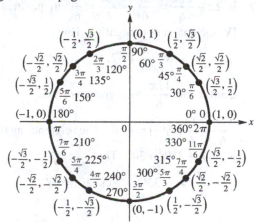

17. $\dfrac{7\pi}{6}$ is in quadrant III, so the reference angle

is $\dfrac{7\pi}{6} - \pi = \dfrac{7\pi}{6} - \dfrac{6\pi}{6} = \dfrac{\pi}{6}$. In quadrant III, the sine is negative. Thus,

$\sin\dfrac{7\pi}{6} = -\sin\dfrac{\pi}{6} = -\dfrac{1}{2}$. Converting $\dfrac{7\pi}{6}$ to

degrees, we have $\dfrac{7\pi}{6} = \dfrac{7}{6}(180°) = 210°$.

The reference angle is $210° - 180° = 30°$.

Thus, $\sin\dfrac{7\pi}{6} = \sin 210° = -\sin 30° = -\dfrac{1}{2}$.

19. $\dfrac{3\pi}{4}$ is in quadrant II, so the reference angle is

$\pi - \dfrac{3\pi}{4} = \dfrac{4\pi}{4} - \dfrac{3\pi}{4} = \dfrac{\pi}{4}$. In quadrant II, the

tangent is negative. Thus,

$\tan\dfrac{3\pi}{4} = -\tan\dfrac{\pi}{4} = -1$.

Converting $\dfrac{3\pi}{4}$ to degrees, we have

$\dfrac{3\pi}{4} = \dfrac{3}{4}(180°) = 135°$. The reference angle is

$180° - 135° = 45°$. Thus,

$\tan\dfrac{3\pi}{4} = \tan 135° = -\tan 45° = -1$.

21. $\dfrac{11\pi}{6}$ is in quadrant IV, so the reference angle

is $2\pi - \dfrac{11\pi}{6} = \dfrac{12\pi}{6} - \dfrac{11\pi}{6} = \dfrac{\pi}{6}$. In quadrant

IV, the cosecant is negative. Thus,

$\csc\dfrac{11\pi}{6} = -\csc\dfrac{\pi}{6} = -2$.

Converting $\dfrac{11\pi}{6}$ to degrees, we have

$\dfrac{11\pi}{6} = \dfrac{11}{6}(180°) = 330°$. The reference angle

is $360° - 330° = 30°$. Thus,

$\csc\dfrac{11\pi}{6} = \csc 330° = -\csc 30° = -2$.

23. $-\dfrac{4\pi}{3}$ is coterminal with

$-\dfrac{4\pi}{3} + 2\pi = -\dfrac{4\pi}{3} + \dfrac{6\pi}{3} = \dfrac{2\pi}{3}$.

$\dfrac{2\pi}{3}$ is in quadrant II, so the reference angle is

$\pi - \dfrac{2\pi}{3} = \dfrac{3\pi}{3} - \dfrac{2\pi}{3} = \dfrac{\pi}{3}$. In quadrant II, the

cosine is negative. Thus,

$\cos\left(-\dfrac{4\pi}{3}\right) = \cos\dfrac{2\pi}{3} = -\cos\dfrac{\pi}{3} = -\dfrac{1}{2}$.

Converting $\dfrac{2\pi}{3}$ to degrees, we have

$\dfrac{2\pi}{3} = \dfrac{2}{3}(180°) = 120°$. The reference angle is

$180° - 120° = 60°$.

Thus,

$\cos\left(-\dfrac{4\pi}{3}\right) = \cos\dfrac{2\pi}{3} = \cos 120°$

$= -\cos 60° = -\dfrac{1}{2}$

25. $\dfrac{7\pi}{4}$ is in quadrant IV, so the reference angle

is $2\pi - \dfrac{7\pi}{4} = \dfrac{8\pi}{4} - \dfrac{7\pi}{4} = \dfrac{\pi}{4}$. In quadrant IV,

the cosine is positive. Thus,

$\cos\dfrac{7\pi}{4} = \cos\dfrac{\pi}{4} = \dfrac{\sqrt{2}}{2}$.

Converting $\dfrac{7\pi}{4}$ to degrees, we have

$\dfrac{7\pi}{4} = \dfrac{7}{4}(180°) = 315°$. The reference angle is

$360° - 315° = 45°$. Thus,

$\cos\dfrac{7\pi}{4} = \cos 315° = \cos 45° = \dfrac{\sqrt{2}}{2}$.

27. $-\dfrac{4\pi}{3}$ is coterminal with

$-\dfrac{4\pi}{3} + 2\pi = -\dfrac{4\pi}{3} + \dfrac{6\pi}{3} = \dfrac{2\pi}{3}$.

$\dfrac{2\pi}{3}$ is in quadrant II, so the reference angle is

$\pi - \dfrac{2\pi}{3} = \dfrac{3\pi}{3} - \dfrac{2\pi}{3} = \dfrac{\pi}{3}$. In quadrant II, the

sine is positive. Thus,

$\sin\left(-\dfrac{4\pi}{3}\right) = \sin\dfrac{2\pi}{3} = \sin\dfrac{\pi}{3} = \dfrac{\sqrt{3}}{2}$.

Converting $\dfrac{2\pi}{3}$ radians to degrees, we have

$\dfrac{2\pi}{3} = \dfrac{2}{3}(180°) = 120°$. The reference angle is

$180° - 120° = 60°$. Thus,

$\sin\left(-\dfrac{4\pi}{3}\right) = \sin\dfrac{2\pi}{3} = \sin 120° = \sin 60° = \dfrac{\sqrt{3}}{2}$

29. $\dfrac{23\pi}{6}$ is coterminal with

$\dfrac{23\pi}{6} - 2\pi = \dfrac{23\pi}{6} - \dfrac{12\pi}{6} = \dfrac{11\pi}{6}$.

$\dfrac{11\pi}{6}$ is in quadrant IV, so the reference angle

is $2\pi - \dfrac{11\pi}{6} = \dfrac{12\pi}{6} - \dfrac{11\pi}{6} = \dfrac{\pi}{6}$.

(continued)

(*continued*)

In quadrant IV, the secant is positive. Thus,

$$\sec\frac{23\pi}{6} = \sec\frac{11\pi}{6} = \sec\frac{\pi}{6} = \frac{2\sqrt{3}}{3}.$$

Converting $\dfrac{11\pi}{6}$ radians to degrees, we have

$\dfrac{11\pi}{6} = \dfrac{11}{6}(180°) = 330°$. The reference angle is

$360° - 330° = 30°$. Thus,

$$\sec\frac{23\pi}{6} = \sec\frac{11\pi}{6} = \sec 330° = \sec 30° = \frac{2\sqrt{3}}{3}.$$

31. $\dfrac{5\pi}{6}$ is in quadrant II, so the reference angle is

$\pi - \dfrac{5\pi}{6} = \dfrac{6\pi}{6} - \dfrac{5\pi}{6} = \dfrac{\pi}{6}$. In quadrant II, the

tangent is negative. Thus,

$$\tan\frac{5\pi}{6} = -\tan\frac{\pi}{6} = -\frac{\sqrt{3}}{3}.$$

Converting $\dfrac{5\pi}{6}$ radians to degrees, we have

$\dfrac{5\pi}{6} = \dfrac{5}{6}(180°) = 150°$. The reference angle is

$180° - 150° = 30°$. Thus,

$$\tan\frac{5\pi}{6} = \tan 150° = -\tan 30° = -\frac{\sqrt{3}}{3}.$$

For Exercises 33–43, 61–65, and 79–87, your calculator must be set in radian mode. Keystroke sequences may vary based on the type and/or model of calculator being used. As in Example 3, we will set the calculator to show four decimal digits.

33. $\sin 0.6109 \approx 0.5736$

35. $\cos(-1.1519) \approx 0.4068$

37. $\tan 4.0203 \approx 1.2065$

39. $\csc(-9.4946) \approx 14.3338$

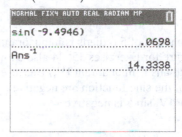

41. $\sec 2.8440 \approx -1.0460$

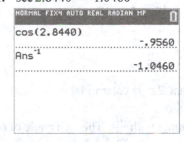

43. $\cot 6.0301 \approx -3.8665$

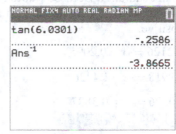

45. $\cos 0.8 \approx 0.7$

47. $\sin 2 \approx 0.9$

49. $\sin 3.8 \approx -0.6$

51. $\cos\theta = -0.65 \Rightarrow x = -0.65 \Rightarrow \theta \approx 2.3$ radians or $\theta \approx 4$ radians

53. $\sin\theta = -0.7 \Rightarrow y = 0.7 \Rightarrow \theta \approx 0.8$ radian or $\theta \approx 2.4$ radians

55. $\cos 2$

$\dfrac{\pi}{2} \approx 1.57$ and $\pi \approx 3.14$, so $\dfrac{\pi}{2} < 2 < \pi$. Thus, an angle of 2 radians is in quadrant II. (The figure for Exercises 35–44 also shows that 2 radians is in quadrant II.) Because values of the cosine function are negative in quadrant II, $\cos 2$ is negative.

57. sin 5

$\frac{3\pi}{2} \approx 4.71$ and $2\pi \approx 6.28$, so $\frac{3\pi}{2} < 5 < 2\pi$.

Thus, an angle of 5 radians is in quadrant IV. (The figure for Exercises 35–44 also shows that 5 radians is in quadrant IV.) Because values of the sine function are negative in quadrant IV, sin 5 is negative.

59. tan 6.29

$2\pi \approx 6.28$ and

$2\pi + \frac{\pi}{2} = \frac{4\pi}{2} + \frac{\pi}{2} = \frac{5\pi}{2} \approx 7.85$, so

$2\pi < 6.29 < \frac{5\pi}{2}$.

Notice that 2π is coterminal with 0 and $\frac{5\pi}{2}$ is coterminal with $\frac{\pi}{2}$. Thus, an angle of 6.29 radians is in quadrant I. Because values of the tangent function are positive in quadrant I, tan 6.29 is positive.

61. $\tan s = 0.2126 \Rightarrow s \approx 0.2095$

63. $\sin s = 0.9918 \Rightarrow s \approx 1.4426$

65. $\sec s = 1.0806 \Rightarrow s \approx 0.3887$

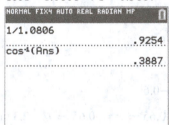

67. $\left[\frac{\pi}{2}, \pi\right]$; $\sin s = \frac{1}{2}$

Recall that $\sin \frac{\pi}{6} = \frac{1}{2}$ and in quadrant II, sin s is positive. Therefore,

$\sin\left(\pi - \frac{\pi}{6}\right) = \sin \frac{5\pi}{6} = \frac{1}{2}$, so $s = \frac{5\pi}{6}$.

69. $\left[\pi, \frac{3\pi}{2}\right]$; $\tan s = \sqrt{3}$

Recall that $\tan \frac{\pi}{3} = \sqrt{3}$ and in quadrant III, tan s is positive. Therefore,

$\tan\left(\pi + \frac{\pi}{3}\right) = \tan \frac{4\pi}{3} = \sqrt{3}$, so $s = \frac{4\pi}{3}$.

71. $\left[\frac{3\pi}{2}, 2\pi\right]$; $\tan s = -1$

Recall that $\tan \frac{\pi}{4} = 1$ and in quadrant IV, tan s is negative. Therefore,

$\tan\left(2\pi - \frac{\pi}{4}\right) = \tan \frac{7\pi}{4} = -1$, so $s = \frac{7\pi}{4}$.

73. $[0, 2\pi)$; $\sin s = -\frac{\sqrt{3}}{2}$

Recall that $\sin \frac{\pi}{3} = \frac{\sqrt{3}}{2}$, and that sin s is negative in quadrants III and IV. Thus, the angles we are seeking have reference angle $\frac{\pi}{3}$ and are located in quadrants III and IV. In quadrant III, $s = \pi + \frac{\pi}{3} = \frac{4\pi}{3}$. In quadrant IV, $s = 2\pi - \frac{\pi}{3} = \frac{5\pi}{3}$. Thus, the angles are $\frac{4\pi}{3}$ and $\frac{5\pi}{3}$.

75. $[0, 2\pi)$; $\cos^2 s = \frac{1}{2} \Rightarrow \cos s = \pm\sqrt{\frac{1}{2}} = \pm\frac{\sqrt{2}}{2}$

Recall that $\cos \frac{\pi}{4} = \frac{\sqrt{2}}{2}$, and that cos s is positive in quadrants I and IV, and negative in quadrants II and III. Thus, the angles we are seeking have reference angle $\frac{\pi}{4}$. In quadrant II, $s = \pi - \frac{\pi}{4} = \frac{3\pi}{4}$. In quadrant III, $s = \pi + \frac{\pi}{4} = \frac{5\pi}{4}$. In quadrant IV, $s = 2\pi - \frac{\pi}{4} = \frac{7\pi}{4}$. Thus, the angles are $\frac{\pi}{4}$, $\frac{3\pi}{4}$, $\frac{5\pi}{4}$, and $\frac{7\pi}{4}$.

77. $[-2\pi, \pi)$; $3\tan^2 s = 1 \Rightarrow \tan^2 s = \frac{1}{3} \Rightarrow$

$\tan s = \pm\sqrt{\frac{1}{3}} = \pm\frac{\sqrt{3}}{3}$

We will split the interval into $[-2\pi, 0)$ and $[0, \pi)$. First we will find the angles in the interval $[0, \pi)$.

(continued on next page)

(continued)

Recall that $\tan\dfrac{\pi}{6} = \dfrac{\sqrt{3}}{3}$, and that $\tan s$ is positive in quadrants I and III, and negative in quadrants II and IV.

In quadrant II, $s = \pi - \dfrac{\pi}{6} = \dfrac{5\pi}{6}$. In quadrant III, $s = \pi + \dfrac{\pi}{6} = \dfrac{7\pi}{6}$. In quadrant IV, $s = 2\pi - \dfrac{\pi}{6} = \dfrac{11\pi}{6}$.

To find the angles in the interval $[-2\pi, 0)$, recall that moving around the unit circle $\dfrac{\pi}{6}$ in the positive

direction yields the same ending point as moving $\dfrac{11\pi}{6}$ units in the negative direction. So $-\dfrac{11\pi}{6}$ is one of the

angles. Moving $\dfrac{5\pi}{6}$ units in the positive direction is the same as moving $\dfrac{7\pi}{6}$ units in the negative direction,

so $-\dfrac{7\pi}{6}$ is another angle. Now we must find the negative angles in quadrants III and IV. Moving $\dfrac{7\pi}{6}$ units

in the positive direction is that same as moving $\dfrac{5\pi}{6}$ units in the negative direction, so $-\dfrac{5\pi}{6}$ is another angle.

Finally, moving $\dfrac{11\pi}{6}$ units in the positive direction is the same as moving $\dfrac{\pi}{6}$ units in the negative direction.

Thus, the angles are $-\dfrac{11\pi}{6}, -\dfrac{7\pi}{6}, -\dfrac{5\pi}{6}, -\dfrac{\pi}{6}, \dfrac{\pi}{6}$, and $\dfrac{5\pi}{6}$.

81. $s = -7.4$

$x = \cos s \Rightarrow x = \cos(-7.4) = 0.4385$

$y = \sin s \Rightarrow y = \sin(-7.4) = -0.8987$

$(x, y) = (0.4385, -0.8987)$

83. $s = 51$

$\cos 51 = 0.7422;\ \sin 51 = 0.6702$

Cosine and sine are both positive, so an angle of 51 radians lies in quadrant I.

85. $s = 65$

$\cos 65 = -0.5625;\ \sin 65 = 0.8268$

Cosine is negative and sine is positive, so an angle of 65 radians lies in quadrant II.

87. $(x, y) = (0.5875, 0.80922417) \Rightarrow$

$\cos s = 0.5875$ and $\sin s = 0.80922417$

Cosine and sine are both positive, so s is in quadrant I.

$s = \cos^{-1} 0.5875 \approx 0.9428$

Verify that $\sin^{-1} 0.80922417 = 0.9428$.

89. (a) New Orleans has latitude $L = 30°$ or 0.5236 radians. The day and time have not changed, so
$D = -0.1425$ and $\omega = 0.7854$.

$\sin\theta = \cos D\cos L\cos\omega + \sin D\sin L$

$\sin\theta = \cos(-0.1425)\cos(0.5236)\cos(0.7854) + \sin(-0.1425)\sin(0.5236)$

$\theta = \sin^{-1}\left[\cos(-0.1425)\cos(0.5236)\cos(0.7854) + \sin(-0.1425)\sin(0.5236)\right]$

≈ 0.5647 radians or $32.4°$

(b) Answers will vary.

91. $t = 60 - 30 \cos \dfrac{x\pi}{6}$

(a) January: $x = 0$;

$$t = 60 - 30 \cos \frac{0 \cdot \pi}{6} = 60 - 30 \cos 0$$
$$= 60 - 30(1) = 60 - 30 = 30°$$

(b) April: $x = 3$;

$$t = 60 - 30 \cos \frac{3 \cdot \pi}{6} = 60 - 30 \cos \frac{\pi}{2}$$
$$= 60 - 30(0) = 60°$$

(c) May: $x = 4$;

$$t = 60 - 30 \cos \frac{4 \cdot \pi}{6} = 60 - 30 \cos \frac{2\pi}{3}$$
$$= 60 - 30 \left(-\frac{1}{2} \right) = 75°$$

(d) June $x = 5$;

$$t = 60 - 30 \cos \frac{5 \cdot \pi}{6} = 60 - 30 \cos \frac{5\pi}{6}$$
$$= 60 - 30 \left(-\frac{\sqrt{3}}{2} \right) = 60 + 15\sqrt{3} \approx 86°$$

(e) August $x = 7$;

$$t = 60 - 30 \cos \frac{7 \cdot \pi}{6} = 60 - 30 \cos \frac{7\pi}{6}$$
$$= 60 - 30 \left(-\frac{\sqrt{3}}{2} \right) = 60 + 15\sqrt{3} \approx 86°$$

(f) October $x = 9$;

$$t = 60 - 30 \cos \frac{9 \cdot \pi}{6} = 60 - 30 \cos \frac{3\pi}{2}$$
$$= 60 - 30(0) = 60°$$

93. Refer to figures 18 and 19 on page 122 in the text.

(a) $OQ = \cos 60° = \dfrac{1}{2}$

(b) $PQ = \sin 60° = \dfrac{\sqrt{3}}{2}$

(c) $VR = \tan 60° = \sqrt{3}$

(d) $OV = \sec 60° = \dfrac{1}{\cos 60°} = \dfrac{1}{\frac{1}{2}} = 2$

(e) $OU = \csc 60° = \dfrac{1}{\sin 60°} = \dfrac{1}{\frac{\sqrt{3}}{2}} = \dfrac{2}{\sqrt{3}} = \dfrac{2\sqrt{3}}{3}$

(f) $US = \cot 60° = \dfrac{1}{\tan 60°} = \dfrac{1}{\sqrt{3}} = \dfrac{\sqrt{3}}{3}$

Chapter 3 Quiz
(Sections 3.1–3.3)

1. $225° = 225 \left(\dfrac{\pi}{180} \text{ radian} \right) = \dfrac{5\pi}{4}$ radians

3. $\dfrac{5\pi}{3} = \dfrac{5\pi}{3} \left(\dfrac{180°}{\pi} \right) = 300°$

5. $r = 300$, $s = 450$

$$s = r\theta \Rightarrow 450 = 300\theta \Rightarrow \theta = \frac{450}{300} = 1.5$$

7. $\dfrac{7\pi}{4}$ is in quadrant IV, so the reference angle

is $2\pi - \dfrac{7\pi}{4} = \dfrac{8\pi}{4} - \dfrac{7\pi}{4} = \dfrac{\pi}{4}$. In quadrant IV, the cosine is positive. Thus,

$$\cos \frac{7\pi}{4} = \cos \frac{\pi}{4} = \frac{\sqrt{2}}{2}.$$

Converting $\dfrac{7\pi}{4}$ to degrees, we have

$\dfrac{7\pi}{4} = \dfrac{7}{4}(180°) = 315°$. The reference angle is

$180° - 315° = 45°$. Thus,

$$\cos \frac{7\pi}{4} = \cos 315° = \cos 45° = \frac{\sqrt{2}}{2}.$$

9. 3π is coterminal with $3\pi - 2\pi = \pi$. So $\tan 3\pi = \tan \pi = 0$
Converting 3π to degrees, we have

$3\pi \left(\dfrac{180}{\pi} \right) = 540°$. The reference angle is

$540° - 360° = 180°$. Thus
$\tan 3\pi = \tan 540° = \tan 180° = 0$.

Section 3.4   Linear and Angular Speed

1. The measure of how fast the position of point P is changing is the <u>linear speed (or linear velocity)</u>.

3. If the angular speed of point P is 1 radian per sec, then P will move around the entire unit circle in <u>2π</u> sec.

5. An angular speed of 1 revolution per min on the unit circle is equivalent to an angular speed, ω, of <u>2π</u> radians per min.

7. $r = 20$ cm, $\omega = \dfrac{\pi}{12}$ radian per sec, $t = 6$ sec

    (a) $\omega = \dfrac{\theta}{t} \Rightarrow \dfrac{\pi}{12} = \dfrac{\theta}{6} \Rightarrow \theta = \dfrac{\pi}{2}$ radians

    (b) $s = r\theta \Rightarrow s = 20 \cdot \dfrac{\pi}{2} = 10\pi$ cm

    (c) $v = \dfrac{r\theta}{t} \Rightarrow v = \dfrac{20 \cdot \frac{\pi}{2}}{6} = \dfrac{10\pi}{6} = \dfrac{5\pi}{3}$ cm per sec

9. $r = 8$ in., $\omega = \dfrac{\pi}{3}$ radian per min, $t = 9$ min

    (a) $\omega = \dfrac{\theta}{t} \Rightarrow \dfrac{\pi}{3} = \dfrac{\theta}{9} \Rightarrow \theta = \dfrac{9\pi}{3} = 3\pi$ radians

    (b) $s = r\theta \Rightarrow s = 8 \cdot 3\pi = 24\pi$ in.

    (c) $v = \dfrac{r\theta}{t} \Rightarrow v = \dfrac{8 \cdot 3\pi}{9} = \dfrac{8\pi}{3}$ in. per min

11. $\omega = \dfrac{2\pi}{3}$ radians per sec, $t = 3$ sec

$\omega = \dfrac{\theta}{t} \Rightarrow \dfrac{2\pi}{3} = \dfrac{\theta}{3} \Rightarrow \theta = 2\pi$ radians

13. $\omega = 0.91$ radian per min, $t = 8.1$ min

$\omega = \dfrac{\theta}{t} \Rightarrow 0.91 = \dfrac{\theta}{8.1} \Rightarrow$

$\theta = (0.91)(8.1) \approx 7.4$ radians

15. $\theta = \dfrac{3\pi}{4}$ radians, $t = 8$ sec

$\omega = \dfrac{\theta}{t} \Rightarrow \theta = \dfrac{\frac{3\pi}{4}}{8} = \dfrac{3\pi}{4} \cdot \dfrac{1}{8} = \dfrac{3\pi}{32}$ radian per sec

17. $\theta = 3.871$ radians, $t = 21.47$ sec

$\omega = \dfrac{\theta}{t}$

$\omega = \dfrac{3.871}{21.47} \approx 0.1803$ radian per sec

19. $\theta = \dfrac{2\pi}{9}$ radian, $\omega = \dfrac{5\pi}{27}$ radian per min

$\omega = \dfrac{\theta}{t} \Rightarrow \dfrac{5\pi}{27} = \dfrac{\frac{2\pi}{9}}{t} \Rightarrow \dfrac{5\pi}{27} = \dfrac{2\pi}{9t} \Rightarrow$

$45\pi t = 54\pi \Rightarrow t = \dfrac{54\pi}{45\pi} = \dfrac{6}{5}$ min

21. $r = 12$ m, $\omega = \dfrac{2\pi}{3}$ radians per sec

$v = r\omega \Rightarrow v = 12\left(\dfrac{2\pi}{3}\right) = 8\pi$ m per sec

23. $v = 9$ m per sec, $r = 5$ m

$v = r\omega \Rightarrow 9 = 5\omega \Rightarrow \omega = \dfrac{9}{5}$ radians per sec

25. $v = 12$ m per sec, $\omega = \dfrac{3\pi}{2}$ radians per sec

$v = r\omega \Rightarrow 12 = \dfrac{3\pi}{2} r \Rightarrow$

$r = 12 \cdot \dfrac{2}{3\pi} = \dfrac{8}{\pi}$ m

27. $r = 6$ cm, $\omega = \dfrac{\pi}{3}$ radians per sec, $t = 9$ sec

$s = r\omega t \Rightarrow s = 6\left(\dfrac{\pi}{3}\right)(9) = 18\pi$ cm

29. $s = 6\pi$ cm, $r = 2$ cm, $\omega = \dfrac{\pi}{4}$ radian per sec

$s = r\omega t \Rightarrow 6\pi = 2\left(\dfrac{\pi}{4}\right)t \Rightarrow 6\pi = \left(\dfrac{\pi}{2}\right)t \Rightarrow$

$t = 6\pi\left(\dfrac{2}{\pi}\right) = 12$ sec

31. $s = \dfrac{3\pi}{4}$ km, $r = 2$ km, $t = 4$ sec

$s = r\omega t \Rightarrow \dfrac{3\pi}{4} = 2\omega \cdot 4 \Rightarrow$

$\dfrac{3\pi}{4} = 8\omega \Rightarrow \omega = \dfrac{3\pi}{4} \cdot \dfrac{1}{8} = \dfrac{3\pi}{32}$ radian per sec

33. The hour hand of a clock moves through an angle of 2π radians (one complete revolution) in 12 hours, so

$\omega = \dfrac{\theta}{t} = \dfrac{2\pi}{12} = \dfrac{\pi}{6}$ radian per hr.

35. The minute hand makes one revolution per hour. Each revolution is 2π radians, so we have $\omega = 2\pi(1) = 2\pi$ radians per hr . There are 60 minutes in 1 hour, so $\omega = \dfrac{2\pi}{60} = \dfrac{\pi}{30}$ radian per min.

37. The minute hand of a clock moves through an angle of 2π radians in 60 min, and at the tip of the minute hand, $r = 7$ cm, so we have

$v = \dfrac{r\theta}{t} \Rightarrow v = \dfrac{7(2\pi)}{60} = \dfrac{7\pi}{30}$ cm per min

39. The flywheel making 42 rotations per min turns through an angle $42(2\pi) = 84\pi$ radians in 1 minute with $r = 2$ m. So,

$$v = \frac{r\theta}{t} \Rightarrow v = \frac{2(84\pi)}{1} = 168\pi \text{ m per min}$$

41. At 500 rotations per min, the propeller turns through an angle of $\theta = 500(2\pi) = 1000\pi$ radians in 1 min with $r = \frac{3}{2} = 1.5$ m, we have

$$v = \frac{r\theta}{t} \Rightarrow v = \frac{1.5(1000\pi)}{1} = 1500\pi \text{ m per min.}$$

43. At 215 revolutions per minute, the bicycle tire is moving $215(2\pi) = 430\pi$ radians per min. This is the angular velocity ω. The linear velocity of the bicycle is

$$v = r\omega = 13.0(430\pi) = 5590\pi \text{ in. per min.}$$

Convert this to miles per hour:

$$v = \frac{5590\pi \text{ in.}}{\text{min}} \cdot \frac{60 \text{ min}}{\text{hr}} \cdot \frac{1 \text{ ft}}{12 \text{ in.}} \cdot \frac{1 \text{ mi}}{5280 \text{ ft}}$$
$$\approx 16.6 \text{ mph}$$

45. (a) $\theta = \frac{1}{365}(2\pi) = \frac{2\pi}{365}$ radian

(b) $\omega = \frac{2\pi}{365}$ radian per day

$$= \frac{2\pi}{365} \cdot \frac{1}{24} \text{ radian per hr}$$

$$= \frac{\pi}{4380} \text{ radian per hr}$$

(c) $v = r\omega$

$$v = (93,000,000)\left(\frac{\pi}{4380}\right) \approx 67,000 \text{ mph}$$

47. (a) Because $s = 56$ cm of belt go around in $t = 18$ sec, the linear velocity is

$$v = \frac{s}{t} \Rightarrow v = \frac{56}{18} = \frac{28}{9} \approx 3.1 \text{ cm per sec}$$

(b) Because the 56-cm belt goes around in 18 sec, we have

$$v = r\omega$$

$$\frac{56}{18} = (12.96)\omega \Rightarrow \frac{28}{9} = (12.96)\omega$$

$$\omega = \frac{\frac{28}{9}}{12.96} \approx 0.24 \text{ radian per sec}$$

49. $\omega = (152)(2\pi) = 304\pi$ radians per min

$$= \frac{304\pi}{60} \text{ radians per sec}$$

$$= \frac{76\pi}{15} \text{ radians per sec}$$

$$v = r\omega \Rightarrow 59.4 = r\left(\frac{76\pi}{15}\right) \Rightarrow$$

$$r = 59.4\left(\frac{15}{76\pi}\right) \approx 3.73 \text{ cm}$$

51. In one minute, the propeller makes 5000 revolutions. Each revolution is 2π radians, so we have $5000(2\pi) = 10,000\pi$ radians per min. There are 60 sec in a minute, so

$$\omega = \frac{10,000\pi}{60} = \frac{500\pi}{3} \approx 523.6 \text{ radians per sec}$$

Chapter 3   Review Exercises

1. A central angle of a circle that intercepts an arc of length 2 times the radius of the circle has a measure of 2 radians.

3. To find a coterminal angle, add or subtract multiples of 2π. Three of the many possible answers are $1 + 2\pi$, $1 + 4\pi$, and $1 + 6\pi$.

5. $45° = 45\left(\frac{\pi}{180}\text{radian}\right) = \frac{\pi}{4}\text{radian}$

7. $175° = 175\left(\frac{\pi}{180}\text{radian}\right) = \frac{35\pi}{36}\text{radians}$

9. $800° = 800\left(\frac{\pi}{180}\text{radian}\right) = \frac{40\pi}{9}\text{radians}$

11. $\frac{5\pi}{4} = \frac{5\pi}{4}\left(\frac{180°}{\pi}\right) = 225°$

13. $\frac{8\pi}{3} = \frac{8\pi}{3}\left(\frac{180°}{\pi}\right) = 480°$

15. $-\frac{11\pi}{18} = -\frac{11\pi}{18}\left(\frac{180°}{\pi}\right) = -110°$

17. $\frac{15}{60} = \frac{1}{4}$ rotation, so we have

$$\theta = \frac{1}{4}(2\pi) = \frac{\pi}{2}. \text{ Thus,}$$

$$s = r\theta \Rightarrow s = 2\left(\frac{\pi}{2}\right) = \pi \text{ in.}$$

19. $\theta = 3(2\pi) = 6\pi$, so we have

$$s = r\theta \Rightarrow s = 2(6\pi) = 12\pi \text{ in.}$$

21. $r = 15.2$ cm, $\theta = \dfrac{3\pi}{4}$

$$s = r\theta \Rightarrow s = 15.2\left(\dfrac{3\pi}{4}\right) = 11.4\pi \approx 35.8 \text{ cm}$$

23. $s = 7.683$, $r = 8.973$ cm

$$s = r\theta \Rightarrow 7.683 = 8.973\theta \Rightarrow$$
$$\theta = \dfrac{7.683}{8.973} \approx 0.8562 \text{ radian}$$

$$0.8592 \text{ radian} = 0.8592\left(\dfrac{180°}{\pi}\right) \approx 49.06°$$

25. $r = 38.0$ m, $\theta = 21°40'$

First convert $\theta = 21°40'$ to radians:

$$\theta = 21°40' = \left(21 + \tfrac{40}{60}\right)\left(\dfrac{\pi}{180}\right)$$
$$= \dfrac{65}{3}\left(\dfrac{\pi}{180}\right) = \dfrac{13\pi}{108}$$

$$\mathscr{A} = \dfrac{1}{2}r^2\theta$$

$$\mathscr{A} = \dfrac{1}{2}(38.0)^2\left(\dfrac{13\pi}{108}\right) \approx 273 \text{ m}^2$$

27. The central angle in degrees measures $0.517°$. Converting to radians, we have

$$0.517° = (0.517)\left(\dfrac{\pi}{180} \text{ radian}\right)$$
$$= \dfrac{0.517\pi}{180} \text{ radian}$$

$$s = r\theta \Rightarrow s = 238,900\left(\dfrac{0.517\pi}{180}\right) \approx 2156$$

Recall from page 112 of your text (above Exercises 43–46), for very small central angles, there is little difference between the arc and the inscribed chord. Thus, the diameter of the moon is approximately 2156 mi. (rounded to four significant digits)

29. (a) The hour hand of a clock moves through an angle of 2π radians in 12 hours, so

$$\omega = \dfrac{\theta}{t} = \dfrac{2\pi}{12} = \dfrac{\pi}{6} \text{ radian per hour}$$

In two hours, the angle would be

$$2\left(\dfrac{\pi}{6}\right) = \dfrac{\pi}{3} \text{ radians.}$$

(b) The distance s the tip of the hour hand travels during the time period from 1 o'clock to 3 o'clock is the arc length for $\theta = \dfrac{\pi}{3}$ and $r = 6$ in. Thus,

$$s = r\theta = 6\left(\dfrac{\pi}{3}\right) = 2\pi \text{ in.}$$

31. The cities are at 28°N and 12°S. 12°S = −12°N, so

$$\theta = 28° - (-12°) = 40° = 40\left(\dfrac{\pi}{180}\right) = \dfrac{2\pi}{9}$$

radians

$$s = r\theta = 6400\left(\dfrac{2\pi}{9}\right) \approx 4500 \text{ km}$$ (rounded to two significant digits)

33. $r = 2$, $s = 1.5$

$$s = r\theta \Rightarrow 1.5 = 2\theta \Rightarrow \theta = \dfrac{1.5}{2} = \dfrac{3}{4} \text{ radian}$$

$$\mathscr{A} = \dfrac{1}{2}r^2\theta \Rightarrow$$

$$\mathscr{A} = \dfrac{1}{2}(2)^2\left(\dfrac{3}{4}\right) = \dfrac{1}{2}(4)\left(\dfrac{3}{4}\right) = \dfrac{3}{2} = 1.5 \text{ sq units}$$

35. $\tan\dfrac{\pi}{3}$

Converting $\dfrac{\pi}{3}$ radians to degrees, we have

$$\dfrac{\pi}{3} = \dfrac{1}{3}(180°) = 60° \Rightarrow \tan\dfrac{\pi}{3} = \tan 60° = \sqrt{3}$$

37. $\sin\left(-\dfrac{5\pi}{6}\right)$

$-\dfrac{5\pi}{6}$ is coterminal with $-\dfrac{5\pi}{6} + 2\pi = \dfrac{7\pi}{6}$.

$\dfrac{7\pi}{6}$ is in quadrant III, so the reference angle is $\dfrac{7\pi}{6} - \pi = \dfrac{\pi}{6}$. In quadrant III, the sine is negative. Thus,

$$\sin\left(-\dfrac{5\pi}{6}\right) = \sin\dfrac{7\pi}{6} = -\sin\dfrac{\pi}{6} = -\dfrac{1}{2}$$

Converting $\dfrac{5\pi}{6}$ to degrees, we have

$\dfrac{7\pi}{6}\left(\dfrac{180°}{\pi}\right) = 210°$. The reference angle is $210° - 180° = 30°$.

(*continued on next page*)

(*continued*)

Thus,

$$\sin\left(-\frac{5\pi}{6}\right) = \sin\frac{7\pi}{6} = \sin 210°$$

$$= -\sin 30° = -\frac{1}{2}$$

39. $\csc\left(-\frac{11\pi}{6}\right)$

$-\frac{11\pi}{6}$ is coterminal with

$-\frac{11\pi}{6} + 2\pi = -\frac{11\pi}{6} + \frac{12\pi}{6} = \frac{\pi}{6}.$

$\frac{\pi}{6}$ is in quadrant I, so

$\csc\left(-\frac{11\pi}{6}\right) = \csc\frac{\pi}{6} = 2.$ Converting $\frac{\pi}{6}$ to

degrees, we have $\frac{\pi}{6} = \frac{1}{6}(180°) = 30°.$ Thus,

$\csc\left(-\frac{11\pi}{6}\right) = \csc\frac{\pi}{6} = \csc 30° = 2.$

41. Because $0 < 1 < \frac{\pi}{2}$, sin 1 and cos 1 are both
positive. Thus, tan 1 is positive. Also, because
$\frac{\pi}{2} < 2 < \pi$, sin 2 is positive and cos 2 is
negative. Thus, tan 2 is negative. Therefore,
tan 1 > tan 2.

43. Because $\frac{\pi}{2} < 2 < \pi$, sin 2 is positive and
cos 2 is negative. Thus, sin 2 > cos 2.

45. $\sin 1.0472 \approx 0.8660$

47. $\cos(-0.2443) \approx 0.9703$

49. $\sec 7.3159 \approx 1.9513$

```
NORMAL FIX4 AUTO REAL RADIAN MP
cos(7.3159)
                          .5125
Ans⁻¹
                          1.9513
```

51. $\cos s = 0.9250 \Rightarrow s \approx 0.3898$

53. $\sin s = 0.4924 \Rightarrow s \approx 0.5148$

55. $\cot s = 0.5022 \Rightarrow s \approx 1.1054$

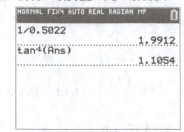

57. $\left[0, \frac{\pi}{2}\right], \cos s = \frac{\sqrt{2}}{2}$

$\cos s = \frac{\sqrt{2}}{2}$, so the reference angle for s must

be $\frac{\pi}{4}$ because $\cos\frac{\pi}{4} = \frac{\sqrt{2}}{2}$. For s to be in the

interval $\left[0, \frac{\pi}{2}\right]$, s must be the reference angle.

Therefore, $s = \frac{\pi}{4}$.

59. $\left[\pi, \frac{3\pi}{2}\right], \sec s = -\frac{2\sqrt{3}}{3}$

$\sec s = -\frac{2\sqrt{3}}{3}$, so the reference angle for s

must be $\frac{\pi}{6}$ because $\sec\frac{\pi}{6} = \frac{2\sqrt{3}}{3}$. For s to be

in the interval $\left[\pi, \frac{3\pi}{2}\right]$, we must add the

reference angle to π. Therefore,

$s = \pi + \frac{\pi}{6} = \frac{7\pi}{6}.$

61. $r = 15$ cm, $\omega = \frac{2\pi}{3}$ radians per sec, $t = 30$ sec

(a) $\omega = \frac{\theta}{t} \Rightarrow \frac{2\pi}{3} = \frac{\theta}{30} \Rightarrow$

$\theta = \frac{2\pi}{3}(30) = 20\pi$ radians

(b) $s = r\theta \Rightarrow s = 15 \cdot 20\pi = 300\pi$ cm

(c) $v = r\omega \Rightarrow v = 15\left(\frac{2\pi}{3}\right) = 10\pi$ cm per sec

63. The flywheel is rotating 90 times per sec or
$90(2\pi) = 180\pi$ radians per sec. Because
$r = 7$ cm, we have
$v = r\omega \Rightarrow v = 7(180\pi) = 1260\pi$ cm per sec

65. In the diagram, $a = 1.4$ in. and $b = 0.2$ in.

$$r = \frac{a^2 + b^2}{2b} = \frac{1.4^2 + 0.2^2}{2(0.2)}$$

$$= \frac{1.96 + 0.04}{0.4} = \frac{2.00}{0.4} = 5$$

Thus, the radius is 5 inches.

Chapter 3   Test

1. $120° = 120\left(\dfrac{\pi}{180} \text{ radian}\right) = \dfrac{2\pi}{3}$ radians

2. $-45° = -45\left(\dfrac{\pi}{180} \text{ radian}\right) = -\dfrac{\pi}{4}$ radian

3. $5° = 5\left(\dfrac{\pi}{180} \text{ radian}\right) = \dfrac{\pi}{36} \approx 0.087$ radian

4. $\dfrac{3\pi}{4} = \dfrac{3\pi}{4}\left(\dfrac{180°}{\pi}\right) = 135°$

5. $-\dfrac{7\pi}{6} = -\dfrac{7\pi}{6}\left(\dfrac{180°}{\pi}\right) = -210°$

6. $4 = 4\left(\dfrac{180°}{\pi}\right) \approx 229.18°$

$= 229° + 0.18(60') = 229°11'$

7. $r = 150$ cm, $s = 200$ cm

(a) $s = r\theta \Rightarrow 200 = 150\theta \Rightarrow \theta = \dfrac{200}{150} = \dfrac{4}{3}$

(b) $\mathcal{A} = \dfrac{1}{2}r^2\theta$

$\mathcal{A} = \dfrac{1}{2}(150)^2\left(\dfrac{4}{3}\right) = \dfrac{1}{2}(22{,}500)\left(\dfrac{4}{3}\right)$

$= 15{,}000$ cm^2

8. $r = \dfrac{1}{2}$ in., $s = 1$ in.

$s = r\theta \Rightarrow 1 = \dfrac{1}{2}\theta \Rightarrow \theta = 2$ radians

For Exercises 9–14, refer to Figure 13 on page 117 of the text.

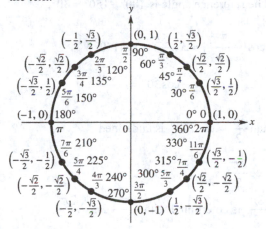

9. $\sin\dfrac{3\pi}{4}$

$\dfrac{3\pi}{4}$ is in quadrant II, so the reference angle is

$\pi - \dfrac{3\pi}{4} = \dfrac{4\pi}{4} - \dfrac{3\pi}{4} = \dfrac{\pi}{4}$. In quadrant II, the

sine is positive. Thus, $\sin\dfrac{3\pi}{4} = \sin\dfrac{\pi}{4} = \dfrac{\sqrt{2}}{2}$.

Converting $\dfrac{3\pi}{4}$ to degrees, we have

$\dfrac{3\pi}{4} = \dfrac{3}{4}(180°) = 135°$. The reference angle is

$180° - 135° = 45°$. Thus,

$\sin\dfrac{3\pi}{4} = \sin 135° = \sin 45° = \dfrac{\sqrt{2}}{2}$.

10. $\cos\left(-\dfrac{7\pi}{6}\right)$

$-\dfrac{7\pi}{6}$ is coterminal with

$-\dfrac{7\pi}{6} + 2\pi = -\dfrac{7\pi}{6} + \dfrac{12\pi}{6} = \dfrac{5\pi}{6}$.

$\dfrac{5\pi}{6}$ is in quadrant II, so the reference angle is

$\pi - \dfrac{5\pi}{6} = \dfrac{6\pi}{6} - \dfrac{5\pi}{6} = \dfrac{\pi}{6}$. In quadrant II, the

cosine is negative. Thus,

$\cos\left(-\dfrac{7\pi}{6}\right) = \cos\dfrac{5\pi}{6} = -\cos\dfrac{\pi}{6} = -\dfrac{\sqrt{3}}{2}$.

Converting $\dfrac{5\pi}{6}$ to degrees, we have

$\dfrac{5\pi}{6} = \dfrac{5}{6}(180°) = 150°$.

(continued on next page)

(*continued*)

The reference angle is $180° - 150° = 30°$. Thus,

$$\cos\left(-\frac{7\pi}{6}\right) = \cos\frac{5\pi}{6} = \cos 150°$$

$$= -\cos 30° = -\frac{\sqrt{3}}{2}.$$

11. $\tan\frac{3\pi}{2} = \tan 270°$ is undefined.

12. $\sec\frac{8\pi}{3}$

$\frac{8\pi}{3}$ is coterminal with $\frac{8\pi}{3} - 2\pi = \frac{2\pi}{3}$.

$\frac{2\pi}{3}$ is in quadrant II, so the reference angles

is $\pi - \frac{2\pi}{3} = \frac{\pi}{3}$. In quadrant II, the secant is negative. Thus,

$$\sec\frac{8\pi}{3} = \sec\frac{2\pi}{3} = -\sec\frac{\pi}{3} = -2.$$

Converting $\frac{2\pi}{3}$ to degrees, we have

$\frac{2\pi}{3} = \frac{2\pi}{3} \cdot \frac{180°}{\pi} = 120°$. The reference angle is $180° - 120° = 60°$. Thus,

$$\sec\frac{8\pi}{3} = \sec\frac{2\pi}{3} = \sec 120° = -\sec 60° = -2.$$

13. $\tan\pi = \tan 180° = 0$

14. $\cos\frac{3\pi}{2} = \cos 270° = 0$

15. $s = \frac{7\pi}{6}$

$\frac{7\pi}{6}$ is in quadrant III, so the reference angle

is $\frac{7\pi}{6} - \pi = \frac{\pi}{6}$. In quadrant III, the sine and cosine are negative.

$$\sin\frac{7\pi}{6} = -\sin\frac{\pi}{6} = -\frac{1}{2}$$

$$\cos\frac{7\pi}{6} = -\cos\frac{\pi}{6} = -\frac{\sqrt{3}}{2}$$

$$\tan\frac{7\pi}{6} = \tan\frac{\pi}{6} = \frac{\sqrt{3}}{3}$$

$$\csc\frac{7\pi}{6} = \frac{1}{\sin\frac{7\pi}{6}} = \frac{1}{-\frac{1}{2}} = -2$$

$$\sec\frac{7\pi}{6} = \frac{1}{\cos\frac{7\pi}{6}} = \frac{1}{-\frac{\sqrt{3}}{2}} = -\frac{2}{\sqrt{3}} = -\frac{2\sqrt{3}}{3}$$

$$\cot\frac{7\pi}{6} = \frac{1}{\tan\frac{7\pi}{6}} = \frac{1}{\frac{\sqrt{3}}{3}} = \frac{3}{\sqrt{3}} = \frac{3\sqrt{3}}{3} = \sqrt{3}$$

16. For any point (x, y) on the unit circle,

$\sin s = \frac{y}{1} = y$ and $\cos s = \frac{x}{1}$, both of which are defined for all values of x and y. Thus, the domains of the sine and cosine functions are both $(-\infty, \infty)$.

$\tan s = \frac{y}{x}$ and $\sec x = \frac{1}{x}$, so these functions are not defined for $x = 0$. This occurs at

$$s = \dots, -\frac{5\pi}{2}, -\frac{3\pi}{2}, -\frac{\pi}{2}, \frac{\pi}{2}, \frac{3\pi}{2}, \frac{5\pi}{2}, \dots$$

Therefore, the domains of the tangent and secant functions are both

$$\left\{ s \mid s \neq (2n+1)\frac{\pi}{2}, \text{ where } n \text{ is any integer} \right\}.$$

$\cot s = \frac{x}{y}$ and $\csc s = \frac{1}{y}$ are not defined for $y = 0$. This occurs at

$$s = \dots, -3\pi, -2\pi, -\pi, 0, \pi, 2\pi, 3\pi, \dots$$

Therefore, the domains of the cotangent and cosecant functions are both

$$\left\{ s \mid s \neq n\pi, \text{ where } n \text{ is any integer} \right\}$$

17. **(a)** $\sin s = 0.8258 \Rightarrow s \approx 0.9716$

(b) $\cos\frac{\pi}{3} = \frac{1}{2}$ and $0 \leq \frac{\pi}{3} \leq \frac{\pi}{2}$, so $s = \frac{\pi}{3}$.

18. **(a)** The speed of ray OP is $\omega = \frac{\pi}{12}$ radian per sec. $\omega = \frac{\theta}{t}$, then in 8 sec,

$$\omega = \frac{\theta}{t} \Rightarrow \frac{\pi}{12} = \frac{\theta}{8} \Rightarrow \theta = \frac{8\pi}{12} = \frac{2\pi}{3}$$

radians

(b) From part (a), P generates an angle of $\frac{2\pi}{3}$ radians in 8 sec. The distance traveled by P along the circle is

$$s = r\theta \Rightarrow s = 60\left(\frac{2\pi}{3}\right) = 40\pi \text{ cm}$$

(c) $v = \frac{s}{t} \Rightarrow \frac{40\pi}{8} = 5\pi$ cm per sec.

19. $r = 483,800,000$ mi

In 11.86 years, Jupiter travels 2π radians.

$$\omega = \frac{\theta}{t} \Rightarrow \omega = \frac{2\pi}{11.86}$$

$$v = r\omega \Rightarrow v = 483,800,000\left(\frac{2\pi}{11.86}\right)$$

$$\approx 256,307,339.9 \text{ miles per year}$$

Now convert this to miles per second:

$$v = \left(\frac{256,307,339.\text{ mi}}{\text{yr}}\right)\left(\frac{1 \text{ yr}}{365 \text{ days}}\right)$$

$$\cdot \left(\frac{1 \text{ day}}{24 \text{ hr}}\right)\left(\frac{1 \text{ hr}}{60 \text{ min}}\right)\left(\frac{1 \text{ min}}{60 \text{ sec}}\right)$$

$$\approx 8.127 \text{ mi per sec}$$

20. (a)

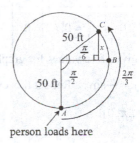

person loads here

Suppose the person takes a seat at point A. The person travels $\frac{\pi}{2}$ radians, the person is 50 ft above the ground. The person travels $\frac{\pi}{6}$ more radians, so let x be the additional vertical distance traveled.

$$\sin\frac{\pi}{6} = \frac{x}{50} \Rightarrow x = 50\sin\frac{\pi}{6} = 50\left(\frac{1}{2}\right) = 25$$

Thus, the person traveled an additional 25 ft above the ground, for a total of 75 ft above the ground.

(b) $t = 30$ sec and $\theta = \frac{2\pi}{3}$ radians, so

$$\omega = \frac{\theta}{t} \Rightarrow \omega = \frac{\frac{2\pi}{3}}{30} = \frac{2\pi}{3} \cdot \frac{1}{30} = \frac{\pi}{45} \text{ radian}$$

per sec. Thus, the Ferris wheel turned at the rate of $\frac{\pi}{45}$ radian per second.

Chapter 4

Graphs of the Circular Functions

Section 4.1 Graphs of the Sine and Cosine Functions

1. The amplitude of the graphs of the sine and cosine function is <u>1</u>, and the period of each is <u>2π</u>.

3. The graph of the sine function crosses the x-axis for all numbers of the form <u>$n\pi$,</u> where n is an integer.

5. The least positive number x for which $\cos x = 0$ is <u>$\frac{\pi}{2}$.</u>

7. $y = -\sin x$
   The graph is a sinusoidal curve with amplitude 1 and period 2π. Bcause $a = -1$, the graph is a reflection of $y = \sin x$ in the x-axis. This matches with graph E.

9. $y = \sin 2x$
   The graph is a sinusoidal curve with amplitude 1 and period π. Because $\sin(2 \cdot 0) = \sin 0 = 0$, the point (0, 0) is on the graph. This matches with graph B.

11. $y = 2\sin x$
    The graph is a sinusoidal curve with amplitude 2 and period 2π. Because $2\sin 0 = 2 \cdot 0 = 0$ and $2\sin \pi = 2 \cdot 1 = 2$, the points (0, 0) and $(\pi, 2)$, are on the graph. This matches with graph F.

13. $y = 2\cos x$
    Amplitude: $|2| = 2$

x	0	$\frac{\pi}{2}$	π	$\frac{3\pi}{2}$	2π
$\cos x$	1	0	-1	0	1
$2\cos x$	2	0	-2	0	2

This table gives five values for graphing one period of the function. Repeat this cycle for the interval $[-2\pi, 0]$.

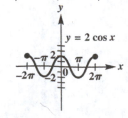

15. $y = \frac{2}{3}\sin x$
    Amplitude: $\left|\frac{2}{3}\right| = \frac{2}{3}$

x	0	$\frac{\pi}{2}$	π	$\frac{3\pi}{2}$	2π
$\sin x$	0	1	0	-1	0
$\frac{2}{3}\sin x$	0	$\frac{2}{3} \approx 0.7$	0	$-\frac{2}{3} \approx -0.7$	0

This table gives five values for graphing one period of $y = \frac{2}{3}\sin x$. Repeat this cycle for the interval $[-2\pi, 0]$.

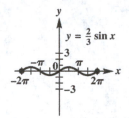

17. $y = -\cos x$
    Amplitude: $|-1| = 1$

x	0	$\frac{\pi}{2}$	π	$\frac{3\pi}{2}$	2π
$\cos x$	1	0	-1	0	1
$-\cos x$	-1	0	1	0	-1

This table gives five values for graphing one period of $y = -\cos x$. Repeat this cycle for the interval $[-2\pi, 0]$.

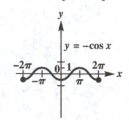

19. $y = -2 \sin x$

Amplitude: $|-2| = 2$

x	0	$\dfrac{\pi}{2}$	π	$\dfrac{3\pi}{2}$	2π
$\sin x$	0	1	0	-1	0
$-2\sin x$	0	-2	0	2	0

This table gives five values for graphing one period of $y = -2\sin x$. Repeat this cycle for the interval $[-2\pi, 0]$.

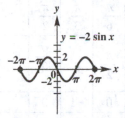

21. $y = \sin(-x)$

Amplitude: 1

x	0	$\dfrac{\pi}{2}$	π	$\dfrac{3\pi}{2}$	2π
$-x$	0	$-\dfrac{\pi}{2}$	$-\pi$	$-\dfrac{3\pi}{2}$	-2π
$\sin(-x)$	0	-1	0	1	0

This table gives five values for graphing one period of $y = \sin(-x)$. Repeat this cycle for the interval $[-2\pi, 0]$.

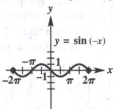

23. $y = \sin \dfrac{1}{2}x$

Period: $\dfrac{2\pi}{\frac{1}{2}} = 4\pi$ and amplitude: $|1| = 1$

Divide the interval $[0, 4\pi]$ into four equal parts to get x-values that will yield minimum and maximum points and x-intercepts. Then make a table. Repeat this cycle for the interval $[-4\pi, 0]$.

x	0	π	2π	3π	4π
$\dfrac{1}{2}x$	0	$\dfrac{\pi}{2}$	π	$\dfrac{3\pi}{2}$	2π
$\sin \dfrac{1}{2}x$	0	1	0	-1	0

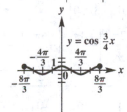

25. $y = \cos \dfrac{3}{4}x$

Period: $\dfrac{2\pi}{\frac{3}{4}} = 2\pi \cdot \dfrac{4}{3} = \dfrac{8\pi}{3}$ and amplitude:

$|1| = 1$

Divide the interval $\left[0, \dfrac{8\pi}{3}\right]$ into four equal parts to get the x-values that will yield minimum and maximum points and x-intercepts. Then make a table. Repeat this cycle for the interval $\left[-\dfrac{8\pi}{3}, 0\right]$.

x	0	$\dfrac{2\pi}{3}$	$\dfrac{4\pi}{3}$	2π	$\dfrac{8\pi}{3}$
$\dfrac{3}{4}x$	0	$\dfrac{\pi}{2}$	π	$\dfrac{3\pi}{2}$	2π
$\cos \dfrac{3}{4}x$	1	0	-1	0	1

27. $y = \sin 3x$

Period: $\dfrac{2\pi}{3}$ and amplitude: $|1| = 1$

Divide the interval $\left[0, \dfrac{2\pi}{3}\right]$ into four equal

parts to get the *x*-values that will yield minimum and maximum points and *x*-intercepts. Then make a table.

Repeat this cycle for the interval $\left[-\dfrac{2\pi}{3}, 0\right]$.

x	0	$\dfrac{\pi}{6}$	$\dfrac{\pi}{3}$	$\dfrac{\pi}{2}$	$\dfrac{2\pi}{3}$
$3x$	0	$\dfrac{\pi}{2}$	π	$\dfrac{3\pi}{2}$	2π
$\sin 3x$	0	1	0	-1	0

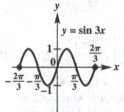

$y = \sin 3x$

29. $y = 2\sin\dfrac{1}{4}x$

Period: $\dfrac{2\pi}{\frac{1}{4}} = 2\pi \cdot \dfrac{4}{1} = 8\pi$ and amplitude:

$|2| = 2$

Divide the interval $[0, 8\pi]$ into four equal

parts to get the *x*-values that will yield minimum and maximum points and *x*-intercepts. Then make a table. Repeat this cycle for the interval $[-8\pi, 0]$.

x	0	2π	4π	6π	8π
$\dfrac{1}{4}x$	0	$\dfrac{\pi}{2}$	π	$\dfrac{3\pi}{2}$	2π
$\sin\dfrac{1}{4}x$	0	1	0	-1	0
$2\sin\dfrac{1}{4}x$	0	2	0	-2	0

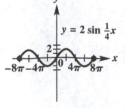

$y = 2\sin\frac{1}{4}x$

31. $y = -2\cos 3x$

Period: $\dfrac{2\pi}{3}$ and amplitude: $|-2| = 2$

Divide the interval $\left[0, \dfrac{2\pi}{3}\right]$ into four equal

parts to get the *x*-values that will yield minimum and maximum points and *x*-intercepts. Then make a table. Repeat this cycle for the interval $\left[-\dfrac{2\pi}{3}, 0\right]$.

x	0	$\dfrac{\pi}{6}$	$\dfrac{\pi}{3}$	$\dfrac{\pi}{2}$	$\dfrac{2\pi}{3}$
$3x$	0	$\dfrac{\pi}{2}$	π	$\dfrac{3\pi}{2}$	2π
$\cos 3x$	1	0	-1	0	1
$-2\cos 3x$	-2	0	2	0	-2

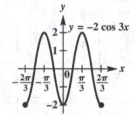

$y = -2\cos 3x$

33. $y = \cos \pi x$

Period: $\dfrac{2\pi}{\pi} = 2$ and amplitude: $|1| = 1$

Divide the interval $[0, 2]$ into four equal parts to get the *x*-values that will yield minimum and maximum points and *x*-intercepts. Then make a table. Repeat this cycle for the interval $[-2, 0]$.

x	0	$\dfrac{1}{2}$	1	$\dfrac{3}{2}$	2
πx	0	$\dfrac{\pi}{2}$	π	$\dfrac{3\pi}{2}$	2π
$\cos \pi x$	1	0	-1	0	1

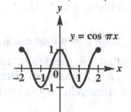

$y = \cos \pi x$

35. $y = -2\sin 2\pi x$

Period: $\dfrac{2\pi}{2\pi} = 1$ and amplitude: $\left|-2\right| = 2$

Divide the interval [0, 1] into four equal parts to get the x-values that will yield minimum and maximum points and x-intercepts. Then make a table. Repeat this cycle for the interval [−1, 0].

x	0	$\dfrac{1}{4}$	$\dfrac{1}{2}$	$\dfrac{3}{4}$	1
$2\pi x$	0	$\dfrac{\pi}{2}$	π	$\dfrac{3\pi}{2}$	2π
$\sin 2\pi x$	0	1	0	−1	0
$-2\sin \pi x$	0	−2	0	2	0

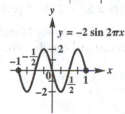

37. $y = \dfrac{1}{2}\cos\dfrac{\pi}{2}x$

Period: $\dfrac{2\pi}{\frac{\pi}{2}} = 2\pi \cdot \dfrac{2}{\pi} = 4$ and amplitude:

$\left|\dfrac{1}{2}\right| = \dfrac{1}{2}$

Divide the interval [0, 4] into four equal parts to get the x-values that will yield minimum and maximum points and x-intercepts. Then make a table. Repeat this cycle for the interval [−4, 0].

x	0	1	2	3	4
$\dfrac{\pi}{2}x$	0	$\dfrac{\pi}{2}$	π	$\dfrac{3\pi}{2}$	2π
$\cos\dfrac{\pi}{2}x$	1	0	−1	0	1
$\dfrac{1}{2}\cos\dfrac{\pi}{2}x$	$\dfrac{1}{2}$	0	$-\dfrac{1}{2}$	0	$\dfrac{1}{2}$

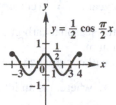

39. $y = \pi \sin \pi x$

Period: $\dfrac{2\pi}{\pi} = 2$ and amplitude: $\left|\pi\right| = \pi$

Divide the interval [0, 2] into four equal parts to get the x-values that will yield minimum and maximum points and x-intercepts. Then make a table. Repeat this cycle for the interval [−2, 0].

x	0	$\dfrac{1}{2}$	1	$\dfrac{3}{2}$	2
πx	0	$\dfrac{\pi}{2}$	π	$\dfrac{3\pi}{2}$	2π
$\sin \pi x$	0	1	0	−1	0
$\pi \sin \pi x$	0	π	0	$-\pi$	0

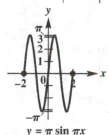

41. The amplitude is $\dfrac{1}{2}\left[2-(-2)\right] = \dfrac{1}{2}(4) = 2$, so $a = 2$. One complete cycle of the graph is achieved in π units, so the period

$\pi = \dfrac{2\pi}{b} \Rightarrow b = \dfrac{2\pi}{\pi} = 2$. Comparing the given

graph with the general sine and cosine curves, we see that this graph is a cosine curve. Substituting $a = 2$ and $b = 2$, the function is $y = 2\cos 2x$. Verify by confirming minimum and maximum points and x-intercepts from the graph:

$(0, 2) \Rightarrow 2 = 2\cos(2\cdot 0) = 2\cos 0 = 2\cdot 1 = 2$

$\left(\dfrac{\pi}{4}, 0\right) \Rightarrow 0 = 2\cos\left(2\cdot\dfrac{\pi}{4}\right) = 2\cos\dfrac{\pi}{2} = 2\cdot 0 = 0$

$\left(\dfrac{\pi}{2}, -2\right) \Rightarrow -2 = 2\cos\left(2\cdot\dfrac{\pi}{2}\right) = 2\cos\pi$
$\qquad\qquad\qquad = 2(-1) = -2$

$\left(\dfrac{3\pi}{4}, 0\right) \Rightarrow 0 = 2\cos\left(2\cdot\dfrac{3\pi}{4}\right)$
$\qquad\qquad\qquad = 2\cos\dfrac{3\pi}{2} = 2\cdot 0 = 0$

$(\pi, 2) \Rightarrow 2 = 2\cos(2\pi) = 2(1) = 2$

43. The amplitude is $\frac{1}{2}[3-(-3)]=\frac{1}{2}(6)=3$, so

$a = 3$. One-half of a cycle of the graph is achieved in 2π units, so the period is

$2 \cdot 2\pi = 4\pi$ and $4\pi = \frac{2\pi}{b} \Rightarrow b = \frac{2\pi}{4\pi} = \frac{1}{2}$.

Comparing the given graph with the general sine and cosine curves, we see that this graph is the reflection of the cosine curve in the x-axis. Thus, $a = -3$. Substituting $a = -3$ and $b = \frac{1}{2}$, the function is $y = -3\cos\frac{1}{2}x$. Verify by confirming minimum and maximum points and x–intercepts from the graph:

$(0, -3) \Rightarrow -3 = -3\cos\left(\frac{1}{2} \cdot 0\right) = -3\cos 0$

$\qquad = -3 \cdot 1 = -3$

$(\pi, 0) \Rightarrow 0 = -3\cos\left(\frac{1}{2} \cdot \pi\right) = -3\cos\frac{\pi}{2}$

$\qquad = -3(0) = 0$

$(2\pi, 3) \Rightarrow 3 = -3\cos\left(\frac{1}{2} \cdot 2\pi\right) = -3\cos\pi$

$\qquad = -3(-1) = 3$

45. The amplitude is $\frac{1}{2}[3-(-3)]=\frac{1}{2}(6)=3$, so

$a = 3$. One complete cycle of the graph is achieved in $\frac{\pi}{2}$ units, so the period

$\frac{\pi}{2} = \frac{2\pi}{b} \Rightarrow b = 2\pi \cdot \frac{2}{\pi} = 4$. Comparing the

given graph with the general sine and cosine curves, we see that this graph is a sine curve. Substituting $a = 3$ and $b = 4$, the function is $y = 3\sin 4x$. Verify by confirming minimum and maximum points and x–intercepts from the graph:

$(0, 0) \Rightarrow 0 = 3\sin(4 \cdot 0) = 3\sin 0 = 3 \cdot 0 = 0$

$\left(\frac{\pi}{8}, 3\right) \Rightarrow 3 = 3\sin\left(4 \cdot \frac{\pi}{8}\right) = 3\sin\frac{\pi}{2}$

$\qquad = 3(1) = 3$

$\left(\frac{\pi}{4}, 0\right) \Rightarrow 0 = 3\sin\left(4 \cdot \frac{\pi}{4}\right) = 3\sin\pi$

$\qquad = 3(0) = 0$

$\left(\frac{3\pi}{8}, -3\right) \Rightarrow -3 = 3\sin\left(4 \cdot \frac{3\pi}{8}\right)$

$\qquad = 3\sin\frac{3\pi}{2} = 3(-1) = -3$

$\left(\frac{\pi}{2}, 0\right) \Rightarrow 0 = 3\sin\left(4 \cdot \frac{\pi}{2}\right) = 3\sin 2\pi = 3(0) = 0$

47. (a) The highest temperature is 80°F; the lowest is 50°F.

(b) The amplitude is

$\frac{1}{2}(80 - 50) = \frac{1}{2}(30) = 15$.

(c) The period is about 35,000 yr.

(d) The trend of the temperature now is downward.

49. The graph repeats each day, so the period is 24 hours.

51. On January 20, low tide was at 6 P.M., with height approximately 0.2 ft.

53. On January 22, high tide was at 2 + 3:18 = 3:18 A.M., with height 2.6 − 0.2 = 2.4 feet.

55. (a) The graph has a general upward trend along with small annual oscillations.

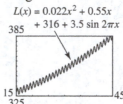

$L(x) = 0.022x^2 + 0.55x + 316 + 3.5\sin 2\pi x$

(b) The seasonal variations are caused by the term $3.5\sin 2\pi x$. The maximums will occur when $2\pi x = \frac{\pi}{2} + 2n\pi$, where n is an integer. Because x cannot be negative, n cannot be negative. This is equivalent to

$2\pi x = \frac{\pi}{2} + 2n\pi,\ n = 0, 1, 2, \ldots$

$2x = \frac{1}{2} + 2n,\ n = 0, 1, 2, \ldots$

$x = \frac{1}{4} + n,\ n = 0, 1, 2, \ldots$

$x = \frac{4n+1}{4},\ n = 0, 1, 2, \ldots$

$x = \frac{1}{4}, \frac{5}{4}, \frac{9}{4}, \ldots$

Because x is in years, $x = \frac{1}{4}$ corresponds to April when the seasonal carbon dioxide levels are maximum.

The minimums will occur when

$2\pi x = \frac{3\pi}{2} + 2n\pi$, where n is an integer.

Because x cannot be negative, n cannot be negative.

(continued on next page)

(*continued*)

This is equivalent to

$$2\pi x = \frac{3\pi}{2} + 2n\pi, \ n = 0, 1, 2, \ldots$$

$$2x = \frac{3}{2} + 2n, \ n = 0, 1, 2, \ldots$$

$$x = \frac{3}{4} + n, \ n = 0, 1, 2, \ldots$$

$$x = \frac{4n+3}{4}, \ n = 0, 1, 2, \ldots$$

$$x = \frac{3}{4}, \frac{7}{4}, \frac{11}{4}, \ldots$$

This is $\frac{1}{2}$ yr later, which corresponds to October.

(c) Answers will vary. Sample answer: The quadratic function provides the general increasing nature of the level, while the sine function provides the fluctuations as the years go by.

57. $T(x) = 37 + 21\sin\left[\frac{2\pi}{365}(x - 91)\right]$

(a) March 15 (day 74)

$$T(74) = 37 + 21\sin\left[\frac{2\pi}{365}(74 - 91)\right]$$

$$\approx 31°F$$

(b) April 5 (day 95)

$$T(95) = 37 + 21\sin\left[\frac{2\pi}{365}(95 - 91)\right]$$

$$\approx 38°F$$

(c) Day 200

$$T(200) = 37 + 21\sin\left[\frac{2\pi}{365}(200 - 91)\right]$$

$$\approx 57.0° \approx 57°F$$

(d) June 25 is day 176.

$(31 + 28 + 31 + 30 + 31 + 25 = 176)$

$$T(176) = 37 + 21\sin\left[\frac{2\pi}{365}(176 - 91)\right]$$

$$\approx 57.9° \approx 58°F$$

(e) October 1 is day 274.

$31 + 28 + 31 + 30 + 31$
$\quad + 30 + 31 + 31 + 30 + 1 = 274$

$$T(274) = 37 + 21\sin\left[\frac{2\pi}{365}(274 - 91)\right]$$

$$\approx 36.8° \approx 37°F$$

(f) December 31 is day 365.

$$T(365) = 37 + 21\sin\left[\frac{2\pi}{365}(365 - 91)\right]$$

$$\approx 16.0° \approx 16°F$$

59. $-1 \leq y \leq 1$

Amplitude: 1

Period: 8 squares $= 8(30°) = 240°$ or $\frac{4\pi}{3}$

61. No, we can't say that $\sin bx = b\sin x$. If b is not zero, then the period of $y = \sin bx$ is $\frac{2\pi}{|b|}$, and the amplitude is 1. The period of $y = b\sin x$ is 2π, and the amplitude is $|b|$.

63. $X \approx -0.4161468$, $Y \approx 0.90929743$. X is cos 2, and Y is sin 2.

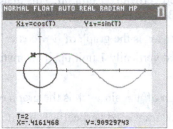

65. $X = 2$, $Y \approx -0.4161468$; $\cos 2 \approx -0.4161468$

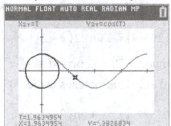

Section 4.2   Translations of the Graphs of the Sine and Cosine Functions

1. The graph of $y = \sin\left(x + \frac{\pi}{4}\right)$ is obtained by shifting the graph of $y = \sin x$ $\frac{\pi}{4}$ units <u>left</u>.

3. The graph of $y = 4\sin x$ is obtained by stretching the graph of $y = \sin x$ vertically by a factor of <u>4</u>.

5. The graph of $y = 6 + 3\sin x$ is obtained by shifting the graph of $y = 3\sin x$ <u>6</u> units <u>up</u>.

7. The graph of $y = 3 + 5\cos\left(x + \frac{\pi}{5}\right)$ is obtained by shifting the graph of $y = \cos x$ $\frac{\pi}{5}$ units horizontally to the <u>left</u>, stretching it vertically by a factor of <u>5</u>, and then shifting it <u>3</u> units vertically in the <u>up</u> direction.

9. $y = \sin\left(x - \frac{\pi}{4}\right)$ is the graph of $y = \sin x$, shifted to the right $\frac{\pi}{4}$ unit. This matches choice D.

11. $y = \cos\left(x - \frac{\pi}{4}\right)$ is the graph of $y = \cos x$, shifted to the right $\frac{\pi}{4}$ unit. This matches choice H.

13. $y = 1 + \sin x$ is the graph of $y = \sin x$, translated vertically 1 unit up. This matches choice B.

15. $y = 1 + \cos x$ is the graph of $y = \cos x$, translated vertically 1 unit up. This matches choice I.

17. The graph of $y = \sin x + 1$ is the graph of $y = \sin x$ translated vertically 1 unit up, while the graph of $y = \sin(x+1)$ is the graph of $y = \sin x$ shifted horizontally 1 unit left.

19. $y = 3\sin(2x - 4) = 3\sin[2(x-2)]$

The amplitude $= |3| = 3$, period $= \frac{2\pi}{2} = \pi$, and phase shift $= 2$. This matches choice B.

21. $y = -4\sin(3x - 2) = -4\sin\left[3\left(x - \frac{2}{3}\right)\right]$

The amplitude $= |-4| = 4$, period $= \frac{2\pi}{3}$, and phase shift $= \frac{2}{3}$. This matches choice C.

23. If the graph of $y = \cos x$ is translated $\frac{\pi}{2}$ units horizontally to the <u>right</u>, it will coincide with the graph of $y = \sin x$.

25. This is a sine curve that has been shifted one unit down, so the equation is $y = -1 + \sin x$. Verify by confirming minimum and maximum points and x–intercepts from the graph:
$(0,-1) \Rightarrow -1 = -1 + \sin 0 = -1 + 0 = -1$
$\left(\frac{\pi}{2}, 0\right) \Rightarrow 0 = -1 + \sin\frac{\pi}{2} = -1 + 1 = 0$
$(\pi, -1) \Rightarrow -1 = -1 + \sin\pi = -1 + 0 = -1$

$\left(\frac{3\pi}{2}, -2\right) \Rightarrow -2 = -1 + \sin\frac{3\pi}{2} = -1 + (-1) = -2$
$(2\pi, -1) \Rightarrow -1 = 1 + \sin(2\pi) = -1 + 0 = -1$

27. The maximum is at $\left(\frac{\pi}{3}, 1\right)$, so this is a cosine curve that has been shifted $\frac{\pi}{3}$ units to the right.

Thus, the equation is $y = \cos\left(x - \frac{\pi}{3}\right)$. Verify by confirming minimum and maximum points and x–intercepts from the graph:
$\left(\frac{\pi}{3}, 1\right) \Rightarrow 1 = \cos\left(\frac{\pi}{3} - \frac{\pi}{3}\right) = \cos 0 = 1$
$\left(\frac{4\pi}{3}, -1\right) \Rightarrow -1 = \cos\left(\frac{4\pi}{3} - \frac{\pi}{3}\right) = \cos\pi = -1$
$\left(\frac{7\pi}{3}, 1\right) \Rightarrow 1 = \cos\left(\frac{7\pi}{3} - \frac{\pi}{3}\right) = \cos 2\pi = 1$
$\left(\frac{10\pi}{3}, -1\right) \Rightarrow -1 = \cos\left(\frac{10\pi}{3} - \frac{\pi}{3}\right) = \cos 3\pi = -1$
$\left(\frac{13\pi}{3}, 1\right) \Rightarrow 1 = \cos\left(\frac{13\pi}{3} - \frac{\pi}{3}\right) = \cos 4\pi = 1$

29. $y = 2\sin(x + \pi)$

amplitude: $|2| = 2$; period: $\frac{2\pi}{1} = 2\pi$; There is no vertical translation. The phase shift is π units to the left.

31. $y = -\frac{1}{4}\cos\left(\frac{1}{2}x + \frac{\pi}{2}\right) = -\frac{1}{4}\cos\left[\frac{1}{2}\left[x - (-\pi)\right]\right]$

amplitude: $\left|-\frac{1}{4}\right| = \frac{1}{4}$; period:
$\frac{2\pi}{\frac{1}{2}} = 2\pi \cdot \frac{2}{1} = 4\pi$
There is no vertical translation. The phase shift is π units to the left.

33. $y = 3\cos\left[\frac{\pi}{2}\left(x - \frac{1}{2}\right)\right]$

amplitude: $|3| = 3$; period: $\frac{2\pi}{\frac{\pi}{2}} = 2\pi \cdot \frac{2}{\pi} = 4$

There is no vertical translation. The phase shift is $\frac{1}{2}$ unit to the right.

35. $y = 2 - \sin\left(3x - \dfrac{\pi}{5}\right) = -\sin\left[3\left(x - \dfrac{\pi}{15}\right)\right] + 2$

amplitude: $|-1| = 1$; period: $\dfrac{2\pi}{3}$

The vertical translation is 2 units up. The

phase shift is $\dfrac{\pi}{15}$ unit to the right

37. $y = \cos\left(x - \dfrac{\pi}{2}\right)$

Step 1: Find the interval whose length is $\dfrac{2\pi}{b}$.

$0 \le x - \dfrac{\pi}{2} \le 2\pi \Rightarrow 0 + \dfrac{\pi}{2} \le x \le 2\pi + \dfrac{\pi}{2} \Rightarrow$
$\dfrac{\pi}{2} \le x \le \dfrac{5\pi}{2}$

Step 2: Divide the period into four equal parts

to get the following *x*-values: $\dfrac{\pi}{2}$, π, $\dfrac{3\pi}{2}$, 2π,

$\dfrac{5\pi}{2}$

Step 3: Evaluate the function for each of the
five *x*-values

x	$\dfrac{\pi}{2}$	π	$\dfrac{3\pi}{2}$	2π	$\dfrac{5\pi}{2}$
$x - \dfrac{\pi}{2}$	0	$\dfrac{\pi}{2}$	π	$\dfrac{3\pi}{2}$	2π
$\cos\left(x - \dfrac{\pi}{2}\right)$	1	0	-1	0	1

Steps 4 and 5: Plot the points found in the
table and join them with a sinusoidal curve.
By graphing an additional period to the right,
we obtain the following graph.

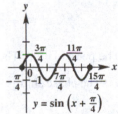

$y = \cos\left(x - \dfrac{\pi}{2}\right)$

The amplitude is 1. The period is 2π. There is

no vertical translation. The phase shift is $\dfrac{\pi}{2}$

unit to the right.

39. $y = \sin\left(x + \dfrac{\pi}{4}\right)$

Step 1: Find the interval whose length is $\dfrac{2\pi}{b}$.

$0 \le x + \dfrac{\pi}{4} \le 2\pi \Rightarrow 0 - \dfrac{\pi}{4} \le x \le 2\pi - \dfrac{\pi}{4} \Rightarrow$
$-\dfrac{\pi}{4} \le x \le \dfrac{7\pi}{4}$

Step 2: Divide the period into four equal parts

to get the following *x*-values: $-\dfrac{\pi}{4}$, $\dfrac{\pi}{4}$,

$\dfrac{3\pi}{4}, \dfrac{5\pi}{4}, \dfrac{7\pi}{4}$

Step 3: Evaluate the function for each of the
five *x*-values

x	$-\dfrac{\pi}{4}$	$\dfrac{\pi}{4}$	$\dfrac{3\pi}{4}$	$\dfrac{5\pi}{4}$	$\dfrac{7\pi}{4}$
$x + \dfrac{\pi}{4}$	0	$\dfrac{\pi}{2}$	π	$\dfrac{3\pi}{2}$	2π
$\sin\left(x + \dfrac{\pi}{4}\right)$	0	1	0	-1	0

Steps 4 and 5: Plot the points found in the
table and join them with a sinusoidal curve.
By graphing an additional period to the right,
we obtain the following graph.

$y = \sin\left(x + \dfrac{\pi}{4}\right)$

The amplitude is 1. The period is 2π. There is

no vertical translation. The phase shift is $\dfrac{\pi}{4}$

unit to the left.

41. $y = 2\cos\left(x - \dfrac{\pi}{3}\right)$

Step 1: Find the interval whose length is $\dfrac{2\pi}{b}$.

$0 \le x - \dfrac{\pi}{3} \le 2\pi \Rightarrow 0 + \dfrac{\pi}{3} \le x \le 2\pi + \dfrac{\pi}{3} \Rightarrow$
$\dfrac{\pi}{3} \le x \le \dfrac{7\pi}{3}$

(*continued on next page*)

(*continued*)

Step 2: Divide the period into four equal parts to get the following *x*-values $\dfrac{\pi}{3}, \dfrac{5\pi}{6}, \dfrac{4\pi}{3}, \dfrac{11\pi}{6}, \dfrac{7\pi}{3}$

Step 3: Evaluate the function for each of the five *x*-values.

x	$\dfrac{\pi}{3}$	$\dfrac{5\pi}{6}$	$\dfrac{4\pi}{3}$	$\dfrac{11\pi}{6}$	$\dfrac{7\pi}{3}$
$x - \dfrac{\pi}{3}$	0	$\dfrac{\pi}{2}$	π	$\dfrac{3\pi}{2}$	2π
$\cos\left(x - \dfrac{\pi}{3}\right)$	1	0	-1	0	1
$2\cos\left(x - \dfrac{\pi}{3}\right)$	2	0	-2	0	2

Steps 4 and 5: Plot the points found in the table and join them with a sinusoidal curve. By graphing an additional period to the right, we obtain the following graph.

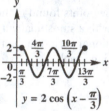

$$y = 2\cos\left(x - \frac{\pi}{3}\right)$$

The amplitude is 2. The period is 2π. There is no vertical translation. The phase shift is $\dfrac{\pi}{3}$ units to the right.

43. $y = \dfrac{3}{2}\sin\left[2\left(x + \dfrac{\pi}{4}\right)\right]$

Step 1: Find the interval whose length is $\dfrac{2\pi}{b}$.

$$0 \le 2\left(x + \frac{\pi}{4}\right) \le 2\pi \Rightarrow 0 \le x + \frac{\pi}{4} \le \pi \Rightarrow$$

$$-\frac{\pi}{4} \le x \le \frac{3\pi}{4}$$

Step 2: Divide the period into four equal parts to get the *x*-values: $-\dfrac{\pi}{4}, 0, \dfrac{\pi}{4}, \dfrac{\pi}{2}, \dfrac{3\pi}{4}$

Step 3: Evaluate the function for each of the five *x*-values

x	$-\dfrac{\pi}{4}$	0	$\dfrac{\pi}{4}$	$\dfrac{\pi}{2}$	$\dfrac{3\pi}{4}$
$2\left(x + \dfrac{\pi}{4}\right)$	0	$\dfrac{\pi}{2}$	π	$\dfrac{3\pi}{2}$	2π
$\sin\left[2\left(x + \dfrac{\pi}{4}\right)\right]$	0	1	0	-1	0
$\dfrac{3}{2}\sin\left[2\left(x + \dfrac{\pi}{4}\right)\right]$	0	$\dfrac{3}{2}$	0	$-\dfrac{3}{2}$	0

Steps 4 and 5: Plot the points found in the table and join them with a sinusoidal curve.

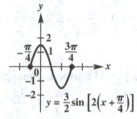

$$y = \frac{3}{2}\sin\left[2\left(x + \frac{\pi}{4}\right)\right]$$

The amplitude is $\dfrac{3}{2}$. The period is $\dfrac{2\pi}{2} = \pi$. There is no vertical translation. The phase shift is $\dfrac{\pi}{4}$ unit to the left.

45. $y = -4\sin(2x - \pi) = -4\sin\left[2\left(x - \dfrac{\pi}{2}\right)\right]$

Step 1: Find the interval whose length is $\dfrac{2\pi}{b}$.

$$0 \le 2\left(x - \frac{\pi}{2}\right) \le 2\pi \Rightarrow 0 \le x - \frac{\pi}{2} \le \frac{2\pi}{2} \Rightarrow$$

$$0 \le x - \frac{\pi}{2} \le \pi \Rightarrow \frac{\pi}{2} \le x \le \frac{3\pi}{2}$$

Step 2: Divide the period into four equal parts to get the following *x*-values: $\dfrac{\pi}{2}, \dfrac{3\pi}{4}, \pi, \dfrac{5\pi}{4}, \dfrac{3\pi}{2}$

Step 3: Evaluate the function for each of the five *x*-values.

x	$\dfrac{\pi}{2}$	$\dfrac{3\pi}{4}$	π	$\dfrac{5\pi}{4}$	$\dfrac{3\pi}{2}$
$2\left(x - \dfrac{\pi}{2}\right)$	0	$\dfrac{\pi}{2}$	π	$\dfrac{3\pi}{2}$	2π
$\sin\left[2\left(x - \dfrac{\pi}{2}\right)\right]$	0	1	0	-1	0

(*continued on next page*)

(*continued*)

x	$\dfrac{\pi}{2}$	$\dfrac{3\pi}{4}$	π	$\dfrac{5\pi}{4}$	$\dfrac{3\pi}{2}$
$-4\sin\left[2\left(x-\dfrac{\pi}{2}\right)\right]$	0	-4	0	4	0

Steps 4 and 5: Plot the points found in the table and join them with a sinusoidal curve.

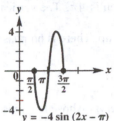

$y = -4\sin(2x - \pi)$

The amplitude is $\left|-4\right|$, which is 4. The

period is $\dfrac{2\pi}{2}$, which is π. There is no

vertical translation. The phase shift is $\dfrac{\pi}{2}$

units to the right

47. $y = \dfrac{1}{2}\cos\left(\dfrac{1}{2}x - \dfrac{\pi}{4}\right) = \dfrac{1}{2}\cos\left[\dfrac{1}{2}\left(x - \dfrac{\pi}{2}\right)\right]$

Step 1: Find the interval whose length is $\dfrac{2\pi}{b}$.

$0 \le \dfrac{1}{2}\left(x - \dfrac{\pi}{2}\right) \le 2\pi \Rightarrow 0 \le x - \dfrac{\pi}{2} \le 4\pi \Rightarrow$

$\dfrac{\pi}{2} \le x \le \dfrac{8\pi}{2} + \dfrac{\pi}{2} \Rightarrow \dfrac{\pi}{2} \le x \le \dfrac{9\pi}{2}$

Step 2: Divide the period into four equal parts

to get the x-values: $\dfrac{\pi}{2}, \dfrac{3\pi}{2}, \dfrac{5\pi}{2}, \dfrac{7\pi}{2}, \dfrac{9\pi}{2}$

Step 3: Evaluate the function for each of the five x-values.

x	$\dfrac{\pi}{2}$	$\dfrac{3\pi}{2}$	$\dfrac{5\pi}{2}$	$\dfrac{7\pi}{2}$	$\dfrac{9\pi}{2}$
$\dfrac{1}{2}\left(x - \dfrac{\pi}{2}\right)$	0	$\dfrac{\pi}{2}$	π	$\dfrac{3\pi}{2}$	2π
$\cos\left[\dfrac{1}{2}\left(x - \dfrac{\pi}{2}\right)\right]$	1	0	-1	0	1
$\dfrac{1}{2}\cos\left[\dfrac{1}{2}\left(x - \dfrac{\pi}{2}\right)\right]$	$\dfrac{1}{2}$	0	$-\dfrac{1}{2}$	0	$\dfrac{1}{2}$

Steps 4 and 5: Plot the points found in the table and join them with a sinusoidal curve.

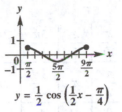

$y = \dfrac{1}{2}\cos\left(\dfrac{1}{2}x - \dfrac{\pi}{4}\right)$

The amplitude is $\dfrac{1}{2}$. The period is $\dfrac{2\pi}{\frac{1}{2}}$,

which is 4π. There is no vertical translation.

The phase shift is $\dfrac{\pi}{2}$ units to the right.

49. $y = -3 + 2\sin x$

Step 1: The period is 2π.

Step 2: Divide the period into four equal parts

to get the x-values: $0, \dfrac{\pi}{2}, \pi, \dfrac{3\pi}{2}, \pi, 2\pi$

Step 3: Evaluate the function for each of the five x-values:

x	0	$\dfrac{\pi}{2}$	π	$\dfrac{3\pi}{2}$	2π
$\sin x$	0	1	0	-1	0
$2\sin x$	0	2	0	-2	0
$-3 + 2\sin x$	-3	-1	-3	-5	-3

Steps 4 and 5: Plot the points found in the table and join them with a sinusoidal curve. By graphing an additional period to the left, we obtain the following graph.

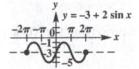

$y = -3 + 2\sin x$

The amplitude is 2. The vertical translation is 3 units down. There is no phase shift.

51. $y = -1 - 2\cos 5x$

Step 1: Find the interval whose length is $\dfrac{2\pi}{b}$.

$0 \le 5x \le 2\pi \Rightarrow 0 \le x \le \dfrac{2\pi}{5}$

Step 2: Divide the period into four equal parts

to get the x-values: $0, \dfrac{\pi}{10}, \dfrac{\pi}{5}, \dfrac{3\pi}{10}, \dfrac{2\pi}{5}$

Step 3: Evaluate the function for each of the five x-values.

(*continued on next page*)

(*continued*)

x	0	$\dfrac{\pi}{10}$	$\dfrac{\pi}{5}$	$\dfrac{3\pi}{10}$	$\dfrac{2\pi}{5}$
$5x$	0	$\dfrac{\pi}{2}$	π	$\dfrac{3\pi}{2}$	2π
$\cos 5x$	1	0	-1	0	1
$-2\cos 5x$	-2	0	2	0	-2
$-1-2\cos 5x$	-3	-1	1	-1	-3

Steps 4 and 5: Plot the points found in the table and join them with a sinusoidal curve. By graphing an additional period to the left, we obtain the following graph.

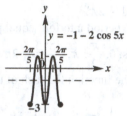

The period is $\dfrac{2\pi}{5}$. The amplitude is $\lvert -2 \rvert$, which is 2. The vertical translation is 1 unit down. There is no phase shift.

53. $y = 1 - 2\cos\dfrac{1}{2}x$

Step 1: Find the interval whose length is $\dfrac{2\pi}{b}$.

$0 \le \dfrac{1}{2}x \le 2\pi \Rightarrow 0 \le x \le 4\pi$

Step 2: Divide the period into four equal parts to get the following x-values: $0,\ \pi,\ 2\pi,\ 3\pi,\ 4\pi$

Step 3: Evaluate the function for each of the five x-values.

x	0	π	2π	3π	4π
$\dfrac{1}{2}x$	0	$\dfrac{\pi}{2}$	π	$\dfrac{3\pi}{2}$	2π
$\cos\dfrac{1}{2}x$	1	0	-1	0	1
$-2\cos\dfrac{1}{2}x$	-2	0	2	0	-2
$1-2\cos\dfrac{1}{2}x$	-1	1	3	1	-1

Steps 4 and 5: Plot the points found in the table and join them with a sinusoidal curve. By graphing an additional period to the left, we obtain the following graph.

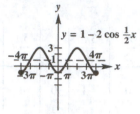

The amplitude is $\lvert -2 \rvert$, which is 2. The period is $\dfrac{2\pi}{\frac{1}{2}}$, which is 4π. The vertical translation is 1 unit up. There is no phase shift.

55. $y = -2 + \dfrac{1}{2}\sin 3x$

Step 1: Find the interval whose length is $\dfrac{2\pi}{b}$.

$0 \le 3x \le 2\pi \Rightarrow 0 \le x \le \dfrac{2\pi}{3}$

Step 2: Divide the period into four equal parts to get the following x-values: $0,\ \dfrac{\pi}{6},\ \dfrac{\pi}{3},\ \dfrac{\pi}{2},\ \dfrac{2\pi}{3}$

Step 3: Evaluate the function for each of the five x-values.

x	0	$\dfrac{\pi}{6}$	$\dfrac{\pi}{3}$	$\dfrac{\pi}{2}$	$\dfrac{2\pi}{3}$
$3x$	0	$\dfrac{\pi}{2}$	π	$\dfrac{3\pi}{2}$	2π
$\sin 3x$	0	1	0	-1	0
$\dfrac{1}{2}\sin 3x$	0	$\dfrac{1}{2}$	0	$-\dfrac{1}{2}$	0
$-2+\dfrac{1}{2}\sin 3x$	-2	$-\dfrac{3}{2}$	-2	$-\dfrac{5}{2}$	-2

Steps 4 and 5: Plot the points found in the table and join them with a sinusoidal curve. By graphing an additional period to the left, we obtain the following graph.

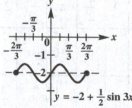

The amplitude is $\left\lvert \dfrac{1}{2} \right\rvert = \dfrac{1}{2}$. The period is $\dfrac{2\pi}{3}$. The vertical translation is 2 units down. There is no phase shift.

57. $y = -3 + 2\sin\left(x + \dfrac{\pi}{2}\right)$

Step 1: Find the interval whose length is $\dfrac{2\pi}{b}$.

$$0 \le x + \dfrac{\pi}{2} \le 2\pi \Rightarrow 0 - \dfrac{\pi}{2} \le x \le 2\pi - \dfrac{\pi}{2} \Rightarrow$$

$$-\dfrac{\pi}{2} \le x \le \dfrac{3\pi}{2}$$

Step 2: Divide the period into four equal parts to get the following *x*-values: $-\dfrac{\pi}{2}, 0, \dfrac{\pi}{2}, \pi,$

$\dfrac{3\pi}{2}$

Step 3: Evaluate the function for each of the five *x*-values

x	$-\dfrac{\pi}{2}$	0	$\dfrac{\pi}{2}$	π	$\dfrac{3\pi}{2}$
$x + \dfrac{\pi}{2}$	0	$\dfrac{\pi}{2}$	π	$\dfrac{3\pi}{2}$	2π
$\sin\left(x + \dfrac{\pi}{2}\right)$	0	1	0	-1	0
$2\sin\left(x + \dfrac{\pi}{2}\right)$	0	2	0	-2	0
$-3 + 2\sin\left(x + \dfrac{\pi}{2}\right)$	-3	-1	-3	-5	-3

Steps 4 and 5: Plot the points found in the table and join them with a sinusoidal curve.

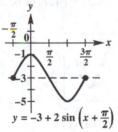

$y = -3 + 2\sin\left(x + \frac{\pi}{2}\right)$

The amplitude is $|2|$, which is 2. The period is 2π. The vertical translation is 3 units down.

The phase shift is $\dfrac{\pi}{2}$ units to the left.

59. $y = \dfrac{1}{2} + \sin\left[2\left(x + \dfrac{\pi}{4}\right)\right]$

Step 1: Find the interval whose length is $\dfrac{2\pi}{b}$.

$$0 \le 2\left(x + \dfrac{\pi}{4}\right) \le 2\pi \Rightarrow 0 \le x + \dfrac{\pi}{4} \le \dfrac{2\pi}{2} \Rightarrow$$

$$0 \le x + \dfrac{\pi}{4} \le \pi \Rightarrow -\dfrac{\pi}{4} \le x \le \dfrac{3\pi}{4}$$

Step 2: Divide the period into four equal parts to get the following *x*-values: $-\dfrac{\pi}{4}, 0, \dfrac{\pi}{4},$

$\dfrac{\pi}{2}, \dfrac{3\pi}{4}$

Step 3: Evaluate the function for each of the five *x*-values.

x	$-\dfrac{\pi}{4}$	0	$\dfrac{\pi}{4}$	$\dfrac{\pi}{2}$	$\dfrac{3\pi}{4}$
$2\left(x + \dfrac{\pi}{4}\right)$	0	$\dfrac{\pi}{2}$	π	$\dfrac{3\pi}{2}$	2π
$\sin\left[2\left(x + \dfrac{\pi}{4}\right)\right]$	0	1	0	-1	0
$\dfrac{1}{2} + \sin\left[2\left(x + \dfrac{\pi}{4}\right)\right]$	$\dfrac{1}{2}$	$\dfrac{3}{2}$	$\dfrac{1}{2}$	$-\dfrac{1}{2}$	$\dfrac{1}{2}$

Steps 4 and 5: Plot the points found in the table and join them with a sinusoidal curve.

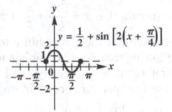

$y = \frac{1}{2} + \sin\left[2\left(x + \frac{\pi}{4}\right)\right]$

The amplitude is $|1|$, which is 1. The period is $\dfrac{2\pi}{2}$, which is π. The vertical translation is $\dfrac{1}{2}$ unit up. The phase shift is $\dfrac{\pi}{4}$ units to the left.

61. (a) Let January correspond to $x = 1$, February to $x = 2$, ..., and December of the second year to $x = 24$. Yes, the data appear to outline the graph of a translated sine graph.

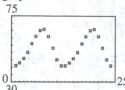

(b) The sine graph is vertically centered around the line $y = 53.5$. This line represents the average annual temperature in Seattle of 53.5°F. (This is also the actual average annual temperature.)

$y = 53.5$

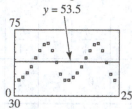

(c) The amplitude of the sine graph is 12.5 because the highest average monthly temperature is 66, the lowest average monthly temperature is 41, and

$\frac{1}{2}(66 - 41) = \frac{1}{2}(25) = 12.5$. The period is 12 because the temperature cycles every twelve months. Let $b = \frac{2\pi}{12} = \frac{\pi}{6}$. One

way to determine the phase shift is to use the following technique. The minimum temperature occurs in January, $x = 1$ and the hottest temperature occurs in August, $x = 8$. Then, $\frac{1}{2}(1 + 8) = 4.5$ gives a good

approximation for the phase shift.

(d) Let $f(x) = a \sin b(x - d) + c$. The amplitude is 12.5, so let $a = 12.5$. The period is equal to 1 yr or 12 mo, so

$b = \frac{\pi}{6}$. The average of the maximum and

minimum temperatures is

$\frac{1}{2}(66 + 41) = \frac{1}{2}(107) = 53.5$. Thus,

$f(x) = 12.5 \sin \left[\frac{\pi}{6}(x - 4.5) \right] + 53.5$

(e) Plotting the data with

$f(x) = 12.5 \sin \left[\frac{\pi}{6}(x - 4.5) \right] + 53.5$ on

the same coordinate axes gives a good fit.

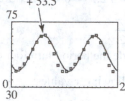

(f)

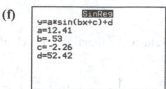

TI-84 Plus fixed to the nearest hundredth

From the sine regression we have

$y \approx 12.41 \sin(0.53x - 2.26) + 52.42$

63. (a) See the graph in part (c)

(b)

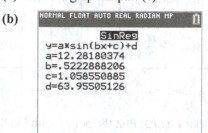

$y = 12.28 \sin(0.52x + 1.06) + 63.96$

(c)

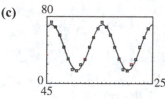

65. (a) See the graph in part (c)

(b)

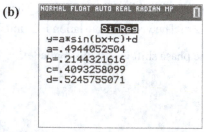

$y = 0.49 \sin(0.21x + 0.41) + 0.52$

(c)

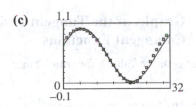

Chapter 4 Quiz

(Section 4.1–4.2)

1. $y = 3 - 4\sin\left(2x + \dfrac{\pi}{2}\right)$

$= 3 - 4\sin\left[2\left(x + \dfrac{\pi}{4}\right)\right] = -4\sin\left[2\left(x + \dfrac{\pi}{4}\right)\right] + 3$

Amplitude: 4; period: $\dfrac{2\pi}{2} = \pi$

Vertical translation: 3 units up

Phase shift: $\dfrac{\pi}{4}$ unit to the left

3. $y = -\dfrac{1}{2}\cos 2x$

Period: $\dfrac{2\pi}{2} = \pi$ and amplitude: $\left|-\dfrac{1}{2}\right| = \dfrac{1}{2}$

Divide the interval $[0, \pi]$ into four equal parts to get the x-values that will yield minimum and maximum points and x-intercepts. Make a table. Repeat this cycle for the interval $[-\pi, 0]$.

x	0	$\dfrac{\pi}{4}$	$\dfrac{\pi}{2}$	$\dfrac{3\pi}{4}$	π
$2x$	0	$\dfrac{\pi}{2}$	π	$\dfrac{3\pi}{2}$	2π
$\cos 2x$	1	0	-1	0	1
$-\dfrac{1}{2}\cos 2x$	$-\dfrac{1}{2}$	0	$\dfrac{1}{2}$	0	$-\dfrac{1}{2}$

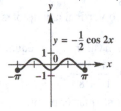

5. $y = -2\cos\left(x + \dfrac{\pi}{4}\right)$

Step 1: Find the interval whose length is $\dfrac{2\pi}{b}$.

$0 \le x + \dfrac{\pi}{4} \le 2\pi \Rightarrow 0 - \dfrac{\pi}{4} \le x \le 2\pi - \dfrac{\pi}{4} \Rightarrow$

$-\dfrac{\pi}{4} \le x \le \dfrac{7\pi}{4}$

Step 2: Divide the period into four equal parts to get the following x-values: $-\dfrac{\pi}{4}, \dfrac{\pi}{4},$

$\dfrac{3\pi}{4}, \dfrac{5\pi}{4}, \dfrac{7\pi}{4}$

Step 3: Evaluate the function for each of the five x-values

x	$-\dfrac{\pi}{4}$	$\dfrac{\pi}{4}$	$\dfrac{3\pi}{4}$	$\dfrac{5\pi}{4}$	$\dfrac{7\pi}{4}$
$x + \dfrac{\pi}{4}$	0	$\dfrac{\pi}{2}$	π	$\dfrac{3\pi}{2}$	2π
$\cos\left(x + \dfrac{\pi}{4}\right)$	1	0	-1	0	1
$-2\cos\left(x + \dfrac{\pi}{4}\right)$	-2	0	2	0	-2

Steps 4 and 5: Plot the points found in the table and join them with a sinusoidal curve. By graphing an additional period to the right, we obtain the following graph.

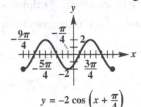

$$y = -2\cos\left(x + \dfrac{\pi}{4}\right)$$

The amplitude is 2. The period is 2π.

7. $y = -1 + \dfrac{1}{2}\sin x$

Period: 2π; amplitude: $\dfrac{1}{2}$

Vertical translation: 1 unit down
No phase shift.
This is the graph of $y = \sin x$ translated one unit down and compressed vertically by a factor of $\dfrac{1}{2}$.

(*continued on next page*)

(*continued*)

x	0	$\dfrac{\pi}{2}$	π	$\dfrac{3\pi}{2}$	2π
$\sin x$	0	1	0	-1	0
$\dfrac{1}{2}\sin x$	0	$\dfrac{1}{2}$	0	$-\dfrac{1}{2}$	0
$-1+\dfrac{1}{2}\sin x$	-1	$-\dfrac{1}{2}$	-1	$-\dfrac{3}{2}$	-1

Plot the points found in the table and join them with a sinusoidal curve. By graphing an additional period to the left, we obtain the following graph.

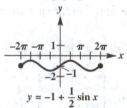

$y = -1 + \dfrac{1}{2}\sin x$

9. The amplitude is $\dfrac{1}{2}\left[1-(-1)\right]=\dfrac{1}{2}(2)=1$, so $a = 1$. One complete cycle of the graph is achieved in π units, so the period

$\pi = \dfrac{2\pi}{b} \Rightarrow b = \dfrac{2\pi}{\pi} = 2$. Comparing the given graph with the general sine and cosine curves, we see that this graph is a cosine curve. Substituting $a = 1$ and $b = 2$, the function is $y = \cos 2x$. Verify by confirming minimum and maximum points and x-intercepts from the graph:

$(0,1) \Rightarrow 1 = \cos(2 \cdot 0) = \cos 0 = 1$

$\left(\dfrac{\pi}{4},0\right) \Rightarrow 0 = \cos\left(2 \cdot \dfrac{\pi}{4}\right) = \cos\dfrac{\pi}{2} = 0$

$\left(\dfrac{\pi}{2},-1\right) \Rightarrow -1 = \cos\left(2 \cdot \dfrac{\pi}{2}\right) = \cos\pi = -1$

$\left(\dfrac{3\pi}{4},0\right) \Rightarrow 0 = \cos\left(2 \cdot \dfrac{3\pi}{4}\right) = \cos\dfrac{3\pi}{2} = 0$

$(\pi,1) \Rightarrow 1 = \cos(2\pi) = 1$

11. $f(x) = 12\sin\left[\dfrac{\pi}{6}(x-3.9)\right]+72$

April is represented by $x = 4$.

$f(4) = 12\sin\left[\dfrac{\pi}{6}(4-3.9)\right]+72 \approx 73°F$

Section 4.3 Graphs of the Tangent and Cotangent Functions

1. The least positive value x for which $\tan x = 0$ is <u>π</u>.

3. Between any two successive vertical asymptotes, the graph of $y = \tan x$ <u>increases</u>.

5. The negative value k with the greatest value for which $x = k$ is a vertical asymptote of the graph of $y = \tan x$ is <u>$-\dfrac{\pi}{2}$</u>.

7. $y = -\tan x$
The graph is the reflection of the graph of $y = \tan x$ about the x-axis. This matches with graph C.

9. $y = \tan\left(x - \dfrac{\pi}{4}\right)$

The graph is the graph of $y = \tan x$ shifted $\dfrac{\pi}{4}$ units to the right. This matches with graph B.

11. $y = \cot\left(x + \dfrac{\pi}{4}\right)$

The graph is the graph of $y = \cot x$ shifted $\dfrac{\pi}{4}$ units to the left. This matches with graph F.

13. $y = \tan 4x$
Step 1: Find the period and locate the vertical asymptotes. The period of tangent is $\dfrac{\pi}{b}$, so the period for this function is $\dfrac{\pi}{4}$. Tangent has asymptotes of the form $bx = -\dfrac{\pi}{2}$ and $bx = \dfrac{\pi}{2}$. Therefore, the asymptotes for $y = \tan 4x$ are

$4x = -\dfrac{\pi}{2} \Rightarrow x = -\dfrac{\pi}{8}$ and $4x = \dfrac{\pi}{2} \Rightarrow x = \dfrac{\pi}{8}$.

Step 2: Sketch the two vertical asymptotes found in Step 1.
Step 3: Divide the interval into four equal parts: $-\dfrac{\pi}{8}, -\dfrac{\pi}{16}, 0, \dfrac{\pi}{16}, \dfrac{\pi}{8}$

Step 4: Finding the first-quarter point, midpoint, and third-quarter point, we have

$\left(-\dfrac{\pi}{16}, -1\right),\ (0, 0),\ \left(\dfrac{\pi}{16}, 1\right)$

Step 5: Join the points with a smooth curve.

(*continued on next page*)

(*continued*)

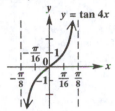

$y = \tan 4x$

15. $y = 2\tan x$

Step 1: Find the period and locate the vertical asymptotes. The period of tangent is $\dfrac{\pi}{b}$, so the period for this function is π. Tangent has asymptotes of the form $bx = -\dfrac{\pi}{2}$ and $bx = \dfrac{\pi}{2}$.

Therefore, the asymptotes for $y = 2\tan x$ are

$$x = -\dfrac{\pi}{2} \text{ and } x = \dfrac{\pi}{2}.$$

Step 2: Sketch the two vertical asymptotes found in Step 1.

Step 3: Divide the interval into four equal

parts: $-\dfrac{\pi}{2}, -\dfrac{\pi}{4}, 0, \dfrac{\pi}{4}, \dfrac{\pi}{2}$

Step 4: Finding the first-quarter point, midpoint, and third-quarter point, we have

$$\left(-\dfrac{\pi}{4}, -2\right), (0,0), \left(\dfrac{\pi}{4}, 2\right)$$

Step 5: Join the points with a smooth curve. The graph is "stretched" because $a = 2$ and $|2| > 1$.

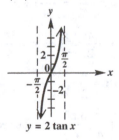

$y = 2\tan x$

17. $y = 2\tan\dfrac{1}{4}x$

Step 1: Find the period and locate the vertical asymptotes. The period of tangent is $\dfrac{\pi}{b}$, so the period for this function is 4π. Tangent has asymptotes of the form $bx = -\dfrac{\pi}{2}$ and $bx = \dfrac{\pi}{2}$.

Therefore, the asymptotes for $y = 2\tan\dfrac{1}{4}x$

are

$$\dfrac{1}{4}x = -\dfrac{\pi}{2} \Rightarrow x = -2\pi \text{ and } \dfrac{1}{4}x = \dfrac{\pi}{2} \Rightarrow x = 2\pi$$

Step 2: Sketch the two vertical asymptotes found in Step 1.

Step 3: Divide the interval into four equal parts: $-2\pi, -\pi, 0, \pi, 2\pi$

Step 4: Finding the first-quarter point, midpoint, and third-quarter point, we have:

$$(-\pi, -2), (0, 0), (\pi, 2)$$

Step 5: Join the points with a smooth curve.

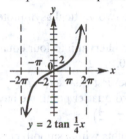

$y = 2\tan\frac{1}{4}x$

19. $y = \cot 3x$

Step 1: Find the period and locate the vertical asymptotes. The period of cotangent is $\dfrac{\pi}{b}$, so the period for this function is $\dfrac{\pi}{3}$. Cotangent has asymptotes of the form $bx = 0$ and $bx = \pi$. The asymptotes for $y = \cot 3x$ are

$$3x = 0 \Rightarrow x = 0 \text{ and } 3x = \pi \Rightarrow x = \dfrac{\pi}{3}$$

Step 2: Sketch the two vertical asymptotes found in Step 1.

Step 3: Divide the interval into four equal

parts: $0, \dfrac{\pi}{12}, \dfrac{\pi}{6}, \dfrac{\pi}{4}, \dfrac{\pi}{3}$

Step 4: Finding the first-quarter point, midpoint, and third-quarter point, we have

$$\left(\dfrac{\pi}{12}, 1\right), \left(\dfrac{\pi}{6}, 0\right), \left(\dfrac{\pi}{4}, -1\right)$$

Step 5: Join the points with a smooth curve.

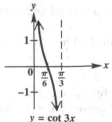

$y = \cot 3x$

21. $y = -2\tan\dfrac{1}{4}x$

Step 1: Find the period and locate the vertical asymptotes. The period of tangent is $\dfrac{\pi}{b}$, so the period for this function is 4π. Tangent has asymptotes of the form $bx = -\dfrac{\pi}{2}$ and $bx = \dfrac{\pi}{2}$.

Therefore, the asymptotes for $y = -2\tan\dfrac{1}{4}x$ are

$\dfrac{1}{4}x = -\dfrac{\pi}{2} \Rightarrow x = -2\pi$ and $\dfrac{1}{4}x = \dfrac{\pi}{2} \Rightarrow x = 2\pi$

Step 2: Sketch the two vertical asymptotes found in Step 1.
Step 3: Divide the interval into four equal parts: $-2\pi, -\pi, 0, \pi, 2\pi$
Step 4: Finding the first-quarter point, midpoint, and third-quarter point, we have $(-\pi, 2)$, $(0, 0)$, $(\pi, -2)$
Step 5: Join the points with a smooth curve.

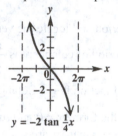

23. $y = \dfrac{1}{2}\cot 4x$

Step 1: Find the period and locate the vertical asymptotes. The period of cotangent is $\dfrac{\pi}{b}$, so the period for this function is $\dfrac{\pi}{4}$. Cotangent has asymptotes of the form $bx = 0$ and $bx = \pi$.

The asymptotes for $y = \dfrac{1}{2}\cot 4x$ are

$4x = 0 \Rightarrow x = 0$ and $4x = \pi \Rightarrow x = \dfrac{\pi}{4}$

Step 2: Sketch the two vertical asymptotes found in Step 1.
Step 3: Divide the interval into four equal parts: $0, \dfrac{\pi}{16}, \dfrac{\pi}{8}, \dfrac{3\pi}{16}, \dfrac{\pi}{4}$
Step 4: Finding the first-quarter point, midpoint, and third-quarter point, we have $\left(\dfrac{\pi}{16}, \dfrac{1}{2}\right)$, $\left(\dfrac{\pi}{8}, 0\right)$, $\left(\dfrac{3\pi}{16}, -\dfrac{1}{2}\right)$

Step 5: Join the points with a smooth curve.

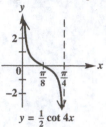

25. $y = \tan(2x - \pi) = \tan\left[2\left(x - \dfrac{\pi}{2}\right)\right]$

Period: $\dfrac{\pi}{b} = \dfrac{\pi}{2}$

Vertical translation: none

Phase shift (horizontal translation): $\dfrac{\pi}{2}$ units to the right
Because the function is to be graphed over a two-period interval, locate three adjacent vertical asymptotes. Asymptotes of the graph $y = \tan x$ occur at $-\dfrac{\pi}{2}, \dfrac{\pi}{2}$, and $\dfrac{3\pi}{2}$, so use the following equations to locate asymptotes:

$2\left(x - \dfrac{\pi}{2}\right) = -\dfrac{\pi}{2}$, $2\left(x - \dfrac{\pi}{2}\right) = \dfrac{\pi}{2}$, and $2\left(x - \dfrac{\pi}{2}\right) = \dfrac{3\pi}{2}$

Solve each of these equations:

$2\left(x - \dfrac{\pi}{2}\right) = -\dfrac{\pi}{2} \Rightarrow x - \dfrac{\pi}{2} = -\dfrac{\pi}{4} \Rightarrow x = \dfrac{\pi}{4}$

$2\left(x - \dfrac{\pi}{2}\right) = \dfrac{\pi}{2} \Rightarrow x - \dfrac{\pi}{2} = \dfrac{\pi}{4} \Rightarrow x = \dfrac{3\pi}{4}$

$2\left(x - \dfrac{\pi}{2}\right) = \dfrac{3\pi}{2} \Rightarrow x - \dfrac{\pi}{2} = \dfrac{3\pi}{4} \Rightarrow x = \dfrac{5\pi}{4}$

Divide the interval $\left(\dfrac{\pi}{4}, \dfrac{3\pi}{4}\right)$ into four equal parts to obtain the following key x-values:

first-quarter value: $\dfrac{3\pi}{8}$; middle value: $\dfrac{\pi}{2}$;

third-quarter value: $\dfrac{5\pi}{8}$

Evaluating the given function at these three key x-values gives the points $\left(\dfrac{3\pi}{8}, -1\right)$, $\left(\dfrac{\pi}{2}, 0\right)$, $\left(\dfrac{5\pi}{8}, 1\right)$

(continued on next page)

(continued)

Connect these points with a smooth curve and continue to graph to approach the asymptote $x = \dfrac{\pi}{4}$ and $x = \dfrac{3\pi}{4}$ to complete one period of the graph. Sketch the identical curve between the asymptotes $x = \dfrac{3\pi}{4}$ and $x = \dfrac{5\pi}{4}$ to complete a second period of the graph.

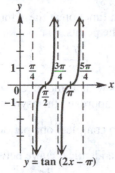

$y = \tan (2x - \pi)$

27. $y = \cot\left(3x + \dfrac{\pi}{4}\right) = \cot\left[3\left(x + \dfrac{\pi}{12}\right)\right]$

Period: $\dfrac{\pi}{b} = \dfrac{\pi}{3}$

Vertical translation: none

Phase shift (horizontal translation): $\dfrac{\pi}{12}$ unit to the left

Because the function is to be graphed over a two-period interval, locate three adjacent vertical asymptotes. Asymptotes of the graph $y = \cot x$ occur at multiples of π, so use the following equations to locate asymptotes:

$3\left(x + \dfrac{\pi}{12}\right) = 0$, $3\left(x + \dfrac{\pi}{12}\right) = \pi$, and

$3\left(x + \dfrac{\pi}{12}\right) = 2\pi$

Solve each of these equations:

$3\left(x + \dfrac{\pi}{12}\right) = 0 \Rightarrow x + \dfrac{\pi}{12} = 0 \Rightarrow x = -\dfrac{\pi}{12}$

$3\left(x + \dfrac{\pi}{12}\right) = \pi \Rightarrow x + \dfrac{\pi}{12} = \dfrac{\pi}{3} \Rightarrow$

$x = \dfrac{\pi}{3} - \dfrac{\pi}{12} = \dfrac{\pi}{4}$

$3\left(x + \dfrac{\pi}{12}\right) = 2\pi \Rightarrow x + \dfrac{\pi}{12} = \dfrac{2\pi}{3} \Rightarrow$

$x = \dfrac{2\pi}{3} - \dfrac{\pi}{12} \Rightarrow x = \dfrac{7\pi}{12}$

Divide the interval $\left(\dfrac{\pi}{4}, \dfrac{7\pi}{12}\right)$ into four equal parts to obtain the following key x-values:

first-quarter value: $\dfrac{\pi}{3}$; middle value: $\dfrac{5\pi}{12}$;

third-quarter value: $\dfrac{\pi}{2}$

Evaluating the given function at these three key x-values gives the points.

$\left(\dfrac{\pi}{3}, 1\right)$, $\left(\dfrac{5\pi}{12}, 0\right)$, $\left(\dfrac{\pi}{2}, -1\right)$

Connect these points with a smooth curve and continue to graph to approach the asymptote $x = \dfrac{\pi}{4}$ and $x = \dfrac{7\pi}{12}$ to complete one period of the graph. Sketch the identical curve between the asymptotes $x = -\dfrac{\pi}{12}$ and $x = \dfrac{\pi}{4}$ to complete a second period of the graph.

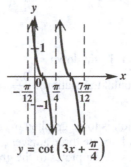

$y = \cot\left(3x + \dfrac{\pi}{4}\right)$

29. $y = 1 + \tan x$

This is the graph of $y = \tan x$ translated vertically 1 unit up.

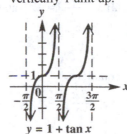

$y = 1 + \tan x$

31. $y = 1 - \cot x$

This is the graph of $y = \cot x$ reflected about the x-axis and then translated vertically 1 unit up.

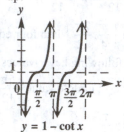

$y = 1 - \cot x$

33. $y = -1 + 2\tan x$

This is the graph of $y = 2\tan x$ translated vertically 1 unit down.

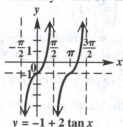

$y = -1 + 2\tan x$

35. $y = -1 + \dfrac{1}{2}\cot(2x - 3\pi) = -1 + \dfrac{1}{2}\cot\left[2\left(x - \dfrac{3\pi}{2}\right)\right]$

Period: $\dfrac{\pi}{b} = \dfrac{\pi}{2}$.

Vertical translation: 1 unit down

Phase shift (horizontal translation): $\dfrac{3\pi}{2}$ units to the right

Because the function is to be graphed over a two-period interval, locate three adjacent vertical asymptotes. Asymptotes of the graph $y = \cot x$ occur at multiples of π, use the following equations to locate asymptotes:

$2\left(x - \dfrac{3\pi}{2}\right) = -2\pi, \ 2\left(x - \dfrac{3\pi}{2}\right) = -\pi,$ and

$2\left(x - \dfrac{3\pi}{2}\right) = 0$

Solve each of these equations:

$2\left(x - \dfrac{3\pi}{2}\right) = -2\pi \Rightarrow x - \dfrac{3\pi}{2} = -\pi \Rightarrow$

$x = -\pi + \dfrac{3\pi}{2} = \dfrac{\pi}{2}$

$2\left(x - \dfrac{3\pi}{2}\right) = -\pi \Rightarrow x - \dfrac{3\pi}{2} = -\dfrac{\pi}{2} \Rightarrow$

$x = -\dfrac{\pi}{2} + \dfrac{3\pi}{2} \Rightarrow x = \dfrac{2\pi}{2} = \pi$

$2\left(x - \dfrac{3\pi}{2}\right) = 0 \Rightarrow x - \dfrac{3\pi}{2} = 0 \Rightarrow x = \dfrac{3\pi}{2}$

Divide the interval $\left(\dfrac{\pi}{2}, \pi\right)$ into four equal parts to obtain the following key x-values:

first-quarter value: $\dfrac{5\pi}{8}$; middle value: $\dfrac{3\pi}{4}$;

third-quarter value: $\dfrac{7\pi}{8}$

Evaluating the given function at these three key x-values gives the points.

$\left(\dfrac{5\pi}{8}, -\dfrac{1}{2}\right), \left(\dfrac{3\pi}{4}, -1\right), \left(\dfrac{7\pi}{8}, -\dfrac{3}{2}\right)$

Connect these points with a smooth curve and continue to graph to approach the asymptote

$x = \dfrac{\pi}{2}$ and $x = \pi$ to complete one period of the graph. Sketch the identical curve between the asymptotes $x = \pi$ and $x = \dfrac{3\pi}{2}$ to complete a second period of the graph.

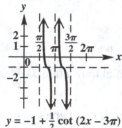

$y = -1 + \frac{1}{2}\cot(2x - 3\pi)$

37. $y = 1 - 2\cot\left[2\left(x + \dfrac{\pi}{2}\right)\right]$

Period: $\dfrac{\pi}{b} = \dfrac{\pi}{2}$

Vertical translation: 1 unit up

Phase shift (horizontal translation): $\dfrac{\pi}{2}$ unit left

Because the function is to be graphed over a two-period interval, locate three adjacent vertical asymptotes. Asymptotes of the graph $y = \cot x$ occur at multiples of π, so use the following equations to locate asymptotes.

$2\left(x + \dfrac{\pi}{2}\right) = 0, \ 2\left(x + \dfrac{\pi}{2}\right) = \pi,$ and

$2\left(x + \dfrac{\pi}{2}\right) = 2\pi$

Solve each of these equations:

$2\left(x + \dfrac{\pi}{2}\right) = 0 \Rightarrow x + \dfrac{\pi}{2} = 0 \Rightarrow x = 0 - \dfrac{\pi}{2} = -\dfrac{\pi}{2}$

(continued on next page)

(continued)

$$2\left(x+\frac{\pi}{2}\right)=\pi \Rightarrow x+\frac{\pi}{2}=\frac{\pi}{2} \Rightarrow$$

$$x=\frac{\pi}{2}-\frac{\pi}{2} \Rightarrow x=0$$

$$2\left(x+\frac{\pi}{2}\right)=2\pi \Rightarrow x+\frac{\pi}{2}=\pi \Rightarrow$$

$$x=\pi-\frac{\pi}{2}=\frac{\pi}{2}$$

Divide the interval $\left(0,\dfrac{\pi}{2}\right)$ into four equal

parts to obtain the following key x-values:

first-quarter value: $\dfrac{\pi}{8}$; middle value: $\dfrac{\pi}{4}$;

third-quarter value: $\dfrac{3\pi}{8}$

Evaluating the given function at these three key x-values gives the points

$$\left(\frac{\pi}{8},-1\right), \left(\frac{\pi}{4},1\right), \left(\frac{3\pi}{8},3\right)$$

Connect these points with a smooth curve and continue to graph to approach the asymptote

$x=0$ and $x=\dfrac{\pi}{2}$ to complete one period of

the graph. Sketch the identical curve between

the asymptotes $x=-\dfrac{\pi}{2}$ and $x=0$ to

complete a second period of the graph.

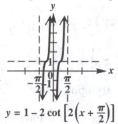

$y = 1 - 2\cot\left[2\left(x+\frac{\pi}{2}\right)\right]$

39. The asymptotes are at $-\dfrac{\pi}{2}, \dfrac{\pi}{2}$, and $\dfrac{3\pi}{2}$, so

this is a tangent function of the form
$y = a\tan x$. The graph passes through the point

$\left(\dfrac{\pi}{4},-2\right)$. Substituting these values into the

generic equation gives

$$y=a\tan x \Rightarrow -2=a\tan\frac{\pi}{4} \Rightarrow -2=a\cdot 1 \Rightarrow$$

$$-2=a$$

Thus, the equation of the graph is $y = -2\tan x$.

41. The asymptotes occur at $0, \dfrac{\pi}{3}$, and $\dfrac{2\pi}{3}$, so

this is a cotangent function of the form

$y = a\cot bx$. The period of the function is $\dfrac{\pi}{3}$,

so we have $\dfrac{\pi}{b}=\dfrac{\pi}{3} \Rightarrow b=3$. The graph passes

through the point $\left(\dfrac{\pi}{12},1\right)$.

Substituting these values into the generic equation gives

$$y=a\cot bx \Rightarrow 1=a\cot\left(3\cdot\frac{\pi}{12}\right)\Rightarrow$$

$$1=a\cot\frac{\pi}{4} \Rightarrow 1=a\cdot 1 \Rightarrow 1=a$$

Thus, the equation of the graph is $y = \cot 3x$.

43. Since the asymptotes occur at $-\pi$ and π, this is a tangent function with period 2π instead of π. Therefore, the coefficient of x is $\frac{1}{2}$. The graph is vertically translated 1 unit up compared to the graph of $y = \tan\frac{1}{2}x$, so an equation for this graph is $y = 1 + \tan\frac{1}{2}x$.

45. True; $\dfrac{\pi}{2}$ is the smallest positive value where

$\cos\dfrac{\pi}{2}=0$. Since $\tan\dfrac{\pi}{2}=\dfrac{\sin\dfrac{\pi}{2}}{\cos\dfrac{\pi}{2}}$, $\dfrac{\pi}{2}$ is the

smallest positive value where the tangent

function is undefined. Thus, $k = \dfrac{\pi}{2}$ is the

smallest positive value for which $x = k$ is an asymptote for the tangent function.

47. False; $\tan(-x)=\dfrac{\sin(-x)}{\cos(-x)}=\dfrac{-\sin x}{\cos x}=-\tan x$

(because $\sin x$ is odd and $\cos x$ is even) for all x in the domain. Moreover, if $\tan(-x)=\tan x$, then the graph would be symmetric about the y-axis, which it is not.

49. The function $\tan x$ has a period of π, so it repeats four times over the interval $(-2\pi, 2\pi]$. Since its range is $(-\infty,\infty)$, $\tan x = c$ has four solutions for every value of c.

51. $\tan(-x) = \dfrac{\sin(-x)}{\cos(-x)} = \dfrac{-\sin x}{\cos x} = -\tan x$,

$\left\{ x \mid x \neq (2n+1)\dfrac{\pi}{4}, \text{ where } n \text{ is any integer} \right\}.$

53. $d = 4\tan\left[2\pi(0)\right] = 4\tan 0 \approx 4(0) = 0 \text{ m}$

55. $d = 4\tan\left[2\pi(1.2)\right] = 4\tan(2.4\pi)$
$\approx 4(3.0777) \approx 12.3 \text{ m}$

57. The least positive number for which $y = \cot x$ is undefined is $x = \pi$.

59. The vertical asymptotes in general occur at

$x = \dfrac{5\pi}{4} + n\pi$, where n is an integer.

61. $0.32175055 + \pi \approx 3.4633432$

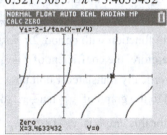

The next positive x-intercept is $(3.46, 0)$.

Section 4.4 Graphs of the Secant and Cosecant Functions

1. A   **3.** D   **5.** C

7. $y = -\csc x$
The graph is the reflection of the graph of $y = \csc x$ about the x-axis. This matches with graph B.

9. $y = \sec\left(x - \dfrac{\pi}{2}\right)$

The graph is the graph of $y = \sec x$ shifted $\dfrac{\pi}{2}$ units to the right. This matches with graph D.

11. $y = 3\sec\dfrac{1}{4}x$

Step 1: Graph the corresponding reciprocal function $y = 3\cos\dfrac{1}{4}x$. The period is

$\dfrac{2\pi}{\frac{1}{4}} = 2\pi \cdot \dfrac{4}{1} = 8\pi$ and its amplitude is $|3| = 3$.

One period is in the interval $0 \leq x \leq 8\pi$.

Dividing the interval into four equal parts gives us the following key points: $(0, 1)$, $(2\pi, 0)$, $(4\pi, -1)$, $(6\pi, 0)$, $(8\pi, 1)$

Step 2: The vertical asymptotes of $y = \sec\dfrac{1}{4}x$

are at the x-intercepts of $y = \cos\dfrac{1}{4}x$, which

are $x = 2\pi$ and $x = 6\pi$. Continuing this pattern to the left, we also have a vertical asymptote of $x = -2\pi$.

Step 3: Sketch the graph.

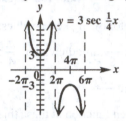

13. $y = -\dfrac{1}{2}\csc\left(x + \dfrac{\pi}{2}\right)$

Step 1: Graph the corresponding reciprocal

function $y = -\dfrac{1}{2}\sin\left(x + \dfrac{\pi}{2}\right)$. The period is

2π and its amplitude is $\left|-\dfrac{1}{2}\right| = \dfrac{1}{2}$. One period

is in the interval $-\dfrac{\pi}{2} \leq x \leq \dfrac{3\pi}{2}$.

Dividing the interval into four equal parts

gives us the following key points: $\left(-\dfrac{\pi}{2}, 0\right)$,

$\left(0, -\dfrac{1}{2}\right)$, $\left(\dfrac{\pi}{2}, 0\right)$, $\left(\pi, \dfrac{1}{2}\right)$, $\left(\dfrac{3\pi}{2}, 0\right)$

Step 2: The vertical asymptotes of

$y = -\dfrac{1}{2}\csc\left(x + \dfrac{\pi}{2}\right)$ are at the x-intercepts of

$y = -\dfrac{1}{2}\sin\left(x + \dfrac{\pi}{2}\right)$, which are $x = -\dfrac{\pi}{2}$,

$x = \dfrac{\pi}{2}$, and $x = \dfrac{3\pi}{2}$.

Step 3: Sketch the graph.

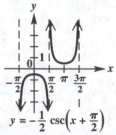

15. $y = \csc\left(x - \dfrac{\pi}{4}\right)$

Step 1: Graph the corresponding reciprocal

function $y = \sin\left(x - \dfrac{\pi}{4}\right)$ The period is 2π

and its amplitude is $|1| = 1$. One period is in

the interval $\dfrac{\pi}{4} \le x \le \dfrac{9\pi}{4}$. Dividing the

interval into four equal parts gives us the

following key points: $\left(\dfrac{\pi}{4}, 0\right)$, $\left(\dfrac{3\pi}{4}, 1\right)$,

$\left(\dfrac{5\pi}{4}, 0\right)$, $\left(\dfrac{7\pi}{4}, -1\right)$, $\left(\dfrac{9\pi}{4}, 0\right)$

Step 2: The vertical asymptotes of

$y = \csc\left(x - \dfrac{\pi}{4}\right)$ are at the x-intercepts of

$y = \sin\left(x - \dfrac{\pi}{4}\right)$, which are $x = \dfrac{\pi}{4}$, $x = \dfrac{5\pi}{4}$,

and $x = \dfrac{9\pi}{4}$.

Step 3: Sketch the graph.

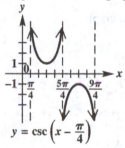

$y = \csc\left(x - \dfrac{\pi}{4}\right)$

17. $y = \sec\left(x + \dfrac{\pi}{4}\right)$

Step 1: Graph the corresponding reciprocal

function $y = \cos\left(x + \dfrac{\pi}{4}\right)$. The period is 2π

and its amplitude is $|1| = 1$. One period is in

the interval $-\dfrac{\pi}{4} \le x \le \dfrac{7\pi}{4}$. Dividing the

interval into four equal parts gives us the

following key points: $\left(-\dfrac{\pi}{4}, 1\right)$, $\left(\dfrac{\pi}{4}, 0\right)$,

$\left(\dfrac{3\pi}{4}, -1\right)$, $\left(\dfrac{5\pi}{4}, 0\right)$, $\left(\dfrac{7\pi}{4}, 1\right)$

Step 2: The vertical asymptotes of

$y = \sec\left(x + \dfrac{\pi}{4}\right)$ are at the x-intercepts of

$y = \cos\left(x + \dfrac{\pi}{4}\right)$, which are $x = \dfrac{\pi}{4}$ and

$x = \dfrac{5\pi}{4}$. Continuing this pattern to the right,

we also have a vertical asymptote of $x = \dfrac{9\pi}{4}$.

Step 3: Sketch the graph.

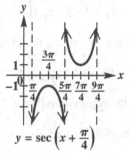

$y = \sec\left(x + \dfrac{\pi}{4}\right)$

19. $y = \csc\left(\dfrac{1}{2}x - \dfrac{\pi}{4}\right) = \csc\left[\dfrac{1}{2}\left(x - \dfrac{\pi}{2}\right)\right]$

Step 1: Graph the corresponding reciprocal

function $y = \sin\left[\dfrac{1}{2}\left(x - \dfrac{\pi}{2}\right)\right]$. The period is

$\dfrac{2\pi}{\frac{1}{2}} = 2\pi \cdot \dfrac{2}{1} = 4\pi$ and its amplitude is $|1| = 1$.

One period is in the interval $\dfrac{\pi}{2} \le x \le \dfrac{9\pi}{2}$.

Dividing the interval into four equal parts

gives us the following key points: $\left(\dfrac{\pi}{2}, 0\right)$,

$\left(\dfrac{3\pi}{2}, 1\right)$, $\left(\dfrac{5\pi}{2}, 0\right)$, $\left(\dfrac{7\pi}{2}, -1\right)$, $\left(\dfrac{9\pi}{2}, 0\right)$

Step 2: The vertical asymptotes of

$y = \csc\left[\dfrac{1}{2}\left(x - \dfrac{\pi}{2}\right)\right]$ are at the x-intercepts of

$y = \sin\left[\dfrac{1}{2}\left(x - \dfrac{\pi}{2}\right)\right]$, which are $x = \dfrac{\pi}{2}$,

$x = \dfrac{5\pi}{2}$, and $x = \dfrac{9\pi}{2}$.

Step 3: Sketch the graph.

(continued on next page)

(*continued*)

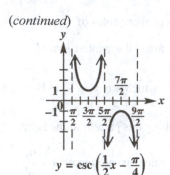

$$y = \csc\left(\frac{1}{2}x - \frac{\pi}{4}\right)$$

21. $y = 2 + 3\sec(2x - \pi) = 2 + 3\sec\left[2\left(x - \frac{\pi}{2}\right)\right]$

Step 1: Graph the corresponding reciprocal

function $y = 2 + 3\cos\left[2\left(x - \frac{\pi}{2}\right)\right]$. The

period is π and its amplitude is $|3| = 3$. One

period is in the interval $\frac{\pi}{2} \le x \le \frac{3\pi}{2}$.

Dividing the interval into four equal parts
gives us the following key points:

$$\left(\frac{\pi}{2}, 5\right), \left(\frac{3\pi}{4}, 2\right), \ (\pi, -1), \left(\frac{5\pi}{4}, 2\right), \left(\frac{3\pi}{2}, 5\right)$$

Step 2: The vertical asymptotes of

$y = 2 + 3\sec\left[2\left(x - \frac{\pi}{2}\right)\right]$ are at the x-

intercepts of $y = 3\cos 2\left(x - \frac{\pi}{2}\right)$, which are

$x = \frac{3\pi}{4}$ and $x = \frac{5\pi}{4}$. Continuing this pattern

to the left, we also have a vertical asymptote

of $x = \frac{\pi}{4}$.

Step 3: Sketch the graph.

$$y = 2 + 3\sec(2x - \pi)$$

23. $y = 1 - \frac{1}{2}\csc\left(x - \frac{3\pi}{4}\right)$

Step 1: Graph the corresponding reciprocal

function $y = 1 - \frac{1}{2}\sin\left(x - \frac{3\pi}{4}\right)$. The period is

2π and its amplitude is $\frac{1}{2}$.

One period is in the interval $\frac{3\pi}{4} \le x \le \frac{11\pi}{4}$.

Dividing the interval into four equal parts
gives us the following key points:

$$\left(\frac{3\pi}{4}, 1\right), \left(\frac{5\pi}{4}, \frac{1}{2}\right),$$

$$\left(\frac{7\pi}{4}, 1\right), \left(\frac{9\pi}{4}, \frac{3}{2}\right), \left(\frac{11\pi}{4}, 1\right)$$

Step 2: The vertical asymptotes of

$y = 1 - \frac{1}{2}\csc\left(x - \frac{3\pi}{4}\right)$ are at the x-intercepts

of $y = -\frac{1}{2}\sin\left(x - \frac{3\pi}{4}\right)$, which are $x = \frac{3\pi}{4}$,

$x = \frac{7\pi}{4}$, and $x = \frac{11\pi}{4}$.

Step 3: Sketch the graph.

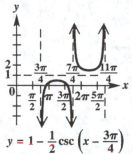

$$y = 1 - \frac{1}{2}\csc\left(x - \frac{3\pi}{4}\right)$$

For exercises 25–29, other answers are possible.

25.   Since the graph crosses the y-axis at $(0, 1)$, this
is a secant graph with $a = 1$. The period is

$$\left|-\frac{\pi}{4} - \frac{\pi}{4}\right| = \left|-\frac{\pi}{2}\right| = \frac{\pi}{2}. \text{ Thus,}$$

$$b = \frac{2\pi}{\frac{\pi}{2}} \Rightarrow b = 4. \text{ The equation of the graph is}$$

$y = \sec 4x$.

27.   This is the graph of $y = \csc x$ translated two
units down. Thus, the equation of the graph is
$y = -2 + \csc x$.

29.   This is the graph of $y = \sec x$, reflected across
the x-axis and translated one unit down. Thus,
the equation of the graph is $y = -1 - \sec x$.

31. True. Because $\tan x = \dfrac{\sin x}{\cos x}$ and

$\sec x = \dfrac{1}{\cos x}$, the tangent and secant

functions will be undefined at the same values.

33. True. $\sec(-x) = \dfrac{1}{\cos(-x)} = \dfrac{1}{\cos(x)} = \sec(x)$

(because $\cos x$ is even) for all x in the domain. Moreover, if $\sec(-x) = \sec x$, then the graph would be symmetric about the y-axis, which it is.

35. None; $|\cos x| \le 1$ for all x, so

$\dfrac{1}{|\cos x|} \ge 1$ and $|\sec x| \ge 1$. Because $|\sec x| \ge 1$,

$\sec x$ has no values in the interval $(-1, 1)$.

37. $\sec(-x) = \dfrac{1}{\cos(-x)} = \dfrac{1}{\cos(x)} = \sec(x)$,

$\left\{ x \mid x \ne (2n+1)\dfrac{\pi}{2}, \text{ where } n \text{ is any integer} \right\}.$

39. $t = 0$

$a = 4\left| \sec 0 \right| = 4\left| 1 \right| = 4(1) = 4$ m

41. $t = 1.24$

$a = 4\left| \sec\left[2\pi(1.24) \right] \right| \approx 4\left| 15.9260 \right|$

$= 4(15.9260) \approx 63.7$ m

43. $y_1 = \sin x$; $y_2 = \sin 2x$; $y_3 = y_1 + y_2$
Graph the functions in the window
$[-2\pi, 2\pi] \times [-3, 3]$.

y_1 is in black, y_2 is in dark grey, and y_3 is in light grey.

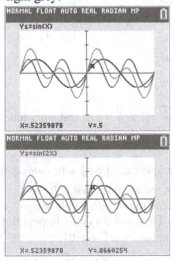

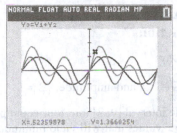

$Y_1\left(\dfrac{\pi}{6}\right) + Y_2\left(\dfrac{\pi}{6}\right) \approx 0.5 + 0.8660254$

$\qquad\qquad = 1.3660254$

$\qquad\qquad = Y_3\left(\dfrac{\pi}{6}\right) = \left(Y_1 + Y_2\right)\left(\dfrac{\pi}{6}\right)$

45. $y_1 = \tan x$; $y_2 = \sec x$; $y_3 = y_1 + y_2$
Graph the functions in the window
$[-2\pi, 2\pi] \times [-3, 3]$.

y_1 is in black, y_2 is in dark grey, and y_3 is in light grey.

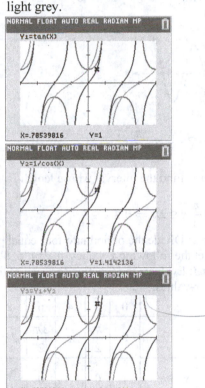

$Y_1\left(\dfrac{\pi}{4}\right) + Y_2\left(\dfrac{\pi}{4}\right) \approx 1 + 1.4142136 = 2.4142136$

$\qquad\qquad = Y_3\left(\dfrac{\pi}{4}\right) = \left(Y_1 + Y_2\right)\left(\dfrac{\pi}{4}\right)$

Summary Exercises on Graphing Circular Functions

1. $y = 2 \sin \pi x$

Period: $\dfrac{2\pi}{\pi} = 2$ and amplitude: $|2| = 2$

Divide the interval [0, 2] into four equal parts to get the x-values that will yield minimum and maximum points and x-intercepts. Then make a table.

x	0	$\dfrac{1}{2}$	1	$\dfrac{3}{2}$	2
πx	0	$\dfrac{\pi}{2}$	π	$\dfrac{3\pi}{2}$	2π
$\sin \pi x$	0	1	0	-1	0
$2 \sin \pi x$	0	2	0	-2	0

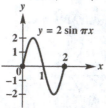

3. $y = -2 + \dfrac{1}{2} \cos \dfrac{\pi}{4} x$

Step 1: Find the interval whose length is $\dfrac{2\pi}{b}$.

$$0 \le \dfrac{\pi}{4} x \le 2\pi \Rightarrow 0 \le x \le 8$$

Step 2: Divide the period into four equal parts to get the following x-values: 0, 2, 4, 6, 8

Step 3: Evaluate the function for each of the five x-values.

x	0	2	4	6	8
$\dfrac{\pi}{4} x$	0	$\dfrac{\pi}{2}$	π	$\dfrac{3\pi}{2}$	2π
$\cos \dfrac{\pi}{4} x$	1	0	-1	0	1
$\dfrac{1}{2} \cos \dfrac{\pi}{4} x$	$\dfrac{1}{2}$	0	$-\dfrac{1}{2}$	0	$\dfrac{1}{2}$
$-2 + \dfrac{1}{2} \cos \dfrac{\pi}{4} x$	$-\dfrac{3}{2}$	-2	$-\dfrac{5}{2}$	-2	$-\dfrac{3}{2}$

Steps 4 and 5: Plot the points found in the table and join them with a sinusoidal curve. By graphing an additional period to the left, we obtain the following graph.

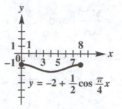

The amplitude is $\left|\dfrac{1}{2}\right| = \dfrac{1}{2}$. The period is 8. The vertical translation is 2 units down. There is no phase shift.

5. $y = -4 \csc \dfrac{1}{2} x$

Step 1: Graph the corresponding reciprocal function $y = -4 \sin \dfrac{1}{2} x$ The period is

$\dfrac{2\pi}{\frac{1}{2}} = 2\pi \cdot \dfrac{2}{1} = 4\pi$ and its amplitude is

$|-4| = 4.$

One period is in the interval $0 \le x \le 4\pi$. Dividing the interval into four equal parts gives us the following key points: $(0, 0)$, $(\pi, -4)$, $(2\pi, 0)$, $(3\pi, 4)$, $(4\pi, 0)$

Step 2: The vertical asymptotes of

$y = -4 \csc \dfrac{1}{2} x$ are at the x-intercepts of

$y = -4 \sin \dfrac{1}{2} x$, which are $x = 0$, $x = 2\pi$, and

$x = 4\pi.$

Step 3: Sketch the graph.

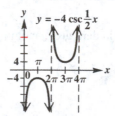

7. $y = -5 \sin \dfrac{x}{3}$

Period: $\dfrac{2\pi}{\frac{1}{3}} = 2\pi \cdot \dfrac{3}{1} = 6\pi$ and amplitude:

$|-5| = 5$

Divide the interval $[0, 6\pi]$ into four equal parts to get the x-values that will yield minimum and maximum points and x-intercepts. Then make a table. Repeat this cycle for the interval $[-6\pi, 0]$.

(continued on next page)

(*continued*)

x	0	$\dfrac{3\pi}{2}$	3π	$\dfrac{9\pi}{2}$	6π
$\dfrac{x}{3}$	0	$\dfrac{\pi}{2}$	π	$\dfrac{3\pi}{2}$	2π
$\sin\dfrac{x}{3}$	0	1	0	-1	0
$-5\sin\dfrac{x}{3}$	0	-5	0	5	0

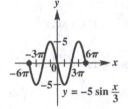

9. $y = 3 - 4\sin\left(\dfrac{5}{2}x + \pi\right) = 3 - 4\sin\left[\dfrac{5}{2}\left(x + \dfrac{2\pi}{5}\right)\right]$

Step 1: Find the interval whose length is $\dfrac{2\pi}{b}$.

$0 \le \dfrac{5}{2}\left(x + \dfrac{2\pi}{5}\right) \le 2\pi \Rightarrow$

$0 \le x + \dfrac{2\pi}{5} \le \dfrac{4\pi}{5} \Rightarrow -\dfrac{2\pi}{5} \le x \le \dfrac{2\pi}{5}$

Step 2: Divide the period into four equal parts to get the following x-values: $-\dfrac{2\pi}{5}, -\dfrac{\pi}{5}, 0,$

$\dfrac{\pi}{5}, \dfrac{2\pi}{5}$

Step 3: Evaluate the function for each of the five x-values

x	$-\dfrac{2\pi}{5}$	$-\dfrac{\pi}{5}$	0	$\dfrac{\pi}{5}$	$\dfrac{2\pi}{5}$
$\dfrac{5}{2}\left(x + \dfrac{2\pi}{5}\right)$	0	$\dfrac{\pi}{2}$	π	$\dfrac{3\pi}{2}$	2π
$\sin\left[\dfrac{5}{2}\left(x + \dfrac{2\pi}{5}\right)\right]$	0	1	0	-1	0
$-4\sin\left[\dfrac{5}{2}\left(x + \dfrac{2\pi}{5}\right)\right]$	0	-4	0	4	0
$3 - 4\sin\left[\dfrac{5}{2}\left(x + \dfrac{2\pi}{5}\right)\right]$	3	-1	3	7	3

Steps 4 and 5: Plot the points found in the table and join them with a sinusoidal curve. By graphing an additional period to the right, we obtain the following graph.

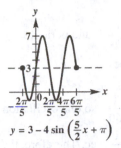

$y = 3 - 4\sin\left(\dfrac{5}{2}x + \pi\right)$

The amplitude is $\left|-4\right|$, which is 4. The period is $\dfrac{2\pi}{\frac{5}{2}}$, which is $\dfrac{4\pi}{5}$. The vertical translation is 3 units up. The phase shift is $\dfrac{2\pi}{5}$ units to the left.

Section 4.5   Harmonic Motion

1. The amplitude is 5.

3. The frequency is $\frac{2}{2\pi} = \frac{1}{\pi}$ oscillation per second.

5. $s\left(\dfrac{\pi}{2}\right) = 5\cos\left(2 \cdot \dfrac{\pi}{2}\right) = 5(-1) = -5$

7. (a) The object is pulled down 4 units, so $s(0) = -4$. Thus, we have
    $s(0) = -4 = a\cos\left[\omega(0)\right] \Rightarrow$
    $-4 = a\cos 0 \Rightarrow -4 = a(1) \Rightarrow a = -4$
    The time it takes to complete one oscillation is 3 sec, so $P = 3$ sec.
    $P = 3 \text{ sec} \Rightarrow 3 = \dfrac{2\pi}{\omega} \Rightarrow 3 = \dfrac{2\pi}{\omega} \Rightarrow$
    $3\omega = 2\pi \Rightarrow \omega = \dfrac{2\pi}{3}$
    Therefore, $s(t) = -4\cos\dfrac{2\pi}{3}t$.

    (b) $s(1) = -4\cos\left[\dfrac{2\pi}{3}(1.25)\right]$
    $= -4\cos\left[\dfrac{2\pi}{3}\left(\dfrac{5}{4}\right)\right] = -4\cos\dfrac{5\pi}{6}$
    $= -4\left(-\dfrac{\sqrt{3}}{2}\right) = 2\sqrt{3} \approx 3.46$ units

    (c) The frequency is the reciprocal of the period, or $\dfrac{1}{3}$ oscillation per second.

9. $E = 5\cos 120\pi t$

    (a) Amplitude: $|5| = 5$ and period:

$$\frac{2\pi}{120\pi} = \frac{1}{60} \text{ sec}$$

    (b) The period is $\dfrac{1}{60}$, so one oscillation is

completed in $\dfrac{1}{60}$ sec. Therefore, the frequency is 60 oscillations per second.

    (c) $t = 0,\ E = 5\cos 120\pi (0) = 5\cos 0 = 5(1) = 5$

$t = 0.03,$
$E = 5\cos 120\pi (0.03) = 5\cos 3.6\pi \approx 1.545$

$t = 0.06,$
$E = 5\cos 120\pi (0.06) = 5\cos 7.2\pi \approx -4.045$

$t = 0.09,$
$E = 5\cos 120\pi (0.09)$
$\quad = 5\cos 10.8\pi \approx -4.045$

$t = 0.12,$
$E = 5\cos 120\pi (0.12) = 5\cos 14.4\pi \approx 1.545$

    (d)

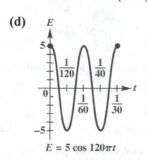

$E = 5\cos 120\pi t$

11. (a) $a = 2, \omega = 2$

$s(t) = a\sin \omega t \Rightarrow s(t) = 2\sin 2t$

amplitude $= |a| = |2| = 2$; period

$= \dfrac{2\pi}{\omega} = \dfrac{2\pi}{2} = \pi$; frequency $= \dfrac{\omega}{2\pi} = \dfrac{1}{\pi}$
rotation per second

(b) $a = 2, \omega = 4$

$s(t) = a\sin \omega t \Rightarrow s(t) = 2\sin 4t$

amplitude $= |a| = |2| = 2$; period

$= \dfrac{2\pi}{\omega} = \dfrac{2\pi}{4} = \dfrac{\pi}{2}$; frequency

$= \dfrac{\omega}{2\pi} = \dfrac{4}{2\pi} = \dfrac{2}{\pi}$ rotation per second

13. $P = 2\pi\sqrt{\dfrac{L}{32}}; L = \dfrac{1}{2}$ ft

The period is

$$P = 2\pi\sqrt{\frac{L}{32}} \Rightarrow P = 2\pi\sqrt{\frac{\frac{1}{2}}{32}} = 2\pi\sqrt{\frac{1}{64}}$$

$$= 2\pi \cdot \frac{1}{8} = \frac{\pi}{4} \text{ sec}$$

The frequency is the reciprocal of the period,

or $\dfrac{4}{\pi}$ oscillations per second.

15. $s(t) = a\sin\sqrt{\dfrac{k}{m}}\,t; k = 4; P = 1$ sec

A period of 1 sec is produced when $\dfrac{2\pi}{\sqrt{\dfrac{k}{m}}} = 1$

Since $k = 4$, we can solve

$$\frac{2\pi}{\sqrt{\dfrac{k}{m}}} = 1 \Rightarrow \frac{2\pi}{\sqrt{\dfrac{4}{m}}} = 1 \Rightarrow 2\pi = \sqrt{\frac{4}{m}} \Rightarrow$$

$$4\pi^2 = \frac{4}{m} \Rightarrow 4\pi^2 m = 4 \Rightarrow m = \frac{4}{4\pi^2} = \frac{1}{\pi^2}$$

17. $s(t) = -5\cos 4\pi t$, $a = |-5| = 5,\ \omega = 4\pi$

    (a) maximum height = amplitude
$$= a = |-5| = 5 \text{ in.}$$

    (b) frequency
$$= \frac{\omega}{2\pi} = \frac{4\pi}{2\pi} = 2 \text{ cycles per sec; period}$$
$$= \frac{2\pi}{\omega} = \frac{1}{2} \text{ sec}$$

    (c) $s(t) = -5\cos 4\pi t = 5 \Rightarrow \cos 4\pi t = -1 \Rightarrow$
$$4\pi t = \pi \Rightarrow t = \frac{1}{4}$$
The weight first reaches its maximum
height after $\dfrac{1}{4}$ sec.

    (d) $s(1.3) = -5\cos\left[4\pi (1.3)\right]$
$$= -5\cos 5.2\pi \approx 4,$$
After 1.3 sec, the weight is about 4.0 in.
above the equilibrium position.

19. $a = -3$

    (a) We will use a model of the form
$s(t) = a \cos \omega t$ with $a = -3$.

$$s(0) = -3\cos\left[\omega(0)\right]$$
$$= -3\cos 0 = -3(1) = -3$$

Using a cosine function rather than a sine function will avoid the need for a phase shift. The frequency $= \dfrac{6}{\pi}$ cycles per sec, so, by definition,

$$\frac{\omega}{2\pi} = \frac{6}{\pi} \Rightarrow \omega\pi = 12\pi \Rightarrow \omega = 12.$$

Therefore, a model for the position of the weight at time t seconds is
$s(t) = -3\cos 12t.$

    (b) The period is the reciprocal of the frequency, or $\dfrac{\pi}{6}$ sec.

21. $s(0) = 2$ in.; $P = 0.5$ sec

    (a) Given $s(t) = a\cos \omega t$, the period is $\dfrac{2\pi}{\omega}$ and the amplitude is $|a|$.

$$P = 0.5 \text{ sec} \Rightarrow 0.5 = \frac{2\pi}{\omega} \Rightarrow \frac{1}{2} = \frac{2\pi}{\omega} \Rightarrow$$
$$\omega = 4\pi$$
$$s(0) = 2 = a\cos\left[\omega(0)\right] \Rightarrow$$
$$2 = a\cos 0 \Rightarrow 2 = a(1) \Rightarrow a = 2$$

Thus, $s(t) = 2\cos 4\pi t.$

    (b) $s(1) = 2\cos\left[4\pi(1)\right] = 2\cos 4\pi$
$$= 2(1) = 2$$

The weight is neither moving upward nor downward. At $t = 1$, the motion of the weight is changing from up to down.

23. $s(0) = -3$ in.; $P = 0.8$ sec

    (a) Given $s(t) = a\cos \omega t$, the period is $\dfrac{2\pi}{\omega}$ and the amplitude is $|a|$.

$$P = 0.8 \text{ sec} \Rightarrow 0.8 = \frac{2\pi}{\omega} \Rightarrow \frac{4}{5} = \frac{2\pi}{\omega} \Rightarrow$$
$$4\omega = 10\pi \Rightarrow \omega = \frac{10\pi}{4} = 2.5\pi$$
$$s(0) = -3 = a\cos\left[\omega(0)\right] \Rightarrow$$
$$-3 = a\cos 0 \Rightarrow -3 = a(1) \Rightarrow a = -3$$

Thus, $s(t) = -3\cos 2.5\pi t.$

    (b) $s(1) = -3\cos\left[2.5\pi(1)\right] = -3\cos\dfrac{5\pi}{2}$
$$= -3(0) = 0$$

The weight is moving upward.

25. The frequency is $\dfrac{\omega}{2\pi}$, so

$$27.5 = \frac{\omega}{2\pi} \Rightarrow \omega = 55\pi. \text{ Because } s(0) = 0.21,$$
$$0.21 = a\cos\left[\omega(0)\right] \Rightarrow 0.21 = a\cos 0 \Rightarrow$$
$$0.21 = a(1) \Rightarrow a = 0.21.$$

Thus, $s(t) = 0.21\cos 55\pi t.$

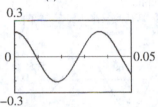

27. The frequency is $\dfrac{\omega}{2\pi}$, so

$$55 = \frac{\omega}{2\pi} \Rightarrow \omega = 110\pi. \text{ Since } s(0) = 0.14,$$
$$0.14 = a\cos\left[\omega(0)\right] \Rightarrow 0.14 = a\cos 0 \Rightarrow$$
$$0.14 = a(1) \Rightarrow a = 0.14. \text{ Thus,}$$
$$s(t) = 0.14\cos 110\pi t.$$

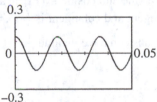

29. The weight was pulled down 11 in. from the equilibrium position before it was released.

31.

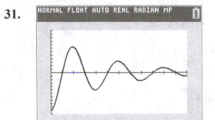

The graph intersects the horizontal axis at $t = 1, 3, 5, 7, 9, 11$.

33. (a)

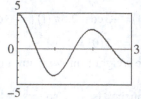

(b) The upper envelope of the graph of $y_3 = 5e^{-0.3x}\cos(\pi x)$ is $y_1 = 5e^{-0.3x}$.

(c) $5e^{-0.3x} = 5e^{-0.3x}\cos(\pi x)$ when $\cos(\pi x) = 1$, or when $x = 0$ and $x = 2$. This can be verified by examining the graphs.

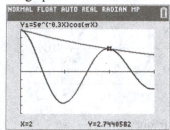

Chapter 4 Review Exercises

1. B; The amplitude is $|4| = 4$ and period is $\dfrac{2\pi}{2} = \pi$.

3. The range of sine and cosine is $[-1, 1]$ and the range of tangent and cotangent is $(-\infty, \infty)$, so those functions can have y-value $\frac{1}{2}$.

5. $y = 2\sin x$
Amplitude: 2
Period: 2π
Vertical translation: none
Phase shift: none

7. $y = -\dfrac{1}{2}\cos 3x$

Amplitude: $\left|-\dfrac{1}{2}\right| = \dfrac{1}{2}$

Period: $\dfrac{2\pi}{3}$

Vertical translation: none
Phase shift: none

9. $y = 1 + 2\sin\dfrac{1}{4}x$

Amplitude: $|2| = 2$

Period: $\dfrac{2\pi}{\frac{1}{4}} = 8\pi$

Vertical translation: up 1 unit
Phase shift: none

11. $y = 3\cos\left(x + \dfrac{\pi}{2}\right) = 3\cos\left[x - \left(-\dfrac{\pi}{2}\right)\right]$

Amplitude: $|3| = 3$
Period: 2π
Vertical translation: none

Phase shift: $\dfrac{\pi}{2}$ units to the left

13. $y = \dfrac{1}{2}\csc\left(2x - \dfrac{\pi}{4}\right) = \dfrac{1}{2}\csc\left[2\left(x - \dfrac{\pi}{8}\right)\right]$

Amplitude: not applicable

Period: $\dfrac{2\pi}{2} = \pi$

Vertical translation: none

Phase shift: $\dfrac{\pi}{8}$ unit to the right

15. $y = \dfrac{1}{3}\tan\left(3x - \dfrac{\pi}{3}\right) = \dfrac{1}{3}\tan\left[3\left(x - \dfrac{\pi}{9}\right)\right]$

Amplitude: not applicable

Period: $\dfrac{\pi}{3}$

Vertical translation: none

Phase shift: $\dfrac{\pi}{9}$ unit to the right

17. The tangent function has a period of π and x-intercepts at integral multiples of π.

19. The cosine function has a period of 2π and has the value 0 when $x = \dfrac{\pi}{2}$.

21. The cotangent function has a period of π and decreases on the interval $(0, \pi)$.

23. Answers will vary. Sample answer: The period is 10, so the value of the function repeats every 10 units. Thus, $f(5) = f(5 + 2\cdot 10) = f(25) = 3$.

25. $y = 3\sin x$

Period: 2π and amplitude: $|3| = 3$

Divide the interval $[0, 2\pi]$ into four equal parts to get x-values that will yield minimum and maximum points and x-intercepts. Then make a table.

x	0	$\dfrac{\pi}{2}$	π	$\dfrac{3\pi}{2}$	2π
$\sin x$	0	1	0	-1	0
$3\sin x$	0	3	0	-3	0

This table gives five values for graphing one period of $y = 3\sin x$.

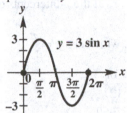

27. $y = -\tan x$

This is a reflection of the graph of $y = \tan x$ over the x-axis. The period is π and vertical asymptotes are $x = -\dfrac{\pi}{2}$ and $x = \dfrac{\pi}{2}$.

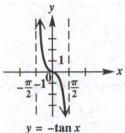

$y = -\tan x$

29. $y = 2 + \cot x$

This is the graph of $y = \cot x$ translated up 2 units. The period is π and the vertical asymptotes are $x = 0$ and $x = \pi$.

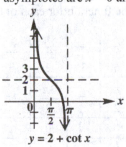

$y = 2 + \cot x$

31. $y = \sin 2x$

Period: $\dfrac{2\pi}{2} = \pi$ and amplitude: $|1| = 1$

Divide the interval $[0, \pi]$ into four equal parts to get the x-values that will yield minimum and maximum points and x-intercepts. Then make a table.

x	0	$\dfrac{\pi}{4}$	$\dfrac{\pi}{2}$	$\dfrac{3\pi}{4}$	π
$2x$	0	$\dfrac{\pi}{2}$	π	$\dfrac{3\pi}{2}$	2π
$\sin 2x$	0	1	0	-1	0

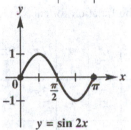

$y = \sin 2x$

33. $y = 3\cos 2x$

Period: $\dfrac{2\pi}{2} = \pi$ and amplitude: $|3| = 3$

Divide the interval $[0, \pi]$ into four equal parts to get the x-values that will yield minimum and maximum points and x-intercepts. Then make a table.

x	0	$\dfrac{\pi}{4}$	$\dfrac{\pi}{2}$	$\dfrac{3\pi}{4}$	π
$2x$	0	$\dfrac{\pi}{2}$	π	$\dfrac{3\pi}{2}$	2π
$\cos 2x$	1	0	-1	0	1
$3\cos 2x$	3	0	-3	0	3

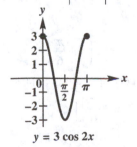

$y = 3\cos 2x$

35. $y = \cos\left(x - \dfrac{\pi}{4}\right)$

Step 1: Find the interval whose length is $\dfrac{2\pi}{b}$.

$0 \le x - \dfrac{\pi}{4} \le 2\pi \Rightarrow 0 + \dfrac{\pi}{4} \le x \le 2\pi + \dfrac{\pi}{4} \Rightarrow$

$\dfrac{\pi}{4} \le x \le \dfrac{9\pi}{4}$

Step 2: Divide the period into four equal parts to get the following x-values: $\dfrac{\pi}{4}$, π, $\dfrac{3\pi}{2}$, 2π,

$\dfrac{9\pi}{4}$

Step 3: Evaluate the function for each of the five x-values.

x	$\dfrac{\pi}{4}$	$\dfrac{3\pi}{4}$	$\dfrac{5\pi}{4}$	$\dfrac{7\pi}{4}$	$\dfrac{9\pi}{4}$
$x - \dfrac{\pi}{4}$	0	$\dfrac{\pi}{2}$	π	$\dfrac{3\pi}{2}$	2π
$\cos\left(x - \dfrac{\pi}{4}\right)$	1	0	-1	0	1

Steps 4 and 5: Plot the points found in the table and join them with a sinusoidal curve.

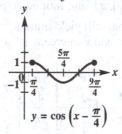

The amplitude is 1. The period is 2π. There is no vertical translation. The phase shift is $\dfrac{\pi}{4}$ unit to the right.

37. $y = \sec\left(2x + \dfrac{\pi}{3}\right) = \sec\left[2\left(x + \dfrac{\pi}{6}\right)\right]$

Step 1: Graph the corresponding reciprocal function $y = \cos\left[2\left(x + \dfrac{\pi}{6}\right)\right]$.

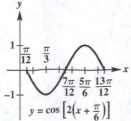

The period is $\dfrac{2\pi}{2} = \pi$, and its amplitude is $|1| = 1$. One period is in the interval

$\dfrac{\pi}{12} \le x \le \dfrac{13\pi}{12}$.

Dividing the interval into four equal parts gives the key points $\left(\dfrac{\pi}{12}, 0\right), \left(\dfrac{\pi}{3}, -1\right),$

$\left(\dfrac{7\pi}{12}, 0\right), \left(\dfrac{5\pi}{6}, 1\right),$ and $\left(\dfrac{13\pi}{12}, 0\right).$

Step 2: The vertical asymptotes of

$y = \sec 2\left(x + \dfrac{\pi}{6}\right)$ are at the x-intercepts of

$y = \cos 2\left(x + \dfrac{\pi}{6}\right),$ which are $x = \dfrac{\pi}{12},$

$x = \dfrac{7\pi}{12},$ and $x = \dfrac{13\pi}{12}.$

Step 3: Sketch the graph.

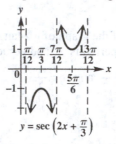

39. $y = 1 + 2\cos 3x$

Step 1: Find the interval whose length is $\dfrac{2\pi}{b}$.

$0 \le 3x \le 2\pi \Rightarrow 0 \le x \le \dfrac{2\pi}{3}$

Step 2: Divide the period into four equal parts to get the following x-values:

$0, \dfrac{\pi}{6}, \dfrac{\pi}{3}, \dfrac{\pi}{2}, \dfrac{2\pi}{3}$

Step 3: Evaluate the function for each of the five x-values.

x	0	$\dfrac{\pi}{6}$	$\dfrac{\pi}{3}$	$\dfrac{\pi}{2}$	$\dfrac{2\pi}{3}$
$3x$	0	$\dfrac{\pi}{2}$	π	$\dfrac{3\pi}{2}$	2π
$\cos 3x$	1	0	-1	0	1
$2\cos 3x$	2	0	-2	0	2
$1 + 2\cos 3x$	3	1	-1	1	3

(continued on next page)

(*continued*)

Steps 4 and 5: Plot the points found in the table and join them with a sinusoidal curve.

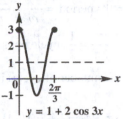

$y = 1 + 2 \cos 3x$

The period is $\dfrac{2\pi}{3}$. The amplitude is $|2|$, which is 2. The vertical translation is 1 unit up. There is no phase shift.

41. $y = 2 \sin \pi x$

Period: $\dfrac{2\pi}{\pi} = 2$ and amplitude: $|2| = 2$

Divide the interval $[0, 2]$ into four equal parts to get the x-values that will yield minimum and maximum points and x-intercepts. Then make a table.

x	0	$\dfrac{1}{2}$	1	$\dfrac{3}{2}$	2
πx	0	$\dfrac{\pi}{2}$	π	$\dfrac{3\pi}{2}$	2π
$\sin \pi x$	0	1	0	-1	0
$2 \sin \pi x$	0	2	0	-2	0

Steps 4 and 5: Plot the points found in the table and join them with a sinusoidal curve.

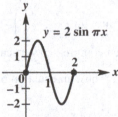

43. (a) See the graph in part (c).

(b)

$y = 8.02 \sin (0.52x + 0.84) + 59.83$

(c)

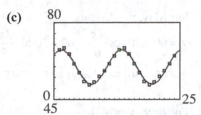

45. This is the graph of $y = \sin x$ reflected across the x-axis and translated 1 unit up. The equation is $y = -\sin x + 1$.

47. This is the graph of $y = \tan x$ with period 2π and stretched vertically by a factor of 2. The equation is $y = 2 \tan \dfrac{1}{2} x$.

49. (a) The shorter leg of the right triangle has length $h_2 - h_1$. Thus, we have

$$\cot \theta = \dfrac{d}{h_2 - h_1} \Rightarrow d = (h_2 - h_1) \cot \theta$$

(b) When $h_2 = 55$ and $h_1 = 5$,

$$d = (55 - 5) \cot \theta = 50 \cot \theta.$$

The period is π, but the graph wanted is d for $0 < \theta < \dfrac{\pi}{2}$. The asymptote is the line $\theta = 0$. Also, when

$$\theta = \dfrac{\pi}{4}, d = 50 \cot \dfrac{\pi}{4} = 50(1) = 50.$$

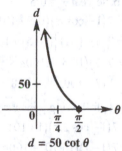

$d = 50 \cot \theta$

51. $t = 60 - 30 \cos \dfrac{x\pi}{6}$

(a) For January, $x = 0$. Thus,

$$t = 60 - 30 \cos \dfrac{0 \cdot \pi}{6} = 60 - 30 \cos 0$$
$$= 60 - 30(1) = 60 - 30 = 30°F$$

(b) For April, $x = 3$. Thus,

$$t = 60 - 30 \cos \dfrac{3\pi}{6} = 60 - 30 \cos \dfrac{\pi}{2}$$
$$= 60 - 30(0) = 60 - 0 = 60°F$$

(c) For May, $x = 4$. Thus,

$$t = 60 - 30\cos\frac{4\pi}{6} = 60 - 30\cos\frac{2\pi}{3}$$

$$= 60 - 30\left(-\frac{1}{2}\right) = 60 + 15 = 75°F$$

(d) For June, $x = 5$. Thus,

$$t = 60 - 30\cos\frac{5\pi}{6} = 60 - 30\left(-\frac{\sqrt{3}}{2}\right)$$

$$= 60 + 15\sqrt{3} \approx 86°F$$

(e) For August, $x = 7$. Thus,

$$t = 60 - 30\cos\frac{7\pi}{6} = 60 - 30\left(-\frac{\sqrt{3}}{2}\right)$$

$$= 60 + 15\sqrt{3} \approx 86°F$$

(f) For October, $x = 9$. Thus,

$$t = 60 - 30\cos\frac{9\pi}{6} = 60 - 30\cos\frac{3\pi}{2}$$

$$= 60 - 30(0) = 60 - 0 = 60°F$$

53. $P(x) = 7(1 - \cos 2\pi x)(x + 10) + 100e^{0.2x}$

(a) January 1, base year $x = 0$

$$P(0) = 7(1 - \cos 0)(10) + 100e^0$$

$$= 7(1 - 1)(10) + 100(1)$$

$$= 7(0)(10) + 100 = 0 + 100 = 100$$

(b) July 1, base year $x = 0.5$

$$P(.5) = 7(1 - \cos\pi)(0.5 + 10) + 100e^{0.2(0.5)}$$

$$= 7[1 - (-1)](10.5) + 100e^{0.1}$$

$$= 7(2)(10.5) + 100e^{0.1}$$

$$= 147 + 100e^{0.1} \approx 258$$

(c) January 1, following year $x = 1$

$$P(1) = 7(1 - \cos 2\pi)(1 + 10) + 100e^{0.2}$$

$$= 7(1 - 1)(1 + 10) + 100e^{0.2}$$

$$= 7(0)(11) + 100e^2 = 0 + 100e^{0.2}$$

$$= 100e^{0.2} \approx 122$$

(d) July 1, following year $x = 1.5$

$$P(1.5) = 7(1 - \cos 3\pi)(1.5 + 10) + 100e^{0.2(1.5)}$$

$$= 7[1 - (-1)](11.5) + 100e^{0.3}$$

$$= 7(2)(11.5) + 100e^{0.3}$$

$$= 161 + +100e^{0.3} \approx 296$$

55. $s(t) = 4\sin\pi t$

$a = 4, \omega = \pi$

amplitude $= |a| = 4$; period $= \dfrac{2\pi}{\omega} = \dfrac{2\pi}{\pi} = 2$

frequency $= \dfrac{\omega}{2\pi} = \dfrac{\pi}{2\pi} = \dfrac{1}{2}$ cycle per sec

57. The frequency is the number of cycles in one unit of time.

$$s(1.5) = 4\sin 1.5\pi = 4\sin\frac{3\pi}{2} = 4(-1) = -4$$

$$s(2) = 4\sin 2\pi = 4(0) = 0$$

$$s(3.25) = 4\sin 3.25\pi = 4\sin\frac{13\pi}{4} = 4\sin\frac{5\pi}{4}$$

$$= 4\left(-\frac{\sqrt{2}}{2}\right) = -2\sqrt{2}$$

Chapter 4   Test

1. (a) $y = \sec x$ (b) $y = \sin x$

   (c) $y = \cos x$ (d) $y = \tan x$

   (e) $y = \csc x$ (e) $y = \cot x$

2. (a) This is a cosine curve with period 4π, so

   $4\pi = \dfrac{2\pi}{b} \Rightarrow b = \dfrac{2\pi}{4\pi} = \dfrac{1}{2}$. The graph has

   been shifted 1 unit up, so an equation is

   $y = 1 + \cos\frac{1}{2}x$.

   (b) This is a cotangent curve that has been reflected across the y-axis. Since the

   graph passes through $\left(\dfrac{\pi}{4}, -\dfrac{1}{2}\right)$, the

   graph has been compressed vertically by

   a factor of $\dfrac{1}{2}$. Thus, an equation of the

   graph is $y = -\dfrac{1}{2}\cot x$.

3. (a) The domain of the cosine function is $(-\infty, \infty)$.

   (b) The range of the sine function is $[-1, 1]$.

   (c) The least positive value for which the

   tangent function is undefined is $\dfrac{\pi}{2}$.

   (d) The range of the secant function is $(-\infty, -1] \cup [1, \infty)$.

4. $y = 3 - 6\sin\left(2x + \dfrac{\pi}{2}\right) = 3 - 6\sin\left[2\left(x + \dfrac{\pi}{4}\right)\right]$

$\qquad = 3 - 6\sin\left[2\left[x - \left(-\dfrac{\pi}{4}\right)\right]\right]$

(a) The period is $\dfrac{2\pi}{2} = \pi$.

(b) The amplitude is 6.

(c) The range is $[-3, 9]$.

(d) The y-intercept occurs when $x = 0$.

$-6\sin\left(2 \cdot 0 + \dfrac{\pi}{2}\right) + 3 = -6\sin\left(0 + \dfrac{\pi}{2}\right) + 3$

$\qquad = -6\sin\left(\dfrac{\pi}{2}\right) + 3$

$\qquad = -6\,(1) + 3 = -3$

The x-intercept is $(0, -3)$.

(e) The phase shift is $\dfrac{\pi}{4}$ unit to the left

$\left(\text{that is}, -\dfrac{\pi}{4}\right)$

5. $y = \sin(2x + \pi) = \sin\left[2\left(x + \dfrac{\pi}{2}\right)\right]$

$\qquad = \sin\left[2\left(x - \left(-\dfrac{\pi}{2}\right)\right)\right]$

Step 1: Find the interval whose length is $\dfrac{2\pi}{b}$.

$0 \le 2\left(x + \dfrac{\pi}{2}\right) \le 2\pi \Rightarrow 0 \le x + \dfrac{\pi}{2} \le \dfrac{2\pi}{2} \Rightarrow$

$0 \le x + \dfrac{\pi}{2} \le \pi \Rightarrow -\dfrac{\pi}{2} \le x \le \dfrac{\pi}{2}$

Step 2: Divide the period into four equal parts

to get the x-values: $-\dfrac{\pi}{2}, -\dfrac{\pi}{4}, 0, \dfrac{\pi}{4}, \dfrac{\pi}{2}$

Step 3: Evaluate the function for each of the five x-values

x	$-\dfrac{\pi}{2}$	$-\dfrac{\pi}{4}$	0	$\dfrac{\pi}{4}$	$\dfrac{\pi}{2}$
$x + \dfrac{\pi}{2}$	0	$\dfrac{\pi}{4}$	$\dfrac{\pi}{2}$	$\dfrac{3\pi}{4}$	π
$2\left(x + \dfrac{\pi}{2}\right)$	0	$\dfrac{\pi}{2}$	π	$\dfrac{3\pi}{2}$	2π
$\sin\left[2\left(x + \dfrac{\pi}{2}\right)\right]$	0	1	0	-1	0

Steps 4 and 5: Plot the points found in the table and join them with a sinusoidal curve.

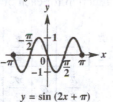

$y = \sin(2x + \pi)$

The period is π. There is no vertical translation. The phase shift is $\dfrac{\pi}{2}$ units to the right.

6. $y = -\cos 2x$

Period: $\dfrac{2\pi}{2} = \pi$ and amplitude: $|-1| = 1$

Divide the interval $[0, \pi]$ into four equal parts to get the x-values that will yield minimum and maximum points and x-intercepts. Then make a table. Repeat this cycle for the interval $[-\pi, 0]$.

x	0	$\dfrac{\pi}{4}$	$\dfrac{\pi}{2}$	$\dfrac{3\pi}{4}$	π
$2x$	0	$\dfrac{\pi}{2}$	π	$\dfrac{3\pi}{2}$	2π
$\cos 2x$	1	0	-1	0	1
$-\cos 2x$	-1	0	1	0	-1

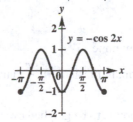

$y = -\cos 2x$

7. $y = 2 + \cos x$

This is the graph of $y = \cos x$ translated vertically 2 units up.

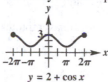

$y = 2 + \cos x$

8. $y = -1 + 2\sin(x + \pi)$

Step 1: Find the interval whose length is $\dfrac{2\pi}{b}$.

$0 \le x + \pi \le 2\pi \Rightarrow -\pi \le x \le \pi$

Step 2: Divide the period into four equal parts

to get the x-values: $-\pi, -\dfrac{\pi}{2}, 0, \dfrac{\pi}{2}, \pi$

(continued on next page)

(*continued*)

Step 3: Evaluate the function for each of the five *x*-values

x	$-\dfrac{\pi}{2}$	$-\dfrac{\pi}{2}$	0	$\dfrac{\pi}{2}$	π
$x + \pi$	0	$\dfrac{\pi}{2}$	π	$\dfrac{3\pi}{2}$	2π
$\sin(x + \pi)$	0	1	0	-1	0
$2\sin(x + \pi)$	0	2	0	-2	0
$-1 + 2\sin(x + \pi)$	-1	1	-1	-3	-1

Steps 4 and 5: Plot the points found in the table and join them with a sinusoidal curve. Repeat this cycle for the interval $[-\pi, 0]$.

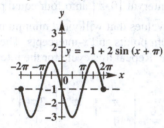

The amplitude is $|2|$, which is 2. The period is 2π. The vertical translation is 1 unit down. The phase shift is π units to the left.

9. $y = \tan\left(x - \dfrac{\pi}{2}\right)$

Period: π
Vertical translation: none

Phase shift (horizontal translation): $\dfrac{\pi}{2}$ units

to the right
Because the function is to be graphed over a two-period interval, locate three adjacent vertical asymptotes. Asymptotes of the graph $y = \tan x$ occur at $-\dfrac{\pi}{2}$, and $\dfrac{\pi}{2}$, so use the

following equations to locate asymptotes:

$x - \dfrac{\pi}{2} = -\dfrac{\pi}{2} \Rightarrow x = 0$ and

$x - \dfrac{\pi}{2} = \dfrac{\pi}{2} \Rightarrow x = \pi.$

Divide the interval $(0, \pi)$ into four equal parts to obtain the key *x*-values: first-quarter value:

$\dfrac{\pi}{4}$; middle value: $\dfrac{\pi}{2}$; third-quarter value:

$\dfrac{3\pi}{4}$

Evaluating the given function at these three

key *x*-values gives the points: $\left(\dfrac{\pi}{4}, -1\right)$,

$\left(\dfrac{\pi}{2}, 0\right)$, $\left(\dfrac{3\pi}{4}, 1\right)$. Connect these points with

a smooth curve and continue to graph to approach the asymptote $x = 0$ and $x = \pi$ to complete one period of the graph. Repeat this cycle for the interval $[-\pi, 0]$.

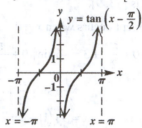

10. $y = -2 - \cot\left(x - \dfrac{\pi}{2}\right)$

Period: $\dfrac{\pi}{b} = \dfrac{\pi}{1} = \pi$

Vertical translation: 2 units down

Phase shift (horizontal translation): $\dfrac{\pi}{2}$ units

to the right
Because the function is to be graphed over a two-period interval, locate three adjacent vertical asymptotes. Asymptotes of the graph $y = \cot x$ occur at multiples of π, so use the following equations to locate asymptotes:

$x - \dfrac{\pi}{2} = -\pi$, $x - \dfrac{\pi}{2} = 0$, and $x - \dfrac{\pi}{2} = \pi$.

Solve each of these equations:

$x - \dfrac{\pi}{2} = -\pi \Rightarrow x = -\dfrac{\pi}{2}$; $x - \dfrac{\pi}{2} = 0 \Rightarrow x = \dfrac{\pi}{2}$;

$x - \dfrac{\pi}{2} = \pi \Rightarrow x = \dfrac{3\pi}{2}$

Divide the interval $\left(-\dfrac{\pi}{2}, \dfrac{\pi}{2}\right)$ into four equal

parts to obtain the following key *x*-values.

first-quarter value: $-\dfrac{\pi}{4}$ middle value: 0;

third-quarter value: $\dfrac{\pi}{4}$

Evaluating the given function at these three

key *x*-values gives the points: $\left(-\dfrac{\pi}{4}, -3\right)$,

$(0, -2)$, $\left(\dfrac{\pi}{4}, -1\right)$

(*continued on next page*)

(*continued*)

Connect these points with a smooth curve and continue to graph to approach the asymptote $x = -\dfrac{\pi}{2}$ and $x = \dfrac{\pi}{2}$ to complete one period of the graph. Sketch the identical curve between the asymptotes $x = \dfrac{\pi}{2}$ and $x = \dfrac{3\pi}{2}$ to complete a second period of the graph.

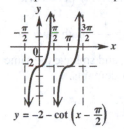

$y = -2 - \cot\left(x - \dfrac{\pi}{2}\right)$

11. $y = -\csc 2x$

Step 1: Graph the corresponding reciprocal function $y = -\sin 2x$ The period is $\dfrac{2\pi}{2} = \pi$ and its amplitude is $|-1| = 1$. One period is in the interval $0 \le x \le \pi$. Dividing the interval into four equal parts gives us the key points:

$(0,0)$, $\left(\dfrac{\pi}{2}, -1\right)$, $\left(\dfrac{\pi}{2}, 0\right)$, $\left(\dfrac{3\pi}{4}, 1\right)$, $(\pi, 0)$

Step 2: The vertical asymptotes of $y = -\csc 2x$ are at the x-intercepts of $y = -\sin 2x$, which are $x = 0$, $x = \dfrac{\pi}{2}$, and $x = \pi$. Continuing this pattern to the right, we also have a vertical asymptotes of $x = \dfrac{3\pi}{2}$ and $x = 2\pi$.

Step 3: Sketch the graph.

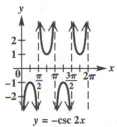

$y = -\csc 2x$

12. $y = 3\csc \pi x$

Step 1: Graph the corresponding reciprocal function $y = 3\sin \pi x$. The period is $\dfrac{2\pi}{\pi} = 2$, and the amplitude is $|3| = 3$. One period is in the interval $0 \le x \le 2$. Dividng the interval into four equal parts give the key points

$(0,0)$, $\left(\dfrac{1}{2}, 1\right)$, $(1,0)$, $\left(\dfrac{3}{2}, -1\right)$, $(2,0)$.

Step 2: The vertical asymptotes of $y = 3\csc \pi x$ are at the x-intercepts of $y = 3\sin \pi x$, namely $x = 0$, $x = 1$, and $x = 2$.

Step 3: Sketch the graph. Repeat this cycle for the interval $[-2, 0]$.

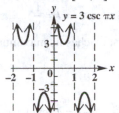

Period: 2.
Amplitude: Not applicable
Phase shift: none
Vertical translation: none

13. (a)

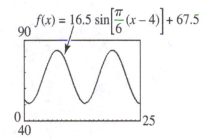

(b) Amplitude: 16.5

Period: $\dfrac{2\pi}{\frac{\pi}{6}} = 2\pi \cdot \dfrac{6}{\pi} = 12$;

Phase shift: 4 units to the right
Vertical translation: 67.5 units up

(c) For the month of December, $x = 12$.

$$f(12) = 16.5\sin\left[\dfrac{\pi}{6}(12 - 4)\right] + 67.5$$

$$= 16.5\sin\left(\dfrac{4\pi}{3}\right) + 67.5$$

$$= 16.5\left(-\dfrac{\sqrt{3}}{2}\right) + 67.5 \approx 53°F$$

(d) Examining the graph shows that a minimum of 51 occurs at $x = 13 = 12 + 1$ implies a minimum average monthly temperature of 51°F in January. A maximum of 84 occurs at $x = 7$ and $x = 19 = 12 + 7$ a maximum average monthly temperature of 84°F in July.

(e) The average annual temperature is about 67.5°F. This is the vertical translation.

14. $s(t) = -4 \cos 8\pi t$, $a = |-4| = 4$, $\omega = 8\pi$

(a) maximum height = amplitude
$$= a = |-4| = 4 \text{ in.}$$

(b) $s(t) = -4 \cos 8\pi t = 4 \Rightarrow \cos 8\pi t = -1 \Rightarrow$
$$8\pi t = \pi \Rightarrow t = \frac{1}{8}$$

The weight first reaches its maximum height after $\frac{1}{8}$ sec.

(c) frequency $= \dfrac{\omega}{2\pi} = \dfrac{8\pi}{2\pi} = 4$ cycles per sec;

period $= \dfrac{2\pi}{\omega} = \dfrac{2\pi}{8\pi} = \dfrac{1}{4}$ sec

15. The functions $y = \sin x$ and $y = \cos x$ both have all real numbers as their domains. The functions $f(x) = \tan x = \dfrac{\sin x}{\cos x}$ and $f(x) = \sec x = \dfrac{1}{\cos x}$ both have cos x in their denominators. Therefore, both the tangent and secnt functions have the same restrictions on their domains. Similarly, $f(x) = \cot x = \dfrac{\cos x}{\sin x}$ and $f(x) = \csc x = \dfrac{1}{\sin x}$ both have sin x in their denominators, and so have the same restrictions on their domains.

Chapter 5

Trigonometric Identities

Section 5.1 Fundamental Identities

1. B; $\dfrac{\cos x}{\sin x} = \cot x$

3. E; $\cos(-x) = \cos x$

5. A; $1 = \sin^2 x + \cos^2 x$

7. By a negative angle identity, $\cos(-\theta) = \cos\theta.$ Thus, if $\cos\theta = -0.65,$ then
$\cos(-\theta) = \underline{-0.65}.$

9. By a quotient identity, $\tan\theta = \dfrac{\sin\theta}{\cos\theta}.$ By a negative angle identity $\sin(-\theta) = -\sin\theta$ and $\cos(-\theta) = \cos\theta.$ Thus, if $\cos x = 0.8$ and $\sin x = 0.6,$ then
$$\tan(-x) = \frac{\sin(-x)}{\cos(-x)} = \frac{-\sin x}{\cos x} = \frac{-0.6}{0.8} = \underline{-0.75}.$$

11. $\cos\theta = \dfrac{3}{4}$, θ is in quadrant I.

An identity that relates sine and cosine is $\sin^2\theta + \cos^2\theta = 1.$
$$\sin^2\theta + \cos^2\theta = 1 \Rightarrow \sin^2\theta + \left(\frac{3}{4}\right)^2 = 1 \Rightarrow$$
$$\sin^2\theta = 1 - \frac{9}{16} = \frac{7}{16} \Rightarrow \sin\theta = \pm\frac{\sqrt{7}}{4}$$
θ is in quadrant I, so $\sin\theta = \dfrac{\sqrt{7}}{4}.$

13. $\cot\theta = -\dfrac{1}{5},$ θ in quadrant IV

Use the identity $1 + \cot^2\theta = \csc^2\theta$ because $\sin\theta = \dfrac{1}{\csc\theta}.$
$$1 + \cot^2\theta = \csc^2\theta \Rightarrow 1 + \left(-\frac{1}{5}\right)^2 = \csc^2\theta \Rightarrow$$
$$1 + \frac{1}{25} = \csc^2\theta \Rightarrow \frac{26}{25} = \csc^2\theta \Rightarrow \csc\theta = \pm\frac{\sqrt{26}}{5}$$
θ is in quadrant IV, so $\csc\theta < 0,$ so
$\csc\theta = -\dfrac{\sqrt{26}}{5}.$

Thus,
$$\sin\theta = \frac{1}{\csc\theta} = -\frac{5}{\sqrt{26}} = -\frac{5}{\sqrt{26}} \cdot \frac{\sqrt{26}}{\sqrt{26}} = -\frac{5\sqrt{26}}{26}$$

15. $\cos(-\theta) = \dfrac{\sqrt{5}}{5},$ $\tan\theta < 0$

Because $\cos(-\theta) = \dfrac{\sqrt{5}}{5},$ we have $\cos\theta = \dfrac{\sqrt{5}}{5}$ by a negative angle identity. An identity that relates sine and cosine is $\sin^2\theta + \cos^2\theta = 1.$
$$\sin^2\theta + \cos^2\theta = 1 \Rightarrow \sin^2\theta + \left(\frac{\sqrt{5}}{5}\right)^2 = 1 \Rightarrow$$
$$\sin^2\theta + \frac{5}{25} = 1 \Rightarrow \sin^2\theta = 1 - \frac{5}{25} = 1 - \frac{1}{5} = \frac{4}{5} \Rightarrow$$
$$\sin\theta = \pm\frac{2}{\sqrt{5}} = \pm\frac{2}{\sqrt{5}} \cdot \frac{\sqrt{5}}{\sqrt{5}} = \pm\frac{2\sqrt{5}}{5}$$
Because $\tan\theta < 0$ and $\cos\theta > 0,$ θ is in quadrant IV and $\sin\theta < 0.$ Thus,
$\sin\theta = -\dfrac{2\sqrt{5}}{5}.$

17. $\tan\theta = -\dfrac{\sqrt{6}}{2},$ $\cos\theta > 0$
$$\tan^2\theta + 1 = \sec^2\theta \Rightarrow \left(-\frac{\sqrt{6}}{2}\right)^2 + 1 = \sec^2\theta \Rightarrow$$
$$\frac{6}{4} + 1 = \frac{10}{4} = \frac{5}{2} = \sec^2\theta \Rightarrow$$
$$\sec\theta = \pm\sqrt{\frac{5}{2}} = \pm\frac{\sqrt{5}}{\sqrt{2}} \cdot \frac{\sqrt{2}}{\sqrt{2}} = \pm\frac{\sqrt{10}}{2}$$
Because $\cos\theta = \dfrac{1}{\sec\theta}$ and $\cos\theta > 0,$
$$\cos\theta = \frac{1}{\frac{\sqrt{10}}{2}} = \frac{2}{\sqrt{10}} = \frac{2}{\sqrt{10}} \cdot \frac{\sqrt{10}}{\sqrt{10}}$$
$$= \frac{2\sqrt{10}}{10} = \frac{\sqrt{10}}{5}.$$
Now, use the identity $\sin^2\theta + \cos^2\theta = 1$:
$$\sin^2\theta + \left(\frac{\sqrt{10}}{5}\right)^2 = 1 \Rightarrow \sin^2\theta + \frac{10}{25} = 1 \Rightarrow$$
$$\sin^2\theta = 1 - \frac{10}{25} = 1 - \frac{2}{5} = \frac{3}{5} \Rightarrow$$

(continued on next page)

(*continued*)

$$\sin\theta = \pm\sqrt{\frac{3}{5}} = \pm\frac{\sqrt{3}}{\sqrt{5}}\cdot\frac{\sqrt{5}}{\sqrt{5}} = \pm\frac{\sqrt{15}}{5}$$

Because $\tan\theta < 0$ and $\cos\theta > 0$, θ is in quadrant IV and $\sin\theta < 0$. Thus,

$$\sin\theta = -\frac{\sqrt{15}}{5}.$$

19. $\sec\theta = \frac{11}{4}, \cot\theta < 0$

Because $\cos\theta = \frac{1}{\sec\theta}, \cos\theta = \frac{1}{\frac{11}{4}} = \frac{4}{11}$. Use

the identity $\sin^2\theta + \cos^2\theta = 1$ to obtain

$$\sin^2\theta + \cos^2\theta = 1 \Rightarrow \sin^2\theta + \left(\frac{4}{11}\right)^2 = 1 \Rightarrow$$

$$\sin^2\theta + \frac{16}{121} = 1 \Rightarrow \sin^2\theta = 1 - \frac{16}{121} \Rightarrow$$

$$\sin^2\theta = \frac{105}{121} \Rightarrow \sin\theta = \pm\frac{\sqrt{105}}{11}$$

Because $\cot\theta < 0$ and $\sec\theta > 0$, θ is in quadrant IV and $\sin\theta < 0$. Thus,

$$\sin\theta = -\frac{\sqrt{105}}{11}.$$

21. $\csc\theta = -\frac{9}{4}$

$$\sin\theta = \frac{1}{\csc\theta}, \text{ so } \sin\theta = \frac{1}{-\frac{9}{4}} = -\frac{4}{9}.$$

23. Because $\sin\theta = \frac{1}{\csc\theta}$, the sign of $\sin\theta$ will

be the same as $\csc\theta$.

25. $f(x) = \frac{\sin x}{x}$

$$f(-x) = \frac{\sin(-x)}{-x} = \frac{-\sin x}{-x} = \frac{\sin x}{x}$$

Because $f(x) = f(-x)$, the function is even.

27. This is the graph of $f(x) = \sec x$. It is symmetric about the *y*-axis.

$$f(-x) = \sec(-x) = \frac{1}{\cos(-x)} = \frac{1}{\cos x}$$
$$= \sec x = f(x)$$

Because $f(x) = f(-x)$, the function is even.

29. This is the graph of $f(x) = \cot x$. It is symmetric about the origin.

$$f(-x) = \cot(-x) = \frac{\cos(-x)}{\sin(-x)} = \frac{\cos x}{-\sin x}$$
$$= -\frac{\cos x}{\sin x} = -\cot x = -f(x)$$

Because $f(x) = -f(x)$, the function is odd.

31. $\sin\theta = \frac{2}{3}, \theta$ in quadrant II

θ is in quadrant II, so the sine and cosecant function values are positive. The cosine, tangent, cotangent, and secant function values are negative.

$$\sin^2\theta + \cos^2\theta = 1 \Rightarrow$$

$$\cos^2\theta = 1 - \sin^2\theta = 1 - \left(\frac{2}{3}\right)^2 = 1 - \frac{4}{9} = \frac{5}{9} \Rightarrow$$

$$\cos\theta = -\frac{\sqrt{5}}{3}, \text{ since } \cos\theta < 0$$

$$\tan\theta = \frac{\sin\theta}{\cos\theta} = \frac{\frac{2}{3}}{-\frac{\sqrt{5}}{3}} = -\frac{2}{\sqrt{5}}$$

$$= -\frac{2}{\sqrt{5}}\cdot\frac{\sqrt{5}}{\sqrt{5}} = -\frac{2\sqrt{5}}{5}$$

$$\cot\theta = \frac{1}{\tan\theta} = \frac{1}{-\frac{2}{\sqrt{5}}} = -\frac{\sqrt{5}}{2}$$

$$\sec\theta = \frac{1}{\cos\theta} = \frac{1}{-\frac{\sqrt{5}}{3}} = -\frac{3}{\sqrt{5}}$$

$$= -\frac{3}{\sqrt{5}}\cdot\frac{\sqrt{5}}{\sqrt{5}} = -\frac{3\sqrt{5}}{5}$$

$$\csc\theta = \frac{1}{\sin\theta} = \frac{1}{\frac{2}{3}} = \frac{3}{2}$$

33. $\tan\theta = -\frac{1}{4}, \theta$ in quadrant IV

θ is in quadrant IV, so the cosine and secant function values are positive. The sine, tangent, cotangent, and cosecant function values are negative.

$$\cot\theta = \frac{1}{\tan\theta} = \frac{1}{-\frac{1}{4}} = -4$$

$$\sec^2\theta = 1 + \tan^2\theta = 1 + \left(-\frac{1}{4}\right)^2$$

$$= 1 + \frac{1}{16} = \frac{17}{16} \Rightarrow$$

$$\sec\theta = \frac{\sqrt{17}}{4}, \text{ since } \sec\theta > 0$$

(*continued on next page*)

(*continued*)

$$\cos\theta = \frac{1}{\sec\theta} = \frac{1}{\frac{\sqrt{17}}{4}} = \frac{4}{\sqrt{17}}$$

$$= \frac{4}{\sqrt{17}} \cdot \frac{\sqrt{17}}{\sqrt{17}} = \frac{4\sqrt{17}}{17}$$

$$\sin^2\theta + \cos^2\theta = 1 \Rightarrow$$

$$\sin^2\theta = 1 - \cos^2\theta = 1 - \left(\frac{4}{\sqrt{17}}\right)^2$$

$$\sin^2\theta = 1 - \frac{16}{17} = \frac{1}{17} \Rightarrow$$

$$\sin\theta = -\frac{1}{\sqrt{17}} = -\frac{1}{\sqrt{17}} \cdot \frac{\sqrt{17}}{\sqrt{17}} = -\frac{\sqrt{17}}{17},$$

since $\sin\theta < 0$

$$\csc\theta = \frac{1}{\sin\theta} = \frac{1}{-\frac{1}{\sqrt{17}}} = -\sqrt{17}$$

35. $\cot\theta = \frac{4}{3}, \sin\theta > 0$

$\cot\theta > 0$ and $\sin\theta > 0$, so θ is in quadrant I and all the function values are positive.

$$\tan = \frac{1}{\cot\theta} = \frac{1}{\frac{4}{3}} = \frac{3}{4}$$

$$\sec^2\theta = 1 + \tan^2\theta = 1 + \left(\frac{3}{4}\right)^2 = 1 + \frac{9}{16} = \frac{25}{16} \Rightarrow$$

$$\sec\theta = \frac{5}{4}, \text{ since } \sec\theta > 0$$

$$\cos\theta = \frac{1}{\sec\theta} = \frac{1}{\frac{5}{4}} = \frac{4}{5}$$

$$\sin^2\theta = 1 - \cos^2\theta = 1 - \left(\frac{4}{5}\right)^2 = 1 - \frac{16}{25} = \frac{9}{25} \Rightarrow$$

$$\sin\theta = \frac{3}{5}, \text{ since } \sin\theta > 0$$

$$\csc\theta = \frac{1}{\sin\theta} = \frac{1}{\frac{3}{5}} = \frac{5}{3}$$

37. $\sec\theta = \frac{4}{3}, \sin\theta < 0$

$\sec\theta > 0$ and $\sin\theta < 0$, so θ is in quadrant IV and the cosine function value is positive. The tangent, cotangent, and cosecant function values are negative.

$$\cos\theta = \frac{1}{\sec\theta} = \frac{1}{\frac{4}{3}} = \frac{3}{4}$$

$$\sin^2\theta = 1 - \cos^2\theta = 1 - \left(\frac{3}{4}\right)^2 = 1 - \frac{9}{16} = \frac{7}{16} \Rightarrow$$

$$\sin\theta = -\frac{\sqrt{7}}{4}, \text{ since } \sin\theta < 0$$

$$\tan\theta = \frac{\sin\theta}{\cos\theta} = \frac{-\frac{\sqrt{7}}{4}}{\frac{3}{4}} = -\frac{\sqrt{7}}{3}$$

$$\cot\theta = \frac{1}{\tan\theta} = -\frac{1}{\frac{\sqrt{7}}{3}} = -\frac{3}{\sqrt{7}} \cdot \frac{\sqrt{7}}{\sqrt{7}} = -\frac{3\sqrt{7}}{7}$$

$$\csc\theta = \frac{1}{\sin\theta} = \frac{1}{-\frac{\sqrt{7}}{4}} = -\frac{4}{\sqrt{7}} \cdot \frac{\sqrt{7}}{\sqrt{7}} = -\frac{4\sqrt{7}}{7}$$

39. C

$$-\tan x \cos x = -\frac{\sin x}{\cos x} \cdot \cos x$$
$$= -\sin x = \sin(-x)$$

41. E; $\dfrac{\sec x}{\csc x} = \dfrac{\frac{1}{\cos x}}{\frac{1}{\sin x}} = \dfrac{\sin x}{\cos x} = \tan x$

43. B; $\cos^2 x = \dfrac{1}{\sec^2 x}$

45. Find $\sin\theta$ if $\cos\theta = \dfrac{x}{x+1}$.

$\sin^2\theta + \cos^2\theta = 1$ and $\cos\theta = \dfrac{x}{x+1}$, so

$$\sin^2\theta = 1 - \cos^2\theta = 1 - \left(\frac{x}{x+1}\right)^2$$

$$= 1 - \frac{x^2}{(x+1)^2} = \frac{(x+1)^2 - x^2}{(x+1)^2}$$

$$= \frac{x^2 + 2x + 1 - x^2}{(x+1)^2} = \frac{2x+1}{(x+1)^2}$$

Thus, $\sin\theta = \dfrac{\pm\sqrt{2x+1}}{x+1}$.

47. $\sin^2 x + \cos^2 x = 1 \Rightarrow \sin^2 x = 1 - \cos^2 x \Rightarrow$
$\sin x = \pm\sqrt{1 - \cos^2 x}$

49. $\tan^2 x + 1 = \sec^2 x \Rightarrow \tan^2 x = \sec^2 x - 1 \Rightarrow$
$\tan x = \pm\sqrt{\sec^2 x - 1}$

51. $\csc x = \dfrac{1}{\sin x} \Rightarrow$

$\csc x = \dfrac{1}{\pm\sqrt{1-\cos^2 x}}$

$= \dfrac{\pm 1}{\sqrt{1-\cos^2 x}} \cdot \dfrac{\sqrt{1-\cos^2 x}}{\sqrt{1-\cos^2 x}}$

$= \dfrac{\pm\sqrt{1-\cos^2 x}}{1-\cos^2 x}$

For exercises 53–77, there may be more than one possible answer.

53. $\cot\theta\sin\theta = \dfrac{\cos\theta}{\sin\theta}\cdot\sin\theta = \cos\theta$

55. $\sec\theta\cot\theta\sin\theta = \dfrac{1}{\cos\theta}\cdot\dfrac{\cos\theta}{\sin\theta}\cdot\dfrac{\sin\theta}{1}$

$= \dfrac{\sin\theta\cos\theta}{\cos\theta\sin\theta} = 1$

57. $\cos\theta\csc\theta = \cos\theta\cdot\dfrac{1}{\sin\theta} = \dfrac{\cos\theta}{\sin\theta} = \cot\theta$

59. $\sin^2\theta\left(\csc^2\theta - 1\right) = \sin^2\theta\left(\dfrac{1}{\sin^2\theta} - 1\right)$

$= \dfrac{\sin^2\theta}{\sin^2\theta} - \sin^2\theta$

$= 1 - \sin^2\theta = \cos^2\theta$

61. $(1-\cos\theta)(1+\sec\theta)$

$= 1 + \sec\theta - \cos\theta - \cos\theta\sec\theta$

$= 1 + \sec\theta - \cos\theta - \cos\theta\left(\dfrac{1}{\cos\theta}\right)$

$= 1 + \sec\theta - \cos\theta - 1 = \sec\theta - \cos\theta$

63. $\dfrac{1+\tan(-\theta)}{\tan(-\theta)} = \dfrac{1-\tan\theta}{-\tan\theta} = \dfrac{1}{-\tan\theta} + \dfrac{-\tan\theta}{-\tan\theta}$

$= -\cot\theta + 1$

65. $\dfrac{1-\cos^2(-\theta)}{1+\tan^2(-\theta)} = \dfrac{\sin^2(-\theta)}{\sec^2(-\theta)} = \dfrac{\sin^2(-\theta)}{\dfrac{1}{\cos^2(-\theta)}}$

$= \sin^2(-\theta)\cos^2(-\theta)$

$= \sin^2\theta\cos^2\theta$

67. $\sec\theta - \cos\theta = \dfrac{1}{\cos\theta} - \cos\theta = \dfrac{1}{\cos\theta} - \dfrac{\cos^2\theta}{\cos\theta}$

$= \dfrac{1-\cos^2\theta}{\cos\theta} = \dfrac{\sin^2\theta}{\cos\theta}$

$= \dfrac{\sin\theta}{\cos\theta}\cdot\sin\theta = \tan\theta\sin\theta$

69. $(\sec\theta + \csc\theta)(\cos\theta - \sin\theta)$

$= \left(\dfrac{1}{\cos\theta} + \dfrac{1}{\sin\theta}\right)(\cos\theta - \sin\theta)$

$= \dfrac{1}{\cos\theta}(\cos\theta) - \dfrac{1}{\cos\theta}(\sin\theta)$

$\qquad + \dfrac{1}{\sin\theta}(\cos\theta) - \dfrac{1}{\sin\theta}(\sin\theta)$

$= 1 - \dfrac{\sin\theta}{\cos\theta} + \dfrac{\cos\theta}{\sin\theta} - 1 = -\tan\theta + \cot\theta$

$= \cot\theta - \tan\theta$

71. $\sin\theta\left(\csc\theta - \sin\theta\right) = \sin\theta\csc\theta - \sin^2\theta$

$= \sin\theta\cdot\dfrac{1}{\sin\theta} - \sin^2\theta$

$= 1 - \sin^2\theta = \cos^2\theta$

73. $\dfrac{1+\tan^2\theta}{1+\cot^2\theta} = \dfrac{\sec^2\theta}{\csc^2\theta} = \dfrac{\dfrac{1}{\cos^2\theta}}{\dfrac{1}{\sin^2\theta}}$

$= \dfrac{1}{\cos^2\theta}\cdot\dfrac{\sin^2\theta}{1} = \dfrac{\sin^2\theta}{\cos^2\theta} = \tan^2\theta$

75. $\dfrac{\csc\theta}{\cot(-\theta)} = \dfrac{\csc\theta}{-\cot\theta} = \dfrac{\dfrac{1}{\sin\theta}}{-\dfrac{\cos\theta}{\sin\theta}} = -\dfrac{1}{\cos\theta}$

$= -\sec\theta$

77. $\sin^2(-\theta) + \tan^2(-\theta) + \cos^2(-\theta)$

$= \left[\sin^2(-\theta) + \cos^2(-\theta)\right] + \tan^2(-\theta)$

$= 1 + \tan^2(-\theta) = 1 + (-\tan\theta)^2$

$= 1 + \tan^2\theta = \sec^2\theta$

79. $\cos x = \dfrac{1}{5}$, so x is in quadrant I or quadrant IV.

$\sin x = \pm\sqrt{1-\cos^2 x} = \pm\sqrt{1-\left(\dfrac{1}{5}\right)^2} = \pm\sqrt{\dfrac{24}{25}}$

$= \pm\dfrac{\sqrt{24}}{5} = \pm\dfrac{2\sqrt{6}}{5}$

$\tan x = \dfrac{\sin x}{\cos x} = \dfrac{\pm\dfrac{2\sqrt{6}}{5}}{\dfrac{1}{5}} = \pm 2\sqrt{6}$

$\sec x = \dfrac{1}{\cos x} = \dfrac{1}{\dfrac{1}{5}} = 5$

(continued on next page)

(continued)

Quadrant I:

$$\frac{\sec x - \tan x}{\sin x} = \frac{5 - 2\sqrt{6}}{\frac{2\sqrt{6}}{5}} = \frac{25 - 10\sqrt{6}}{2\sqrt{6}}$$

$$= \frac{25 - 10\sqrt{6}}{2\sqrt{6}} \cdot \frac{\sqrt{6}}{\sqrt{6}} = \frac{25\sqrt{6} - 60}{12}$$

Quadrant IV:

$$\frac{\sec x - \tan x}{\sin x} = \frac{5 - \left(-2\sqrt{6}\right)}{-\frac{2\sqrt{5}}{5}} = \frac{25 + 10\sqrt{6}}{-2\sqrt{6}}$$

$$= \frac{25 + 10\sqrt{6}}{-2\sqrt{6}} \cdot \frac{-\sqrt{6}}{-\sqrt{6}} = \frac{-25\sqrt{6} - 60}{12}$$

In Exercises 81–83, the functions are graphed in the window $[-2\pi, -2\pi] \times [-4, 4]$.

81. The equation $\cos 2x = 1 - 2\sin^2 x$ is an identity. $y_1 = \cos 2x,\ y_2 = 1 - 2\sin^2 x$

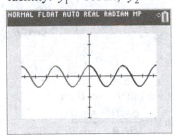

83. The equation $\sin x = \sqrt{1 - \cos^2 x}$ is not an identity. $y_1 = \sin x,\ y_2 = \sqrt{1 - \cos^2 x}$

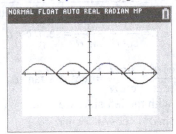

85. $y = \sin(-2x) \Rightarrow y = -\sin(2x)$

87. $y = \cos(-4x) \Rightarrow y = \cos(4x)$

89. (a) $y = \sin(-4x) \Rightarrow y = -\sin(4x)$

(b) $y = \cos(-2x) \Rightarrow y = \cos(2x)$

(c) $y = -5\sin(-3x) \Rightarrow y = -5\left[-\sin(3x)\right] \Rightarrow$
$y = 5\sin(3x)$

Section 5.2 Verifying Trigonometric Identities

1. B   3. A

5. $\sin^2\theta + \cos^2\theta = \underline{1}$

7. $\sin(-\theta) = \underline{-\sin\theta}$

9. $\tan\theta = \dfrac{1}{\cot\theta} = \dfrac{\sin\theta}{\cos\theta}$

11. $\cot\theta + \dfrac{1}{\cot\theta} = \cot\theta + \tan\theta$

$$= \frac{\cos\theta}{\sin\theta} + \frac{\sin\theta}{\cos\theta}$$

$$= \frac{\cos^2\theta + \sin^2\theta}{\sin\theta\cos\theta}$$

$$= \frac{1}{\sin\theta\cos\theta} \text{ or } \csc\theta\sec\theta$$

13. $\tan s(\cot x + \csc x) = \dfrac{\sin x}{\cos x}\left(\dfrac{\cos x}{\sin x} + \dfrac{1}{\sin x}\right)$

$$= 1 + \frac{1}{\cos x}$$

$$= 1 + \sec x$$

15. $\dfrac{1}{\csc^2\theta} + \dfrac{1}{\sec^2\theta} = \sin^2\theta + \cos^2\theta = 1$

17. $(\sin\alpha - \cos\alpha)^2$

$$= \sin^2\alpha - 2\sin\alpha\cos\alpha + \cos^2\alpha$$

$$= (\sin^2\alpha + \cos^2\alpha) - 2\sin\alpha\cos\alpha$$

$$= 1 - 2\sin\alpha\cos\alpha$$

19. $(1 + \sin t)^2 + \cos^2 t = 1 + 2\sin t + \sin^2 t + \cos^2 t$

$$= 1 + 2\sin t + (\sin^2 t + \cos^2 t)$$

$$= 1 + 2\sin t + 1 = 2 + 2\sin t$$

21. $\dfrac{1}{1 + \cos x} - \dfrac{1}{1 - \cos x}$

$$= \frac{1 - \cos x}{(1 + \cos x)(1 - \cos x)} - \frac{1 + \cos x}{(1 + \cos x)(1 - \cos x)}$$

$$= \frac{(1 - \cos x) - (1 + \cos x)}{(1 + \cos x)(1 - \cos x)}$$

$$= \frac{1 - \cos x - 1 - \cos x}{1 - \cos^2 x} = -\frac{2\cos x}{\sin^2 x} \text{ or}$$

$$-\frac{2\cos x}{\sin^2 x} = -\frac{2\cos x}{\sin x \sin x} = -2\left(\frac{\cos x}{\sin x}\right)\left(\frac{1}{\sin x}\right)$$

$$= -2\cot x\csc x$$

23. $\sin^2\theta - 1 = (\sin\theta + 1)(\sin\theta - 1)$

25. $(\sin x + 1)^2 - (\sin x - 1)^2$
$= \left[(\sin x + 1) + (\sin x - 1)\right]$
$\qquad \cdot \left[(\sin x + 1) - (\sin x - 1)\right]$
$= (\sin x + 1 + \sin x - 1)(\sin x + 1 - \sin x + 1)$
$= (2\sin x)(2) = 4\sin x$

27. $2\sin^2 x + 3\sin x + 1$
Let $a = \sin x$.
$2\sin^2 x + 3\sin x + 1 = 2a^2 + 3a + 1$
$\qquad\qquad = (2a + 1)(a + 1)$
$\qquad\qquad = (2\sin x + 1)(\sin x + 1)$

29. $\cos^4 x + 2\cos^2 x + 1$
Let $\cos^2 x = a$.
$\cos^4 x + 2\cos^2 x + 1 = a^2 + 2a + 1$
$\qquad\qquad = (a + 1)^2 = \left(\cos^2 x + 1\right)^2$

31. $\sin^3 x - \cos^3 x$
Let $\sin x = a$ and $\cos x = b$.
$\sin^3 x - \cos^3 x$
$= a^3 - b^3 = (a - b)\left(a^2 + ab + b^2\right)$
$= (\sin x - \cos x)\left(\sin^2 x + \sin x \cos x + \cos^2 x\right)$
$= (\sin x - \cos x)\left[\left(\sin^2 x + \cos^2 x\right) + \sin x \cos x\right]$
$= (\sin x - \cos x)(1 + \sin x \cos x)$

33. $\tan\theta \cos\theta = \dfrac{\sin\theta}{\cos\theta}\cos\theta = \sin\theta$

35. $\sec r \cos r = \dfrac{1}{\cos r} \cdot \cos r = 1$

37. $\dfrac{\sin\beta \tan\beta}{\cos\beta} = \tan\beta \tan\beta = \tan^2\beta$

39. $\sec^2 x - 1 = \dfrac{1}{\cos^2 x} - 1 = \dfrac{1}{\cos^2 x} - \dfrac{\cos^2 x}{\cos^2 x}$
$= \dfrac{1 - \cos^2 x}{\cos^2 x} = \dfrac{\sin^2 x}{\cos^2 x} = \tan^2 x$

41. $\dfrac{\sin^2 x}{\cos^2 x} + \sin x \csc x = \tan^2 x + \sin x \cdot \dfrac{1}{\sin x}$
$= \tan^2 x + 1 = \sec^2 x$

43. $1 - \dfrac{1}{\csc^2 x} = 1 - \sin^2 x = \cos^2 x$

45. Verify $\dfrac{\cot\theta}{\csc\theta} = \cos\theta$.
$\dfrac{\cot\theta}{\csc\theta} = \dfrac{\dfrac{\cos\theta}{\sin\theta}}{\dfrac{1}{\sin\theta}} = \dfrac{\cos\theta}{\sin\theta} \cdot \dfrac{\sin\theta}{1} = \cos\theta$

47. Verify $\dfrac{1 - \sin^2\beta}{\cos\beta} = \cos\beta$.
$\dfrac{1 - \sin^2\beta}{\cos\beta} = \dfrac{\cos^2\beta}{\cos\beta} = \cos\beta$

49. Verify $\cos^2\theta(\tan^2\theta + 1) = 1$.
$\cos^2\theta(\tan^2\theta + 1) = \cos^2\theta\left(\dfrac{\sin^2\theta}{\cos^2\theta} + 1\right)$
$= \cos^2\theta\left(\dfrac{\sin^2\theta}{\cos^2\theta} + \dfrac{\cos^2\theta}{\cos^2\theta}\right)$
$= \cos^2\theta\left(\dfrac{\sin^2\theta + \cos^2\theta}{\cos^2\theta}\right)$
$= \cos^2\theta\left(\dfrac{1}{\cos^2\theta}\right) = 1$

51. Verify $\cot\theta + \tan\theta = \sec\theta \csc\theta$.
$\cot\theta + \tan\theta = \dfrac{\cos\theta}{\sin\theta} + \dfrac{\sin\theta}{\cos\theta}$
$= \dfrac{\cos^2\theta}{\sin\theta\cos\theta} + \dfrac{\sin^2\theta}{\sin\theta\cos\theta}$
$= \dfrac{\cos^2\theta + \sin^2\theta}{\cos\theta\sin\theta} = \dfrac{1}{\cos\theta\sin\theta}$
$= \dfrac{1}{\cos\theta} \cdot \dfrac{1}{\sin\theta} = \sec\theta \csc\theta$

53. Verify $\dfrac{\cos\alpha}{\sec\alpha} + \dfrac{\sin\alpha}{\csc\alpha} = \sec^2\alpha - \tan^2\alpha$.
Working with the left side, we have
$\dfrac{\cos\alpha}{\sec\alpha} + \dfrac{\sin\alpha}{\csc\alpha} = \dfrac{\cos\alpha}{\dfrac{1}{\cos\alpha}} + \dfrac{\sin\alpha}{\dfrac{1}{\sin\alpha}}$
$= \cos^2\alpha + \sin^2\alpha = 1$
Working with the right side, we have
$\sec^2\alpha - \tan^2\alpha = 1$.
$\dfrac{\cos\alpha}{\sec\alpha} + \dfrac{\sin\alpha}{\csc\alpha} = 1 = \sec^2\alpha - \tan^2\alpha,$ so the statement has been verified.

55. Verify $\sin^4\theta - \cos^4\theta = 2\sin^2\theta - 1$.

$\sin^4\theta - \cos^4\theta$

$= \left(\sin^2\theta + \cos^2\theta\right)\left(\sin^2\theta - \cos^2\theta\right)$

$= 1\cdot\left(\sin^2\theta - \cos^2\theta\right) = \sin^2\theta - \cos^2\theta$

$= \sin^2\theta - \left(1 - \sin^2\theta\right) = 2\sin^2\theta - 1$

57. Verify $\dfrac{1-\cos x}{1+\cos x} = (\cot x - \csc x)^2$.

Work with the left side.

$\dfrac{1-\cos x}{1+\cos x} = \dfrac{(1-\cos x)(1-\cos x)}{(1+\cos x)(1-\cos x)}$

$= \dfrac{1 - 2\cos x + \cos^2 x}{1 - \cos^2 x}$

$= \dfrac{1 - 2\cos x + \cos^2 x}{\sin^2 x}$

Work with the right side.

$(\cot x - \csc x)^2 = \left(\dfrac{\cos x}{\sin x} - \dfrac{1}{\sin x}\right)^2 = \left(\dfrac{\cos x - 1}{\sin x}\right)^2$

$= \dfrac{\cos^2 x - 2\cos x + 1}{\sin^2 x}$

$\dfrac{1-\cos x}{1+\cos x} = \dfrac{\cos^2 x - 2\cos x + 1}{\sin^2 x} = (\cot x - \csc x)^2$

Thus, the statement has been verified.

59. Verify $\dfrac{\cos\theta + 1}{\tan^2\theta} = \dfrac{\cos\theta}{\sec\theta - 1}$.

Work with the left side.

$\dfrac{\cos\theta + 1}{\tan^2\theta} = \dfrac{\cos\theta + 1}{\sec^2\theta - 1} = \dfrac{\cos\theta + 1}{\dfrac{1}{\cos^2\theta} - 1}$

$= \dfrac{(\cos\theta + 1)\cos^2\theta}{\left(\dfrac{1}{\cos^2\theta} - 1\right)\cos^2\theta}$

$= \dfrac{\cos^2\theta(\cos\theta + 1)}{1 - \cos^2\theta}$

$= \dfrac{\cos^2\theta(\cos\theta + 1)}{(1+\cos\theta)(1-\cos\theta)} = \dfrac{\cos^2\theta}{1 - \cos\theta}$

Now work with the right side.

$\dfrac{\cos\theta}{\sec\theta - 1} = \dfrac{\cos\theta}{\dfrac{1}{\cos\theta} - 1} = \dfrac{\cos\theta}{\dfrac{1}{\cos\theta} - 1}\cdot\dfrac{\cos\theta}{\cos\theta}$

$= \dfrac{\cos^2\theta}{1 - \cos\theta}$

$\dfrac{\cos\theta + 1}{\tan^2\theta} = \dfrac{\cos^2\theta}{1 - \cos\theta} = \dfrac{\cos\theta}{\sec\theta - 1}$

Thus, the statement has been verified.

61. Verify $\dfrac{1}{1 - \sin\theta} + \dfrac{1}{1 + \sin\theta} = 2\sec^2\theta$.

$\dfrac{1}{1 - \sin\theta} + \dfrac{1}{1 + \sin\theta}$

$= \dfrac{1 + \sin\theta}{(1+\sin\theta)(1-\sin\theta)} + \dfrac{1 - \sin\theta}{(1+\sin\theta)(1-\sin\theta)}$

$= \dfrac{(1+\sin\theta) + (1-\sin\theta)}{(1+\sin\theta)(1-\sin\theta)} = \dfrac{1 + \sin\theta + 1 - \sin\theta}{(1+\sin\theta)(1-\sin\theta)}$

$= \dfrac{2}{1 - \sin^2\theta} = \dfrac{2}{\cos^2\theta} = 2\sec^2\theta$

63. Verify $\dfrac{\cot\alpha + 1}{\cot\alpha - 1} = \dfrac{1 + \tan\alpha}{1 - \tan\alpha}$.

$\dfrac{\cot\alpha + 1}{\cot\alpha - 1} = \dfrac{\dfrac{\cos\alpha}{\sin\alpha} + 1}{\dfrac{\cos\alpha}{\sin\alpha} - 1} = \dfrac{\dfrac{\cos\alpha}{\sin\alpha} + 1}{\dfrac{\cos\alpha}{\sin\alpha} - 1}\cdot\dfrac{\sin\alpha}{\sin\alpha}$

$= \dfrac{\cos\alpha + \sin\alpha}{\cos\alpha - \sin\alpha}$

$= \dfrac{\cos\alpha + \sin\alpha}{\cos\alpha - \sin\alpha}\cdot\dfrac{\dfrac{1}{\cos\alpha}}{\dfrac{1}{\cos\alpha}}$

$= \dfrac{\dfrac{\cos\alpha}{\cos\alpha} + \dfrac{\sin\alpha}{\cos\alpha}}{\dfrac{\cos\alpha}{\cos\alpha} - \dfrac{\sin\alpha}{\cos\alpha}} = \dfrac{1 + \tan\alpha}{1 - \tan\alpha}$

65. Verify $\dfrac{\cos\theta}{\sin\theta\cot\theta} = 1$.

$\dfrac{\cos\theta}{\sin\theta\cot\theta} = \dfrac{\cos\theta}{\sin\theta\cdot\dfrac{\cos\theta}{\sin\theta}} = \dfrac{\cos\theta}{\cos\theta} = 1$

67. Verify $\dfrac{\sec^4\theta - \tan^4\theta}{\sec^2\theta + \tan^2\theta} = \sec^2\theta - \tan^2\theta$.

$\dfrac{\sec^4\theta - \tan^4\theta}{\sec^2\theta + \tan^2\theta}$

$= \dfrac{\left(\sec^2\theta + \tan^2\theta\right)\left(\sec^2\theta - \tan^2\theta\right)}{\sec^2\theta + \tan^2\theta}$

$= \sec^2\theta - \tan^2\theta$

69. Verify $\dfrac{\tan^2 t - 1}{\sec^2 t} = \dfrac{\tan t - \cot t}{\tan t + \cot t}$.

Simplify the right side

$$\dfrac{\tan t - \cot t}{\tan t + \cot t} = \dfrac{\tan t - \dfrac{1}{\tan t}}{\tan t + \dfrac{1}{\tan t}} = \dfrac{\tan t - \dfrac{1}{\tan t}}{\tan t + \dfrac{1}{\tan t}} \cdot \dfrac{\tan t}{\tan t} = \dfrac{\tan^2 t - 1}{\tan^2 t + 1} = \dfrac{\tan^2 t - 1}{\sec^2 t}$$

71. Verify $\sin^2 \alpha \sec^2 \alpha + \sin^2 \alpha \csc^2 \alpha = \sec^2 \alpha$.

$$\sin^2 \alpha \sec^2 \alpha + \sin^2 \alpha \csc^2 \alpha = \sin^2 \alpha \cdot \dfrac{1}{\cos^2 \alpha} + \sin^2 \alpha \cdot \dfrac{1}{\sin^2 \alpha} = \dfrac{\sin^2 \alpha}{\cos^2 \alpha} + 1 = \tan^2 \alpha + 1 = \sec^2 \alpha$$

73. Verify $\dfrac{\tan x}{1 + \cos x} + \dfrac{\sin x}{1 - \cos x} = \cot x + \sec x \csc x$.

$$\dfrac{\tan x}{1 + \cos x} + \dfrac{\sin x}{1 - \cos x} = \dfrac{\tan x(1 - \cos x)}{(1 + \cos x)(1 - \cos x)} + \dfrac{\sin x(1 + \cos x)}{(1 + \cos x)(1 - \cos x)} = \dfrac{\tan x(1 - \cos x) + \sin x(1 + \cos x)}{(1 + \cos x)(1 - \cos x)}$$

$$= \dfrac{\tan x - \sin x + \sin x + \sin x \cos x}{1 - \cos^2 x} = \dfrac{\tan x + \sin x \cos x}{\sin^2 x} = \dfrac{\tan x}{\sin^2 x} + \dfrac{\sin x \cos x}{\sin^2 x}$$

$$= \tan x \cdot \dfrac{1}{\sin^2 x} + \dfrac{\cos x}{\sin x} = \dfrac{\sin x}{\cos x} \cdot \dfrac{1}{\sin^2 x} + \cot x = \dfrac{1}{\cos x} \cdot \dfrac{1}{\sin x} + \cot x$$

$$= \sec x \csc x + \cot x$$

75. Verify $\dfrac{1 + \cos x}{1 - \cos x} - \dfrac{1 - \cos x}{1 + \cos x} = 4 \cot x \csc x$

$$\dfrac{1 + \cos x}{1 - \cos x} - \dfrac{1 - \cos x}{1 + \cos x} = \dfrac{(1 + \cos x)^2}{(1 + \cos x)(1 - \cos x)} - \dfrac{(1 - \cos x)^2}{(1 + \cos x)(1 - \cos x)}$$

$$= \dfrac{1 + 2\cos x + \cos^2 x}{(1 + \cos x)(1 - \cos x)} - \dfrac{1 - 2\cos x + \cos^2 x}{(1 + \cos x)(1 - \cos x)}$$

$$= \dfrac{1 + 2\cos x + \cos^2 x - 1 + 2\cos x - \cos^2 x}{(1 + \cos x)(1 - \cos x)}$$

$$= \dfrac{4\cos x}{1 - \cos^2 x} = \dfrac{4\cos x}{\sin^2 x} = 4 \cdot \dfrac{\cos x}{\sin x} \cdot \dfrac{1}{\sin x} = 4 \cot x \csc x$$

77. Verify $\dfrac{1 - \sin \theta}{1 + \sin \theta} = \sec^2 \theta - 2 \sec \theta \tan \theta + \tan^2 \theta$

Simplify the right side

$$\sec^2 \theta - 2 \sec \theta \tan \theta + \tan^2 \theta = \dfrac{1}{\cos^2 \theta} - 2 \cdot \dfrac{1}{\cos \theta} \cdot \dfrac{\sin \theta}{\cos \theta} + \dfrac{\sin^2 \theta}{\cos^2 \theta} = \dfrac{1 - 2\sin \theta + \sin^2 \theta}{\cos^2 \theta} = \dfrac{(1 - \sin \theta)^2}{1 - \sin^2 \theta}$$

$$= \dfrac{(1 - \sin \theta)^2}{(1 + \sin \theta)(1 - \sin \theta)} = \dfrac{1 - \sin \theta}{1 + \sin \theta}$$

79. Verify $\dfrac{-1}{\tan \alpha - \sec \alpha} + \dfrac{-1}{\tan \alpha + \sec \alpha} = 2 \tan \alpha$.

$$\dfrac{-1}{\tan \alpha - \sec \alpha} + \dfrac{-1}{\tan \alpha + \sec \alpha} = \dfrac{-\tan \alpha - \sec \alpha}{(\tan \alpha + \sec \alpha)(\tan \alpha - \sec \alpha)} + \dfrac{-\tan \alpha + \sec \alpha}{(\tan \alpha + \sec \alpha)(\tan \alpha - \sec \alpha)}$$

$$= \dfrac{-\tan \alpha - \sec \alpha - \tan \alpha + \sec \alpha}{(\tan \alpha + \sec \alpha)(\tan \alpha - \sec \alpha)} = \dfrac{-2\tan \alpha}{\tan^2 \alpha - \sec^2 \alpha} = \dfrac{-2\tan \alpha}{\tan^2 \alpha - (\tan^2 \alpha + 1)}$$

$$= \dfrac{-2\tan \alpha}{\tan^2 \alpha - \tan^2 \alpha - 1} = \dfrac{-2\tan \alpha}{-1} = 2 \tan \alpha$$

81. Verify $\left(1-\cos^2\alpha\right)\left(1+\cos^2\alpha\right)=2\sin^2\alpha-\sin^4\alpha$.

$$\left(1-\cos^2\alpha\right)\left(1+\cos^2\alpha\right)=\sin^2\alpha\left(1+\cos^2\alpha\right)=\sin^2\alpha\left(2-\sin^2\alpha\right)=2\sin^2\alpha-\sin^4\alpha$$

83. Verify $\dfrac{1-\cos x}{1+\cos x}=\csc^2 x-2\csc x\cot x+\cot^2 x$

Work with the left side:

$$\frac{1-\cos x}{1+\cos x}=\frac{1-\cos x}{1+\cos x}\cdot\frac{1-\cos x}{1-\cos x}=\frac{1-2\cos x+\cos^2 x}{1-\cos^2 x}=\frac{1-2\cos x+\cos^2 x}{\sin^2 x}$$

Work with the right side:

$$\csc^2 x-2\csc x\cot x+\cot^2 x=\frac{1}{\sin^2 x}-\frac{2\cos x}{\sin^2 x}+\frac{\cos^2 x}{\sin^2 x}=\frac{1-2\cos x+\cos^2 x}{\sin^2 x}$$

$$\frac{1-\cos x}{1+\cos x}=\frac{1-2\cos x+\cos^2 x}{\sin^2 x}=\csc^2 x-2\csc x\cot x+\cot^2 x,\ \text{so the statement has been verified.}$$

85. Verify $\left(2\sin x+\cos x\right)^2+\left(2\cos x-\sin x\right)^2=5$

$$\left(2\sin x+\cos x\right)^2+\left(2\cos x-\sin x\right)^2=\left(4\sin^2 x+4\sin x\cos x+\cos^2 x\right)+\left(4\cos^2 x-4\sin x\cos x+\sin^2 x\right)$$

$$=4\left(\sin^2 x+\cos^2 x\right)+\left(\cos^2 x+\sin^2 x\right)=4+1=5$$

87. Verify $\sec x-\cos x+\csc x-\sin x-\sin x\tan x=\cos x\cot x$

$$\sec x-\cos x+\csc x-\sin x-\sin x\tan x=\frac{1}{\cos x}-\cos x+\frac{1}{\sin x}-\sin x-\sin x\left(\frac{\sin x}{\cos x}\right)$$

$$=\left(\frac{1}{\cos x}-\cos x\right)+\left(\frac{1}{\sin x}-\sin x\right)-\frac{\sin^2 x}{\cos x}$$

$$=\frac{1-\cos^2 x}{\cos x}+\frac{1-\sin^2 x}{\sin x}-\frac{\sin^2 x}{\cos x}$$

$$=\left(\frac{1-\cos^2 x}{\cos x}-\frac{\sin^2 x}{\cos x}\right)+\frac{1-\sin^2 x}{\sin x}=\frac{1-\cos^2 x-\sin^2 x}{\cos x}+\frac{\cos^2 x}{\sin x}$$

$$=\frac{1-\left(\cos^2 x+\sin^2 x\right)}{\cos x}+\frac{\cos^2 x}{\sin x}=\frac{1-1}{\cos x}+\cos x\cdot\frac{\cos x}{\sin x}=\cos x\cot x$$

In Exercises 89–95, the functions are graphed in the window $\left[-2\pi,\ 2\pi\right]\times\left[-4,\ 4\right]$.

89. $\left(\sec\theta+\tan\theta\right)\left(1-\sin\theta\right)$ appears to be equivalent to $\cos\theta$.

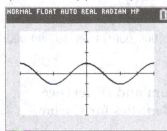

$$\left(\sec\theta+\tan\theta\right)\left(1-\sin\theta\right)=\left(\frac{1}{\cos\theta}+\frac{\sin\theta}{\cos\theta}\right)\left(1-\sin\theta\right)=\left(\frac{1+\sin\theta}{\cos\theta}\right)\left(1-\sin\theta\right)$$

$$=\frac{\left(1+\sin\theta\right)\left(1-\sin\theta\right)}{\cos\theta}=\frac{1-\sin^2\theta}{\cos\theta}=\frac{\cos^2\theta}{\cos\theta}=\cos\theta$$

91. $\dfrac{\cos\theta+1}{\sin\theta+\tan\theta}$ appears to be equivalent to $\cot\theta$.

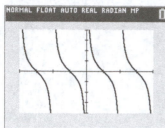

$$\frac{\cos\theta+1}{\sin\theta+\tan\theta}=\frac{1+\cos\theta}{\sin\theta+\dfrac{\sin\theta}{\cos\theta}}=\frac{1+\cos\theta}{\sin\theta\left(1+\dfrac{1}{\cos\theta}\right)}$$

$$=\frac{1+\cos\theta}{\sin\theta\left(1+\dfrac{1}{\cos\theta}\right)}\cdot\frac{\cos\theta}{\cos\theta}$$

$$=\frac{(1+\cos\theta)\cos\theta}{\sin\theta(\cos\theta+1)}=\frac{\cos\theta}{\sin\theta}=\cot\theta.$$

93. Is $\dfrac{2+5\cos x}{\sin x}=2\csc x+5\cot x$ an identity?

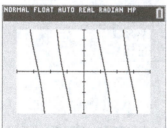

The graphs of $y_1=\dfrac{2+5\cos x}{\sin x}$ and
$y_2=2\csc x+5\cot x$ appear to be the same.

$$\frac{2+5\cos x}{\sin x}=\frac{2}{\sin x}+\frac{5\cos x}{\sin x}=2\csc x+5\cot x$$

Thus, the given statement is an identity.

95. Is $\dfrac{\tan x-\cot x}{\tan x+\cot x}=2\sin^2 x$ an identity?

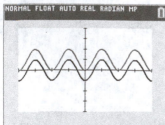

The graphs of
$y=\dfrac{\tan x-\cot x}{\tan x+\cot x}$ and $y=2\sin^2 x$ are not the same. The given statement is not an identity.

97. Show that $\sin(\csc t)=1$ is not an identity.

We need to find only one value for which the statement is false. Let $t=2$. Use a calculator to find that $\sin(\csc 2)\approx 0.891094$, which is not equal to 1. $\sin(\csc t)=1$ does not hold true for *all* real numbers t. Thus, it is not an identity.

99. Show that $\csc t=\sqrt{1+\cot^2 t}$ is not an identity.

Let $t=\dfrac{\pi}{4}$. We have $\csc\dfrac{\pi}{4}=\sqrt{2}$ and

$$\sqrt{1+\cot^2\frac{\pi}{4}}=\sqrt{1+1^2}=\sqrt{1+1}=\sqrt{2}.\text{ But let}$$

$t=-\dfrac{\pi}{4}$. We have $\csc\left(-\dfrac{\pi}{4}\right)=-\sqrt{2}$ and

$$\sqrt{1+\cot^2\left(-\frac{\pi}{4}\right)}=\sqrt{1+(-1)^2}=\sqrt{1+1}=\sqrt{2}.$$

$\csc t=\sqrt{1+\cot^2 t}$ does not hold true for *all* real numbers t. Thus, it is not an identity.

101. (a) $I=k\cos^2\theta=k\left(1-\sin^2\theta\right)$

(b) When $\theta=0$, $\cos\theta=1$, its maximum value. Thus, $\cos^2\theta$ will be a maximum and, as a result, I will be maximized if k is a positive constant.

103. The sum of L and C equals 3.

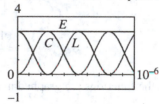

105. $E(t)=L(t)+C(t)$
$$=3\cos^2(6,000,000t)+3\sin^2(6,000,000t)$$
$$=3\left[\cos^2(6,000,000t)+\sin^2(6,000,000t)\right]$$
$$=3\cdot 1=3$$

Section 5.3   Sum and Difference Identities for Cosine

1. F; $\cos(x+y)=\cos x\cos y-\sin x\sin y$

3. E; $\cos\left(\dfrac{\pi}{2}-x\right)=\sin x$

5. E

$$\cos\left(x - \frac{\pi}{2}\right) = \cos\left[-\left(\frac{\pi}{2} - x\right)\right]$$
$$= \cos\left(\frac{\pi}{2} - x\right) = \sin x$$

7. H; $\tan\left(\frac{\pi}{2} - x\right) = \cot x$

9. $\cos 75° = \cos(30° + 45°)$
$$= \cos 30° \cos 45° - \sin 30° \sin 45°$$
$$= \frac{\sqrt{3}}{2} \cdot \frac{\sqrt{2}}{2} - \frac{1}{2} \cdot \frac{\sqrt{2}}{2}$$
$$= \frac{\sqrt{6}}{4} - \frac{\sqrt{2}}{4} = \frac{\sqrt{6} - \sqrt{2}}{4}$$

11. $\cos(-105°) = \cos\left[-60° + (-45°)\right]$
$$= \cos(-60°)\cos(-45°) - \sin(-60°)\sin(-45°)$$
$$= \frac{1}{2} \cdot \frac{\sqrt{2}}{2} - \left(-\frac{\sqrt{3}}{2}\right)\left(-\frac{\sqrt{2}}{2}\right)$$
$$= \frac{\sqrt{2}}{4} - \frac{\sqrt{6}}{4} = \frac{\sqrt{2} - \sqrt{6}}{4}$$

13. $\cos\left(\frac{7\pi}{12}\right) = \cos\left(\frac{4\pi}{12} + \frac{3\pi}{12}\right) = \cos\left(\frac{\pi}{3} + \frac{\pi}{4}\right)$
$$= \cos\frac{\pi}{3}\cos\frac{\pi}{4} - \sin\frac{\pi}{3}\sin\frac{\pi}{4}$$
$$= \frac{1}{2} \cdot \frac{\sqrt{2}}{2} - \frac{\sqrt{3}}{2} \cdot \frac{\sqrt{2}}{2}$$
$$= \frac{\sqrt{2}}{4} - \frac{\sqrt{6}}{4} = \frac{\sqrt{2} - \sqrt{6}}{4}$$

15. $\cos\left(-\frac{\pi}{12}\right) = \cos\left(\frac{2\pi}{12} - \frac{3\pi}{12}\right) = \cos\left(\frac{\pi}{6} - \frac{\pi}{4}\right)$
$$= \cos\frac{\pi}{6}\cos\frac{\pi}{4} + \sin\frac{\pi}{6}\sin\frac{\pi}{4}$$
$$= \frac{\sqrt{3}}{2} \cdot \frac{\sqrt{2}}{2} + \frac{1}{2} \cdot \frac{\sqrt{2}}{2}$$
$$= \frac{\sqrt{6}}{4} + \frac{\sqrt{2}}{4} = \frac{\sqrt{6} + \sqrt{2}}{4}$$

17. $\cos 40° \cos 50° - \sin 40° \sin 50°$
$$= \cos(40° + 50°) = \cos 90° = 0$$

19. $\tan 87° = \cot(90° - 87°) = \cot 3°$

21. $\cos\frac{\pi}{12} = \sin\left(\frac{\pi}{2} - \frac{\pi}{12}\right) = \sin\frac{5\pi}{12}$

23. $\csc 14°24' = \sec(90° - 14°24') = \sec 75°36'$

25. $\sin\frac{5\pi}{8} = \cos\left(\frac{\pi}{2} - \frac{5\pi}{8}\right) = \cos\left(\frac{4\pi}{8} - \frac{5\pi}{8}\right)$
$$= \cos\left(-\frac{\pi}{8}\right)$$

27. $\sec 146°42' = \csc(90° - 146°42')$
$$= \csc\left[-(146°42' - 90°)\right]$$
$$= \csc(-56°42')$$

29. $\cot 176.9814° = \tan(90° - 176.9814°)$
$$= \tan(-86.9814°)$$

31. Because $\frac{\pi}{6} = \frac{\pi}{2} - \frac{\pi}{3}$, $\cot\frac{\pi}{3} = \tan\frac{\pi}{6}$.
$$\cot\frac{\pi}{3} = \tan\frac{\pi}{6}$$

33. Because $90° - 57° = 33°$, $\sin 57° = \cos 33°$.
$$\sin 57° = \cos 33°$$

35. Because $90° - 70° = 20°$, and $\sin x = \frac{1}{\csc x}$,
$$\cos 70° = \sin 20° = \frac{1}{\csc 20°}.$$
$$\cos 70° = \frac{1}{\csc 20°}$$

For exercises 37–41, other answers are possible.

37. $\tan\theta = \cot(45° + 2\theta)$
   Because $\tan\theta = \cot(90° - \theta)$,
$$90° - \theta = 45° + 2\theta \Rightarrow 90° = 45° + 3\theta \Rightarrow$$
$$3\theta = 45° \Rightarrow \theta = 15°$$

39. $\sec x = \csc\frac{2\pi}{3}$
   By a cofunction identity, $\sec x = \csc\left(\frac{\pi}{2} - x\right)$.
   Thus,
$$\csc\frac{2\pi}{3} = \csc\left(\frac{\pi}{2} - x\right) \Rightarrow \frac{2\pi}{3} = \frac{\pi}{2} - x \Rightarrow$$
$$x = -\frac{\pi}{6}$$

41. $\sin(3\theta - 15°) = \cos(\theta + 25°)$

Because $\sin\theta = \cos(90° - \theta)$, we have

$$\sin(3\theta - 15°) = \cos\left[90° - (3\theta - 15°)\right]$$
$$= \cos(90° - 3\theta + 15°)$$
$$= \cos(105° - 3\theta)$$

Solve $\cos(105° - 3\theta) = \cos(\theta + 25°)$.

$$\cos(105° - 3\theta) = \cos(\theta + 25°)$$
$$105° - 3\theta = \theta + 25°$$
$$105° = 4\theta + 25°$$
$$4\theta = 80° \Rightarrow \theta = 20°$$

43. $\cos(0° - \theta) = \cos 0° \cos\theta + \sin 0° \sin\theta$
$$= (1)\cos\theta + (0)\sin\theta = \cos\theta$$

45. $\cos(\theta - 180°) = \cos\theta\cos 180° + \sin\theta\sin 180°$
$$= \cos\theta(-1) + \sin\theta(0)$$
$$= -\cos\theta + 0 = -\cos\theta$$

47. $\cos(0° + \theta) = \cos 0° \cos\theta - \sin 0° \sin\theta$
$$= (1)\cos\theta - (0)\sin\theta = \cos\theta$$

49. $\cos(180° + \theta) = \cos 180° \cos\theta - \sin 180° \sin\theta$
$$= (-1)\cos\theta - (0)\sin\theta$$
$$= -\cos\theta - 0 = -\cos\theta$$

51. $\sin s = \dfrac{3}{5}$ and $\sin t = -\dfrac{12}{13}$, s is in quadrant I and t is in quadrant III.

$\sin s = \dfrac{y}{r} = \dfrac{3}{5} \Rightarrow y = 3, r = 5$. Substituting into the Pythagorean theorem, we have

$x^2 + 3^2 = 5^2 \Rightarrow x^2 = 16 \Rightarrow x = 4$, because $\cos s > 0$. Thus, $\cos s = \dfrac{x}{r} = \dfrac{4}{5}$. We will use a Pythagorean identity to find the value of $\cos t$.

$$\cos t = -\sqrt{1 - \left(-\frac{12}{13}\right)^2} = -\sqrt{1 - \frac{144}{169}}$$
$$= -\sqrt{\frac{25}{169}} = -\frac{5}{13}$$

$$\cos(s + t) = \cos s \cos t - \sin s \sin t$$
$$= \left(\frac{4}{5}\right)\left(-\frac{5}{13}\right) - \left(\frac{3}{5}\right)\left(-\frac{12}{13}\right)$$
$$= -\frac{20}{65} + \frac{36}{65} = \frac{16}{65}$$

$$\cos(s - t) = \cos s \cos t + \sin s \sin t$$
$$= \left(\frac{4}{5}\right)\left(-\frac{5}{13}\right) + \left(\frac{3}{5}\right)\left(-\frac{12}{13}\right)$$
$$= -\frac{20}{65} - \frac{36}{65} = -\frac{56}{65}$$

53. $\cos s = -\dfrac{1}{5}$, $\sin t = \dfrac{3}{5}$, s and t are in quadrant II.

$\cos s = \dfrac{x}{r} \Rightarrow \cos s = -\dfrac{1}{5} = \dfrac{-1}{5} \Rightarrow x = -1, r = 5$.

Substituting into the Pythagorean theorem, we have $(-1)^2 + y^2 = 5^2 \Rightarrow y^2 = 24 \Rightarrow y = \sqrt{24}$,

because $\sin x > 0$. Thus, $\sin s = \dfrac{y}{r} = \dfrac{\sqrt{24}}{5}$.

We will use a Pythagorean identity to find the value of $\cos t$.

$$\cos t = -\sqrt{1 - \sin^2 t} = -\sqrt{1 - \left(\frac{3}{5}\right)^2}$$
$$= -\sqrt{1 - \frac{9}{25}} = -\sqrt{\frac{16}{25}} = -\frac{4}{5}$$

$$\cos(s + t) = \cos s \cos t - \sin s \sin t$$
$$= \left(-\frac{1}{5}\right)\left(-\frac{4}{5}\right) - \left(\frac{\sqrt{24}}{5}\right)\left(\frac{3}{5}\right)$$
$$= \frac{4}{25} - \frac{3\sqrt{24}}{25} = \frac{4}{25} - \frac{6\sqrt{6}}{25}$$
$$= \frac{4 - 6\sqrt{6}}{25}$$

$$\cos(s - t) = \cos s \cos t + \sin s \sin t$$
$$= \left(-\frac{1}{5}\right)\left(-\frac{4}{5}\right) + \left(\frac{\sqrt{24}}{5}\right)\left(\frac{3}{5}\right)$$
$$= \frac{4}{25} + \frac{3\sqrt{24}}{25} = \frac{4}{25} + \frac{6\sqrt{6}}{25}$$
$$= \frac{4 + 6\sqrt{6}}{25}$$

55. $\sin s = \dfrac{\sqrt{5}}{7}$ and $\sin t = \dfrac{\sqrt{6}}{8}$, s and t are in quadrant I.

$\sin s = \dfrac{y}{r} = \dfrac{\sqrt{5}}{7} \Rightarrow y = \sqrt{5}, r = 7$. Substituting into the Pythagorean theorem, we have

$x^2 + \left(\sqrt{5}\right)^2 = 7^2 \Rightarrow x^2 = 44 \Rightarrow x = \sqrt{44}$,

because $\cos s > 0$. Thus, $\cos s = \dfrac{x}{r} = \dfrac{\sqrt{44}}{7}$.

(*continued on next page*)

(*continued*)

We will use a Pythagorean identity to find the value of cos t.

$$\cos t = \sqrt{1 - \left(\frac{\sqrt{6}}{8}\right)^2} = \sqrt{1 - \frac{6}{64}} = \sqrt{\frac{58}{64}} = \frac{\sqrt{58}}{8}$$

$$\cos(s+t) = \cos s \cos t - \sin s \sin t$$
$$= \left(\frac{\sqrt{44}}{7}\right)\left(\frac{\sqrt{58}}{8}\right) - \left(\frac{\sqrt{5}}{7}\right)\left(\frac{\sqrt{6}}{8}\right)$$
$$= \frac{2\sqrt{638}}{56} - \frac{\sqrt{30}}{56} = \frac{2\sqrt{638} - \sqrt{30}}{56}$$

$$\cos(s+t) = \cos s \cos t - \sin s \sin t$$
$$= \left(\frac{\sqrt{44}}{7}\right)\left(\frac{\sqrt{58}}{8}\right) + \left(\frac{\sqrt{5}}{7}\right)\left(\frac{\sqrt{6}}{8}\right)$$
$$= \frac{2\sqrt{638}}{56} + \frac{\sqrt{30}}{56} = \frac{2\sqrt{638} + \sqrt{30}}{56}$$

57. True or false: $\cos 42° = \cos(30° + 12°)$

$42° = 30° + 12°$. Thus, the given statement is true.

59. True or false:

$\cos 74° = \cos 60° \cos 14° + \sin 60° \sin 14°$
$\cos 74° = \cos(60° + 14°)$
$\qquad = \cos 60° \cos 14° - \sin 60° \sin 14°$
$\qquad \ne \cos 60° \cos 14° + \sin 60° \sin 14°$

Thus, the given statement is false.

61. True or false:

$$\cos\frac{\pi}{3} = \cos\frac{\pi}{12}\cos\frac{\pi}{4} - \sin\frac{\pi}{12}\sin\frac{\pi}{4}.$$

$$\cos\frac{\pi}{3} = \cos\left(\frac{\pi}{12} + \frac{\pi}{4}\right)$$
$$= \cos\frac{\pi}{12}\cos\frac{\pi}{4} - \sin\frac{\pi}{12}\sin\frac{\pi}{4}$$

Thus, the given statement is true.

63. True or false:

$\cos 70° \cos 20° - \sin 70° \sin 20° = 0$.
$\cos 70° \cos 20° - \sin 70° \sin 20°$
$\quad = \cos(70° + 20°) = \cos 90° = 0$

Thus, the given statement is true.

65. True or false: $\tan\left(x - \dfrac{\pi}{2}\right) = \cot x$.

$$\tan\left(x - \frac{\pi}{2}\right) = -\tan\left[-\left(x - \frac{\pi}{2}\right)\right]$$
$$= -\tan\left(\frac{\pi}{2} - x\right) = -\cot x \ne \cot x$$

Thus, the given statement is false.

67. Verify $\cos\left(\dfrac{\pi}{2} + x\right) = -\sin x$.

$$\cos\left(\frac{\pi}{2} + x\right) = \cos\frac{\pi}{2}\cos x - \sin\frac{\pi}{2}\sin x$$
$$= (0)\cos x - (1)\sin x = -\sin x$$

69. Verify $\cos 2x = \cos^2 x - \sin^2 x$

$\cos 2x = \cos(x + x)$
$\qquad = \cos x \cos x - \sin x \sin x$
$\qquad = \cos^2 x - \sin^2 x$

71. Verify $\cos 2x = 1 - 2\sin^2 x$

From exercise 69, we have

$\cos 2x = \cos^2 x - \sin^2 x$
$\cos 2x = \cos^2 x - \sin^2 x$
$\qquad = \left(1 - \sin^2 x\right) - \sin^2 x$
$\qquad = 1 - 2\sin^2 x$

73. Verify $\cos 2x = \dfrac{\cot^2 x - 1}{\cot^2 x + 1}$

$$\frac{\cot^2 x - 1}{\cot^2 x + 1} = \frac{\cot^2 x - 1}{\csc^2 x}$$
$$= \left(\cot^2 x - 1\right)\left(\sin^2 x\right)$$
$$= \left(\frac{\cos^2 x}{\sin^2 x} - 1\right)\left(\sin^2 x\right)$$
$$= \cos^2 x - \sin^2 x$$
$$= \cos 2x \text{ (from exercise 69)}$$

75. (a) There are 60 cycles per sec, so the number of cycles in 0.05 sec is given by (0.05 sec)(60 cycles per sec) = 3 cycles.

(b) Because $V = 163 \sin \omega t$ and the maximum value of $\sin \omega t$ is 1, the maximum voltage is 163. Similarly, because the minimum value of $\sin \omega t$ is –1, the minimum voltage is –163.

(c) Because $V = 163 \sin \omega t$ and the minimum value of $\sin \omega t$ is –1, the minimum voltage is –163. Therefore, the voltage is not always equal to 115.

77. Because 90° is a quadrantal angle whose terminal side lies along the y-axis, follow the reasoning in Case 2 in the text. If θ is a small positive angle, then $90° + \theta$ lies in quadrant II and $\sin \theta$ is positive while $\cos \theta$ is negative. Thus, $\cos(90° + \theta) = -\sin \theta$.

79. Because $180°$ is a quadrantal angle whose terminal side lies along the x-axis, follow the reasoning in Case 1 in the text. If θ is a small positive angle, then $180° + \theta$ lies in quadrant III and $\cos\theta$ is negative. Thus,
$$\cos(180° + \theta) = -\cos\theta.$$

81. Because $180°$ is a quadrantal angle whose terminal side lies along the x-axis, follow the reasoning in Case 1 in the text. If θ is a small positive angle, then $180° + \theta$ lies in quadrant III and $\sin\theta$ is negative. Thus,
$$\sin(180° - \theta) = -\sin\theta.$$

Section 5.4   Sum and Difference Identities for Sine and Tangent

1. D; $\sin(A + B) = \sin A \cos B + \cos A \sin B$

3. B; $\tan(A + B) = \dfrac{\tan A + \tan B}{1 - \tan A \tan B}$

5. C
$$\sin 60° \cos 45° + \cos 45° \sin 60°$$
$$= \sin(60° + 45°)$$
$$= \sin 105°$$

7. A
$$\frac{\tan\dfrac{\pi}{3} + \tan\dfrac{\pi}{4}}{1 - \tan\dfrac{\pi}{3}\tan\dfrac{\pi}{4}} = \tan\left(\frac{\pi}{3} + \frac{\pi}{4}\right) = \tan\frac{7\pi}{12}$$

9. $\sin 165° = \sin(180° - 15°)$
$$= \sin 180° \cos 15° - \cos 180° \sin 15°$$
$$= (0)\cos 15° - (-1)\sin 15° = 0 + \sin 15°$$
$$= \sin 15°$$
Now use a difference identity to find $\sin 15°$.
$$\sin 15° = \sin(45° - 30°)$$
$$= \sin 45° \cos 30° - \cos 45° \sin 30°$$
$$= \frac{\sqrt{2}}{2}\cdot\frac{\sqrt{3}}{2} - \frac{\sqrt{2}}{2}\cdot\frac{1}{2}$$
$$= \frac{\sqrt{6}}{4} - \frac{\sqrt{2}}{4} = \frac{\sqrt{6} - \sqrt{2}}{4}$$

11. $\tan 165° = \tan(180° - 15°)$
$$= \frac{\tan 180° - \tan 15°}{1 + \tan 180° \tan 15°}$$
$$= \frac{0 - \tan 15°}{1 + 0\cdot\tan 15°} = -\tan 15°$$
Now use a difference identity to find $\tan 15°$.

$$\tan 15° = \tan(45° - 30°) = \frac{\tan 45° - \tan 30°}{1 + \tan 45° \tan 30°}$$
$$= \frac{1 - \dfrac{\sqrt{3}}{3}}{1 + 1\cdot\dfrac{\sqrt{3}}{3}} = \frac{3 - \sqrt{3}}{3 + \sqrt{3}}$$
$$= \frac{3 - \sqrt{3}}{3 + \sqrt{3}}\cdot\frac{3 - \sqrt{3}}{3 - \sqrt{3}} = \frac{9 - 3\sqrt{3} - 3\sqrt{3} + 3}{9 - 3}$$
$$= \frac{12 - 6\sqrt{3}}{6} = 2 - \sqrt{3}$$
Thus,
$$\tan 165° = -\tan 15° = -\left(2 - \sqrt{3}\right) = -2 + \sqrt{3}.$$

13. $\sin\dfrac{5\pi}{12} = \sin\left(\dfrac{\pi}{4} + \dfrac{\pi}{6}\right)$
$$= \sin\frac{\pi}{4}\cos\frac{\pi}{6} + \cos\frac{\pi}{4}\sin\frac{\pi}{6}$$
$$= \frac{\sqrt{2}}{2}\cdot\frac{\sqrt{3}}{2} + \frac{\sqrt{2}}{2}\cdot\frac{1}{2}$$
$$= \frac{\sqrt{6}}{4} + \frac{\sqrt{2}}{4} = \frac{\sqrt{6} + \sqrt{2}}{4}$$

15. $\tan\dfrac{\pi}{12} = \tan\left(\dfrac{\pi}{4} - \dfrac{\pi}{6}\right) = \dfrac{\tan\dfrac{\pi}{4} - \tan\dfrac{\pi}{6}}{1 + \tan\dfrac{\pi}{4}\tan\dfrac{\pi}{6}}$
$$= \frac{1 - \dfrac{\sqrt{3}}{3}}{1 + \dfrac{\sqrt{3}}{3}} = \frac{1 - \dfrac{\sqrt{3}}{3}}{1 + \dfrac{\sqrt{3}}{3}}\cdot\frac{3}{3} = \frac{3 - \sqrt{3}}{3 + \sqrt{3}}$$
$$= \frac{3 - \sqrt{3}}{3 + \sqrt{3}}\cdot\frac{3 - \sqrt{3}}{3 - \sqrt{3}} = \frac{\left(3 - \sqrt{3}\right)^2}{3^2 - \left(\sqrt{3}\right)^2}$$
$$= \frac{9 - 6\sqrt{3} + 3}{9 - 3} = \frac{12 - 6\sqrt{3}}{6} = 2 - \sqrt{3}$$

17. $\sin\dfrac{7\pi}{12} = \sin\left(\dfrac{\pi}{4} + \dfrac{\pi}{3}\right)$
$$= \sin\frac{\pi}{4}\cos\frac{\pi}{3} + \cos\frac{\pi}{4}\sin\frac{\pi}{3}$$
$$= \frac{\sqrt{2}}{2}\cdot\frac{1}{2} + \frac{\sqrt{2}}{2}\cdot\frac{\sqrt{3}}{2} = \frac{\sqrt{2} + \sqrt{6}}{4}$$

19. $\sin\left(-\dfrac{7\pi}{12}\right) = \sin\left(-\dfrac{\pi}{3} - \dfrac{\pi}{4}\right)$

$\qquad = \sin\left(-\dfrac{\pi}{3}\right)\cos\dfrac{\pi}{4} - \cos\left(-\dfrac{\pi}{3}\right)\sin\dfrac{\pi}{4}$

$\qquad = -\sin\dfrac{\pi}{3}\cos\dfrac{\pi}{4} - \cos\dfrac{\pi}{3}\sin\dfrac{\pi}{4}$

$\qquad = -\dfrac{\sqrt{3}}{2}\cdot\dfrac{\sqrt{2}}{2} - \dfrac{1}{2}\cdot\dfrac{\sqrt{2}}{2}$

$\qquad = -\dfrac{\sqrt{6}}{4} - \dfrac{\sqrt{2}}{4} = \dfrac{-\sqrt{6} - \sqrt{2}}{4}$

21. $\sin\left(-\dfrac{5\pi}{12}\right) = \tan\left(-\dfrac{\pi}{6} - \dfrac{\pi}{4}\right)$

$\qquad = \dfrac{\tan\left(-\dfrac{\pi}{6}\right) + \tan\left(-\dfrac{\pi}{4}\right)}{1 - \tan\left(-\dfrac{\pi}{6}\right)\tan\left(-\dfrac{\pi}{4}\right)}$

$\qquad = \dfrac{-\tan\dfrac{\pi}{6} - \tan\dfrac{\pi}{4}}{1 - \tan\dfrac{\pi}{6}\tan\dfrac{\pi}{4}}$

$\qquad = \dfrac{-\dfrac{\sqrt{3}}{3} - 1}{1 - \left(\dfrac{\sqrt{3}}{3}\right)(1)} = \dfrac{-\sqrt{3} - 3}{3 - \sqrt{3}}$

$\qquad = \dfrac{-\sqrt{3} - 3}{3 - \sqrt{3}}\cdot\dfrac{3 + \sqrt{3}}{3 + \sqrt{3}}$

$\qquad = \dfrac{-3\sqrt{3} - 3 - 9 - 3\sqrt{3}}{3^2 - \left(\sqrt{3}\right)^2} = \dfrac{-12 - 6\sqrt{3}}{9 - 3}$

$\qquad = \dfrac{-12 - 6\sqrt{3}}{6} = -2 - \sqrt{3}$

23. $\tan\dfrac{11\pi}{12} = \tan\left(\pi - \dfrac{\pi}{12}\right)$

$\qquad = \dfrac{\tan\pi - \tan\dfrac{\pi}{12}}{1 + \tan\pi\tan\dfrac{\pi}{12}} = -\tan\dfrac{\pi}{12}$

Now use a difference identity to find $\tan\dfrac{\pi}{12}$.

$\tan\dfrac{\pi}{12} = \tan\left(\dfrac{\pi}{4} - \dfrac{\pi}{6}\right) = \dfrac{\tan\dfrac{\pi}{4} - \tan\dfrac{\pi}{6}}{1 + \tan\dfrac{\pi}{4}\tan\dfrac{\pi}{6}}$

$\qquad = \dfrac{1 - \dfrac{\sqrt{3}}{3}}{1 + 1\cdot\dfrac{\sqrt{3}}{3}} = \dfrac{1 - \dfrac{\sqrt{3}}{3}}{1 + \dfrac{\sqrt{3}}{3}} = \dfrac{3 - \sqrt{3}}{3 + \sqrt{3}}$

$= \dfrac{3 - \sqrt{3}}{3 + \sqrt{3}} = \dfrac{3 - \sqrt{3}}{3 + \sqrt{3}}\cdot\dfrac{3 - \sqrt{3}}{3 - \sqrt{3}}$

$= \dfrac{9 - 6\sqrt{3} + 3}{9 - 3} = \dfrac{12 - 6\sqrt{3}}{6}$

$= \dfrac{6\left(2 - \sqrt{3}\right)}{6} = 2 - \sqrt{3}$

Thus,

$\tan\dfrac{11\pi}{12} = -\tan\dfrac{\pi}{12} = -\left(2 - \sqrt{3}\right) = -2 + \sqrt{3}.$

25. $\sin 76°\cos 31° - \cos 76°\sin 31° = \sin\left(76° - 31°\right)$

$\qquad\qquad\qquad\qquad = \sin 45° = \dfrac{\sqrt{2}}{2}$

27. $\sin\dfrac{\pi}{5}\cos\dfrac{3\pi}{10} + \cos\dfrac{\pi}{5}\sin\dfrac{3\pi}{10}$

$\qquad = \sin\left(\dfrac{\pi}{5} + \dfrac{3\pi}{10}\right) = \sin\left(\dfrac{\pi}{2}\right) = 1$

29. $\dfrac{\tan 80° + \tan 55°}{1 - \tan 80°\tan 55°} = \tan\left(80° + 55°\right)$

$\qquad\qquad\qquad\qquad = \tan 135° = -1$

31. $\dfrac{\tan\frac{5\pi}{9} + \tan\frac{4\pi}{9}}{1 - \tan\frac{5\pi}{9}\tan\frac{4\pi}{9}} = \tan\left(\dfrac{5\pi}{9} + \dfrac{4\pi}{9}\right)$

$\qquad\qquad\qquad\qquad = \tan\pi = 0$

33. $\cos\left(30° + \theta\right) = \cos 30°\cos\theta - \sin 30°\sin\theta$

$\qquad = \dfrac{\sqrt{3}}{2}\cos\theta - \dfrac{1}{2}\sin\theta$

$\qquad = \dfrac{1}{2}\left(\sqrt{3}\cos\theta - \sin\theta\right)$

$\qquad = \dfrac{\sqrt{3}\cos\theta - \sin\theta}{2}$

35. $\cos\left(60° + \theta\right) = \cos 60°\cos\theta - \sin 60°\sin\theta$

$\qquad = \dfrac{1}{2}\cos\theta - \dfrac{\sqrt{3}}{2}\sin\theta$

$\qquad = \dfrac{1}{2}\left(\cos\theta - \sqrt{3}\sin\theta\right)$

$\qquad = \dfrac{\cos\theta - \sqrt{3}\sin\theta}{2}$

37. $\cos\left(\dfrac{3\pi}{4} - x\right) = \cos\dfrac{3\pi}{4}\cos x + \sin\dfrac{3\pi}{4}\sin x$

$\qquad = \left(-\dfrac{\sqrt{2}}{2}\right)\cos x + \left(\dfrac{\sqrt{2}}{2}\right)\sin x$

$\qquad = \dfrac{\sqrt{2}}{2}\left(-\cos x + \sin x\right)$

$\qquad = \dfrac{\sqrt{2}\left(\sin x - \cos x\right)}{2}$

39. $\tan(\theta + 30°) = \dfrac{\tan\theta + \tan 30°}{1 - \tan\theta\tan 30°}$

$= \dfrac{\tan\theta + \dfrac{1}{\sqrt{3}}}{1 - \left(\dfrac{1}{\sqrt{3}}\right)\tan\theta}$

$= \dfrac{\sqrt{3}\tan\theta + 1}{\sqrt{3} - \tan\theta}$

41. $\sin\left(\dfrac{\pi}{4} + x\right) = \sin\dfrac{\pi}{4}\cos x + \cos\dfrac{\pi}{4}\sin x$

$= \dfrac{\sqrt{2}}{2}\cos x + \dfrac{\sqrt{2}}{2}\sin x$

$= \dfrac{\sqrt{2}(\cos x + \sin x)}{2}$

43. $\sin(270° - \theta) = \sin 270°\cos\theta - \cos 270°\sin\theta$

$= (-1)(\cos\theta) - (0)(\sin\theta)$

$= -\cos\theta$

45. $\tan(2\pi - x) = \dfrac{\tan 2\pi - \tan x}{1 + \tan 2\pi\tan x}$

$= \dfrac{0 - \tan x}{1 + 0\cdot\tan x} = -\tan x$

47. $\tan(\pi - x) = \dfrac{\tan\pi - \tan x}{1 + \tan\pi\tan x} = \dfrac{0 - \tan x}{1 + 0\cdot\tan x}$

$= -\tan x$

49. Answers will vary. Sample answer: Cotangent, secant, and cosecant formulas can be written using their reciprocal functions tangent, cosine, and sine.

51. $\cos s = \dfrac{3}{5}$, $\sin t = \dfrac{5}{13}$, and s and t are in quadrant I.

First find the values of $\sin s$, $\tan s$, $\cos t$, and $\tan t$. Because s and t are both in quadrant I, the values of $\sin s$ and $\cos t$, $\tan s$, and $\tan t$ will be positive.

$\sin s = \sqrt{1 - \left(\dfrac{3}{5}\right)^2} = \sqrt{1 - \dfrac{9}{25}} = \sqrt{\dfrac{16}{25}} = \dfrac{4}{5}$

$\cos t = \sqrt{1 - \left(\dfrac{5}{13}\right)^2} = \sqrt{1 - \dfrac{25}{169}} = \sqrt{\dfrac{144}{169}} = \dfrac{12}{13}$

$\tan s = \dfrac{\sin s}{\cos s} = \dfrac{\frac{4}{5}}{\frac{3}{5}} = \dfrac{4}{3}$

$\tan t = \dfrac{\sin t}{\cos t} = \dfrac{\frac{5}{13}}{\frac{12}{13}} = \dfrac{5}{12}$

(a) $\sin(s + t) = \sin s\cos t + \cos s\sin t$

$= \left(\dfrac{4}{5}\right)\left(\dfrac{12}{13}\right) + \left(\dfrac{3}{5}\right)\left(\dfrac{5}{13}\right)$

$= \dfrac{48}{65} + \dfrac{15}{65} = \dfrac{63}{65}$

(b) $\tan(s + t) = \dfrac{\tan s + \tan t}{1 - \tan s\tan t} = \dfrac{\frac{4}{3} + \frac{5}{12}}{1 - \left(\frac{4}{3}\right)\left(\frac{5}{12}\right)}$

$= \dfrac{48 + 15}{36 - 20} = \dfrac{63}{16}$

(c) From parts (a) and (b), $\sin(s + t) > 0$ and $\tan(s + t) > 0$. The only quadrant in which the values of both the sine and the tangent are positive is quadrant I, so $s + t$ is in quadrant I.

53. $\cos s = -\dfrac{8}{17}$ and $\cos t = -\dfrac{3}{5}$, s and t are in quadrant III

First find the values of $\sin s$, $\sin t$, $\tan s$, and $\tan t$. Because s and t are both in quadrant III, the values of $\sin s$ and $\sin t$ will be negative, while $\tan s$ and $\tan t$ will be positive.

$\sin s = -\sqrt{1 - \cos^2 s} = -\sqrt{1 - \left(-\dfrac{8}{17}\right)^2}$

$= -\sqrt{1 - \dfrac{64}{289}} = -\sqrt{\dfrac{225}{289}} = -\dfrac{15}{17}$

$\sin t = -\sqrt{1 - \cos^2 t} = -\sqrt{1 - \left(-\dfrac{3}{5}\right)^2}$

$= -\sqrt{1 - \dfrac{9}{25}} = -\sqrt{\dfrac{16}{25}} = -\dfrac{4}{5}$

$\tan s = \dfrac{\sin s}{\cos s} = \dfrac{-\frac{15}{17}}{-\frac{8}{17}} = \dfrac{15}{8}$

$\tan t = \dfrac{\sin t}{\cos t} = \dfrac{-\frac{4}{5}}{-\frac{3}{5}} = \dfrac{4}{3}$

(a) $\sin(s + t) = \sin s\cos t + \cos s\sin t$

$= \left(-\dfrac{15}{17}\right)\left(-\dfrac{3}{5}\right) + \left(-\dfrac{8}{17}\right)\left(-\dfrac{4}{5}\right)$

$= \dfrac{45}{85} + \dfrac{32}{85} = \dfrac{77}{85}$

(b) $\tan(s + t) = \dfrac{\frac{15}{8} + \frac{4}{3}}{1 - \left(\frac{15}{8}\right)\left(\frac{4}{3}\right)} = \dfrac{45 + 32}{24 - 60}$

$= \dfrac{77}{-36} = -\dfrac{77}{36}$

(c) From parts (a) and (b), $\sin (s + t) > 0$ and $\tan (s + t) < 0$. The only quadrant in which the value of the sine is positive and the value of the tangent is negative is quadrant II, so $s + t$ is in quadrant II.

55. $\sin s = \dfrac{2}{3}$ and $\sin t = -\dfrac{1}{3}$, s is in quadrant II and t is in quadrant IV.

First find the values of $\cos s$, $\cos t$, $\tan s$, and $\tan t$. Because s is in quadrant II and t is in quadrant IV, the values of $\cos s$, $\tan s$, and $\tan t$ will be negative, while $\cos t$ will be positive.

$$\cos s = -\sqrt{1 - \left(\dfrac{2}{3}\right)^2} = -\sqrt{1 - \dfrac{4}{9}} = -\sqrt{\dfrac{5}{9}} = -\dfrac{\sqrt{5}}{3}$$

$$\cos t = \sqrt{1 - \left(-\dfrac{1}{3}\right)^2} = \sqrt{1 - \dfrac{1}{9}} = \sqrt{\dfrac{8}{9}}$$
$$= \dfrac{\sqrt{8}}{3} = \dfrac{2\sqrt{2}}{3}$$

$$\tan s = \dfrac{\sin s}{\cos s} = \dfrac{\frac{2}{3}}{-\frac{\sqrt{5}}{3}} = -\dfrac{2}{\sqrt{5}} = -\dfrac{2\sqrt{5}}{5}$$

$$\tan t = \dfrac{\sin t}{\cos t} = \dfrac{-\frac{1}{3}}{\frac{2\sqrt{2}}{3}} = -\dfrac{1}{2\sqrt{2}} = -\dfrac{\sqrt{2}}{4}$$

(a) $\sin (s + t) = \sin s \cos t + \cos s \sin t$

$$= \left(\dfrac{2}{3}\right)\left(\dfrac{2\sqrt{2}}{3}\right) + \left(-\dfrac{\sqrt{5}}{3}\right)\left(-\dfrac{1}{3}\right)$$

$$= \dfrac{4\sqrt{2}}{9} + \dfrac{\sqrt{5}}{9} = \dfrac{4\sqrt{2} + \sqrt{5}}{9}$$

(b) Different forms of $\tan (s + t)$ will be obtained depending on whether $\tan s$ and $\tan t$ are written with rationalized denominators.

$\tan (s + t)$

$$= \dfrac{-\frac{2\sqrt{5}}{5} + \left(-\frac{\sqrt{2}}{4}\right)}{1 - \left(-\frac{2\sqrt{5}}{5}\right)\left(-\frac{\sqrt{2}}{4}\right)} = \dfrac{-8\sqrt{5} - 5\sqrt{2}}{20 - 2\sqrt{10}}$$

$$= \dfrac{-8\sqrt{5} - 5\sqrt{2}}{20 - 2\sqrt{10}} \cdot \dfrac{20 + 2\sqrt{10}}{20 + 2\sqrt{10}}$$

$$= \dfrac{-160\sqrt{5} - 16\sqrt{50} - 100\sqrt{2} - 10\sqrt{20}}{400 - 40}$$

$$= \dfrac{-160\sqrt{5} - 80\sqrt{2} - 100\sqrt{2} - 20\sqrt{5}}{360}$$

$$= \dfrac{-180\sqrt{5} - 180\sqrt{2}}{360} = \dfrac{-\sqrt{5} - \sqrt{2}}{2}$$

or

$$\tan (s + t) = \dfrac{-\frac{2}{\sqrt{5}} + \left(-\frac{1}{2\sqrt{2}}\right)}{1 - \left(-\frac{2}{\sqrt{5}}\right)\left(-\frac{1}{2\sqrt{2}}\right)}$$

$$= \dfrac{-4\sqrt{2} - \sqrt{5}}{2\sqrt{10} - 2} = \dfrac{4\sqrt{2} + \sqrt{5}}{2 - 2\sqrt{10}}$$

$$= \dfrac{4\sqrt{2} + \sqrt{5}}{2 - 2\sqrt{10}} \cdot \dfrac{2 + 2\sqrt{10}}{2 + 2\sqrt{10}}$$

$$= \dfrac{8\sqrt{2} + 8\sqrt{20} + 2\sqrt{5} + 2\sqrt{50}}{4 - 40}$$

$$= \dfrac{8\sqrt{2} + 16\sqrt{5} + 2\sqrt{5} + 10\sqrt{2}}{-36}$$

$$= \dfrac{18\sqrt{2} + 18\sqrt{5}}{-36} = \dfrac{-\sqrt{2} - \sqrt{5}}{2}$$

(c) To find the quadrant of $s + t$, notice from the preceding that $\sin (s + t) = \dfrac{4\sqrt{2} + \sqrt{5}}{9} > 0$

and $\tan (s + t) = \dfrac{-8\sqrt{5} - 5\sqrt{2}}{20 - 2\sqrt{10}} \approx -1.8 < 0$.

The only quadrant in which the values of sine are positive and tangent is negative is quadrant II. Therefore, $s + t$ is in quadrant II.

The graphs in exercises 57–59 are shown in the following window $[-2\pi, 2\pi] \times [-4, 4]$.

57. $\sin\left(\dfrac{\pi}{2} + \theta\right)$ appears to be equivalent to $\cos \theta$.

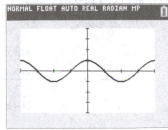

$$\sin\left(\dfrac{\pi}{2} + \theta\right) = \sin\dfrac{\pi}{2}\cos\theta + \sin\theta\cos\dfrac{\pi}{2}$$

$$= 1 \cdot \cos\theta + \sin\theta \cdot 0$$

$$= \cos\theta + 0 = \cos\theta$$

59. $\tan\left(\dfrac{\pi}{2}+\theta\right)$ appears to be equivalent to $-\cot\theta$.

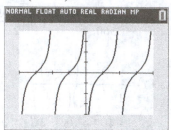

$$\tan\left(\frac{\pi}{2}+\theta\right)=\frac{\sin\left(\dfrac{\pi}{2}+\theta\right)}{\cos\left(\dfrac{\pi}{2}+\theta\right)}=\frac{\sin\dfrac{\pi}{2}\cos\theta+\cos\dfrac{\pi}{2}\sin\theta}{\cos\dfrac{\pi}{2}\cos\theta-\sin\dfrac{\pi}{2}\sin\theta}=\frac{1\cdot\cos\theta+0\cdot\sin\theta}{0\cdot\cos\theta-1\cdot\sin\theta}=\frac{\cos\theta}{-\sin\theta}=-\cot\theta$$

61. Verify $\sin 2x = 2\sin x \cos x$ is an identity.

$\sin 2x = \sin(x+x)=\sin x\cos x+\cos x\sin x=2\sin x\cos x$

63. Verify $\sin\left(\dfrac{7\pi}{6}+x\right)-\cos\left(\dfrac{2\pi}{3}+x\right)=0$ is an identity.

$$\sin\left(\frac{7\pi}{6}+x\right)-\cos\left(\frac{2\pi}{3}+x\right)=\left(\sin\frac{7\pi}{6}\cos x+\cos\frac{7\pi}{6}\sin x\right)-\left(\cos\frac{2\pi}{3}\cos x-\sin\frac{2\pi}{3}\sin x\right)$$

$$=\left(-\frac{1}{2}\cos x-\frac{\sqrt{3}}{2}\sin x\right)-\left(-\frac{1}{2}\cos x-\frac{\sqrt{3}}{2}\sin x\right)=0$$

65. Verify $\dfrac{\cos(\alpha-\beta)}{\cos\alpha\sin\beta}=\tan\alpha+\cot\beta$ is an identity.

$$\frac{\cos(\alpha-\beta)}{\cos\alpha\sin\beta}=\frac{\cos\alpha\cos\beta+\sin\alpha\sin\beta}{\cos\alpha\sin\beta}=\frac{\cos\alpha\cos\beta}{\cos\alpha\sin\beta}+\frac{\sin\alpha\sin\beta}{\cos\alpha\sin\beta}=\frac{\cos\beta}{\sin\beta}+\frac{\sin\alpha}{\cos\alpha}=\cot\beta+\tan\alpha$$

67. Verify that $\dfrac{\sin(x-y)}{\sin(x+y)}=\dfrac{\tan x-\tan y}{\tan x+\tan y}$ is an identity

$$\frac{\sin(x-y)}{\sin(x+y)}=\frac{\sin x\cos y-\cos x\sin y}{\sin x\cos y+\cos x\sin y}=\frac{\dfrac{\sin x\cos y}{\cos x\cos y}-\dfrac{\cos x\sin y}{\cos x\cos y}}{\dfrac{\sin x\cos y}{\cos x\cos y}+\dfrac{\cos x\sin y}{\cos x\cos y}}=\frac{\dfrac{\sin x}{\cos x}\cdot\dfrac{\cos y}{\cos y}-\dfrac{\cos x}{\cos x}\cdot\dfrac{\sin y}{\cos y}}{\dfrac{\sin x}{\cos x}\cdot\dfrac{\cos y}{\cos y}+\dfrac{\cos x}{\cos x}\cdot\dfrac{\sin y}{\cos y}}$$

$$=\frac{\dfrac{\sin x}{\cos x}\cdot 1-1\cdot\dfrac{\sin y}{\cos y}}{\dfrac{\sin x}{\cos x}\cdot 1+1\cdot\dfrac{\sin y}{\cos y}}=\frac{\dfrac{\sin x}{\cos x}-\dfrac{\sin y}{\cos y}}{\dfrac{\sin x}{\cos x}+\dfrac{\sin y}{\cos y}}=\frac{\tan x-\tan y}{\tan x+\tan y}$$

69. Verify $\dfrac{\sin(s-t)}{\sin t}+\dfrac{\cos(s-t)}{\cos t}=\dfrac{\sin s}{\sin t\cos t}$ is an identity.

$$\frac{\sin(s-t)}{\sin t}+\frac{\cos(s-t)}{\cos t}=\frac{\sin s\cos t-\sin t\cos s}{\sin t}+\frac{\cos s\cos t+\sin t\sin s}{\cos t}$$

$$=\frac{\sin s\cos^2 t-\sin t\cos t\cos s}{\sin t\cos t}+\frac{\sin t\cos t\cos s+\sin^2 t\sin s}{\sin t\cos t}=\frac{\sin s\cos^2 t+\sin s\sin^2 t}{\sin t\cos t}$$

$$=\frac{\sin s\left(\cos^2 t+\sin^2 t\right)}{\sin t\cos t}=\frac{\sin s}{\sin t\cos t}$$

71. (a) $F = \dfrac{0.6W \sin(\theta + 90°)}{\sin 12°}$

$= \dfrac{0.6(170)\sin(30 + 90)°}{\sin 12°}$

$= \dfrac{102 \sin 120°}{\sin 12°} \approx 425 \text{ lb}$

(This is a good reason why people frequently have back problems.)

(b) $F = \dfrac{0.6W \sin(\theta + 90°)}{\sin 12°}$

$= \dfrac{0.6W(\sin\theta \cos 90° + \sin 90° \cos\theta)}{\sin 12°}$

$= \dfrac{0.6W(\sin\theta \cdot 0 + 1 \cdot \cos\theta)}{\sin 12°}$

$= \dfrac{0.6W(0 + \cos\theta)}{\sin 12°}$

$= \dfrac{0.6}{\sin 12°}W \cos\theta \approx 2.9W \cos\theta$

(c) F will be maximum when $\cos\theta = 1$ or $\theta = 0°$. ($\theta = 0°$ corresponds to the back being horizontal which gives a maximum force on the back muscles. This agrees with intuition because stress on the back increases as one bends farther until the back is parallel with the ground.)

73. $E = 20 \sin\left(\dfrac{\pi t}{4} - \dfrac{\pi}{2}\right)$

$= 20\left(\sin\dfrac{\pi t}{4}\cos\dfrac{\pi}{2} - \cos\dfrac{\pi t}{4}\sin\dfrac{\pi}{2}\right)$

$= 20\left(\sin\dfrac{\pi t}{4}(0) - \cos\dfrac{\pi t}{4}(1)\right)$

$= 20\left(0 - \cos\dfrac{\pi t}{4}\right) = -20\cos\dfrac{\pi t}{4}$

75. $y' = r\cos(\theta + R) = r[\cos\theta\cos R - \sin\theta\sin R]$

$= (r\cos\theta)\cos R - (r\sin\theta)\sin R$

$= y\cos R - z\sin R$

77. $\angle\beta$ and $\angle ABC$ are supplementary, so $m\angle ABC = 180° - \beta$.

79. $\tan\theta = \tan(\beta - \alpha) = \dfrac{\tan\beta - \tan\alpha}{1 + \tan\beta\tan\alpha}$

81. $x + y = 9,\ 2x + y = -1$

Convert each equation to slope-intercept form to find the slopes:

$x + y = 9 \Rightarrow y = -x + 9 \Rightarrow m_1 = -1$

$2x + y = -1 \Rightarrow y = -2x - 1 \Rightarrow m_2 = -2$

$\tan\theta = \dfrac{m_2 - m_1}{1 + m_1 m_2} = \dfrac{-2 - (-1)}{1 + (-1)(-2)} = -\dfrac{1}{3} \Rightarrow$

$\theta = \tan^{-1}\left(-\dfrac{1}{3}\right) \approx -18.4°$

Note that the angle must be positive, so $\theta \approx 18.4°$

Chapter 5 Quiz
(Sections 5.1−5.4)

1. $\sin\theta = -\dfrac{7}{25}$, θ is in quadrant IV

In quadrant IV, the cosine and secant function values are positive. The tangent, cotangent, and cosecant function values are negative.

$\cos\theta = \sqrt{1 - \sin^2\theta} = \sqrt{1 - \left(-\dfrac{7}{25}\right)^2}$

$= \sqrt{1 - \dfrac{49}{625}} = \sqrt{\dfrac{576}{625}} = \dfrac{24}{25}$

$\tan\theta = \dfrac{\sin\theta}{\cos\theta} = \dfrac{-\frac{7}{25}}{\frac{24}{25}} = -\dfrac{7}{24}$

$\cot\theta = \dfrac{1}{\tan\theta} = \dfrac{1}{-\frac{7}{24}} = -\dfrac{24}{7}$

$\sec\theta = \dfrac{1}{\cos\theta} = \dfrac{1}{\frac{24}{25}} = \dfrac{25}{24}$

$\csc\theta = \dfrac{1}{\sin\theta} = \dfrac{1}{-\frac{7}{25}} = -\dfrac{25}{7}$

3. $\sin\left(-\dfrac{7\pi}{12}\right) = -\sin\left(\dfrac{7\pi}{12}\right) = -\sin\left(\dfrac{\pi}{3} + \dfrac{\pi}{4}\right)$

$= -\left(\sin\dfrac{\pi}{3}\cos\dfrac{\pi}{4} + \cos\dfrac{\pi}{3}\sin\dfrac{\pi}{4}\right)$

$= -\left[\dfrac{\sqrt{3}}{2}\left(\dfrac{\sqrt{2}}{2}\right) + \dfrac{1}{2}\left(\dfrac{\sqrt{2}}{2}\right)\right]$

$= -\left(\dfrac{\sqrt{6}}{4} + \dfrac{\sqrt{2}}{4}\right) = -\left(\dfrac{\sqrt{6} + \sqrt{2}}{4}\right)$

$= \dfrac{-\sqrt{6} - \sqrt{2}}{4}$

5. $\cos A = \dfrac{3}{5}, \sin B = -\dfrac{5}{13}, 0 < A < \dfrac{\pi}{2},$ and $\pi < B < \dfrac{3\pi}{2}$

$\cos A = \dfrac{3}{5} = \dfrac{y}{r} \Rightarrow y = 3, r = 5$. Substituting into the Pythagorean theorem, we have $x^2 + 3^3 = 5^2 \Rightarrow x = 4$,

because $\sin A > 0$. Thus, $\sin A = \dfrac{4}{5}$.

We will use a Pythagorean identity to find the value of $\cos B$.

$\cos B = -\sqrt{1 - \left(-\dfrac{5}{13}\right)^2} = -\sqrt{1 - \dfrac{25}{169}} = -\sqrt{\dfrac{144}{169}} = -\dfrac{12}{13}$

Note that $\cos B$ is negative because B is in quadrant III.

(a) $\cos(A + B) = \cos A \cos B - \sin A \sin B = \dfrac{3}{5}\left(-\dfrac{12}{13}\right) - \dfrac{4}{5}\left(-\dfrac{5}{13}\right) = -\dfrac{36}{65} + \dfrac{20}{65} = -\dfrac{16}{65}$

(b) $\sin(A + B) = \sin A \cos B + \cos A \sin B = \dfrac{4}{5}\left(-\dfrac{12}{13}\right) + \dfrac{3}{5}\left(-\dfrac{5}{13}\right) = -\dfrac{48}{65} - \dfrac{15}{65} = -\dfrac{63}{65}$

(c) Both $\cos(A + B)$ and $\sin(A + B)$ are negative. Thus $(A + B)$ is in quadrant III.

7. Verify $\dfrac{1 + \sin\theta}{\cot^2\theta} = \dfrac{\sin\theta}{\csc\theta - 1}$ is an identity.

Working with the right side, we have $\dfrac{\sin\theta}{\csc\theta - 1} = \dfrac{\sin\theta}{\csc\theta - 1} \cdot \dfrac{\csc\theta + 1}{\csc\theta + 1} = \dfrac{\sin\theta\csc\theta + \sin\theta}{\csc^2\theta - 1} = \dfrac{1 + \sin\theta}{\cot^2\theta}$

9. Verify $\dfrac{\sin^2\theta - \cos^2\theta}{\sin^4\theta - \cos^4\theta} = 1$ is an identity.

$\dfrac{\sin^2\theta - \cos^2\theta}{\sin^4\theta - \cos^4\theta} = \dfrac{\sin^2\theta - \cos^2\theta}{\left(\sin^2\theta - \cos^2\theta\right)\left(\sin^2\theta + \cos^2\theta\right)} = \dfrac{1}{1} = 1$

Section 5.5   Double-Angle Identities

1. C. $2\cos^2 15° - 1 = \cos(2 \cdot 15°) = \cos 30° = \dfrac{\sqrt{3}}{2}$

3. B. $2\sin 22.5° \cos 22.5° = \sin(2 \cdot 22.5°) = \sin 45° = \dfrac{\sqrt{2}}{2}$

5. F. $4\sin\dfrac{\pi}{3}\cos\dfrac{\pi}{3} = 2\sin\left(2 \cdot \dfrac{\pi}{3}\right) = 2\sin\dfrac{2\pi}{3} = 2 \cdot \dfrac{\sqrt{3}}{2} = \sqrt{3}$

7. $\sin\theta = \dfrac{2}{5}, \cos\theta < 0$

$\cos 2\theta = 1 - 2\sin^2\theta \Rightarrow \cos 2\theta = 1 - 2\left(\dfrac{2}{5}\right)^2 1 - 2 \cdot \dfrac{4}{25} = 1 - \dfrac{8}{25} = \dfrac{17}{25}$

$\cos^2 2\theta + \sin^2 2\theta = 1 \Rightarrow \sin^2 2\theta = 1 - \cos^2 2\theta \Rightarrow \sin^2 2\theta = 1 - \left(\dfrac{17}{25}\right)^2 = 1 - \dfrac{289}{625} = \dfrac{336}{625}$

$\cos\theta < 0$, so $\sin 2\theta < 0$ because $\sin 2\theta = 2\sin\theta\cos\theta < 0$ and $\sin\theta > 0$.

$\sin 2\theta = -\sqrt{\dfrac{336}{625}} = -\dfrac{\sqrt{336}}{25} = -\dfrac{4\sqrt{21}}{25}$

9. $\tan x = 2, \cos x > 0$

$$\tan 2x = \frac{2\tan x}{1-\tan^2 x} = \frac{2(2)}{1-2^2} = \frac{4}{1-4} = -\frac{4}{3}$$

Both $\tan x$ and $\cos x$ are positive, so x must be in quadrant I. Because $0° < x < 90°$, then $0° < 2x < 180°$. Thus, $2x$ must be in either quadrant I or quadrant II. However, $\tan 2x < 0$, so $2x$ is in quadrant II, and $\sec 2x$ is negative.

$$\sec^2 2x = 1 + \tan^2 2x = 1 + \left(-\frac{4}{3}\right)^2 = 1 + \frac{16}{9} = \frac{25}{9}$$

$$\sec 2x = -\frac{5}{3} \Rightarrow \cos 2x = \frac{1}{\sec 2x} = -\frac{3}{5}$$

$$\cos^2 2x + \sin^2 2x = 1 \Rightarrow \sin^2 2x = 1 - \cos^2 2x \Rightarrow$$

$$\sin^2 2x = 1 - \left(-\frac{3}{5}\right)^2 \Rightarrow \sin^2 2x = 1 - \frac{9}{25} = \frac{16}{25}$$

In quadrants I and II, $\sin 2x > 0$. Thus, we have $\sin 2x = \sqrt{\frac{16}{25}} = \frac{4}{5}$.

11. $\sin\theta = -\frac{\sqrt{5}}{7}, \cos\theta > 0$

$$\cos^2\theta = 1 - \sin^2\theta = 1 - \left(-\frac{\sqrt{5}}{7}\right)^2 = 1 - \frac{5}{49} = \frac{44}{49}$$

$$\cos\theta > 0, \text{ so } \cos\theta = \sqrt{\frac{44}{49}} = \frac{\sqrt{44}}{7} = \frac{2\sqrt{11}}{7}.$$

$$\cos 2\theta = 1 - 2\sin^2\theta = 1 - 2\left(-\frac{\sqrt{5}}{7}\right)^2$$

$$= 1 - 2 \cdot \frac{5}{49} = 1 - \frac{10}{49} = \frac{39}{49}$$

$$\sin 2\theta = 2\sin\theta\cos\theta = 2\left(-\frac{\sqrt{5}}{7}\right)\left(\frac{2\sqrt{11}}{7}\right)$$

$$= -\frac{4\sqrt{55}}{49}$$

13. $\cos 2\theta = \frac{3}{5}, \theta$ is in quadrant I.

$$\cos 2\theta = 2\cos^2\theta - 1 \Rightarrow \frac{3}{5} = 2\cos^2\theta - 1 \Rightarrow$$

$$2\cos^2\theta = \frac{3}{5} + 1 = \frac{8}{5} \Rightarrow \cos^2\theta = \frac{8}{10} = \frac{4}{5}$$

θ is in quadrant I, so $\cos\theta > 0$. Thus,

$$\cos\theta = \sqrt{\frac{4}{5}} = \frac{2}{\sqrt{5}} = \frac{2\sqrt{5}}{5}.$$

θ is in quadrant I, so $\sin\theta > 0$.

$$\sin\theta = \sqrt{1-\cos^2\theta} = \sqrt{1-\left(\frac{2\sqrt{5}}{5}\right)^2} = \sqrt{1-\frac{20}{25}}$$

$$= \sqrt{\frac{5}{25}} = \sqrt{\frac{1}{5}} = \frac{1}{\sqrt{5}} = \frac{1}{\sqrt{5}} \cdot \frac{\sqrt{5}}{\sqrt{5}} = \frac{\sqrt{5}}{5}$$

15. $\cos 2\theta = -\frac{5}{12}, 90° < \theta < 180°$

$$\cos 2\theta = 2\cos^2\theta - 1 \Rightarrow$$

$$2\cos^2\theta = \cos 2\theta + 1 = -\frac{5}{12} + 1 = \frac{7}{12} \Rightarrow$$

$$\cos^2\theta = \frac{7}{24}$$

θ is in quadrant II, so $\cos\theta < 0$. Thus,

$$\cos\theta = -\sqrt{\frac{7}{24}} = -\frac{\sqrt{7}}{\sqrt{24}} = -\frac{\sqrt{7}}{2\sqrt{6}}$$

$$= -\frac{\sqrt{7}}{2\sqrt{6}} \cdot \frac{\sqrt{6}}{\sqrt{6}} = -\frac{\sqrt{42}}{12}$$

$$\sin^2\theta = 1 - \cos^2\theta = 1 - \left(-\sqrt{\frac{7}{24}}\right)^2$$

$$= 1 - \frac{7}{24} = \frac{17}{24}$$

θ is in quadrant II, so $\sin\theta > 0$. Thus,

$$\sin\theta = \sqrt{\frac{17}{24}} = \frac{\sqrt{17}}{\sqrt{24}} = \frac{\sqrt{17}}{2\sqrt{6}}$$

$$= \frac{\sqrt{17}}{2\sqrt{6}} \cdot \frac{\sqrt{6}}{\sqrt{6}} = \frac{\sqrt{102}}{12}$$

17. Verify $(\sin x + \cos x)^2 = \sin 2x + 1$ is an identity.

$$(\sin x + \cos x)^2 = \sin^2 x + 2\sin x \cos x + \cos^2 x$$

$$= \left(\sin^2 x + \cos^2 x\right) + 2\sin x \cos x$$

$$= 1 + \sin 2x$$

19. Verify $(\cos 2x + \sin 2x)^2 = 1 + \sin 4x$ is an identity.

$$(\cos 2x + \sin 2x)^2$$

$$= \cos^2 2x + 2\cos 2x \sin 2x + \sin^2 2x$$

$$= \left(\cos^2 2x + \sin^2 2x\right) + 2\cos 2x \sin 2x$$

$$= 1 + \sin 4x$$

21. Verify $\tan 8\theta - \tan 8\theta \tan^2 4\theta = 2\tan 4\theta$ is an identity.

$$\tan 8\theta - \tan 8\theta \tan^2 4\theta$$

$$= \tan 8\theta\left(1 - \tan^2 4\theta\right)$$

$$= \frac{2\tan 4\theta}{1 - \tan^2 4\theta}\left(1 - \tan^2 4\theta\right) = 2\tan 4\theta$$

23. Verify $\cos 2\theta = \dfrac{2-\sec^2\theta}{\sec^2\theta}$ is an identity. Working with the right side, we have

$$\dfrac{2-\sec^2\theta}{\sec^2\theta} = \dfrac{2-\dfrac{1}{\cos^2\theta}}{\dfrac{1}{\cos^2\theta}} = \dfrac{2-\dfrac{1}{\cos^2\theta}}{\dfrac{1}{\cos^2\theta}}\cdot\dfrac{\cos^2\theta}{\cos^2\theta} = \dfrac{2\cos^2\theta-1}{1} = \cos 2\theta$$

25. Verify that $\sin 4x = 4\sin x\cos x\cos 2x$ is an identity.

$$\sin 4x = \sin 2(2x) = 2\sin 2x\cos 2x = 2(2\sin x\cos x)\cos 2x = 4\sin x\cos x\cos 2x$$

27. Verify $\dfrac{2\cos 2\theta}{\sin 2\theta} = \cot\theta - \tan\theta$ is an identity.
     Work with the right side.

$$\cot\theta-\tan\theta = \dfrac{\cos\theta}{\sin\theta}-\dfrac{\sin\theta}{\cos\theta} = \dfrac{\cos\theta}{\sin\theta}\cdot\dfrac{\cos\theta}{\cos\theta}-\dfrac{\sin\theta}{\cos\theta}\cdot\dfrac{\sin\theta}{\sin\theta} = \dfrac{\cos^2\theta-\sin^2\theta}{\sin\theta\cos\theta} = \dfrac{2(\cos^2\theta-\sin^2\theta)}{2\sin\theta\cos\theta} = \dfrac{2\cos 2\theta}{\sin 2\theta}$$

29. Verify $\tan x + \cot x = 2\csc 2x$ is an identity.

$$\tan x + \cot x = \dfrac{\sin x}{\cos x}\cdot\dfrac{\sin x}{\sin x}+\dfrac{\cos x}{\sin x}\cdot\dfrac{\cos x}{\cos x} = \dfrac{\sin^2 x+\cos^2 x}{\cos x\sin x} = \dfrac{1}{\cos x\sin x} = \dfrac{2}{2\cos x\sin x} = \dfrac{2}{\sin 2x} = 2\csc 2x$$

31. Verify $1 + \tan x\tan 2x = \sec 2x$ is an identity.

$$1+\tan x\tan 2x = 1+\tan x\left(\dfrac{2\tan x}{1-\tan^2 x}\right) = 1+\dfrac{2\tan^2 x}{1-\tan^2 x} = \dfrac{(1-\tan^2 x)+2\tan^2 x}{1-\tan^2 x}$$

$$= \dfrac{1+\tan^2 x}{1-\tan^2 x} = \dfrac{1+\dfrac{\sin^2 x}{\cos^2 x}}{1-\dfrac{\sin^2 x}{\cos^2 x}} = \dfrac{1+\dfrac{\sin^2 x}{\cos^2 x}}{1-\dfrac{\sin^2 x}{\cos^2 x}}\cdot\dfrac{\cos^2 x}{\cos^2 x} = \dfrac{\cos^2 x+\sin^2 x}{\cos^2 x-\sin^2 x} = \dfrac{1}{\cos 2x}$$

$$= \sec 2x$$

33. Verify $\sin 2A\cos 2A = \sin 2A - 4\sin^3 A\cos A$ is an identity.

$$\sin 2A\cos 2A = (2\sin A\cos A)(1-2\sin^2 A) = 2\sin A\cos A - 4\sin^3 A\cos A = \sin 2A - 4\sin^3 A\cos A$$

35. Verify $\tan(\theta-45°)+\tan(\theta+45°) = 2\tan 2\theta$ is an identity.

$$\tan(\theta-45°)+\tan(\theta+45°) = \dfrac{\tan\theta-\tan 45°}{1+\tan\theta\tan 45°}+\dfrac{\tan\theta+\tan 45°}{1-\tan\theta\tan 45°} = \dfrac{\tan\theta-1}{1+\tan\theta}+\dfrac{\tan\theta+1}{1-\tan\theta} = \dfrac{\tan\theta-1}{\tan\theta+1}-\dfrac{\tan\theta+1}{\tan\theta-1}$$

$$= \dfrac{(\tan\theta-1)^2-(\tan\theta+1)^2}{(\tan\theta+1)(\tan\theta-1)} = \dfrac{(\tan^2\theta-2\tan\theta+1)-(\tan^2\theta+2\tan\theta+1)}{\tan^2\theta-1}$$

$$= \dfrac{-4\tan\theta}{\tan^2\theta-1} = \dfrac{4\tan\theta}{1-\tan^2\theta} = \dfrac{2(2\tan\theta)}{1-\tan^2\theta} = 2\tan 2\theta$$

37. $\cos^2 15° - \sin^2 15° = \cos\left[2(15°)\right] = \cos 30° = \dfrac{\sqrt{3}}{2}$

39. $1-2\sin^2 15° = \cos\left[2(15°)\right] = \cos 30° = \dfrac{\sqrt{3}}{2}$

41. $2\cos^2 67\dfrac{1}{2}° - 1 = \cos^2 67\dfrac{1}{2}° - \sin^2 67\dfrac{1}{2}° = \cos 2\left(67\dfrac{1}{2}°\right) = \cos 135° = -\dfrac{\sqrt{2}}{2}$

43. $\dfrac{\tan 51°}{1 - \tan^2 51°}$

$\dfrac{2\tan A}{1 - \tan^2 A} = \tan 2A$, so we have

$\dfrac{1}{2}\left(\dfrac{2\tan A}{1 - \tan^2 A}\right) = \dfrac{1}{2}\tan 2A \Rightarrow$

$\dfrac{\tan A}{1 - \tan^2 A} = \dfrac{1}{2}\tan 2A \Rightarrow$

$\dfrac{\tan 51°}{1 - \tan^2 51°} = \dfrac{1}{2}\tan\left[2(51°)\right] = \dfrac{1}{2}\tan 102°$

45. $\dfrac{1}{4} - \dfrac{1}{2}\sin^2 47.1° = \dfrac{1}{4}\left(1 - 2\sin^2 47.1°\right)$

$= \dfrac{1}{4}\cos\left[2(47.1°)\right]$

$= \dfrac{1}{4}\cos 94.2°$

47. $\sin^2 \dfrac{2\pi}{5} - \cos^2 \dfrac{2\pi}{5} = -\left(\cos^2 \dfrac{2\pi}{5} - \sin^2 \dfrac{2\pi}{5}\right)$

$\cos^2 A - \sin^2 A = \cos 2A \Rightarrow$

$-\left(\cos^2 \dfrac{2\pi}{5} - \sin^2 \dfrac{2\pi}{5}\right) = -\cos\left(2 \cdot \dfrac{2\pi}{5}\right)$

$= -\cos \dfrac{4\pi}{5}$

49. $\sin 4x = \sin\left[2(2x)\right] = 2\sin 2x \cos 2x$

$= 2\left(2\sin x \cos x\right)\left(\cos^2 x - \sin^2 x\right)$

$= 4\sin x \cos^3 x - 4\sin^3 x \cos x$

51. $\tan 3x = \tan\left(2x + x\right) = \dfrac{\tan 2x + \tan x}{1 - \tan 2x \tan x}$

$= \dfrac{\dfrac{2\tan x}{1 - \tan^2 x} + \tan x}{1 - \dfrac{2\tan x}{1 - \tan^2 x}\cdot \tan x}$

$= \dfrac{\dfrac{2\tan x}{1 - \tan^2 x} + \dfrac{\left(1 - \tan^2 x\right)\tan x}{1 - \tan^2 x}}{\dfrac{1 - \tan^2 x}{1 - \tan^2 x} - \dfrac{2\tan^2 x}{1 - \tan^2 x}}$

$= \dfrac{\dfrac{2\tan x}{1 - \tan^2 x} + \dfrac{\tan x - \tan^3 x}{1 - \tan^2 x}}{\dfrac{1 - \tan^2 x}{1 - \tan^2 x} - \dfrac{2\tan^2 x}{1 - \tan^2 x}} \cdot \dfrac{1 - \tan^2 x}{1 - \tan^2 x}$

$= \dfrac{2\tan x + \tan x - \tan^3 x}{1 - \tan^2 x - 2\tan^2 x}$

$= \dfrac{3\tan x - \tan^3 x}{1 - 3\tan^2 x}$

Exercises 53–55 are graphed in the window $\left[-2\pi, 2\pi\right]$ by $\left[-4, 4\right]$.

53. $\cos^4 x - \sin^4 x$ appears to be equivalent to $\cos 2x$.

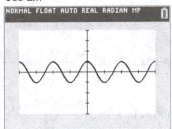

$\cos^4 x - \sin^4 x$

$= \left(\cos^2 x + \sin^2 x\right)\left(\cos^2 x - \sin^2 x\right)$

$= 1 \cdot \cos 2x = \cos 2x$

55. $\dfrac{2\tan x}{2 - \sec^2 x}$ appears to be equivalent to $\tan 2x$.

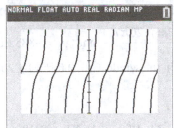

$\dfrac{2\tan x}{2 - \sec^2 x} = \dfrac{2\tan x}{1 - \left(\sec^2 x - 1\right)} = \dfrac{2\tan x}{1 - \tan^2 x}$

$= \tan 2x$

57. $2\sin 58° \cos 102°$

$= 2\left(\dfrac{1}{2}\left[\sin\left(58° + 102°\right) + \sin\left(58° - 102°\right)\right]\right)$

$= \sin 160° + \sin\left(-44°\right) = \sin 160° - \sin 44°$

59. $2\sin \dfrac{\pi}{6}\cos \dfrac{\pi}{3}$

$= 2 \cdot \dfrac{1}{2}\left[\sin\left(\dfrac{\pi}{6} + \dfrac{\pi}{3}\right) + \sin\left(\dfrac{\pi}{6} - \dfrac{\pi}{3}\right)\right]$

$= \sin \dfrac{\pi}{2} + \sin\left(-\dfrac{\pi}{6}\right) = \sin \dfrac{\pi}{2} - \sin \dfrac{\pi}{6}$

61. $6\sin 4x \sin 5x$

$= 6 \cdot \dfrac{1}{2}\left[\cos\left(4x - 5x\right) - \cos\left(4x + 5x\right)\right]$

$= 3\cos\left(-x\right) - 3\cos 9x = 3\cos x - 3\cos 9x$

63. $\cos 4x - \cos 2x = -2 \sin\left(\dfrac{4x+2x}{2}\right)\sin\left(\dfrac{4x-2x}{2}\right) = -2\sin\dfrac{6x}{2}\sin\dfrac{2x}{2} = -2\sin 3x \sin x$

65. $\sin 25° + \sin(-48°) = 2\sin\left(\dfrac{25°+(-48°)}{2}\right)\cos\left(\dfrac{25°-(-48°)}{2}\right) = 2\sin\dfrac{-23°}{2}\cos\dfrac{73°}{2}$

$\qquad = 2\sin(-11.5°)\cos 36.5° = -2\sin 11.5° \cos 36.5°$

67. $\cos 4x + \cos 8x = 2\cos\left(\dfrac{4x+8x}{2}\right)\cos\left(\dfrac{4x-8x}{2}\right) = 2\cos\dfrac{12x}{2}\cos\dfrac{-4x}{2} = 2\cos 6x\cos(-2x) = 2\cos 6x\cos 2x$

69. From Example 6, $W = \dfrac{(163\sin 120\pi t)^2}{15} \Rightarrow W \approx 1771.3(\sin 120\pi t)^2$. Thus, we have

$1771.3(\sin 120\pi t)^2 = 1771.3\sin 120\pi t \cdot \sin 120\pi t$

$\qquad = (1771.3)\left(\dfrac{1}{2}\right)\left[\cos(120\pi t - 120\pi t) - \cos(120\pi t + 120\pi t)\right]$

$\qquad = 885.6(\cos 0 - \cos 240\pi t) = 885.6(1 - \cos 240\pi t) = -885.6\cos 240\pi t + 885.6$

If we compare this to $W = a\cos(\omega t) + c$, then $a = -885.6$, $c = 885.6$, and $\omega = 240\pi$. The graph shows

$y_1 = \dfrac{(163\sin 120\pi t)^2}{15}$ and $y_2 = -885.6\cos 240\pi t + 885.6$

graphed in the window $[0, 0.05] \times [0, 2000]$.

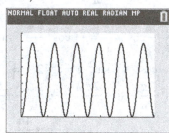

Section 5.6   Half-Angle Identities

1. 195° lies in quadrant III, and the sine is negative in quadrant III, so choose the negative square root.

3. 225° lies in quadrant III, and the tangent is positive in quadrant III, so choose the positive square root.

5. C.

$\sin 15° = \sin\dfrac{30°}{2} = \sqrt{\dfrac{1-\cos 30°}{2}} = \sqrt{\dfrac{1-\frac{\sqrt{3}}{2}}{2}}$

$\qquad = \sqrt{\dfrac{2-\sqrt{3}}{2\cdot 2}} = \sqrt{\dfrac{2-\sqrt{3}}{4}} = \dfrac{\sqrt{2-\sqrt{3}}}{2}$

7. D.

$\cos\dfrac{\pi}{8} = \cos\dfrac{\frac{\pi}{4}}{2} = \sqrt{\dfrac{1+\cos\frac{\pi}{4}}{2}} = \sqrt{\dfrac{1+\frac{\sqrt{2}}{2}}{2}}$

$\qquad = \sqrt{\dfrac{2+\sqrt{2}}{2\cdot 2}} = \dfrac{\sqrt{2+\sqrt{2}}}{2}$

9. F.

$\tan 67.5° = \tan\dfrac{135°}{2} = \dfrac{1-\cos 135°}{\sin 135°}$

$\qquad = \dfrac{1-\left(-\frac{\sqrt{2}}{2}\right)}{\frac{\sqrt{2}}{2}} = \dfrac{1-\left(-\frac{\sqrt{2}}{2}\right)}{\frac{\sqrt{2}}{2}} \cdot \dfrac{2}{2}$

$\qquad = \dfrac{2+\sqrt{2}}{\sqrt{2}} = \dfrac{2+\sqrt{2}}{\sqrt{2}} \cdot \dfrac{\sqrt{2}}{\sqrt{2}}$

$\qquad = \dfrac{2\sqrt{2}+2}{2} = \sqrt{2}+1$

11. $\sin 67.5° = \sin\left(\dfrac{135°}{2}\right)$

67.5° is in quadrant I, so $\sin 67.5° > 0$.

$\sin 67.5° = \sqrt{\dfrac{1-\cos 135°}{2}} = \sqrt{\dfrac{1-(-\cos 45°)}{2}}$

$\qquad = \sqrt{\dfrac{1+\frac{\sqrt{2}}{2}}{2}} = \sqrt{\dfrac{1+\frac{\sqrt{2}}{2}}{2} \cdot \dfrac{2}{2}}$

$\qquad = \dfrac{\sqrt{2+\sqrt{2}}}{\sqrt{4}} = \dfrac{\sqrt{2+\sqrt{2}}}{2}$

13. $\tan 195° = \tan\left(\dfrac{390°}{2}\right) = \dfrac{\sin 390°}{1+\cos 390°}$

$= \dfrac{\sin 30°}{1+\cos 30°} = \dfrac{\frac{1}{2}}{1+\frac{\sqrt{3}}{2}} = \dfrac{\frac{1}{2}}{1+\frac{\sqrt{3}}{2}} \cdot \dfrac{2}{2}$

$= \dfrac{1}{2+\sqrt{3}} = \dfrac{1}{2+\sqrt{3}} \cdot \dfrac{2-\sqrt{3}}{2-\sqrt{3}} = \dfrac{2-\sqrt{3}}{4-3}$

$= \dfrac{2-\sqrt{3}}{1} = 2-\sqrt{3}$

15. $\cos 165° = \cos\left(\dfrac{330°}{2}\right)$

$165°$ is in quadrant II, so $\cos 165° < 0$.

$\cos 165° = -\sqrt{\dfrac{1+\cos 330°}{2}} = -\sqrt{\dfrac{1+\cos 30°}{2}}$

$= -\sqrt{\dfrac{1+\frac{\sqrt{3}}{2}}{2}} = -\sqrt{\dfrac{2+\sqrt{3}}{4}}$

$= \dfrac{-\sqrt{2+\sqrt{3}}}{2}$

17. To find $\sin 7.5°$, you could use the half-angle formulas for sine and cosine as follows:

$\sin 7.5° = \sqrt{\dfrac{1-\cos 15°}{2}}$ and

$\cos 15° = \sqrt{\dfrac{1+\cos 30°}{2}} = \sqrt{\dfrac{1+\frac{\sqrt{3}}{2}}{2}}$

$= \sqrt{\dfrac{2+\sqrt{3}}{4}} = \dfrac{\sqrt{2+\sqrt{3}}}{2}$

Thus,

$\sin 7.5° = \sqrt{\dfrac{1-\cos 15°}{2}} = \sqrt{\dfrac{1-\frac{\sqrt{2+\sqrt{3}}}{2}}{2}}$

$= \sqrt{\dfrac{2-\sqrt{2+\sqrt{3}}}{4}} = \dfrac{\sqrt{2-\sqrt{2+\sqrt{3}}}}{2}$

19. Find $\cos\dfrac{x}{2}$, given $\cos x = \dfrac{1}{4}$, with

$0 < x < \dfrac{\pi}{2}$.

$0 < x < \dfrac{\pi}{2} \Rightarrow 0 < \dfrac{x}{2} < \dfrac{\pi}{4}$, so $\cos\dfrac{x}{2} > 0$.

$\cos\dfrac{x}{2} = \sqrt{\dfrac{1+\cos x}{2}} = \sqrt{\dfrac{1+\frac{1}{4}}{2}} = \sqrt{\dfrac{4+1}{8}} = \sqrt{\dfrac{5}{8}}$

$= \dfrac{\sqrt{5}}{\sqrt{8}} = \dfrac{\sqrt{5}}{2\sqrt{2}} = \dfrac{\sqrt{5}}{2\sqrt{2}} \cdot \dfrac{\sqrt{2}}{\sqrt{2}} = \dfrac{\sqrt{10}}{4}$

21. Find $\tan\dfrac{\theta}{2}$, given $\sin\theta = \dfrac{3}{5}$, with $90° < \theta < 180°$.

To find $\tan\dfrac{\theta}{2}$, we need the values of $\sin\theta$

and $\cos\theta$. We know $\sin\theta = \dfrac{3}{5}$.

$\cos\theta = \pm\sqrt{1-\left(\dfrac{3}{5}\right)^2} = \pm\sqrt{1-\dfrac{9}{25}} = \pm\sqrt{\dfrac{16}{25}} = \pm\dfrac{4}{5}$

Because $90° < \theta < 180°$ (θ is in quadrant II),

$\cos\theta < 0$. Thus, $\cos\theta = -\dfrac{4}{5}$.

$\tan\dfrac{\theta}{2} = \dfrac{\sin\theta}{1+\cos\theta} = \dfrac{\frac{3}{5}}{1+\left(-\frac{4}{5}\right)} = \dfrac{\frac{3}{5}}{1+\left(-\frac{4}{5}\right)} \cdot \dfrac{5}{5}$

$= \dfrac{3}{5-4} = \dfrac{3}{1} = 3$

23. Find $\sin\dfrac{x}{2}$, given $\tan x = 2$, with $0 < x < \dfrac{\pi}{2}$.

Because x is in quadrant I, $\sec x > 0$.

$\sec^2 x = \tan^2 x + 1 \Rightarrow$

$\sec^2 x = 2^2 + 1 = 4+1 = 5 \Rightarrow \sec x = \sqrt{5}$

$\cos x = \dfrac{1}{\sec x} = \dfrac{1}{\sqrt{5}} = \dfrac{1}{\sqrt{5}} \cdot \dfrac{\sqrt{5}}{\sqrt{5}} = \dfrac{\sqrt{5}}{5}$

Because $0 < x < \dfrac{\pi}{2} \Rightarrow 0 < \dfrac{x}{2} < \dfrac{\pi}{4}$, $\dfrac{x}{2}$ is in

quadrant I. Thus, $\sin\dfrac{x}{2} > 0$.

$\sin\dfrac{x}{2} = \sqrt{\dfrac{1-\cos x}{2}} = \sqrt{\dfrac{1-\frac{\sqrt{5}}{5}}{2}} = \dfrac{\sqrt{50-10\sqrt{5}}}{10}$

25. Find $\tan\dfrac{\theta}{2}$, given $\tan\theta = \dfrac{\sqrt{7}}{3}$, with $180° < \theta < 270°$.

$\sec^2\theta = \tan^2\theta + 1 \Rightarrow$

$\sec^2\theta = \left(\dfrac{\sqrt{7}}{3}\right)^2 + 1 = \dfrac{7}{9} + 1 = \dfrac{16}{9}$

θ is in quadrant III, so $\sec\theta < 0$ and $\sin\theta < 0$.

$\sec\theta = -\sqrt{\dfrac{16}{9}} = -\dfrac{4}{3}$ and

$\cos\theta = \dfrac{1}{\sec\theta} = \dfrac{1}{-\frac{4}{3}} = -\dfrac{3}{4}$

(*continued on next page*)

(continued)

$$\sin\theta = -\sqrt{1-\cos^2\theta}$$

$$= -\sqrt{1-\left(-\frac{3}{4}\right)^2} - \sqrt{1-\frac{9}{16}} = -\frac{\sqrt{7}}{4}$$

$$\tan\frac{\theta}{2} = \frac{\sin\theta}{1+\cos\theta} = \frac{-\frac{\sqrt{7}}{4}}{1+\left(-\frac{3}{4}\right)} = \frac{-\sqrt{7}}{4-3} = -\sqrt{7}$$

27. Find $\sin\theta$, given $\cos 2\theta = \frac{3}{5}$, θ is in quadrant I. θ is in quadrant I, so $\sin\theta > 0$.

$$\sin\theta = \sqrt{\frac{1-\cos 2\theta}{2}} \Rightarrow$$

$$\sin\theta = \sqrt{\frac{1-\frac{3}{5}}{2}} = \sqrt{\frac{\frac{2}{5}}{2}} = \sqrt{\frac{2}{10}} = \sqrt{\frac{1}{5}}$$

$$= \frac{1}{\sqrt{5}} \cdot \frac{\sqrt{5}}{\sqrt{5}} = \frac{\sqrt{5}}{5}$$

29. Find $\cos x$, given $\cos 2x = -\frac{5}{12}, \frac{\pi}{2} < x < \pi$.

Because $\frac{\pi}{2} < x < \pi, \cos x < 0$.

$$\cos x = -\sqrt{\frac{1+\cos 2x}{2}} \Rightarrow$$

$$\cos x = -\sqrt{\frac{1+\left(-\frac{5}{12}\right)}{2}} = -\sqrt{\frac{\frac{7}{12}}{2}} = -\sqrt{\frac{7}{24}}$$

$$= -\frac{\sqrt{7}}{2\sqrt{6}} \cdot \frac{\sqrt{6}}{\sqrt{6}} = -\frac{\sqrt{42}}{12}$$

31. $\cos x \approx 0.9682$ and $\sin x \approx 0.25$

$$\tan\frac{x}{2} \approx \frac{\sin x}{1+\cos x} = \frac{0.25}{1+0.9682} = \frac{0.25}{1.9682}$$

$$\approx 0.1270$$

(0.127 to three significant digits)

33. $\sqrt{\dfrac{1-\cos 40°}{2}} = \sin\dfrac{40°}{2} = \sin 20°$

35. $\sqrt{\dfrac{1-\cos 147°}{1+\cos 147°}} = \tan\dfrac{147°}{2} = \tan 73.5°$

37. $\dfrac{1-\cos 59.74°}{\sin 59.74°} = \tan\dfrac{59.74°}{2} = \tan 29.87°$

39. $\pm\sqrt{\dfrac{1+\cos 18x}{2}} = \cos\dfrac{18x}{2} = \cos 9x$

41. $\pm\sqrt{\dfrac{1-\cos 8\theta}{1+\cos 8\theta}} = \tan\dfrac{8\theta}{2} = \tan 4\theta$

43. $\pm\sqrt{\dfrac{1+\cos\frac{x}{4}}{2}} = \cos\dfrac{\frac{x}{4}}{2} = \cos\dfrac{x}{8}$

45. Verify $\sec^2\dfrac{x}{2} = \dfrac{2}{1+\cos x}$ is an identity.

$$\sec^2\frac{x}{2} = \frac{1}{\cos^2\frac{x}{2}} = \frac{1}{\left(\pm\sqrt{\frac{1+\cos x}{2}}\right)^2}$$

$$= \frac{1}{\frac{1+\cos x}{2}} = \frac{2}{1+\cos x}$$

47. Verify $\sin^2\dfrac{x}{2} = \dfrac{\tan x - \sin x}{2\tan x}$ is an identity.
Work with the left side.

$$\sin^2\frac{x}{2} = \left(\pm\sqrt{\frac{1-\cos x}{2}}\right)^2 = \frac{1-\cos x}{2}$$

Work with the right side.

$$\frac{\tan x - \sin x}{2\tan x} = \frac{\frac{\sin x}{\cos x} - \sin x}{2\cdot\frac{\sin x}{\cos x}}$$

$$= \frac{\frac{\sin x}{\cos x} - \sin x}{2\cdot\frac{\sin x}{\cos x}} \cdot \frac{\cos x}{\cos x}$$

$$= \frac{\sin x - \cos x\sin x}{2\sin x}$$

$$= \frac{\sin x(1-\cos x)}{2\sin x} = \frac{1-\cos x}{2}$$

$$\sin^2\frac{x}{2} = \frac{1-\cos x}{2} = \frac{\tan x - \sin x}{2\tan x}, \text{ so the}$$

statement has been verified.

49. Verify $\dfrac{2}{1+\cos x} - \tan^2\dfrac{x}{2} = 1$ is an identity.

$$\frac{2}{1+\cos x} - \tan^2\frac{x}{2} = \frac{2}{1+\cos x} - \left(\pm\sqrt{\frac{1-\cos x}{1+\cos x}}\right)^2$$

$$= \frac{2}{1+\cos x} - \frac{1-\cos x}{1+\cos x}$$

$$= \frac{2-1+\cos x}{1+\cos x}$$

$$= \frac{1+\cos x}{1+\cos x} = 1$$

51. Verify $1 - \tan^2 \dfrac{\theta}{2} = \dfrac{2\cos\theta}{1+\cos\theta}$ is an identity.

$$1 - \tan^2 \frac{\theta}{2} = 1 - \left(\frac{\sin\theta}{1+\cos\theta}\right)^2 = 1 - \frac{\sin^2\theta}{(1+\cos\theta)^2}$$

$$= \frac{(1+\cos\theta)^2 - \sin^2\theta}{(1+\cos\theta)^2}$$

$$= \frac{1 + 2\cos\theta + \cos^2\theta - \sin^2\theta}{(1+\cos\theta)^2}$$

$$= \frac{1 + 2\cos\theta + \cos^2\theta - (1 - \cos^2\theta)}{(1+\cos\theta)^2}$$

$$= \frac{1 + 2\cos\theta + 2\cos^2\theta - 1}{(1+\cos\theta)^2}$$

$$= \frac{2\cos^2\theta + 2\cos\theta}{(1+\cos\theta)^2}$$

$$= \frac{2\cos\theta(1+\cos\theta)}{(1+\cos\theta)^2} = \frac{2\cos\theta}{1+\cos\theta}$$

53. $\tan \dfrac{A}{2} = \dfrac{\sin A}{1+\cos A} = \dfrac{\sin A}{1+\cos A} \cdot \dfrac{1-\cos A}{1-\cos A}$

$$= \frac{\sin A(1-\cos A)}{1-\cos^2 A} = \frac{\sin A(1-\cos A)}{\sin^2 A}$$

$$= \frac{1-\cos A}{\sin A}$$

Exercises 55–57 are graphed in the window $[-2\pi, 2\pi] \times [-4, 4]$.

55. $\dfrac{\sin x}{1+\cos x}$ appears to be equivalent to $\tan \dfrac{x}{2}$.

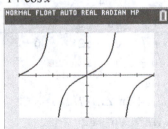

$$\frac{\sin x}{1+\cos x} = \frac{\sin 2\left(\frac{x}{2}\right)}{1+\cos 2\left(\frac{x}{2}\right)} = \frac{2\sin\left(\frac{x}{2}\right)\cos\left(\frac{x}{2}\right)}{1+\left[2\cos^2\left(\frac{x}{2}\right)-1\right]}$$

$$= \frac{2\sin\left(\frac{x}{2}\right)\cos\left(\frac{x}{2}\right)}{2\cos^2\left(\frac{x}{2}\right)} = \frac{\sin\left(\frac{x}{2}\right)}{\cos\left(\frac{x}{2}\right)} = \tan\left(\frac{x}{2}\right)$$

57. $\dfrac{\tan\frac{x}{2} + \cot\frac{x}{2}}{\cot\frac{x}{2} - \tan\frac{x}{2}}$ appears to be equivalent to $\sec x$.

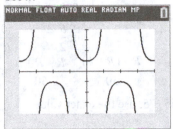

$$\frac{\tan\frac{x}{2} + \cot\frac{x}{2}}{\cot\frac{x}{2} - \tan\frac{x}{2}} = \frac{\dfrac{\sin\frac{x}{2}}{\cos\frac{x}{2}} + \dfrac{\cos\frac{x}{2}}{\sin\frac{x}{2}}}{\dfrac{\cos\frac{x}{2}}{\sin\frac{x}{2}} - \dfrac{\sin\frac{x}{2}}{\cos\frac{x}{2}}}$$

$$= \frac{\dfrac{\sin\frac{x}{2}}{\cos\frac{x}{2}} + \dfrac{\cos\frac{x}{2}}{\sin\frac{x}{2}}}{\dfrac{\cos\frac{x}{2}}{\sin\frac{x}{2}} - \dfrac{\sin\frac{x}{2}}{\cos\frac{x}{2}}} \cdot \frac{\sin\frac{x}{2}\cos\frac{x}{2}}{\sin\frac{x}{2}\cos\frac{x}{2}}$$

$$= \frac{\sin^2\frac{x}{2} + \cos^2\frac{x}{2}}{\cos^2\frac{x}{2} - \sin^2\frac{x}{2}} = \frac{1}{\cos 2\left(\frac{x}{2}\right)}$$

$$= \frac{1}{\cos x} = \sec x$$

59. $\sin\dfrac{\theta}{2} = \dfrac{1}{m}$, $m = \dfrac{5}{4}$

Because $\sin\dfrac{\theta}{2} = \dfrac{1}{\frac{5}{4}} = \dfrac{4}{5}$ and

$\sin^2\dfrac{\theta}{2} = \dfrac{1-\cos\theta}{2}$, we have

$$\left(\frac{4}{5}\right)^2 = \frac{1-\cos\theta}{2} \Rightarrow \frac{16}{25} = \frac{1-\cos\theta}{2} \Rightarrow$$

$$\frac{32}{25} = 1 - \cos\theta \Rightarrow \frac{7}{25} = -\cos\theta \Rightarrow$$

$$\cos\theta = -\frac{7}{25}$$

Thus, we have $\theta = \cos^{-1}\left(-\dfrac{7}{25}\right) \approx 106°$.

61. $\sin\dfrac{\theta}{2} = \dfrac{1}{m}$, $\theta = 60°$

$$\sin\frac{60°}{2} = \frac{1}{m} \Rightarrow \sin 30° = \frac{1}{m}$$

$$\frac{1}{2} = \frac{1}{m} \Rightarrow m = 2$$

63. **(a)** $\cos\dfrac{\theta}{2} = \dfrac{R-b}{R}$

(b) $\tan\dfrac{\theta}{4} = \tan\dfrac{\dfrac{\theta}{2}}{2} = \dfrac{1-\cos\dfrac{\theta}{2}}{\sin\dfrac{\theta}{2}} = \dfrac{1-\dfrac{R-b}{R}}{\dfrac{50}{R}}$

$= \dfrac{R-(R-b)}{50} = \dfrac{R-R+b}{50} = \dfrac{b}{50}$

For exercises 65–75, use the exact value

$\sin 18° = \dfrac{\sqrt{5}-1}{4}$. Other solutions are possible.

65. $\sin^2 18° + \cos^2 18° = 1 \Rightarrow$

$\left(\dfrac{\sqrt{5}-1}{4}\right)^2 + \cos^2 18° = 1 \Rightarrow$

$\cos^2 18° = 1 - \left(\dfrac{\sqrt{5}-1}{4}\right)^2 \Rightarrow$

$\cos 18° = \sqrt{1 - \left(\dfrac{\sqrt{5}-1}{4}\right)^2}$

$= \sqrt{1 - \dfrac{6-2\sqrt{5}}{16}} = \sqrt{\dfrac{10+2\sqrt{5}}{16}}$

$= \dfrac{\sqrt{10+2\sqrt{5}}}{4}$

67. Use the result from exercise 65.

$\cot 18° = \dfrac{\cos 18°}{\sin 18°} = \dfrac{\dfrac{\sqrt{10+2\sqrt{5}}}{4}}{\dfrac{\sqrt{5}-1}{4}} = \dfrac{\sqrt{10+2\sqrt{5}}}{\sqrt{5}-1}$

$= \dfrac{\sqrt{10+2\sqrt{5}}}{\sqrt{5}-1} \cdot \dfrac{\sqrt{5}+1}{\sqrt{5}+1}$

$= \dfrac{\left(\sqrt{10+2\sqrt{5}}\right)\left(\sqrt{5}+1\right)}{5-1}$

$= \dfrac{\left(\sqrt{10+2\sqrt{5}}\right)\left(\sqrt{5}+1\right)}{4}$

69. $\csc 18° = \dfrac{1}{\sin 18°} = \dfrac{1}{\dfrac{\sqrt{5}-1}{4}} = \dfrac{4}{\sqrt{5}-1}$

$= \dfrac{4}{\sqrt{5}-1} \cdot \dfrac{\sqrt{5}+1}{\sqrt{5}+1} = \dfrac{4\left(\sqrt{5}+1\right)}{4} = \sqrt{5}+1$

71. $\sin 72° = \cos 18° = \dfrac{\sqrt{10+2\sqrt{5}}}{4}$

73. Using the result from exercise 66, we have

$\tan 72° = \cot 18° = \dfrac{\left(-5+3\sqrt{5}\right)\left(\sqrt{10-2\sqrt{5}}\right)}{20}$ or

$\tan 72° = \cot 18° = \dfrac{20}{\left(5-\sqrt{5}\right)\left(\sqrt{10-2\sqrt{5}}\right)}$

75. $\sec 72° = \csc 18° = 1+\sqrt{5}$

77. $AB = BD$ because they are both radii of the circle.

79. The sum of the measures of angles DAB and ADB is $180° - 150° = 30°$.
$m\angle DAB = m\angle ADB$, so the measure of each is 15°.

81. $AD^2 = AC^2 + CD^2 \Rightarrow$
$AD^2 = 1^2 + \left(2+\sqrt{3}\right)^2 = 1 + 4 + 4\sqrt{3} + 3$
$= 8 + 4\sqrt{3} = \left(\sqrt{6}+\sqrt{2}\right)^2 \Rightarrow$
$AD = \sqrt{6}+\sqrt{2}$

83. $m\angle EAD = 90°$ (an angle inscribed in a semicircle is a right angle), so
$m\angle EAC + m\angle CAD = 90° \Rightarrow m\angle EAC = 15°$.

$\cos 15° = \dfrac{AC}{AE} = \dfrac{1}{AE} \Rightarrow AE = \dfrac{1}{\cos 15°}$

$AE = \dfrac{1}{\cos 15°} = \dfrac{1}{\dfrac{\sqrt{6}+\sqrt{2}}{4}} = \dfrac{4}{\sqrt{6}+\sqrt{2}}$

$= \dfrac{4}{\sqrt{6}+\sqrt{2}} \cdot \dfrac{\sqrt{6}-\sqrt{2}}{\sqrt{6}-\sqrt{2}} = \dfrac{4\left(\sqrt{6}-\sqrt{2}\right)}{6-2}$

$= \dfrac{4\left(\sqrt{6}-\sqrt{2}\right)}{4} = \sqrt{6}-\sqrt{2}$

$ED = 2BD = 4$. In $\triangle AED$,

$\sin 15° = \dfrac{AE}{ED} = \dfrac{\sqrt{6}-\sqrt{2}}{4}$

Summary Exercises on Verifying Trigonometric Identities

For the following exercises, other solutions are possible.

1. Verify $\tan\theta + \cot\theta = \sec\theta\csc\theta$ is an identity.

$$\tan\theta + \cot\theta = \frac{\sin\theta}{\cos\theta} + \frac{\cos\theta}{\sin\theta} = \frac{\sin^2\theta}{\cos\theta\sin\theta} + \frac{\cos^2\theta}{\cos\theta\sin\theta} = \frac{\sin^2\theta + \cos^2\theta}{\cos\theta\sin\theta} = \frac{1}{\cos\theta\sin\theta}$$

$$= \frac{1}{\cos\theta}\cdot\frac{1}{\sin\theta} = \sec\theta\csc\theta$$

3. Verify $\tan\dfrac{x}{2} = \csc x - \cot x$ is an identity.

Starting on the right side, we have

$$\csc x - \cot x = \frac{1}{\sin x} - \frac{\cos x}{\sin x} = \frac{1-\cos x}{\sin x} = \tan\frac{x}{2}$$

5. Verify $\dfrac{\sin t}{1+\cos t} = \dfrac{1-\cos t}{\sin t}$ is an identity.

$$\frac{\sin t}{1+\cos t} = \frac{\sin t}{1+\cos t}\cdot\frac{1-\cos t}{1-\cos t} = \frac{\sin t(1-\cos t)}{1-\cos^2 t} = \frac{\sin t(1-\cos t)}{\sin^2 t} = \frac{1-\cos t}{\sin t}$$

7. Verify $\sin 2\theta = \dfrac{2\tan\theta}{1+\tan^2\theta}$ is an identity.

Starting on the right side, we have

$$\frac{2\tan\theta}{1+\tan^2\theta} = \frac{2\tan\theta}{\sec^2\theta} = \frac{2\cdot\dfrac{\sin\theta}{\cos\theta}}{\dfrac{1}{\cos^2\theta}} = 2\cdot\frac{\sin\theta}{\cos\theta}\cdot\frac{\cos^2\theta}{1} = 2\sin\theta\cos\theta = \sin 2\theta$$

9. Verify $\cot\theta - \tan\theta = \dfrac{2\cos^2\theta - 1}{\sin\theta\cos\theta}$ is an identity.

$$\cot\theta - \tan\theta = \frac{\cos\theta}{\sin\theta} - \frac{\sin\theta}{\cos\theta} = \frac{\cos^2\theta}{\sin\theta\cos\theta} - \frac{\sin^2\theta}{\sin\theta\cos\theta} = \frac{\cos^2\theta - \sin^2\theta}{\sin\theta\cos\theta} = \frac{\cos^2\theta - (1-\cos^2\theta)}{\sin\theta\cos\theta}$$

$$= \frac{\cos^2\theta - 1 + \cos^2\theta}{\sin\theta\cos\theta} = \frac{2\cos^2\theta - 1}{\sin\theta\cos\theta}$$

11. Verify $\dfrac{\sin(x+y)}{\cos(x-y)} = \dfrac{\cot x + \cot y}{1+\cot x\cot y}$ is an identity.

$$\frac{\sin(x+y)}{\cos(x-y)} = \frac{\sin x\cos y + \cos x\sin y}{\cos x\cos y + \sin x\sin y} = \frac{\sin x\cos y + \cos x\sin y}{\cos x\cos y + \sin x\sin y}\cdot\frac{\dfrac{1}{\cos x\cos y}}{\dfrac{1}{\cos x\cos y}} = \frac{\dfrac{\sin x\cos y}{\cos x\cos y} + \dfrac{\cos x\sin y}{\cos x\cos y}}{\dfrac{\cos x\cos y}{\cos x\cos y} + \dfrac{\sin x\sin y}{\cos x\cos y}}$$

$$= \frac{\dfrac{\sin x}{\cos x} + \dfrac{\sin y}{\cos y}}{1 + \dfrac{\sin x}{\cos x}\cdot\dfrac{\sin y}{\cos y}} = \frac{\cot x + \cot y}{1+\cot x\cot y}$$

13. Verify $\dfrac{\sin\theta + \tan\theta}{1+\cos\theta} = \tan\theta$ is an identity.

$$\frac{\sin\theta + \tan\theta}{1+\cos\theta} = \frac{\sin\theta + \dfrac{\sin\theta}{\cos\theta}}{1+\cos\theta} = \frac{\sin\theta + \dfrac{\sin\theta}{\cos\theta}}{1+\cos\theta}\cdot\frac{\cos\theta}{\cos\theta} = \frac{\sin\theta\cos\theta + \sin\theta}{\cos\theta(1+\cos\theta)} = \frac{\sin\theta(\cos\theta+1)}{\cos\theta(1+\cos\theta)} = \frac{\sin\theta}{\cos\theta} = \tan\theta$$

15. Verify $\cos x = \dfrac{1 - \tan^2 \frac{x}{2}}{1 + \tan^2 \frac{x}{2}}$ is an identity.

$$\frac{1 - \tan^2 \frac{x}{2}}{1 + \tan^2 \frac{x}{2}} = \frac{1 - \left(\frac{1 - \cos x}{\sin x}\right)^2}{1 + \left(\frac{1 - \cos x}{\sin x}\right)^2} = \frac{1 - \frac{(1 - \cos x)^2}{\sin^2 x}}{1 + \frac{(1 - \cos x)^2}{\sin^2 x}} = \frac{1 - \frac{(1 - \cos x)^2}{\sin^2 x}}{1 + \frac{(1 - \cos x)^2}{\sin^2 x}} \cdot \frac{\sin^2 x}{\sin^2 x} = \frac{\sin^2 x - (1 - \cos x)^2}{\sin^2 x + (1 - \cos x)^2}$$

$$= \frac{\sin^2 x - (1 - 2\cos x + \cos^2 x)}{\sin^2 x + (1 - 2\cos x + \cos^2 x)} = \frac{\sin^2 x - 1 + 2\cos x - \cos^2 x}{\sin^2 x + 1 - 2\cos x + \cos^2 x} = \frac{(1 - \cos^2 x) - 1 + 2\cos x - \cos^2 x}{(\sin^2 x + \cos^2 x) + 1 - 2\cos x}$$

$$= \frac{1 - \cos^2 x - 1 + 2\cos x - \cos^2 x}{1 + 1 - 2\cos x} = \frac{2\cos x - 2\cos^2 x}{2 - 2\cos x} = \frac{2\cos x(1 - \cos x)}{2(1 - \cos x)} = \cos x$$

17. Verify $\dfrac{\tan^2 t + 1}{\tan t \csc^2 t} = \tan t$ is an identity.

$$\frac{\tan^2 t + 1}{\tan t \csc^2 t} = \frac{\frac{\sin^2 t}{\cos^2 t} + 1}{\frac{\sin t}{\cos t} \cdot \frac{1}{\sin^2 t}} = \frac{\frac{\sin^2 t}{\cos^2 t} + 1}{\frac{1}{\cos t \sin t}} = \frac{\frac{\sin^2 t}{\cos^2 t} + 1}{\frac{1}{\cos t \sin t}} \cdot \frac{\cos^2 t \sin t}{\cos^2 t \sin t} = \frac{\sin^3 t + \cos^2 t \sin t}{\cos t}$$

$$= \frac{\sin t (\sin^2 t + \cos^2 t)}{\cos t} = \frac{\sin t (1)}{\cos t} = \frac{\sin t}{\cos t} = \tan t$$

19. Verify $\tan 4\theta = \dfrac{2\tan 2\theta}{2 - \sec^2 2\theta}$ is an identity.

$$\frac{2\tan 2\theta}{2 - \sec^2 2\theta} = \frac{2 \cdot \frac{\sin 2\theta}{\cos 2\theta}}{2 - \frac{1}{\cos^2 2\theta}} = \frac{2 \cdot \frac{\sin 2\theta}{\cos 2\theta}}{2 - \frac{1}{\cos^2 2\theta}} \cdot \frac{\cos^2 2\theta}{\cos^2 2\theta} = \frac{2\sin 2\theta \cos 2\theta}{2\cos^2 2\theta - 1} = \frac{\sin[2(2\theta)]}{\cos[2(2\theta)]} = \frac{\sin 4\theta}{\cos 4\theta} = \tan 4\theta$$

21. Verify $\dfrac{\cot s - \tan s}{\cos s + \sin s} = \dfrac{\cos s - \sin s}{\sin s \cos s}$ is an identity.

$$\frac{\cot s - \tan s}{\cos s + \sin s} = \frac{\frac{\cos s}{\sin s} - \frac{\sin s}{\cos s}}{\cos s + \sin s} = \frac{\frac{\cos s}{\sin s} - \frac{\sin s}{\cos s}}{\cos s + \sin s} \cdot \frac{\sin s \cos s}{\sin s \cos s} = \frac{\cos^2 s - \sin^2 s}{(\cos s + \sin s)\sin s \cos s}$$

$$= \frac{(\cos s + \sin s)(\cos s - \sin s)}{(\cos s + \sin s)\sin s \cos s} = \frac{\cos s - \sin s}{\sin s \cos s}$$

23. Verify $\dfrac{\tan(x + y) - \tan y}{1 + \tan(x + y)\tan y} = \tan x$ is an identity.

$$\frac{\tan(x + y) - \tan y}{1 + \tan(x + y)\tan y} = \frac{\frac{\tan x + \tan y}{1 - \tan x \tan y} - \tan y}{1 + \frac{\tan x + \tan y}{1 - \tan x \tan y} \cdot \tan y} = \frac{\frac{\tan x + \tan y}{1 - \tan x \tan y} - \tan y}{1 + \frac{\tan x + \tan y}{1 - \tan x \tan y} \cdot \tan y} \cdot \frac{1 - \tan x \tan y}{1 - \tan x \tan y}$$

$$= \frac{\tan x + \tan y - \tan y(1 - \tan x \tan y)}{1 - \tan x \tan y + (\tan x + \tan y)\tan y} = \frac{\tan x + \tan x \tan^2 y}{1 - \tan x \tan y + \tan x \tan y + \tan^2 y}$$

$$= \frac{\tan x (1 + \tan^2 y)}{1 + \tan^2 y} = \tan x$$

25. Verify $\dfrac{\cos^4 x - \sin^4 x}{\cos^2 x} = 1 - \tan^2 x$ is an identity.

$$\frac{\cos^4 x - \sin^4 x}{\cos^2 x} = \frac{\left(\cos^2 x + \sin^2 x\right)\left(\cos^2 x - \sin^2 x\right)}{\cos^2 x} = \frac{(1)\left(\cos^2 x - \sin^2 x\right)}{\cos^2 x} = \frac{\cos^2 x - \sin^2 x}{\cos^2 x}$$

$$= \frac{\cos^2 x}{\cos^2 x} - \frac{\sin^2 x}{\cos^2 x} = 1 - \tan^2 x$$

27. Verify $\dfrac{2\left(\sin x - \sin^3 x\right)}{\cos x} = \sin 2x$ is an identity.

$$\frac{2\left(\sin x - \sin^3 x\right)}{\cos x} = \frac{2\sin x\left(1 - \sin^2 x\right)}{\cos x} = \frac{2\sin x \cos^2 x}{\cos x} = 2\sin x \cos x = \sin 2x$$

29. Verify $\dfrac{\cos(x+y) + \cos(y-x)}{\sin(x+y) - \sin(y-x)} = \cot x$ is an identity.

$$\frac{\cos(x+y)+\cos(y-x)}{\sin(x+y)-\sin(y-x)} = \frac{(\cos x \cos y - \sin x \sin y)+(\cos y \cos x + \sin y \sin x)}{(\sin x \cos y + \cos x \sin y)-(\sin y \cos x - \cos y \sin x)} = \frac{2\cos x \cos y}{2\cos y \sin x} = \frac{\cos x}{\sin x} = \cot x$$

31. Verify $\sin(60° + x) + \sin(60° - x) = \sqrt{3}\cos x$ is an identity.

$$\sin(60°+x)+\sin(60°-x) = (\sin 60° \cos x + \cos 60° \sin x)+(\sin 60° \cos x - \cos 60° \sin x)$$

$$= 2\sin 60° \cos x = 2\left(\frac{\sqrt{3}}{2}\right)\cos x = \sqrt{3}\cos x$$

33. Verify $\sin^3 \theta + \cos^3 \theta + \sin \theta \cos^2 \theta + \sin^2 \theta \cos \theta = \sin \theta + \cos \theta$ is an identity.

$$\sin^3 \theta + \cos^3 \theta + \sin \theta \cos^2 \theta + \sin^2 \theta \cos \theta = \left(\sin^3 \theta + \sin \theta \cos^2 \theta\right)+\left(\cos^3 \theta + \sin^2 \theta \cos \theta\right)$$

$$= \sin \theta\left(\sin^2 \theta + \cos^2 \theta\right)+\cos \theta\left(\cos^2 \theta + \sin^2 \theta\right)$$

$$= (\sin \theta + \cos \theta)\left(\sin^2 \theta + \cos^2 \theta\right) = \sin \theta + \cos \theta$$

Chapter 5   Review Exercises

1. B   **3.** C   **5.** D

7. $\sec^2 \theta - \tan^2 \theta = \dfrac{1}{\cos^2 \theta} - \dfrac{\sin^2 \theta}{\cos^2 \theta}$

$$= \frac{1 - \sin^2 \theta}{\cos^2 \theta} = \frac{\cos^2 \theta}{\cos^2 \theta} = 1$$

9. $\tan^2 \theta\left(1 + \cot^2 \theta\right) = \dfrac{\sin^2 \theta}{\cos^2 \theta}\left(1 + \dfrac{\cos^2 \theta}{\sin^2 \theta}\right)$

$$= \frac{\sin^2 \theta}{\cos^2 \theta}\left(\frac{\sin^2 \theta + \cos^2 \theta}{\sin^2 \theta}\right)$$

$$= \frac{\sin^2 \theta}{\cos^2 \theta}\left(\frac{1}{\sin^2 \theta}\right) = \frac{1}{\cos^2 \theta}$$

$$= \sec^2 \theta$$

11. $\tan \theta - \sec \theta \csc \theta = \dfrac{\sin \theta}{\cos \theta} - \dfrac{1}{\cos \theta}\cdot\dfrac{1}{\sin \theta}$

$$= \frac{\sin \theta}{\cos \theta} - \frac{1}{\sin \theta \cos \theta}$$

$$= \frac{\sin^2 \theta}{\sin \theta \cos \theta} - \frac{1}{\sin \theta \cos \theta}$$

$$= \frac{\sin^2 \theta - 1}{\sin \theta \cos \theta} = \frac{\left(1 - \cos^2 \theta\right)-1}{\sin \theta \cos \theta}$$

$$= \frac{-\cos^2 \theta}{\sin \theta \cos \theta} = -\frac{\cos \theta}{\sin \theta}$$

$$= -\cot \theta$$

13. $\cos x = \dfrac{3}{5}$, x is in quadrant IV.

$$\sin^2 x = 1 - \cos^2 x = 1 - \left(\dfrac{3}{5}\right)^2 = 1 - \dfrac{9}{25} = \dfrac{16}{25}$$

Because x is in quadrant IV, $\sin x < 0$.

$$\sin x = -\sqrt{\dfrac{16}{25}} = -\dfrac{4}{5}$$

$$\tan x = \dfrac{\sin x}{\cos x} = \dfrac{-\frac{4}{5}}{\frac{3}{5}} = -\dfrac{4}{3}$$

$$\cot(-x) = -\cot x = \dfrac{1}{-\tan x} = \dfrac{1}{-\left(-\frac{4}{3}\right)} = \dfrac{3}{4}$$

15. Use the fact that $165° = 180° - 15°$.

$$\sin 165° = \sin 180° \cos 15° - \cos 180° \sin 15°$$
$$= 0 \cdot \cos 15° - 1 \cdot \sin 15° = \sin 15°$$
$$= \sin(45° - 30°)$$
$$= \sin 45° \cos 30° - \cos 45° \sin 30°$$
$$= \dfrac{\sqrt{2}}{2} \cdot \dfrac{\sqrt{3}}{2} - \dfrac{\sqrt{2}}{2} \cdot \dfrac{1}{2} = \dfrac{\sqrt{6} - \sqrt{2}}{4}$$

$$\cos 165° = \cos 180° \cos 15° + \sin 180° \sin 15°$$
$$= -1 \cdot \cos 15° + 0 \cdot \sin 15° = -\cos 15°$$
$$= -\cos(45° - 30°)$$
$$= -(\cos 45° \cos 30° + \sin 45° \sin 30°)$$
$$= -\left(\dfrac{\sqrt{2}}{2} \cdot \dfrac{\sqrt{3}}{2} + \dfrac{\sqrt{2}}{2} \cdot \dfrac{1}{2}\right)$$
$$= -\left(\dfrac{\sqrt{6} + \sqrt{2}}{4}\right) = \dfrac{-\sqrt{6} - \sqrt{2}}{4}$$

$$\tan 165° = \dfrac{\tan 180° - \tan 15°}{1 + \tan 180° \tan 15°} = \dfrac{0 - \tan 15°}{1 + 0 \cdot \tan 15°}$$
$$= -\tan 15° = -\dfrac{\tan 45° - \tan 30°}{1 + \tan 45° \tan 30°}$$
$$= -\dfrac{1 - \frac{\sqrt{3}}{3}}{1 + \frac{\sqrt{3}}{3}} = -\dfrac{1 - \frac{\sqrt{3}}{3}}{1 + \frac{\sqrt{3}}{3}} \cdot \dfrac{3}{3}$$
$$= -\dfrac{3 - \sqrt{3}}{3 + \sqrt{3}} = -\dfrac{3 - \sqrt{3}}{3 + \sqrt{3}} \cdot \dfrac{3 - \sqrt{3}}{3 - \sqrt{3}}$$
$$= -\dfrac{9 - 3\sqrt{3} - 3\sqrt{3} + 3}{9 - 3} = -\dfrac{12 - 6\sqrt{3}}{6}$$
$$= -\left(2 - \sqrt{3}\right) = -2 + \sqrt{3}$$

$$\cot 165° = \dfrac{1}{\tan 165°} = \dfrac{1}{-2 + \sqrt{3}}$$
$$= \dfrac{1}{-2 + \sqrt{3}} \cdot \dfrac{-2 - \sqrt{3}}{-2 - \sqrt{3}} = \dfrac{-2 - \sqrt{3}}{4 - 3}$$
$$= -2 - \sqrt{3}$$

$$\sec 165° = \dfrac{1}{\cos 165°} = \dfrac{1}{\frac{-\sqrt{6} - \sqrt{2}}{4}} = \dfrac{4}{-\sqrt{6} - \sqrt{2}}$$
$$= \dfrac{4}{-\sqrt{6} - \sqrt{2}} \cdot \dfrac{-\sqrt{6} + \sqrt{2}}{-\sqrt{6} + \sqrt{2}}$$
$$= \dfrac{4\left(-\sqrt{6} + \sqrt{2}\right)}{6 - 2} = -\sqrt{6} + \sqrt{2}$$

$$\csc 165° = \dfrac{1}{\sin 165°} = \dfrac{1}{\frac{\sqrt{6} - \sqrt{2}}{4}} = \dfrac{4}{\sqrt{6} - \sqrt{2}}$$
$$= \dfrac{4}{\sqrt{6} - \sqrt{2}} \cdot \dfrac{\sqrt{6} + \sqrt{2}}{\sqrt{6} + \sqrt{2}} = \dfrac{4\left(\sqrt{6} + \sqrt{2}\right)}{6 - 2}$$
$$= \sqrt{6} + \sqrt{2}$$

17. I. $\cos 210° = \cos(150° + 60°)$
$$= \cos 150° \cos 60° - \sin 150° \sin 60°$$

19. H. $\tan(-35°) = \cot\left[90° - (-35°)\right] = \cot 125°$

21. G. $\cos 35° = \cos(-35°)$

23. J. $\sin 75° = \sin(15° + 60°)$
$$= \sin 15° \cos 60° + \cos 15° \sin 60°$$

25. F. $\cos 300° = \cos 2(150°)$
$$= \cos^2 150° - \sin^2 150°$$

27. Find $\sin(x + y)$, $\cos(x - y)$, and $\tan(x + y)$, given $\sin x = -\dfrac{3}{5}$, $\cos y = -\dfrac{7}{25}$, x and y are in quadrant III. Because x and y are in quadrant III, $\cos x$ and $\sin y$ are negative.

$$\cos x = -\sqrt{1 - \sin^2 x} = -\sqrt{1 - \left(-\dfrac{3}{5}\right)^2}$$
$$= -\sqrt{1 - \dfrac{9}{25}} = -\sqrt{\dfrac{16}{25}} = -\dfrac{4}{5}$$

$$\sin y = -\sqrt{1 - \cos^2 y} = -\sqrt{1 - \left(-\dfrac{7}{25}\right)^2}$$
$$= -\sqrt{1 - \dfrac{49}{625}} = -\sqrt{\dfrac{576}{625}} = -\dfrac{24}{25}$$

$$\sin(x + y) = \sin x \cos y + \cos x \sin y$$
$$= \left(-\dfrac{3}{5}\right)\left(-\dfrac{7}{25}\right) + \left(-\dfrac{4}{5}\right)\left(-\dfrac{24}{25}\right)$$
$$= \dfrac{21}{125} + \dfrac{96}{125} = \dfrac{117}{125}$$

(continued on next page)

(continued)

$$\cos(x-y) = \cos x \cos y + \sin x \sin y$$
$$= \left(-\frac{4}{5}\right)\left(-\frac{7}{25}\right) + \left(-\frac{3}{5}\right)\left(-\frac{24}{25}\right)$$
$$= \frac{28}{125} + \frac{72}{125} = \frac{100}{125} = \frac{4}{5}$$

To find $\tan(x+y)$, first find $\cos(x+y)$.

$$\cos(x+y) = \cos x \cos y - \sin x \sin y$$
$$= \left(-\frac{4}{5}\right)\left(-\frac{7}{25}\right) - \left(-\frac{3}{5}\right)\left(-\frac{24}{25}\right)$$
$$= \frac{28}{125} - \frac{72}{125} = -\frac{44}{125}$$

$$\tan(x+y) = \frac{\sin(x+y)}{\cos(x+y)} = \frac{\frac{117}{125}}{-\frac{44}{5}} \cdot \frac{125}{125} = -\frac{117}{44}$$

Note that using the formula
$$\tan(x+y) = \frac{\tan x + \tan y}{1 - \tan x \tan y}, \text{ we have}$$
$$\tan(x+y) = \frac{\frac{3}{4} + \frac{24}{7}}{1 - \left(\frac{3}{4}\right)\left(\frac{24}{7}\right)} = \frac{21+96}{28-72} = -\frac{117}{44}$$

To find the quadrant of $x+y$, notice that $\sin(x+y) > 0$, which implies $x+y$ is in quadrant I or II. Also $\tan(x+y) < 0$, which implies that $x+y$ is in quadrant II or IV. Therefore, $x+y$ is in quadrant II.

29. Find $\sin(x+y)$, $\cos(x-y)$, and $\tan(x+y)$, given $\sin x = -\frac{1}{2}$, $\cos y = -\frac{2}{5}$, x and y are in quadrant III.

Because x and y are in quadrant III, $\cos x$ and $\sin y$ are negative.

$$\cos x = -\sqrt{1 - \sin^2 x} = -\sqrt{1 - \left(-\frac{1}{2}\right)^2}$$
$$= -\sqrt{1 - \frac{1}{4}} = -\sqrt{\frac{3}{4}} = -\frac{\sqrt{3}}{2}$$

$$\sin y = -\sqrt{1 - \cos^2 y} = -\sqrt{1 - \left(-\frac{2}{5}\right)^2}$$
$$= -\sqrt{1 - \frac{4}{25}} = -\sqrt{\frac{21}{25}} = -\frac{\sqrt{21}}{5}$$

$$\sin(x+y) = \sin x \cos y + \cos x \sin y$$
$$= \left(-\frac{1}{2}\right)\left(-\frac{2}{5}\right) + \left(-\frac{\sqrt{3}}{2}\right)\left(-\frac{\sqrt{21}}{5}\right)$$
$$= \frac{2}{10} + \frac{\sqrt{63}}{10} = \frac{2 + 3\sqrt{7}}{10}$$

$$\cos(x-y) = \cos x \cos y + \sin x \sin y$$
$$= \left(-\frac{\sqrt{3}}{2}\right)\left(-\frac{2}{5}\right) + \left(-\frac{1}{2}\right)\left(-\frac{\sqrt{21}}{5}\right)$$
$$= \frac{2\sqrt{3}}{10} + \frac{\sqrt{21}}{10} = \frac{2\sqrt{3} + \sqrt{21}}{10}$$

To find $\tan(x+y)$, first find $\cos(x+y)$.

$$\cos(x+y) = \cos x \cos y - \sin x \sin y$$
$$= \left(-\frac{2}{5}\right)\left(-\frac{\sqrt{3}}{2}\right) - \left(-\frac{1}{2}\right)\left(-\frac{\sqrt{21}}{5}\right)$$
$$= \frac{2\sqrt{3}}{10} - \frac{\sqrt{21}}{10} = \frac{2\sqrt{3} - \sqrt{21}}{10}$$

$$\tan(x+y) = \frac{\sin(x+y)}{\cos(x+y)} = \frac{\frac{2+3\sqrt{7}}{10}}{\frac{2\sqrt{3}-\sqrt{21}}{10}}$$
$$= \frac{2 + 3\sqrt{7}}{2\sqrt{3} - \sqrt{21}}$$

Using the formula $\tan(x+y) = \frac{\tan x + \tan y}{1 - \tan x \tan y}$, we have

$$\tan(x+y) = \frac{\frac{\sqrt{3}}{3} + \frac{\sqrt{21}}{2}}{1 - \left(\frac{\sqrt{3}}{3}\right)\left(\frac{\sqrt{21}}{2}\right)} = \frac{2\sqrt{3} + 3\sqrt{21}}{6 - 3\sqrt{7}}$$
$$= -\frac{75\sqrt{3} + 24\sqrt{21}}{27} = \frac{-25\sqrt{3} - 8\sqrt{21}}{9}$$

The two forms of $\tan(x+y)$ are equal.

To find the quadrant of $x+y$, notice that $\sin(x+y) > 0$, which implies $x+y$ is in quadrant I or II. Also $\tan(x+y) < 0$, which implies that $x+y$ is in quadrant II or IV. Therefore, $x+y$ is in quadrant II.

31. Find $\sin(x+y)$, $\cos(x-y)$, and $\tan(x+y)$, given $\sin x = \frac{1}{10}$, $\cos y = \frac{4}{5}$, x is in quadrant I and y is in quadrant IV.

Because x is in quadrant I, $\cos x$ is positive.

$$\cos x = \sqrt{1 - \sin^2 x} = \sqrt{1 - \left(\frac{1}{10}\right)^2}$$
$$= \sqrt{1 - \frac{1}{100}} = \sqrt{\frac{99}{100}} = \frac{3\sqrt{11}}{10}$$

Because y is in quadrant IV, $\sin y$ is negative.

$$\sin y = -\sqrt{1 - \cos^2 y} = -\sqrt{1 - \left(\frac{4}{5}\right)^2}$$
$$= -\sqrt{1 - \frac{16}{25}} = -\sqrt{\frac{9}{25}} = -\frac{3}{5}$$

(continued on next page)

(*continued*)

$$\sin(x+y) = \sin x \cos y + \cos x \sin y$$

$$= \left(\frac{1}{10}\right)\left(\frac{4}{5}\right) + \left(\frac{3\sqrt{11}}{10}\right)\left(-\frac{3}{5}\right)$$

$$= \frac{4}{50} - \frac{9\sqrt{11}}{50} = \frac{4 - 9\sqrt{11}}{50}$$

$$\cos(x-y) = \cos x \cos y + \sin x \sin y$$

$$= \left(\frac{3\sqrt{11}}{10}\right)\left(\frac{4}{5}\right) + \left(\frac{1}{10}\right)\left(-\frac{3}{5}\right)$$

$$= \frac{12\sqrt{11}}{50} - \frac{3}{50} = \frac{12\sqrt{11} - 3}{50}$$

To find $\tan(x+y)$, first find $\cos(x+y)$.

$$\cos(x+y) = \cos x \cos y - \sin x \sin y$$

$$= \left(\frac{3\sqrt{11}}{10}\right)\left(\frac{4}{5}\right) - \left(\frac{1}{10}\right)\left(-\frac{3}{5}\right)$$

$$= \frac{12\sqrt{11}}{50} + \frac{3}{50} = \frac{12\sqrt{11} + 3}{50}$$

$$\tan(x+y) = \frac{\sin(x+y)}{\cos(x+y)} = \frac{\frac{4-9\sqrt{11}}{50}}{\frac{12\sqrt{11}+3}{50}}$$

$$= \frac{4 - 9\sqrt{11}}{12\sqrt{11} + 3}$$

To find $\tan(x+y)$ using the formula

$$\tan(x+y) = \frac{\tan x + \tan y}{1 - \tan x \tan y}, \text{ we have}$$

$$\tan x = \frac{\sin x}{\cos x} = \frac{\frac{1}{10}}{\frac{3\sqrt{11}}{10}} = \frac{1}{3\sqrt{11}} \cdot \frac{\sqrt{11}}{\sqrt{11}} = \frac{\sqrt{11}}{33}$$

$$\tan y = \frac{\sin y}{\cos y} = \frac{-\frac{3}{5}}{\frac{4}{5}} = -\frac{3}{4}$$

$$\tan(x+y) = \frac{\tan x + \tan y}{1 - \tan x \tan y} = \frac{\frac{\sqrt{11}}{33} + \left(-\frac{3}{4}\right)}{1 - \left(\frac{\sqrt{11}}{33}\right)\left(-\frac{3}{4}\right)}$$

$$= \frac{4\sqrt{11} - 99}{132 + 3\sqrt{11}} \cdot \frac{132 - 3\sqrt{11}}{132 - 3\sqrt{11}}$$

$$= \frac{528\sqrt{11} - 132 - 13068 + 297\sqrt{11}}{17424 - 99}$$

$$= \frac{825\sqrt{11} - 13200}{17325} = \frac{\sqrt{11} - 16}{21}$$

The two forms of $\tan(x+y)$ are equal.
To find the quadrant of $x+y$, notice that
$\sin(x+y) < 0$, which implies $x+y$ is in
quadrant III or IV. Also $\tan(x+y) < 0$, which
implies that $x+y$ is in quadrant II or IV.
Therefore, $x+y$ is in quadrant IV.

33. Find $\sin\theta$ and $\cos\theta$, given $\cos 2\theta = -\frac{3}{4}$,
$90° < 2\theta < 180°$.
$90° < 2\theta < 180° \Rightarrow 45° < \theta < 90° \Rightarrow \theta$ is in
quadrant I, so $\sin\theta$ and $\cos\theta$ are both
positive.

$$\cos 2\theta = 1 - 2\sin^2\theta \Rightarrow -\frac{3}{4} = 1 - 2\sin^2\theta \Rightarrow$$

$$-\frac{7}{4} = -2\sin^2\theta \Rightarrow \frac{7}{8} = \sin^2\theta \Rightarrow$$

$$\sin\theta = \sqrt{\frac{7}{8}} = \frac{\sqrt{7}}{2\sqrt{2}} = \frac{\sqrt{14}}{4}$$

$$\cos\theta = \sqrt{1 - \sin^2\theta} = \sqrt{1 - \frac{7}{8}} = \sqrt{\frac{1}{8}} = \frac{1}{\sqrt{8}}$$

$$= \frac{1}{2\sqrt{2}} = \frac{\sqrt{2}}{4}$$

35. Find $\sin 2x$ and $\cos 2x$, given $\tan x = 3$,
$\sin x < 0$.
Because $\tan x > 0$ and $\sin x < 0$, x is in
quadrant III, and $2x$ is in quadrant I or II.

$$\tan 2x = \frac{2\tan x}{1 - \tan^2 x} = \frac{2(3)}{1 - 3^2} = \frac{6}{8} = -\frac{3}{4}$$

Because $\tan 2x < 0$, $2x$ is in quadrant II. Thus,
$\sin 2x > 0$ and $\cos 2x < 0$.

$$\sec 2x = -\sqrt{1 + \tan^2 x} = -\sqrt{1 + \left(-\frac{3}{4}\right)^2}$$

$$= -\sqrt{1 + \frac{9}{16}} = -\sqrt{\frac{25}{16}} = -\frac{5}{4} \Rightarrow$$

$$\cos 2x = \frac{1}{-\frac{5}{4}} = -\frac{4}{5}$$

$$\sin 2x = \sqrt{1 - \cos^2(2x)} = \sqrt{1 - \left(-\frac{4}{5}\right)^2}$$

$$= \sqrt{1 - \frac{16}{25}} = \sqrt{\frac{9}{25}} = \frac{3}{5}$$

37. Find $\cos\frac{\theta}{2}$, given $\cos\theta = -\frac{1}{2}$,
$90° < \theta < 180°$.

Because $90° < \theta < 180° \Rightarrow 45° < \frac{\theta}{2} < 90°$, $\frac{\theta}{2}$

is in quadrant I and $\cos\frac{\theta}{2} > 0$.

$$\cos\frac{\theta}{2} = \sqrt{\frac{1 + \left(-\frac{1}{2}\right)}{2}} = \sqrt{\frac{2-1}{4}} = \sqrt{\frac{1}{4}} = \frac{1}{2}$$

39. Find $\tan x$, given $\tan 2x = 2$, with $\pi < x < \frac{3\pi}{2}$.

$$\tan 2x = \frac{2\tan x}{1-\tan^2 x} \Rightarrow 2 = \frac{2\tan x}{1-\tan^2 x} \Rightarrow$$

$$2\tan x = 2\left(1-\tan^2 x\right), \text{ if } \tan x \neq \pm 1$$

Thus, $2\left(\tan^2 x + 2\tan x - 1\right) = 0 \Rightarrow$

$\tan^2 x + 2\tan x - 1 = 0$, so we can use the quadratic formula to solve for $\tan x$.

$$x = \frac{-b \pm \sqrt{b^2 - 4ac}}{2a} \Rightarrow$$

$$\tan x = \frac{-1 \pm \sqrt{1^2 - 4(1)(-1)}}{2} = \frac{-1 \pm \sqrt{5}}{2}$$

Because x is in quadrant III, $\tan x > 0$, so

$$\tan x = \frac{-1 + \sqrt{5}}{2}$$

41. Find $\tan \frac{x}{2}$, given $\sin x = 0.8$, with $0 < x < \frac{\pi}{2}$.

$$\cos x = \pm\sqrt{1 - \sin^2 x} = \pm\sqrt{1 - 0.8^2} = \sqrt{1 - 0.64}$$

$$= \pm\sqrt{0.36} = \pm 0.6$$

Because x is in quadrant I, $\cos x > 0$, so $\cos x = 0.6$

$$\tan \frac{x}{2} = \frac{1 - \cos x}{\sin x} = \frac{1 - 0.6}{0.8} = \frac{0.4}{0.8} = 0.5$$

Exercises 43–47 are graphed in the window $[-2\pi, 2\pi]$ by $[-4, 4]$.

43. $\dfrac{\sin 2x + \sin x}{\cos 2x - \cos x}$ appears to be equivalent to $\cot \dfrac{x}{2}$.

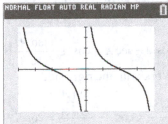

$$\frac{\sin 2x + \sin x}{\cos x - \cos 2x} = \frac{2\sin x \cos x + \sin x}{\cos x - \left(2\cos^2 x - 1\right)}$$

$$= \frac{\sin x\left(2\cos x + 1\right)}{-2\cos^2 x + \cos x + 1}$$

$$= \frac{\sin x\left(2\cos x + 1\right)}{\left(2\cos x + 1\right)\left(-\cos x + 1\right)}$$

$$= \frac{\sin x}{-\cos x + 1} = \frac{\sin x}{1 - \cos x}$$

$$= \frac{1}{\dfrac{1-\cos x}{\sin x}} = \frac{1}{\tan \dfrac{x}{2}} = \cot \frac{x}{2}$$

45. $\dfrac{\sin x}{1 - \cos x}$ appears to be equivalent to $\cot \dfrac{x}{2}$.

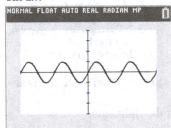

$$\frac{\sin x}{1 - \cos x} = \frac{1}{\dfrac{1 - \cos x}{\sin x}} = \frac{1}{\tan \dfrac{x}{2}} = \cot \frac{x}{2}$$

47. $\dfrac{2\left(\sin x - \sin^3 x\right)}{\cos x}$ appears to be equivalent to $\sin 2x$.

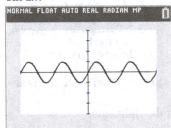

$$\frac{2\left(\sin x - \sin^3 x\right)}{\cos x} = \frac{2\sin x\left(1 - \sin^2 x\right)}{\cos x}$$

$$= \frac{2\sin x \cos^2 x}{\cos x}$$

$$= 2\sin x \cos x = \sin 2x$$

49. Verify $\sin^2 x - \sin^2 y = \cos^2 y - \cos^2 x$ is an identity.

$$\sin^2 x - \sin^2 y = \left(1 - \cos^2 x\right) - \left(1 - \cos^2 y\right)$$

$$= 1 - \cos^2 x - 1 + \cos^2 y$$

$$= \cos^2 y - \cos^2 x$$

51. Verify $\dfrac{\sin^2 x}{2 - 2\cos x} = \cos^2 \dfrac{x}{2}$ is an identity.

$$\dfrac{\sin^2 x}{2 - 2\cos x} = \dfrac{1 - \cos^2 x}{2(1 - \cos x)}$$

$$= \dfrac{(1 - \cos x)(1 + \cos x)}{2(1 - \cos x)}$$

$$= \dfrac{1 + \cos x}{2} = \cos^2 \dfrac{x}{2}$$

53. Verify $2\cos A - \sec A = \cos A - \dfrac{\tan A}{\csc A}$ is an identity. Work with the right side.

$$\cos A - \dfrac{\tan A}{\csc A} = \cos A - \dfrac{\dfrac{\sin A}{\cos A}}{\dfrac{1}{\sin A}} = \cos A - \dfrac{\sin^2 A}{\cos A}$$

$$= \dfrac{\cos^2 A}{\cos A} - \dfrac{\sin^2 A}{\cos A}$$

$$= \dfrac{\cos^2 A - \sin^2 A}{\cos A}$$

$$= \dfrac{\cos^2 A - (1 - \cos^2 A)}{\cos A}$$

$$= \dfrac{2\cos^2 A - 1}{\cos A} = 2\cos A - \dfrac{1}{\cos A}$$

$$= 2\cos A - \sec A$$

55. Verify $1 + \tan^2 \alpha = 2\tan \alpha \csc 2\alpha$ is an identity.
Work with the right side.

$$2\tan \alpha \csc 2\alpha = \dfrac{2\tan \alpha}{\sin 2\alpha} = \dfrac{2 \cdot \dfrac{\sin \alpha}{\cos \alpha}}{2\sin \alpha \cos \alpha}$$

$$= \dfrac{2\sin \alpha}{2\sin \alpha \cos^2 \alpha} = \dfrac{1}{\cos^2 \alpha}$$

$$= \sec^2 \alpha = 1 + \tan^2 \alpha$$

57. Verify $\tan \theta \sin 2\theta = 2 - 2\cos^2 \theta$ is an identity.

$$\tan \theta \sin 2\theta = \tan \theta (2\sin \theta \cos \theta)$$

$$= \dfrac{\sin \theta}{\cos \theta}(2\sin \theta \cos \theta) = 2\sin^2 \theta$$

$$= 2(1 - \cos^2 \theta) = 2 - 2\cos^2 \theta$$

59. Verify $2\tan x \csc 2x - \tan^2 x = 1$ is an identity.

$$2\tan x \csc 2x - \tan^2 x$$

$$= 2\tan x \dfrac{1}{\sin 2x} - \tan^2 x$$

$$= 2 \cdot \dfrac{\sin x}{\cos x} \cdot \dfrac{1}{2\sin x \cos x} - \dfrac{\sin^2 x}{\cos^2 x}$$

$$= \dfrac{1}{\cos^2 x} - \dfrac{\sin^2 x}{\cos^2 x} = \dfrac{1 - \sin^2 x}{\cos^2 x} = \dfrac{\cos^2 x}{\cos^2 x} = 1$$

61. Verify $\tan \theta \cos^2 \theta = \dfrac{2\tan \theta \cos^2 \theta - \tan \theta}{1 - \tan^2 \theta}$ is an identity.
Work with the right side.

$$\dfrac{2\tan \theta \cos^2 \theta - \tan \theta}{1 - \tan^2 \theta}$$

$$= \dfrac{\tan \theta (2\cos^2 \theta - 1)}{1 - \tan^2 \theta}$$

$$= \dfrac{\tan \theta (2\cos^2 \theta - 1)}{1 - \dfrac{\sin^2 \theta}{\cos^2 \theta}} \cdot \dfrac{\cos^2 \theta}{\cos^2 \theta}$$

$$= \dfrac{\tan \theta \cos^2 \theta (2\cos^2 \theta - 1)}{\cos^2 \theta - \sin^2 \theta}$$

$$= \dfrac{\tan \theta \cos^2 \theta (2\cos^2 \theta - 1)}{2\cos^2 \theta - 1} = \tan \theta \cos^2 \theta$$

63. Verify $\dfrac{\sin^2 x - \cos^2 x}{\csc x} = 2\sin^3 x - \sin x$ is an identity.

$$\dfrac{\sin^2 x - \cos^2 x}{\csc x} = \dfrac{\sin^2 x - (1 - \sin^2 x)}{\dfrac{1}{\sin x}}$$

$$= \dfrac{2\sin^2 x - 1}{\dfrac{1}{\sin x}} \cdot \dfrac{\sin x}{\sin x}$$

$$= (2\sin^2 x - 1)\sin x$$

$$= 2\sin^3 x - \sin x$$

65. Verify $\tan 4\theta = \dfrac{2\tan 2\theta}{2 - \sec^2 2\theta}$ is an identity.

$$\tan 4\theta = \tan[2(2\theta)] = \dfrac{2\tan 2\theta}{1 - \tan^2 2\theta}$$

$$= \dfrac{2\tan 2\theta}{1 - (\sec^2 2\theta - 1)} = \dfrac{2\tan 2\theta}{2 - \sec^2 2\theta}$$

67. Verify $\tan\left(\dfrac{x}{2}+\dfrac{\pi}{4}\right)=\sec x+\tan x$ is an identity. Working with the left side, we have

$$\tan\left(\dfrac{x}{2}+\dfrac{\pi}{4}\right)=\dfrac{\tan\frac{x}{2}+\tan\frac{\pi}{4}}{1-\tan\frac{x}{2}\tan\frac{\pi}{4}}=\dfrac{\tan\frac{x}{2}+1}{1-\tan\frac{x}{2}}$$

Working with the right side, we have

$$\sec x+\tan x=\dfrac{1}{\cos x}+\dfrac{\sin x}{\cos x}=\dfrac{1+\sin x}{\cos x}$$

$$=\dfrac{\left(\cos^2\frac{x}{2}+\sin^2\frac{x}{2}\right)+\sin\left[2\left(\frac{x}{2}\right)\right]}{\cos\left[2\left(\frac{x}{2}\right)\right]}$$

$$=\dfrac{\cos^2\frac{x}{2}+\sin^2\frac{x}{2}+2\sin\frac{x}{2}\cos\frac{x}{2}}{\cos^2 x-\sin^2 x}$$

$$=\dfrac{\left(\cos\frac{x}{2}+\sin\frac{x}{2}\right)^2}{\left(\cos\frac{x}{2}+\sin\frac{x}{2}\right)\left(\cos\frac{x}{2}-\sin\frac{x}{2}\right)}$$

$$=\dfrac{\cos\frac{x}{2}+\sin\frac{x}{2}}{\cos\frac{x}{2}-\sin\frac{x}{2}}\cdot\dfrac{\cos\frac{x}{2}}{1}\cdot\dfrac{1}{\cos\frac{x}{2}}$$

$$=\dfrac{\dfrac{\cos\frac{x}{2}}{\cos\frac{x}{2}}+\dfrac{\sin\frac{x}{2}}{\cos\frac{x}{2}}}{\dfrac{\cos\frac{x}{2}}{\cos\frac{x}{2}}-\dfrac{\sin\frac{x}{2}}{\cos\frac{x}{2}}}=\dfrac{1+\tan\frac{x}{2}}{1-\tan\frac{x}{2}}$$

$$\tan\left(\dfrac{x}{2}+\dfrac{\pi}{4}\right)=\dfrac{\tan\frac{x}{2}+1}{1-\tan\frac{x}{2}}=\sec x+\tan x,\text{ so the}$$

statement is verified.

69. Verify $-\cot\dfrac{x}{2}=\dfrac{\sin 2x+\sin x}{\cos 2x-\cos x}$ is an identity.

Work with the right side.

$$\dfrac{\sin 2x+\sin x}{\cos 2x-\cos x}=\dfrac{2\sin x\cos x+\sin x}{\left(2\cos^2 x-1\right)-\cos x}$$

$$=\dfrac{\sin x\left(2\cos x+1\right)}{2\cos^2 x-\cos x-1}$$

$$=\dfrac{\sin x\left(2\cos x+1\right)}{\left(2\cos x+1\right)\left(\cos x-1\right)}$$

$$=\dfrac{\sin x}{1-\cos x}=-\dfrac{\sin x}{\cos x-1}$$

$$=-\dfrac{1}{\dfrac{\sin x}{\cos x-1}}=-\dfrac{1}{\tan\frac{x}{2}}=-\cot\dfrac{x}{2}$$

71. (a) When $h=0$,

$$D=\dfrac{v^2\sin\theta\cos\theta+v\cos\theta\sqrt{\left(v\sin\theta\right)^2+64\cdot 0}}{32}$$

$$=\dfrac{v^2\sin\theta\cos\theta+v\cos\theta\sqrt{\left(v\sin\theta\right)^2}}{32}$$

$$=\dfrac{v^2\sin\theta\cos\theta+v^2\sin\theta\cos\theta}{32}$$

$$=\dfrac{2v^2\sin\theta\cos\theta}{32}=\dfrac{v^2\sin 2\theta}{32}$$

This is dependent on both the velocity and the angle at which the object is thrown.

(b) Using the result from part (a), we have

$$D=\dfrac{v^2\sin 2\theta}{32}=\dfrac{36^2\sin\left(2\cdot 30°\right)}{32}$$

$$=\dfrac{1296\sin\left(60°\right)}{32}=\dfrac{1296\cdot\frac{\sqrt{3}}{2}}{32}$$

$$=\dfrac{81\sqrt{3}}{4}\approx 35\text{ ft}$$

Chapter 5 Test

1. $\cos\theta=\dfrac{24}{25}$, θ is in quadrant IV.

$$\sin^2\theta=1-\cos^2\theta=1-\left(\dfrac{24}{25}\right)^2=1-\dfrac{576}{625}=\dfrac{49}{625}$$

θ is in quadrant IV, so $\sin\theta<0$.

$$\sin\theta=-\sqrt{\dfrac{49}{625}}=-\dfrac{7}{25}$$

$$\tan\theta=\dfrac{\sin\theta}{\cos\theta}=\dfrac{-\frac{7}{25}}{\frac{24}{25}}=-\dfrac{7}{24}$$

$$\cot\theta=\dfrac{1}{\tan\theta}=\dfrac{1}{-\frac{7}{24}}=-\dfrac{24}{7}$$

$$\sec\theta=\dfrac{1}{\cos\theta}=\dfrac{1}{\frac{24}{25}}=\dfrac{25}{24}$$

$$\csc\theta=\dfrac{1}{\sin\theta}=\dfrac{1}{-\frac{7}{25}}=-\dfrac{25}{7}$$

2. $\sec\theta-\sin\theta\tan\theta=\dfrac{1}{\cos\theta}-\sin\theta\cdot\dfrac{\sin\theta}{\cos\theta}$

$$=\dfrac{1-\sin^2\theta}{\cos\theta}=\dfrac{\cos^2\theta}{\cos\theta}=\cos\theta$$

3. $\tan^2 x - \sec^2 x = \dfrac{\sin^2 x}{\cos^2 x} - \dfrac{1}{\cos^2 x}$

$= \dfrac{\sin^2 x - 1}{\cos^2 x} = -\dfrac{1 - \sin^2 x}{\cos^2 x}$

$= -\dfrac{\cos^2 x}{\cos^2 x} = -1$

4. $\cos \dfrac{5\pi}{12} = \cos\left(\dfrac{\pi}{6} + \dfrac{\pi}{4}\right)$

$= \cos \dfrac{\pi}{6} \cos \dfrac{\pi}{4} - \sin \dfrac{\pi}{6} \sin \dfrac{\pi}{4}$

$= \dfrac{\sqrt{3}}{2}\left(\dfrac{\sqrt{2}}{2}\right) - \dfrac{1}{2}\left(\dfrac{\sqrt{2}}{2}\right) = \dfrac{\sqrt{6} - \sqrt{2}}{4}$

5. (a) $\cos(270° - x)$

$= \cos 270° \cos x + \sin 270° \sin x$

$= 0 \cdot \cos x + (-1)\sin x = 0 - \sin x = -\sin x$

(b) $\tan(\pi + x) = \dfrac{\tan \pi + \tan x}{1 - \tan \pi \tan x} = \tan x$

6. $\sin(-22.5°) = \pm\sqrt{\dfrac{1 - \cos(-45°)}{2}} = \pm\sqrt{\dfrac{1 - \dfrac{\sqrt{2}}{2}}{2}}$

$= \pm\sqrt{\dfrac{2 - \sqrt{2}}{4}} = \pm\dfrac{\sqrt{2 - \sqrt{2}}}{2}$

Because $-22.5°$ is in quadrant IV, $\sin(-22.5°)$

is negative. Thus, $\sin(-22.5°) = -\dfrac{\sqrt{2 - \sqrt{2}}}{2}$.

7. $\cot \dfrac{x}{2} - \cot x$ appears to be equivalent to

$\csc x$.

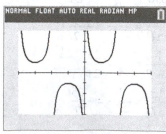

$\cot \dfrac{x}{2} - \cot x = \dfrac{1}{\tan \dfrac{x}{2}} - \dfrac{\cos x}{\sin x} = \dfrac{1}{\dfrac{\sin x}{1 + \cos x}} - \dfrac{\cos x}{\sin x}$

$= \dfrac{1 + \cos x}{\sin x} - \dfrac{\cos x}{\sin x}$

$= \dfrac{1 + \cos x - \cos x}{\sin x} = \dfrac{1}{\sin x} = \csc x$

8. Find $\sin(A + B)$, $\cos(A + B)$, and $\tan(A - B)$, given $\sin A = \dfrac{5}{13}$, $\cos B = -\dfrac{3}{5}$, A is in quadrant I and B is in quadrant II. Thus, $\cos A > 0$ and $\sin B > 0$.

$\cos A = \sqrt{1 - \sin^2 A} = \sqrt{1 - \left(\dfrac{5}{13}\right)^2}$

$= \sqrt{1 - \dfrac{25}{169}} = \sqrt{\dfrac{144}{169}} = \dfrac{12}{13}$

$\sin B = \sqrt{1 - \cos^2 B} = \sqrt{1 - \left(-\dfrac{3}{5}\right)^2}$

$= \sqrt{1 - \dfrac{9}{25}} = \sqrt{\dfrac{16}{25}} = \dfrac{4}{5}$

(a) $\sin(A + B) = \sin A \cos B + \cos A \sin B$

$= \left(\dfrac{5}{13}\right)\left(-\dfrac{3}{5}\right) + \left(\dfrac{12}{13}\right)\left(\dfrac{4}{5}\right)$

$= -\dfrac{15}{65} + \dfrac{48}{65} = \dfrac{33}{65}$

(b) $\cos(A + B) = \cos A \cos B - \sin A \sin B$

$= \left(\dfrac{12}{13}\right)\left(-\dfrac{3}{5}\right) - \left(\dfrac{5}{13}\right)\left(\dfrac{4}{5}\right)$

$= -\dfrac{36}{65} - \dfrac{20}{65} = -\dfrac{56}{65}$

(c) To use the formula

$\tan(A + B) = \dfrac{\tan A + \tan B}{1 - \tan A \tan B}$, first find

$\tan A$ and $\tan B$:

$\tan A = \dfrac{\sin A}{\cos A} = \dfrac{\frac{5}{13}}{\frac{12}{13}} = \dfrac{5}{12}$

$\tan B = \dfrac{\sin B}{\cos B} = \dfrac{\frac{4}{5}}{-\frac{3}{5}} = -\dfrac{4}{3}$

$\tan(A - B) = \dfrac{\dfrac{5}{12} - \left(-\dfrac{4}{3}\right)}{1 + \left(\dfrac{5}{12}\right)\left(-\dfrac{4}{3}\right)}$

$= \dfrac{15 + 48}{36 - 20} = \dfrac{63}{16}$

(d) To find the quadrant of $A + B$, notice that $\sin(A + B) > 0$, which implies $x + y$ is in quadrant I or II. Also $\cos(A + B) < 0$, which implies that $A + B$ is in quadrant II or III. Therefore, $A + B$ is in quadrant II.

9. Given $\cos\theta = -\dfrac{3}{5}$, $90° < \theta < 180°$

θ is in quadrant II, so $\sin\theta > 0$, 2θ is in quadrant III or quadrant IV, and

$\dfrac{\pi}{4} < \dfrac{\theta}{2} < \dfrac{\pi}{2} \Rightarrow \dfrac{\theta}{2}$ is in quadrant I. Also

$\sin\theta = \sqrt{1-\cos^2\theta} = \sqrt{1-\left(-\dfrac{3}{5}\right)^2} = \sqrt{\dfrac{16}{25}} = \dfrac{4}{5}$

$\tan\theta = \dfrac{\sin\theta}{\cos\theta} = \dfrac{\frac{4}{5}}{-\frac{3}{5}} = -\dfrac{4}{3}$

(a) $\cos 2\theta = 2\cos^2\theta - 1 = 2\left(-\dfrac{3}{5}\right)^2 - 1 = -\dfrac{7}{25}$

Note that 2θ is in quadrant III because $\cos 2\theta < 0$.

(b) $\sin 2\theta = 2\sin\theta\cos\theta = 2\left(\dfrac{4}{5}\right)\left(-\dfrac{3}{5}\right) = -\dfrac{24}{25}$

(c) $\tan 2\theta = \dfrac{2\tan\theta}{1-\tan^2\theta} = \dfrac{2\left(-\frac{4}{3}\right)}{1-\left(\frac{4}{3}\right)^2} = \dfrac{-\frac{8}{3}}{1-\frac{16}{9}}$

$= \dfrac{-24}{9-16} = \dfrac{24}{7}$

(d) $\cos\frac{1}{2}\theta = \sqrt{\dfrac{1+\cos\theta}{2}} = \sqrt{\dfrac{1+\left(-\frac{3}{5}\right)}{2}}$

$= \sqrt{\dfrac{5-3}{10}} = \sqrt{\dfrac{2}{10}} = \dfrac{1}{\sqrt{5}} = \dfrac{\sqrt{5}}{5}$

(e) $\tan\frac{1}{2}\theta = \dfrac{\sin\theta}{1+\cos\theta} = \dfrac{\frac{4}{5}}{1-\frac{3}{5}} = \dfrac{4}{5-3} = 2$

10. Verify $\sec^2 B = \dfrac{1}{1-\sin^2 B}$ is an identity.
Work with the right side.

$\dfrac{1}{1-\sin^2 B} = \dfrac{1}{\cos^2 B} = \sec^2 B$

11. Verify $\cos 2A = \dfrac{\cot A - \tan A}{\csc A \sec A}$ is an identity.
Work with the right side.

$\dfrac{\cot A - \tan A}{\csc A \sec A} = \dfrac{\dfrac{\cos A}{\sin A} - \dfrac{\sin A}{\cos A}}{\left(\dfrac{1}{\sin A}\right)\left(\dfrac{1}{\cos A}\right)} \cdot \dfrac{\sin A\cos A}{\sin A\cos A}$

$= \cos^2 A - \sin^2 A = \cos 2A$

12. Verify $\dfrac{\sin 2x}{\cos 2x + 1} = \tan x$ is an identity.

$\dfrac{\sin 2x}{\cos 2x + 1} = \dfrac{2\sin x\cos x}{\left(2\cos^2 x - 1\right) + 1}$

$= \dfrac{2\sin x\cos x}{2\cos^2 x} = \dfrac{\sin x}{\cos x} = \tan x$

13. Verify $\tan^2 x - \sin^2 x = \left(\tan x\sin x\right)^2$ is an identity.

$\tan^2 x - \sin^2 x = \dfrac{\sin^2 x}{\cos^2 x} - \sin^2 x$

$= \dfrac{\sin^2 x - \sin^2 x\cos^2 x}{\cos^2 x}$

$= \dfrac{\sin^2 x\left(1-\cos^2 x\right)}{\cos^2 x} = \dfrac{\sin^2 x\sin^2 x}{\cos^2 x}$

$= \tan^2 x\sin^2 x = \left(\tan x\sin x\right)^2$

14. Verify $\dfrac{\tan x - \cot x}{\tan x + \cot x} = 2\sin^2 x - 1$ is an identity.

$\dfrac{\tan x - \cot x}{\tan x + \cot x} = \dfrac{\dfrac{\sin x}{\cos x} - \dfrac{\cos x}{\sin x}}{\dfrac{\sin x}{\cos x} + \dfrac{\cos x}{\sin x}}$

$= \dfrac{\dfrac{\sin x}{\cos x} - \dfrac{\cos x}{\sin x}}{\dfrac{\sin x}{\cos x} + \dfrac{\cos x}{\sin x}} \cdot \dfrac{\cos x\sin x}{\cos x\sin x}$

$= \dfrac{\sin^2 x - \cos^2 x}{\sin^2 + \cos^2 x}$

$= \sin^2 x - \cos^2 x$

$= \sin^2 x - \left(1-\sin^2 x\right) = 2\sin^2 x - 1$

15. (a) $V = 163\sin\omega t$. $\sin x = \cos\left(\dfrac{\pi}{2} - x\right) \Rightarrow$

$V = 163\cos\left(\dfrac{\pi}{2} - \omega t\right)$.

(b) $V = 163 \sin \omega t = 163 \sin 120 \pi t$

$= 163 \cos\left(\dfrac{\pi}{2} - 120 \pi t\right) \Rightarrow$ the

maximum voltage occurs when

$\cos\left(\dfrac{\pi}{2} - 120 \pi t\right) = 1$. Thus, the

maximum voltage is $V = 163$ volts.

$\cos\left(\dfrac{\pi}{2} - 120 \pi t\right) = 1$ when

$\dfrac{\pi}{2} - 120 \pi t = 2k\pi$, where k is any

integer. The first maximum occurs when

$\dfrac{\pi}{2} - 120 \pi t = 0 \Rightarrow$

$\dfrac{\pi}{2} = 120 \pi t \Rightarrow \dfrac{1}{120 \pi} \cdot \dfrac{\pi}{2} = t \Rightarrow t = \dfrac{1}{240}$

The maximum voltage will first occur at

$\dfrac{1}{240}$ sec .

Chapter 6

Inverse Circular Functions and Trigonometric Equations

Section 6.1   Inverse Circular Functions

1. For a function to have an inverse, it must be one-to-one.

3. $y = \cos^{-1} x$ means that $x = \cos y$ for $0 \le y \le \pi$.

5. If a function f has an inverse and $f(\pi) = -1$, then $f^{-1}(-1) = \pi$.

7. (a) $[-1, 1]$

   (b) $\left[-\dfrac{\pi}{2}, \dfrac{\pi}{2}\right]$

   (c) increasing

   (d) -2 is not in the domain.

9. (a) $(-\infty, \infty)$

   (b) $\left(-\dfrac{\pi}{2}, \dfrac{\pi}{2}\right)$

   (c) increasing

   (d) no

11. The interval must be chosen so that the function is one-to-one. The sine and cosine functions are not one-to-one on the same intervals.

13. $y = \sin^{-1} 0$

   $\sin y = 0, -\dfrac{\pi}{2} \le y \le \dfrac{\pi}{2}$

   $\sin 0 = 0$, so $y = 0$.

15. $y = \cos^{-1}(-1)$

   $\cos y = -1, 0 \le y \le \pi$

   $\cos \pi = -1$, so $y = \pi$.

17. $y = \tan^{-1} 1$

   $\tan y = 1, -\dfrac{\pi}{2} < y < \dfrac{\pi}{2}$

   $\tan \dfrac{\pi}{4} = 1$, so $y = \dfrac{\pi}{4}$.

19. $y = \arctan 0$

   $\tan y = 0, -\dfrac{\pi}{2} < y < \dfrac{\pi}{2}$

   $\tan 0 = 0$, so $y = 0$.

21. $y = \arcsin\left(-\dfrac{\sqrt{3}}{2}\right)$

   $\sin y = -\dfrac{\sqrt{3}}{2}, -\dfrac{\pi}{2} \le y \le \dfrac{\pi}{2}$

   $\sin\left(-\dfrac{\pi}{3}\right) = -\dfrac{\sqrt{3}}{2}$, so $y = -\dfrac{\pi}{3}$.

23. $y = \arccos\left(-\dfrac{\sqrt{3}}{2}\right)$

   $\cos y = -\dfrac{\sqrt{3}}{2}, 0 \le y \le \pi$

   $\cos \dfrac{5\pi}{6} = -\dfrac{\sqrt{3}}{2}$, so $y = \dfrac{5\pi}{6}$.

25. $y = \sin^{-1} \sqrt{3}$

   $\sin y = \sqrt{3}, -\dfrac{\pi}{2} \le y \le \dfrac{\pi}{2}$

   $\sqrt{3} > 1$, so there is no angle θ such that $\sin \theta = \sqrt{3}$. Thus, $\sin^{-1} \sqrt{3}$ does not exist.

27. $y = \cot^{-1}(-1)$

   $\cot y = -1, 0 < y < \pi$

   y is in quadrant II. The reference angle is $\dfrac{\pi}{4}$.

   $\cot \dfrac{3\pi}{4} = 1$, so $y = \dfrac{3\pi}{4}$.

29. $y = \csc^{-1}(-2)$

   $\csc y = -2, -\dfrac{\pi}{2} \le y \le \dfrac{\pi}{2}, y \ne 0$

   y is in quadrant IV. The reference angle is $\dfrac{\pi}{6}$.

   $\csc\left(-\dfrac{\pi}{6}\right) = -2$, so $y = -\dfrac{\pi}{6}$.

31. $y = \text{arc sec}\left(\dfrac{2\sqrt{3}}{3}\right)$

$\sec y = \dfrac{2\sqrt{3}}{3},\ 0 \le y \le \pi,\ y \ne \dfrac{\pi}{2}$

$\sec\dfrac{\pi}{6} = \dfrac{2\sqrt{3}}{3},\ \text{so } y = \dfrac{\pi}{6}.$

33. $y = \sec^{-1} 1$

$\sec y = 1,\ 0 \le y \le \pi,\ y \ne \dfrac{\pi}{2}$

$\sec 0 = 1,\ \text{so } y = 0.$

35. $y = \csc^{-1}\left(\dfrac{\sqrt{2}}{2}\right)$

$\csc y = \dfrac{\sqrt{2}}{2},\ -\dfrac{\pi}{2} \le y \le \dfrac{\pi}{2},\ y \ne 0$

There is no angle θ such that $\csc\theta = \dfrac{\sqrt{2}}{2}$.

Thus, $\csc^{-1}\left(\dfrac{\sqrt{2}}{2}\right)$ does not exist.

37. $\theta = \arctan(-1)$

$\tan\theta = -1,\ -90° < \theta < 90°$

θ is in quadrant IV. The reference angle is $45°$. Thus, $\theta = -45°$.

39. $\theta = \arcsin\left(-\dfrac{\sqrt{3}}{2}\right)$

$\sin\theta = -\dfrac{\sqrt{3}}{2},\ -90° \le \theta \le 90°$

θ is in quadrant IV. The reference angle is $60°$. $\theta = -60°$.

41. $\theta = \arccos\left(-\dfrac{1}{2}\right)$

$\cos\theta = -\dfrac{1}{2},\ 0° \le \theta \le 180°$

θ is in quadrant II. The reference angle is $60°$. Thus, $\theta = 180° - 60° = 120°$.

43. $\theta = \cot^{-1}\left(-\dfrac{\sqrt{3}}{3}\right)$

$\cot\theta = -\dfrac{\sqrt{3}}{3},\ 0° < \theta < 180°$

θ is in quadrant II. The reference angle is $60°$. $\theta = 180° - 60° = 120°$

45. $\theta = \csc^{-1}(-2)$

$\csc\theta = -2$ and $-90° < \theta < 90°,\ \theta \ne 0°$

θ is in quadrant IV. The reference angle is $30°$. $\theta = -30°$

47. $\theta = \sin^{-1} 2$

$\sin\theta = 2,\ 0° \le \theta \le 180°$

There is no angle θ such that $\sin\theta = 2$.

For Exercises 49–57, be sure that your calculator is in degree mode.

49. $\theta = \sin^{-1}(-0.13349122) = -7.6713835°$

51. $\theta = \arccos(-0.39876459) \approx 113.500970°$

53. $\theta = \csc^{-1} 1.9422833 \approx 30.987961°$

55. $\theta = \cot^{-1}(-0.60724226)$

$= \tan^{-1}\left(-\dfrac{1}{0.60724226}\right) \approx -58.7321071°$

Note that θ is in quadrant II because $\cot^{-1}\theta$ is defined for $0° < \theta < 180°$. So,
$\theta = -58.7321071° + 180° = 121.267893°$

57. $\theta = \tan^{-1}(-7.7828641) \approx -82.67832938°$

Note that $\tan^{-1} y$ is defined for $-90° < y < 90°$, so

$\tan^{-1}(-7.7828641) \approx -82.678329°$

For Exercises 59–67, be sure that your calculator is in radian mode.

59. $y = \arcsin 0.92837781 \approx 1.1900238$

61. $y = \cos^{-1}(-0.32647891) \approx 1.9033723$

63. $y = \arctan 1.1111111 \approx 0.83798122$

65. $y = \cot^{-1}(-0.92170128)$

$= \tan^{-1}\left(\dfrac{1}{-0.92170128}\right) \approx 2.3154725$

67. $y = \sec^{-1}(-1.2871684)$

$= \cos^{-1}\left(-\dfrac{1}{1.2871684}\right) \approx 2.4605221$

69.

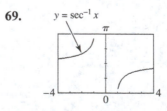

$y = \sec^{-1} x$

71.

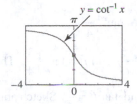

73.

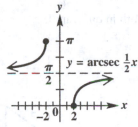

75. $\tan\left(\arccos\dfrac{3}{4}\right)$

Let $\omega = \arccos\dfrac{3}{4}$, so that $\cos\omega = \dfrac{3}{4}$.

Because arccos is defined only in quadrants I

and II, and $\dfrac{3}{4}$ is positive, ω is in quadrant I.

Sketch ω and label a triangle with the side opposite ω equal to

$$\sqrt{4^2 - 3^2} = \sqrt{16-9} = \sqrt{7}.$$

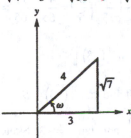

$$\tan\left(\arccos\dfrac{3}{4}\right) = \tan\omega = \dfrac{\sqrt{7}}{3}$$

77. $\cos(\tan^{-1}(-2))$

Let $\omega = \tan^{-1}(-2)$, so that $\tan\omega = -2$.

Because $\tan^{-1}$ is defined only in quadrants I and IV, and -2 is negative, ω is in quadrant IV. Sketch ω and label a triangle with the hypotenuse equal to

$$\sqrt{(-2)^2 + 1} = \sqrt{4+1} = \sqrt{5}.$$

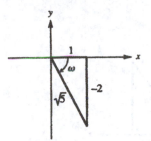

$$\cos(\tan^{-1}(-2)) = \cos\omega = \dfrac{\sqrt{5}}{5}$$

79. $\sin\left(2\tan^{-1}\dfrac{12}{5}\right)$

Let $\omega = \tan^{-1}\dfrac{12}{5}$, so that $\tan\omega = \dfrac{12}{5}$.

Because $\tan^{-1}\omega$ is defined only in quadrants

I and IV, and $\dfrac{12}{5}$ is positive, ω is in quadrant

I.

Sketch ω and label a right triangle with the hypotenuse equal to

$$\sqrt{12^2 + 5^2} = \sqrt{144+25} = \sqrt{169} = 13.$$

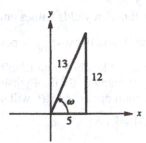

$$\sin\omega = \dfrac{12}{13}; \cos\omega = \dfrac{5}{13}$$

$$\sin\left(2\tan^{-1}\dfrac{12}{5}\right) = \sin(2\omega) = 2\sin\omega\cos\omega$$

$$= 2\left(\dfrac{12}{13}\right)\left(\dfrac{5}{13}\right) = \dfrac{120}{169}$$

81. $\cos\left(2\arctan\dfrac{4}{3}\right)$

Let $\omega = \arctan\dfrac{4}{3}$, so that $\tan\omega = \dfrac{4}{3}$.

Because arctan is defined only in quadrants I

and IV, and $\dfrac{4}{3}$ is positive, ω is in quadrant I.

Sketch ω and label a triangle with the hypotenuse equal to

$$\sqrt{4^2 + 3^2} = \sqrt{16+9} = \sqrt{25} = 5.$$

(continued on next page)

(*continued*)

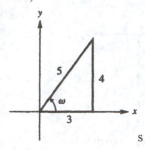

$$\cos\omega = \frac{3}{5}; \sin\omega = \frac{4}{5}$$

$$\cos\left(2\arctan\frac{4}{3}\right) = \cos(2\omega) = \cos^2\omega - \sin^2\omega$$

$$= \left(\frac{3}{5}\right)^2 - \left(\frac{4}{5}\right)^2$$

$$= \frac{9}{25} - \frac{16}{25} = -\frac{7}{25}$$

83. $\sin\left(2\cos^{-1}\frac{1}{5}\right)$

Let $\theta = \cos^{-1}\frac{1}{5}$, so that $\cos\theta = \frac{1}{5}$. The inverse cosine function yields values only in quadrants I and II, and because $\frac{1}{5}$ is positive, θ is in quadrant I. Sketch θ and label the sides of the right triangle. By the Pythagorean theorem, the length opposite to θ will be

$$\sqrt{5^2 - 1^2} = \sqrt{24} = 2\sqrt{6}.$$

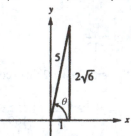

From the figure,

$\sin\theta = \dfrac{2\sqrt{6}}{5}$. Then,

$$\sin\left(2\cos^{-1}\frac{1}{5}\right) = \sin 2\theta = 2\sin\theta\,\cos\theta$$

$$= 2\left(\frac{2\sqrt{6}}{5}\right)\left(\frac{1}{5}\right) = \frac{4\sqrt{6}}{25}$$

85. $\sec(\sec^{-1} 2)$

Secant and inverse secant are inverse functions, so $\sec\left(\sec^{-1} 2\right) = 2$.

87. $\cos\left(\tan^{-1}\frac{5}{12} - \tan^{-1}\frac{3}{4}\right)$

Let $\alpha = \tan^{-1}\frac{5}{12}$ and $\beta = \tan^{-1}\frac{3}{4}$. Then $\tan\alpha = \frac{5}{12}$ and $\tan\beta = \frac{3}{4}$. Sketch angles α and β, both in quadrant I.

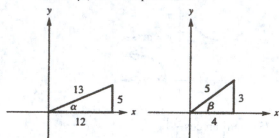

We have $\sin\alpha = \frac{5}{13}$, $\cos\alpha = \frac{12}{13}$, $\sin\beta = \frac{3}{5}$, and $\cos\beta = \frac{4}{5}$.

$$\cos\left(\tan^{-1}\frac{5}{12} - \tan^{-1}\frac{3}{4}\right)$$

$$= \cos(\alpha - \beta) = \cos\alpha\cos\beta + \sin\alpha\sin\beta$$

$$= \left(\frac{12}{13}\right)\left(\frac{4}{5}\right) + \left(\frac{5}{13}\right)\left(\frac{3}{5}\right) = \frac{48}{65} + \frac{15}{65} = \frac{63}{65}$$

89. $\sin\left(\sin^{-1}\frac{1}{2} + \tan^{-1}(-3)\right)$

Let $\sin^{-1}\frac{1}{2} = A$ and $\tan^{-1}(-3) = B$.

Then $\sin A = \frac{1}{2}$ and $\tan B = -3$. Sketch angle A in quadrant I and angle B in quadrant IV.

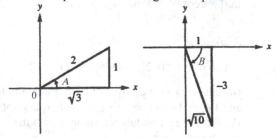

We have $\cos A = \frac{\sqrt{3}}{2}$, $\sin A = \frac{1}{2}$,

$\cos B = \frac{1}{\sqrt{10}} = \frac{\sqrt{10}}{10}$, and

$\sin B = \frac{-3}{\sqrt{10}} = -\frac{3\sqrt{10}}{10}$.

(continued on next page)

(*continued*)

$$\sin\left(\sin^{-1}\frac{1}{2}+\tan^{-1}(-3)\right)$$
$$=\sin(A+B)=\sin A\cos B+\cos A\sin B$$
$$=\frac{1}{2}\cdot\frac{1}{\sqrt{10}}+\frac{\sqrt{3}}{2}\cdot\frac{-3}{\sqrt{10}}$$
$$=\frac{1-3\sqrt{3}}{2\sqrt{10}}=\frac{\sqrt{10}-3\sqrt{30}}{20}$$

For Exercises 91–93, your calculator could be in either degree or radian mode.

91. $\cos(\tan^{-1}0.5)\approx0.894427191$

93. $\tan(\arcsin 0.12251014)\approx0.1234399811$

95. $\sin(\arccos u)$

Let $\theta=\arccos u$, so $\cos\theta=u=\dfrac{u}{1}$. Because

$u>0,\ 0<\theta<\dfrac{\pi}{2}.$

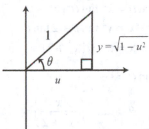

Because $y>0$, from the Pythagorean theorem,
$$y=\sqrt{1^2-u^2}=\sqrt{1-u^2}.$$
Therefore, $\sin\theta=\dfrac{\sqrt{1-u^2}}{1}=\sqrt{1-u^2}$. Thus,
$$\sin(\arccos u)=\sqrt{1-u^2}.$$

97. $\cos(\arcsin u)$

Let $\theta=\arcsin u$, so $\sin\theta=u=\dfrac{u}{1}$. Because

$u>0,\ 0<\theta<\dfrac{\pi}{2}.$

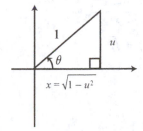

Because $x>0$, from the Pythagorean theorem,
$$x=\sqrt{1^2-u^2}=\sqrt{1-u^2}.$$
Therefore, $\cos\theta=\dfrac{\sqrt{1-u^2}}{1}=\sqrt{1-u^2}$. Thus,
$$\cos(\arcsin u)=\sqrt{1-u^2}$$

99. $\sin\left(2\sec^{-1}\dfrac{u}{2}\right)$

Let $\theta=\sec^{-1}\dfrac{u}{2}$, so $\sec\theta=\dfrac{u}{2}$. Because

$u>0,\ 0<\theta<\dfrac{\pi}{2}.$

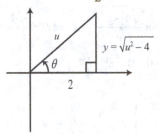

Because $y>0$, from the Pythagorean theorem,
$$y=\sqrt{u^2-2^2}=\sqrt{u^2-4}.$$
Now find $\sin 2\theta$.
$$\sin\theta=\frac{\sqrt{u^2-4}}{u}\text{ and }\cos\theta=\frac{2}{u}\text{ Thus,}$$
$$\sin\left(2\sec^{-1}\frac{u}{2}\right)=\sin 2\theta=2\sin\theta\cos\theta$$
$$=2\left(\frac{\sqrt{u^2-4}}{u}\right)\left(\frac{2}{u}\right)$$
$$=\frac{4\sqrt{u^2-4}}{u^2}$$

101. $\tan\left(\sin^{-1}\dfrac{u}{\sqrt{u^2+2}}\right)$

Let $\theta=\sin^{-1}\dfrac{u}{\sqrt{u^2+2}}$, so $\sin\theta=\dfrac{u}{\sqrt{u^2+2}}$.

Because $u>0,\ 0<\theta<\dfrac{\pi}{2}.$

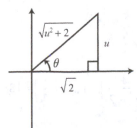

(*continued on next page*)

(*continued*)

Because $x > 0$, from the Pythagorean theorem,

$$x = \sqrt{\left(\sqrt{u^2 + 2}\right)^2 - u^2} = \sqrt{u^2 + 2 - u^2} = \sqrt{2}.$$

Therefore, $\tan \theta = \dfrac{u}{\sqrt{2}} = \dfrac{u\sqrt{2}}{2}$. Thus,

$$\tan\left(\sin^{-1} \frac{u}{\sqrt{u^2 + 2}}\right) = \frac{u\sqrt{2}}{2}.$$

103. $\sec\left(\text{arc cot } \dfrac{\sqrt{4 - u^2}}{u}\right)$

Let $\theta = \text{arc cot } \dfrac{\sqrt{4 - u^2}}{u}$, so $\cot \theta = \dfrac{\sqrt{4 - u^2}}{u}$.

Because $u > 0$, $0 < \theta < \dfrac{\pi}{2}$.

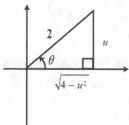

From the Pythagorean theorem,

$$r = \sqrt{\left(\sqrt{4 - u^2}\right)^2 + u^2} = \sqrt{4 - u^2 + u^2}$$

$$= \sqrt{4} = 2.$$

Therefore, $\sec \theta = \dfrac{2}{\sqrt{4 - u^2}} = \dfrac{2\sqrt{4 - u^2}}{4 - u^2}$.

Thus, $\sec\left(\text{arc cot } \dfrac{\sqrt{4 - u^2}}{u}\right) = \dfrac{2\sqrt{4 - u^2}}{4 - u^2}$.

105. From Example 8 in the text, we have

$$\theta = \arcsin \sqrt{\frac{v^2}{2v^2 + 64h}}.$$

$$\theta = \arcsin \sqrt{\frac{32^2}{2\left(32^2\right) + 64(5.0)}} \approx 41°$$

107. $\theta = \tan^{-1}\left(\dfrac{x}{x^2 + 2}\right)$

(a) $x = 1$,

$$\theta = \tan^{-1}\left(\frac{1}{1^2 + 2}\right) = \tan^{-1}\left(\frac{1}{3}\right) \approx 18°$$

(b) $x = 2$,

$$\theta = \tan^{-1}\left(\frac{2}{2^2 + 2}\right) = \tan^{-1}\frac{2}{6}$$

$$= \tan^{-1}\frac{1}{3} \approx 18°$$

(c) $x = 3$,

$$\theta = \tan^{-1}\left(\frac{3}{3^2 + 2}\right) = \tan^{-1}\frac{3}{11} \approx 15°$$

(d) $\tan(\theta + \alpha) = \dfrac{1 + 1}{x} = \dfrac{2}{x}$ and $\tan \alpha = \dfrac{1}{x}$

$$\tan(\theta + \alpha) = \frac{\tan \theta + \tan \alpha}{1 - \tan \theta \tan \alpha}$$

$$\frac{2}{x} = \frac{\tan \theta + \dfrac{1}{x}}{1 - \tan \theta \left(\dfrac{1}{x}\right)}$$

$$\frac{2}{x} = \frac{x \tan \theta + 1}{x - \tan \theta}$$

$$2(x - \tan \theta) = x(x \tan \theta + 1)$$

$$2x - 2 \tan \theta = x^2 \tan \theta + x$$

$$2x - x = x^2 \tan \theta + 2 \tan \theta$$

$$x = \tan \theta \left(x^2 + 2\right)$$

$$\tan \theta = \frac{x}{x^2 + 2}$$

$$\theta = \tan^{-1}\left(\frac{x}{x^2 + 2}\right)$$

(e) If we graph $y_1 = \tan^{-1}\left(\dfrac{x}{x^2 + 2}\right)$ using a graphing calculator, the maximum value of the function occurs when x is 1.4142123 m. (Note: Due to the computational routine, there may be a discrepancy in the last few decimal places.)

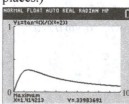

(f) $x = \sqrt{(1)(2)} = \sqrt{2}$

109. The diameter of the earth is 7927 miles at the equator, so the radius of the earth is 3963.5 miles. Then

$$\cos\theta = \frac{3963.5}{20,000+3963.5} = \frac{3963.5}{23,963.5} \text{ and}$$

$$\theta = \arccos\left(\frac{3963.5}{23,963.5}\right) \approx 80.48°.$$

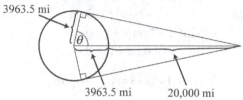

The percent of the equator that can be seen by the satellite is $\dfrac{2\theta}{360}\cdot 100 = \dfrac{2(80.48)}{360} \approx 44.7\%.$

111. $f(x) = 3x-2;\ f^{-1}(x) = \dfrac{1}{3}x + \dfrac{2}{3}$

$$f\left[f^{-1}(x)\right] = f\left[\frac{1}{3}x+\frac{2}{3}\right] = 3\left(\frac{1}{3}x+\frac{2}{3}\right)-2$$
$$= x+2-2 = x$$

$$f^{-1}\left[f(x)\right] = f^{-1}[3x-2] = \frac{3x-2}{3}+\frac{2}{3}$$
$$= \frac{3x-2+2}{3} = x$$

In each case, the result is x.

113. It is the graph of $y = x$.

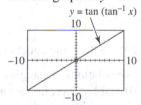

$y = \tan(\tan^{-1} x)$

Section 6.2   Trigonometric Equations I

Use the unit circle below to solve each equation.

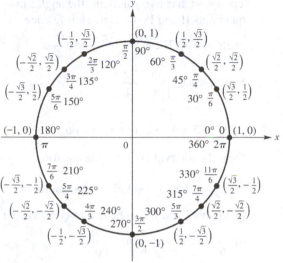

The unit circle $x^2 + y^2 = 1$

1. $\cos x = \dfrac{1}{2}$

Solution set: $\left\{\dfrac{\pi}{3}, \dfrac{5\pi}{3}\right\}$

3. $\sin x = -\dfrac{1}{2}$

Solution set: $\left\{\dfrac{7\pi}{6}, \dfrac{11\pi}{6}\right\}$

5. $\cos x = -1$

Solution set: $\{\pi\}$

7. $\sin\theta = 0$

Solution set: $\{0°, 180°\}$

9. $\cos\theta = -\dfrac{1}{2}$

Solution set: $\{120°, 240°\}$

11. $\sin\theta = -\dfrac{\sqrt{2}}{2}$

Solution set: $\{225°, 315°\}$

13. $-30°$ is not in the interval $[0°, 360°)$.

15. $2\cot x+1=-1\Rightarrow 2\cot x=-2\Rightarrow \cot x=-1$

Over the interval $[0,2\pi)$, the equation $\cot x=-1$ has two solutions, the angles in quadrants II and IV that have a reference angle of $\frac{\pi}{4}$. These are $\frac{3\pi}{4}$ and $\frac{7\pi}{4}$.

Solution set: $\left\{\frac{3\pi}{4},\frac{7\pi}{4}\right\}$

17. $2\sin x+3=4\Rightarrow 2\sin x=1\Rightarrow \sin x=\frac{1}{2}$

Over the interval $[0,2\pi)$, the equation $\sin x=\frac{1}{2}$ has two solutions, the angles in quadrants I and II that have a reference angle of $\frac{\pi}{6}$. These are $\frac{\pi}{6}$ and $\frac{5\pi}{6}$.

Solution set: $\left\{\frac{\pi}{6},\frac{5\pi}{6}\right\}$

19. $\tan^2 x+3=0\Rightarrow \tan^2 x=-3$

The square of a real number cannot be negative, so this equation has no solution.
Solution set: $\varnothing$

21. $(\cot x-1)(\sqrt{3}\cot x+1)=0$

$\cot x-1=0\Rightarrow \cot x=1$ or

$\sqrt{3}\cot x+1=0\Rightarrow \sqrt{3}\cot x=-1\Rightarrow$

$\cot x=-\frac{1}{\sqrt{3}}\Rightarrow \cot x=-\frac{\sqrt{3}}{3}$

Over the interval $[0,2\pi)$, the equation $\cot x=1$ has two solutions, the angles in quadrants I and III that have a reference angle of $\frac{\pi}{4}$. These are $\frac{\pi}{4}$ and $\frac{5\pi}{4}$. In the same interval, $\cot x=-\frac{\sqrt{3}}{3}$ also has two solutions. The angles in quadrants II and IV that have a reference angle of $\frac{\pi}{3}$ are $\frac{2\pi}{3}$ and $\frac{5\pi}{3}$.

Solution set: $\left\{\frac{\pi}{4},\frac{2\pi}{3},\frac{5\pi}{4},\frac{5\pi}{3}\right\}$

23. $\cos^2 x+2\cos x+1=0$

$\cos^2 x+2\cos x+1=0\Rightarrow (\cos x+1)^2=0\Rightarrow$

$\cos x+1=0\Rightarrow \cos x=-1$

Over the interval $[0,2\pi)$, the equation $\cos x=-1$ has one solution. This solution is π. Solution set: $\{\pi\}$

25. $-2\sin^2 x=3\sin x+1$

$2\sin^2 x+3\sin x+1=0$

$(2\sin x+1)(\sin x+1)=0$

$2\sin x+1=0\Rightarrow \sin x=-\frac{1}{2}$ or

$\sin x+1=0\Rightarrow \sin x=-1$

Over the interval $[0,2\pi)$, the equation $\sin x=-\frac{1}{2}$ has two solutions. The angles in quadrants III and IV that have a reference angle of $\frac{\pi}{6}$ are $\frac{7\pi}{6}$ and $\frac{11\pi}{6}$.

In the same interval, $\sin x=-1$ when the angle is $\frac{3\pi}{2}$. Solution set: $\left\{\frac{7\pi}{6},\frac{3\pi}{2},\frac{11\pi}{6}\right\}$

27. $(\cot\theta-\sqrt{3})(2\sin\theta+\sqrt{3})=0$

$\cot\theta-\sqrt{3}=0\Rightarrow \cot\theta=\sqrt{3}$ or

$2\sin\theta+\sqrt{3}=0\Rightarrow 2\sin\theta=-\sqrt{3}\Rightarrow$

$\sin\theta=-\frac{\sqrt{3}}{2}$

Over the interval $[0°,360°)$, the equation $\cot\theta=\sqrt{3}$ has two solutions, the angles in quadrants I and III that have a reference angle of $30°$ These are $30°$ and $210°$. In the same interval, the equation $\sin\theta=-\frac{\sqrt{3}}{2}$ has two solutions, the angles in quadrants III and IV that have a reference angle of $60°$. These are $240°$ and $300°$.

Solution set: $\{30°, 210°, 240°, 300°\}$

29. $2\sin\theta-1=\csc\theta\Rightarrow 2\sin\theta-1=\frac{1}{\sin\theta}\Rightarrow$

$2\sin^2\theta-\sin\theta=1\Rightarrow$

$2\sin^2\theta-\sin\theta-1=0\Rightarrow$

$(2\sin\theta+1)(\sin\theta-1)=0$

$2\sin\theta+1=0\Rightarrow \sin\theta=-\frac{1}{2}$ or

$\sin\theta-1=0\Rightarrow \sin\theta=1$

(continued on next page)

(*continued*)

Over the interval $[0°, 360°)$, the equation

$\sin \theta = -\dfrac{1}{2}$ has two solutions, the angles in

quadrants III and IV that have a reference angle of 30° These are 210° and 330°. In the same interval, the only angle θ for which $\sin \theta = 1$ is 90°.

Solution set: $\{90°, 210°, 330°\}$

31. $\tan \theta - \cot \theta = 0$

$\tan \theta - \cot \theta = 0 \Rightarrow \tan \theta - \dfrac{1}{\tan \theta} = 0 \Rightarrow$

$\tan^2 \theta - 1 = 0 \Rightarrow \tan^2 \theta = 1 \Rightarrow \tan \theta = \pm 1$

Over the interval $[0°, 360°)$, the equation

$\tan \theta = 1$ has two solutions, the angles in quadrants I and III that have a reference angle of 45° These are 45° and 225°.

In the same interval, the equation $\tan \theta = -1$ has two solutions, the angles in quadrants II and IV that have a reference angle of 45°. These are 135° and 315°.

Solution set: $\{45°, 135°, 225°, 315°\}$

33. $\csc^2 \theta - 2 \cot \theta = 0$

$\csc^2 \theta - 2 \cot \theta = 0$

$\left(1 + \cot^2 \theta\right) - 2 \cot \theta = 0$

$\cot^2 \theta - 2 \cot \theta + 1 = 0$

$\left(\cot \theta - 1\right)^2 = 0$

$\cot \theta - 1 = 0 \Rightarrow \cot \theta = 1$

Over the interval $[0°, 360°)$, the equation

$\cot \theta = 1$ has two solutions, the angles in quadrants I and III that have a reference angle of 45°. These are 45° and 225°

Solution set: $\{45°, 225°\}$

35. $2 \tan^2 \theta \sin \theta - \tan^2 \theta = 0$

$2 \tan^2 \theta \sin \theta - \tan^2 \theta = 0$

$\tan^2 \theta (2 \sin \theta - 1) = 0$

$\tan^2 \theta = 0$

$\tan \theta = 0$ or $2 \sin \theta - 1 = 0 \Rightarrow$

$2 \sin \theta = 1 \Rightarrow \sin \theta = \dfrac{1}{2}$

Over the interval $[0°, 360°)$, the equation

$\tan \theta = 0$ has two solutions. These are 0° and 180°. In the same interval, the

equation $\sin \theta = \dfrac{1}{2}$ has two solutions, the

angles in quadrants I and II that have a reference angle of 30°. These are 30° and 150°.

Solution set: $\{0°, 30°, 150°, 180°\}$

37. $\sec^2 \theta \tan \theta = 2 \tan \theta$

$\sec^2 \theta \tan \theta = 2 \tan \theta$

$\sec^2 \theta \tan \theta - 2 \tan \theta = 0$

$\tan \theta \left(\sec^2 \theta - 2\right) = 0$

$\tan \theta = 0$ or $\sec^2 \theta - 2 = 0 \Rightarrow$

$\sec^2 \theta = 2 \Rightarrow \sec \theta = \pm \sqrt{2}$

Over the interval $[0°, 360°)$, the equation

$\tan \theta = 0$ has two solutions. These are 0° and 180°. In the same interval, the

equation $\sec \theta = \sqrt{2}$ has two solutions, the angles in quadrants I and IV that have a reference angle of 45° These are 45° and 315°. Finally, the equation $\sec \theta = -\sqrt{2}$ has two solutions, the angles in quadrants II and III that have a reference angle of 45°. These are 135° and 225°.

Solution set:

$\{0°, 45°, 135°, 180°, 225°, 315°\}$

For Exercises 39–45, make sure your calculator is in degree mode.

39. $9 \sin^2 \theta - 6 \sin \theta = 1$

$9 \sin^2 \theta - 6 \sin \theta = 1 \Rightarrow 9 \sin^2 \theta - 6 \sin \theta - 1 = 0$

We use the quadratic formula with $a = 9$, $b = -6$, and $c = -1$.

$\sin \theta = \dfrac{6 \pm \sqrt{36 - 4(9)(-1)}}{2(9)} = \dfrac{6 \pm \sqrt{36 + 36}}{18}$

$= \dfrac{6 \pm \sqrt{72}}{18} = \dfrac{6 \pm 6\sqrt{2}}{18} = \dfrac{1 \pm \sqrt{2}}{3}$

Because $\sin \theta = \dfrac{1 + \sqrt{2}}{3} > 0$ (and less than 1),

we will obtain two angles. One angle will be in quadrant I and the other will be in quadrant II. Using a calculator, if

$\sin \theta = \dfrac{1 + \sqrt{2}}{3} \approx 0.80473787$, the quadrant I

angle will be approximately 53.6°.

(*continued on next page*)

(*continued*)

The quadrant II angle will be approximately $180° - 53.6° = 126.4°$. Because

$$\sin\theta = \frac{1-\sqrt{2}}{3} < 0 \text{ (and greater than } -1\text{), we}$$

will obtain two angles. One angle will be in quadrant III and the other will be in quadrant IV. Using a calculator, if

$$\sin\theta = \frac{1-\sqrt{2}}{3} \approx -0.13807119, \text{ then}$$

$\theta \approx -7.9°$. This solution is not in the interval $[0°, 360°)$, so we must use it as a reference angle to find angles in the interval. Our reference angle will be $7.9°$. The angle in quadrant III will be approximately $180° + 7.9° = 187.9°$. The angle in quadrant IV will be approximately $360° - 7.9° = 352.1°$.

Solution set: $\{53.6°, 126.4°, 187.9°, 352.1°\}$

41. $\tan^2\theta + 4\tan\theta + 2 = 0$

We use the quadratic formula with $a = 1$, $b = 4$, and $c = 2$.

$$\tan\theta = \frac{-4 \pm \sqrt{16 - 4(1)(2)}}{2(1)} = \frac{-4 \pm \sqrt{16-8}}{2}$$

$$= \frac{-4 \pm \sqrt{8}}{2} = \frac{-4 \pm 2\sqrt{2}}{2} = -2 \pm \sqrt{2}$$

Because $\tan\theta = -2 + \sqrt{2} < 0$, we will obtain two angles. One angle will be in quadrant II and the other will be in quadrant IV. Using a calculator, if $\tan\theta = -2 + \sqrt{2} = -0.5857864$, then $\theta \approx -30.4°$. This solution is not in the interval $[0°, 360°)$, so we must use it as a reference angle to find angles in the interval. Our reference angle will be $30.4°$. The angle in quadrant II will be approximately $180° - 30.4° = 149.6°$. The angle in quadrant IV will be approximately $360° - 30.4° = 329.6°$.

Because $\tan\theta = -2 - \sqrt{2} < 0$, we will obtain two angles. One angle will be in quadrant II and the other will be in quadrant IV. Using a calculator, if $\tan\theta = -2 - \sqrt{2} = -3.4142136$, then $\theta \approx -73.7°$. This solution is not in the interval $[0°, 360°)$, so we must use it as a reference angle to find angles in the interval. Our reference angle will be $73.7°$. The angle in quadrant II will be approximately $180° - 73.7° = 106.3°$

The angle in quadrant IV will be approximately $360° - 73.7° = 286.3°$.

Solution set: $\{106.3°, 149.6°, 286.3°, 329.6°\}$

43. $\sin^2\theta - 2\sin\theta + 3 = 0$

We use the quadratic formula with $a = 1$, $b = -2$, and $c = 3$.

$$\sin\theta = \frac{2 \pm \sqrt{4 - (4)(1)(3)}}{2(1)} = \frac{2 \pm \sqrt{4-12}}{2}$$

$$= \frac{2 \pm \sqrt{-8}}{2} = \frac{2 \pm 2i\sqrt{2}}{2} = 1 \pm i\sqrt{2}$$

Because $1 \pm i\sqrt{2}$ is not a real number, the equation has no real solutions.

Solution set: $\varnothing$

45. $\cot\theta + 2\csc\theta = 3$

$$\cot\theta + 2\csc\theta = 3 \Rightarrow \frac{\cos\theta}{\sin\theta} + \frac{2}{\sin\theta} = 3$$

$$\cos\theta + 2 = 3\sin\theta$$

$$(\cos\theta + 2)^2 = (3\sin\theta)^2$$

$$\cos^2\theta + 4\cos\theta + 4 = 9\sin^2\theta$$

$$\cos^2\theta + 4\cos\theta + 4 = 9(1 - \cos^2\theta)$$

$$\cos^2\theta + 4\cos\theta + 4 = 9 - 9\cos^2\theta$$

$$10\cos^2\theta + 4\cos\theta - 5 = 0$$

We use the quadratic formula with $a = 10$, $b = 4$, and $c = -5$.

$$\cos\theta = \frac{-4 \pm \sqrt{4^2 - 4(10)(-5)}}{2(10)}$$

$$= \frac{-4 \pm \sqrt{16 + 200}}{20} = \frac{-4 \pm \sqrt{216}}{20}$$

$$= \frac{-4 \pm 6\sqrt{6}}{20} = \frac{-2 \pm 3\sqrt{6}}{10}$$

Because $\cos\theta = \dfrac{-2 + 3\sqrt{6}}{10} > 0$ (and less

than 1), we will obtain two angles. One angle will be in quadrant I and the other will be in quadrant IV. Using a calculator, if

$$\cos\theta = \frac{-2 + 3\sqrt{6}}{10} \approx 0.53484692, \text{ the quadrant}$$

I angle will be approximately $57.7°$.
The quadrant IV angle will be approximately $360° - 57.7° = 302.3°$.

Because $\cos\theta = \dfrac{-2 - 3\sqrt{6}}{10} < 0$ (and greater

than -1), we will obtain two angles. One angle will be in quadrant II and the other will be in quadrant III.

(*continued on next page*)

(continued)

Using a calculator, if

$$\cos\theta = \frac{-2-3\sqrt{6}}{10} \approx -0.93484692, \text{ the}$$

quadrant II angle will be approximately 159.2°. The reference angle is 180° − 159.2° = 20.8°. Thus, the quadrant III angle will be approximately 180° + 20.8° = 200.8°. The solution was found by squaring both sides of an equation, so we must check that each proposed solution is a solution of the original equation. 302.3° and 200.8° do not satisfy our original equation. Thus, they are not elements of the solution set.
Solution set: {57.7°, 159.2°}

In Exercises 47–61, if you are using a calculator, make sure it is in radian mode if you are solving for x and in degree mode if you are solving for θ.

47. $\cos\theta + 1 = 0 \Rightarrow \cos\theta = -1 \Rightarrow \theta = 180°$ in the interval $[0, 2\pi)$. The solution set is $\{180° + 360°n, \text{ where } n \text{ is any integer}\}$.

49. $3\csc x - 2\sqrt{3} = 0 \Rightarrow 3\csc x = 2\sqrt{3} \Rightarrow$

$$\csc x = \frac{2\sqrt{3}}{3} \Rightarrow x = \frac{\pi}{3}, \frac{2\pi}{3} \text{ in the interval}$$

$[0, 2\pi)$. The solution set is

$$\left\{\frac{\pi}{3} + 2n\pi, \frac{2\pi}{3} + 2n\pi, \text{ where } n \text{ is any}\right.$$

$$\text{integer}\Big\}.$$

51. $6\sin^2\theta + \sin\theta = 1 \Rightarrow 6\sin^2\theta + \sin\theta - 1 = 0 \Rightarrow$
$(3\sin\theta - 1)(2\sin\theta + 1) = 0 \Rightarrow$

$$\sin\theta = \frac{1}{3} \Rightarrow \theta \approx 19.5° \text{ or}$$

$$\theta \approx 180° - 19.5° = 160.5° \text{ or}$$

$$\sin\theta = -\frac{1}{2} \Rightarrow \theta = 210° \text{ or } \theta = 330°$$

The solution set is $\{19.5° + 360°n,$
$160.5° + 360°n, 210° + 360°n, 330° + 360°n,$
where n is any integer$\}$.

53. $2\cos^2 x + \cos x - 1 = 0$
$(2\cos x - 1)(\cos x + 1) = 0$

$$2\cos x - 1 = 0 \Rightarrow \cos x = \frac{1}{2} \text{ or}$$

$$\cos x + 1 = 0 \Rightarrow \cos x = -1$$

Over the interval $[0, 2\pi)$, the equation

$$\cos x = \frac{1}{2} \text{ has two solutions.}$$

The angles in quadrants I and IV that have a reference angle of $\frac{\pi}{3}$ are $\frac{\pi}{3}$ and $\frac{5\pi}{3}$. In the same interval, $\cos x = -1$ when the angle is π. Thus, the solution set is

$$\left\{\frac{\pi}{3} + 2n\pi, \ \pi + 2n\pi,\right.$$

$$\text{and } \frac{5\pi}{3} + 2n\pi, \text{ where } n \text{ is any integer}\Big\}.$$

55. $\sin\theta\cos\theta - \sin\theta = 0 \Rightarrow \sin\theta(\cos\theta - 1) = 0 \Rightarrow$
$\sin\theta = 0 \Rightarrow \theta = 0° \text{ or } \theta = 180° \text{ or}$
$\cos\theta = 1 \Rightarrow \theta = 0°$
The solution set is $\{180°n, \text{ where } n \text{ is any integer}\}$.

57. $\sin x(3\sin x - 1) = 1 \Rightarrow$
$3\sin^2 x - \sin x - 1 = 0$
Use the quadratic formula with $a = 3$, $b = -1$, and $c = -1$.

$$\sin x = \frac{-(-1) \pm \sqrt{(-1)^2 - 4(3)(-1)}}{2(3)} = \frac{1 \pm \sqrt{13}}{6}$$

Because $\sin x = \dfrac{1 + \sqrt{13}}{6} > 0$ (and less than 1),

we will obtain two angles. One angle will be in quadrant I and the other will be in quadrant II. Using a calculator, if

$$\sin x = \frac{1 + \sqrt{13}}{6} \approx 0.76759188, \text{ the quadrant I}$$

angle will be approximately 0.8751. The quadrant II angle will be approximately $\pi - 0.88 \approx 2.2665$. Because

$$\sin x = \frac{1 - \sqrt{13}}{6} < 0 \text{ (and greater than } -1), \text{ we}$$

will obtain two angles. One angle will be in quadrant III and the other will be in quadrant IV. Using a calculator, if

$$\sin x = \frac{1 - \sqrt{13}}{6} \approx -0.43425855, \text{ then}$$

$x \approx -0.4492$. This solution is not in the interval $[0, 2\pi)$, so we must use it as a reference angle to find angles in the interval. Our reference angle will be 0.4492. The angle in quadrant III will be approximately $\pi + 0.4492 \approx 3.5908$. The angle in quadrant IV will be approximately $2\pi - 0.4492 \approx 5.8340$. Thus, the solution set is
$\{0.8751 + 2n\pi, \ 2.2665 + 2n\pi, \ 3.5908 + 2n\pi,$
and $5.8340 + 2n\pi$, where n is any integer$\}$.

59. $5 + 5\tan^2\theta = 6\sec\theta \Rightarrow 5\left(1 + \tan^2\theta\right) = 6\sec\theta \Rightarrow$

$5\sec^2\theta = 6\sec\theta \Rightarrow 5\sec^2\theta - 6\sec\theta = 0 \Rightarrow$
$\sec\theta\left(5\sec\theta - 6\right) = 0$

$\sec\theta = 0$ or $5\sec\theta - 6 = 0 \Rightarrow \sec\theta = \dfrac{6}{5}$

$\sec\theta = 0$ is an impossible value because the secant function must be either ≥ 1 or ≤ -1.

Because $\sec\theta = \dfrac{6}{5} > 1$, we will obtain two angles.

One angle will be in quadrant I and the other will be in quadrant IV. Using a calculator, if $\sec\theta = \dfrac{6}{5} = 1.2$, the quadrant I angle will be approximately $33.6°$. The quadrant IV angle will be approximately $360° - 33.6° = 326.4°$. Thus, the solution set is $\{33.6° + 360°n$ and $326.4° + 360°n$, where n is any integer$\}$.

61.
$$\dfrac{2\tan\theta}{3 - \tan^2\theta} = 1$$
$$2\tan\theta = 3 - \tan^2\theta$$
$$\tan^2\theta + 2\tan\theta - 3 = 0$$
$$\left(\tan\theta - 1\right)\left(\tan\theta + 3\right) = 0$$

$\tan\theta - 1 = 0 \Rightarrow \tan\theta = 1$ or
$\tan\theta + 3 = 0 \Rightarrow \tan\theta = -3$
Over the interval $[0°, 360°)$, the equation $\tan\theta = 1$ has two solutions $45°$ and $225°$.
Over the same interval, the equation $\tan\theta = -3$ has two solutions that are approximately $-71.6° + 180° = 108.4°$ and $-71.6° + 360° = 288.4°$.

Thus, the solutions are
$45° + 360°n$, $108.4° + 360°n$, $225° + 360°n$
and $288.4° + 360°n$, where n is any integer.
The period of the tangent function is $180°$, so the solution set can also be written as
$\{45° + 180°n$ and $108.4° + 180°n$, where n is any integer$\}$.

63. The x-intercept method is shown below.
$y_1 = x^2 + \sin x - x^3 - \cos x$ is graphed in the window $[0, 2\pi] \times [-1, 1]$.

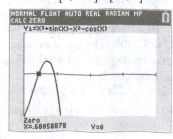

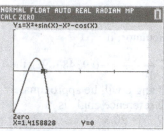

Solution set: $\{0.6806, 1.4159\}$

65. $P = A\sin\left(2\pi ft + \phi\right)$

(a) $0 = 0.004\sin\left[2\pi(261.63)t + \dfrac{\pi}{7}\right]$

$0 = \sin\left(1643.87t + 0.45\right)$

$1643.87t + 0.45 = n\pi$, so $t = \dfrac{n\pi - 0.45}{1643.87}$,

where n is any integer.
If $n = 0$, then $t \approx 0.000274$. If $n = 1$, then $t \approx 0.00164$. If $n = 2$, then $t \approx 0.00355$. If $n = 3$, then $t \approx 0.00546$. The only solutions for t in the interval $[0, 0.005]$ are 0.00164 and 0.00355.

(b)

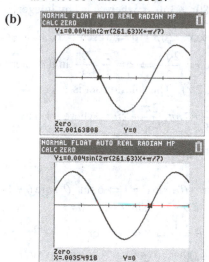

From the graphs we can estimate that $P \leq 0$ on the interval $[0.00164, 0.00355]$.

(c) $P < 0$ implies that there is a decrease in pressure so an eardrum would be vibrating outward.

67. $V = \cos 2\pi t$, $0 \leq t \leq \dfrac{1}{2}$

(a) $V = 0$, $\cos 2\pi t = 0 \Rightarrow 2\pi t = \cos^{-1} 0 \Rightarrow$

$2\pi t = \dfrac{\pi}{2} \Rightarrow t = \dfrac{\dfrac{\pi}{2}}{2\pi} = \dfrac{1}{4}$ sec

(b) $V = 0.5, \cos 2\pi t = 0.5 \Rightarrow$

$$2\pi t = \cos^{-1}(0.5) \Rightarrow 2\pi t = \frac{\pi}{3} \Rightarrow$$

$$t = \frac{\dfrac{\pi}{3}}{2\pi} = \frac{1}{6} \sec$$

(c) $V = 0.25, \cos 2\pi t = 0.25 \Rightarrow$

$$2\pi t = \cos^{-1}(0.25) \Rightarrow$$
$$2\pi t \approx 1.3181161 \Rightarrow$$
$$t \approx \frac{1.3181161}{2\pi} \approx 0.21 \sec$$

Section 6.3 Trigonometric Equations II

1. $\cos 2x = \dfrac{1}{2}$

$2x = \dfrac{\pi}{3}$	$2x = \dfrac{\pi}{3} + 2\pi$	$2x = \dfrac{5\pi}{3}$	$2x = \dfrac{5\pi}{3} + 2\pi$
$x = \dfrac{\pi}{6}$	$2x = \dfrac{7\pi}{3}$	$x = \dfrac{5\pi}{6}$	$2x = \dfrac{11\pi}{3}$
	$x = \dfrac{7\pi}{6}$		$x = \dfrac{11\pi}{6}$

Solution set: $\left\{\dfrac{\pi}{6}, \dfrac{5\pi}{6}, \dfrac{7\pi}{6}, \dfrac{11\pi}{6}\right\}$

3. $\sin 2x = -\dfrac{1}{2}$

$2x = \dfrac{7\pi}{6}$	$2x = \dfrac{7\pi}{6} + 2\pi$	$2x = \dfrac{11\pi}{6}$	$2x = \dfrac{11\pi}{6} + 2\pi$
$x = \dfrac{7\pi}{12}$	$2x = \dfrac{19\pi}{6}$	$x = \dfrac{11\pi}{12}$	$2x = \dfrac{23\pi}{6}$
	$x = \dfrac{19\pi}{12}$		$x = \dfrac{23\pi}{12}$

Solution set: $\left\{\dfrac{7\pi}{12}, \dfrac{11\pi}{12}, \dfrac{19\pi}{12}, \dfrac{23\pi}{12}\right\}$

5. $\cos 2x = -1$

$2x = \pi$	$2x = \pi + 2\pi$
$x = \dfrac{\pi}{2}$	$2x = 3\pi$
	$x = \dfrac{3\pi}{2}$

Solution set: $\left\{\dfrac{\pi}{2}, \dfrac{3\pi}{2}\right\}$

7. $\sin \dfrac{\theta}{2} = 0$

$\dfrac{\theta}{2} = 0°$	$\dfrac{\theta}{2} = 180°$
$\theta = 0°$	$\theta = 360°$

Note that the solution 360° is not included in the given interval. So, the solution set is $\{0°\}$.

9. $\cos \dfrac{\theta}{2} = -\dfrac{1}{2}$

$\dfrac{\theta}{2} = 120°$	$\dfrac{\theta}{2} = 240°$
$\theta = 240°$	$\theta = 480°$

Note that the solution 480° is not included in the given interval. So, the solution set is $\{240°\}$.

11. $\sin \dfrac{\theta}{2} = -\dfrac{\sqrt{2}}{2}$

$\dfrac{\theta}{2} = 225°$	$\dfrac{\theta}{2} = 315°$
$\theta = 450°$	$\theta = 630°$

Neither solution is in the given interval, so the solution set is $\varnothing$.

13. Because $2x = \dfrac{2\pi}{3}, 2\pi, \dfrac{8\pi}{3} \Rightarrow$

$x = \dfrac{2\pi}{6}, \dfrac{2\pi}{2}, \dfrac{8\pi}{6} \Rightarrow x = \dfrac{\pi}{3}, \pi, \dfrac{4\pi}{3}$, the

solution set is $\left\{\dfrac{\pi}{3}, \pi, \dfrac{4\pi}{3}\right\}$.

15. Because $3\theta = 180°, 630°, 720°, 930° \Rightarrow$
$\theta = 60°, 210°, 240°, 310°$, the solution set is $\{60°, 210°, 240°, 310°\}$.

17. $2\cos 2x = \sqrt{3} \Rightarrow \cos 2x = \dfrac{\sqrt{3}}{2}$

Because $0 \le x < 2\pi, 0 \le 2x < 4\pi$. Thus,

$2x = \dfrac{\pi}{6}, \dfrac{11\pi}{6}, \dfrac{13\pi}{6}, \dfrac{23\pi}{6} \Rightarrow$

$x = \dfrac{\pi}{12}, \dfrac{11\pi}{12}, \dfrac{13\pi}{12}, \dfrac{23\pi}{12}$.

Solution set: $\dfrac{\pi}{12}, \dfrac{11\pi}{12}, \dfrac{13\pi}{12}, \dfrac{23\pi}{12}$

19. $\sin 3\theta = -1$
Because $0° \le \theta < 360°, 0° \le 3\theta < 1080°$.
Thus, $3\theta = 270°, 630°, 990° \Rightarrow$
$x = 90°, 210°, 330°$
Solution set: $\{90°, 210°, 330°\}$

21. $3 \tan 3x = \sqrt{3} \Rightarrow \tan 3x = \dfrac{\sqrt{3}}{3}$

Because $0 \le x < 2\pi$, $0 \le 3x < 6\pi$.

Thus, $3x = \dfrac{\pi}{6}, \dfrac{7\pi}{6}, \dfrac{13\pi}{6}, \dfrac{19\pi}{6}, \dfrac{25\pi}{6}, \dfrac{31\pi}{6}$

implies $x = \dfrac{\pi}{18}, \dfrac{7\pi}{18}, \dfrac{13\pi}{18}, \dfrac{19\pi}{18}, \dfrac{25\pi}{18}, \dfrac{31\pi}{18}$.

Solution set:

$$\left\{ \dfrac{\pi}{18}, \dfrac{7\pi}{18}, \dfrac{13\pi}{18}, \dfrac{19\pi}{18}, \dfrac{25\pi}{18}, \dfrac{31\pi}{18} \right\}$$

23. $\sqrt{2} \cos 2\theta = -1 \Rightarrow \cos 2\theta = \dfrac{-1}{\sqrt{2}} = -\dfrac{\sqrt{2}}{2}$

Because $0° \le \theta < 360°$, $0° \le 2\theta < 720°$. Thus,

$2\theta = 135°, 225°, 495°, 585° \Rightarrow$

$\theta = 67.5°, 112.5°, 247.5°, 292.5°$

Solution set: $\{67.5°, 112.5°, 247.5°, 292.5°\}$

25. $\sin \dfrac{x}{2} = \sqrt{2} - \sin \dfrac{x}{2}$

$\sin \dfrac{x}{2} = \sqrt{2} - \sin \dfrac{x}{2} \Rightarrow \sin \dfrac{x}{2} + \sin \dfrac{x}{2} = \sqrt{2} \Rightarrow$

$2 \sin \dfrac{x}{2} = \sqrt{2} \Rightarrow \sin \dfrac{x}{2} = \dfrac{\sqrt{2}}{2}$

Because $0 \le x < 2\pi$, $0 \le \dfrac{x}{2} < \pi$. Thus,

$\dfrac{x}{2} = \dfrac{\pi}{4}, \dfrac{3\pi}{4} \Rightarrow x = \dfrac{\pi}{2}, \dfrac{3\pi}{2}$.

Solution set: $\left\{ \dfrac{\pi}{2}, \dfrac{3\pi}{2} \right\}$

27. $\sin x = \sin 2x$

$\sin x = \sin 2x \Rightarrow \sin x = 2 \sin x \cos x \Rightarrow$

$\sin x - 2 \sin x \cos x = 0 \Rightarrow \sin x (1 - 2 \cos x) = 0$

Over the interval $[0, 2\pi)$, we have

$1 - 2 \cos x = 0 \Rightarrow -2 \cos x = -1 \Rightarrow$

$\cos x = \dfrac{1}{2} \Rightarrow x = \dfrac{\pi}{3}$ or $\dfrac{5\pi}{3}$

$\sin x = 0 \Rightarrow x = 0$ or π

Solution set: $\left\{ 0, \dfrac{\pi}{3}, \pi, \dfrac{5\pi}{3} \right\}$

29. $8 \sec^2 \dfrac{x}{2} = 4 \Rightarrow \sec^2 \dfrac{x}{2} = \dfrac{1}{2} \Rightarrow \sec \dfrac{x}{2} = \pm \dfrac{\sqrt{2}}{2}$

$-\dfrac{\sqrt{2}}{2}$ is not in the interval $(-\infty, -1]$ and $\dfrac{\sqrt{2}}{2}$

is not in the interval $[1, \infty)$, so this equation has no solution. Solution set: $\varnothing$

31. $\sin \dfrac{\theta}{2} = \csc \dfrac{\theta}{2}$

$\sin \dfrac{\theta}{2} = \csc \dfrac{\theta}{2} \Rightarrow \sin \dfrac{\theta}{2} = \dfrac{1}{\sin \dfrac{\theta}{2}} \Rightarrow$

$\sin^2 \dfrac{\theta}{2} = 1 \Rightarrow \sin \dfrac{\theta}{2} = \pm 1$

$0° \le \theta < 360° \Rightarrow 0 \le \dfrac{\theta}{2} < 180°$.

If $\sin \dfrac{\theta}{2} = 1$, $\dfrac{\theta}{2} = 90° \Rightarrow \theta = 180°$.

If $\sin \dfrac{\theta}{2} = -1$, $\dfrac{\theta}{2} = 270° \Rightarrow \theta = 540°$, which is

not in the interval $[0, 360°]$.

Solution set: $\{180°\}$

33. $\cos 2x + \cos x = 0$

We choose an identity for $\cos 2x$ that involves only the cosine function.

$$\cos 2x + \cos x = 0$$
$$\left(2 \cos^2 x - 1 \right) + \cos x = 0$$
$$2 \cos^2 x + \cos x - 1 = 0$$
$$\left(2 \cos x - 1 \right)\left(\cos x + 1 \right) = 0 \Rightarrow$$
$$\cos x = \dfrac{1}{2} \text{ or } \cos x = -1$$

$\cos x = \dfrac{1}{2} \Rightarrow x = \dfrac{\pi}{3}$ or $\dfrac{5\pi}{3}$

$\cos x = -1 \Rightarrow x = \pi$

Solution set: $\left\{ \dfrac{\pi}{3}, \pi, \dfrac{5\pi}{3} \right\}$

35. $\sqrt{2} \sin 3x - 1 = 0$

$\sqrt{2} \sin 3x - 1 = 0 \Rightarrow \sqrt{2} \sin 3x = 1 \Rightarrow$

$\sin 3x = \dfrac{1}{\sqrt{2}} \Rightarrow \sin 3x = \dfrac{\sqrt{2}}{2}$

In quadrant I and II, sine is positive. Thus,

$3x = \dfrac{\pi}{4} + 2n\pi, \dfrac{3\pi}{4} + 2n\pi \Rightarrow$

$x = \dfrac{\pi}{12} + \dfrac{2n\pi}{3}, \dfrac{\pi}{4} + \dfrac{2n\pi}{3}$

Solution set:

$$\left\{ \dfrac{\pi}{12} + \dfrac{2n\pi}{3}, \dfrac{\pi}{4} + \dfrac{2n\pi}{3}, \text{ where } n \text{ is any integer} \right\}$$

37. $\cos \dfrac{\theta}{2} = 1$

$\dfrac{\theta}{2} = 0° + 360°n \Rightarrow \theta = 720°n$.

Solution set: $\{720°n, \text{ where } n \text{ is any integer}\}$

39. $2\sqrt{3}\sin\dfrac{x}{2}=3\Rightarrow\sin\dfrac{x}{2}=\dfrac{3}{2\sqrt{3}}\Rightarrow$

$\sin\dfrac{x}{2}=\dfrac{3\sqrt{3}}{6}\Rightarrow\sin\dfrac{x}{2}=\dfrac{\sqrt{3}}{2}$

$0\le\theta<360°\Rightarrow0°\le\dfrac{\theta}{2}<180°.$

Thus, $\dfrac{x}{2}=\dfrac{\pi}{3}+2n\pi,\ \dfrac{2\pi}{3}+2n\pi\Rightarrow$

$x=\dfrac{2\pi}{3}+4n\pi,\ \dfrac{4\pi}{3}+4n\pi$

Solution set:

$\left\{\dfrac{2\pi}{3}+4n\pi,\ \dfrac{4\pi}{3}+4n\pi,\text{ where }n\text{ is any}\right.$

integer$\Big\}$

41. $2\sin\theta=2\cos2\theta$

$2\sin\theta=2\cos2\theta\Rightarrow\sin\theta=\cos2\theta\Rightarrow$

$\sin\theta=1-2\sin^2\theta\Rightarrow2\sin^2\theta+\sin\theta-1=0\Rightarrow$

$(2\sin\theta-1)(\sin\theta+1)=0\Rightarrow$

$2\sin\theta-1=0$ or $\sin\theta+1=0$

Over the interval $[0°,360°)$, we have

$2\sin\theta-1=0\Rightarrow2\sin\theta=1\Rightarrow\sin\theta=\dfrac{1}{2}\Rightarrow$

$\theta=30°$ or $150°$

$\sin\theta+1=0\Rightarrow\sin\theta=-1\Rightarrow\theta=270°$

Solution set: $\{30°+360°n,\ 150°+360°n,$

$270°+360°n$, where n is any integer$\}$

43. $1-\sin x=\cos2x\Rightarrow1-\sin x=1-2\sin^2x\Rightarrow$

$2\sin^2x-\sin x=0\Rightarrow\sin x(2\sin x-1)=0$

Over the interval $[0,2\pi)$, we have

$\sin x=0\Rightarrow x=0$ or $\pi.$

$2\sin x-1=0\Rightarrow\sin x=\dfrac{1}{2}\Rightarrow x=\dfrac{\pi}{6}$ or $\dfrac{5\pi}{6}$

Solution set:

$\left\{n\pi,\ \dfrac{\pi}{6}+2n\pi,\ \dfrac{5\pi}{6}+2n\pi,\text{ where}\right.$

n is any integer$\Big\}$

45. $3\csc^2\dfrac{x}{2}=2\sec x\Rightarrow\dfrac{3}{\sin^2\dfrac{x}{2}}=\dfrac{2}{\cos x}\Rightarrow$

$\sin^2\dfrac{x}{2}=\dfrac{3}{2}\cos x\Rightarrow\dfrac{1-\cos x}{2}=\dfrac{3}{2}\cos x\Rightarrow$

$1-\cos x=3\cos x\Rightarrow1=4\cos x\Rightarrow\dfrac{1}{4}=\cos x$

Over the interval $[0,2\pi)$, we have

$\cos x=\dfrac{1}{4}\Rightarrow x=1.3181$ or $x=4.9651$

Solution set:

$\{1.3181+2n\pi,\ 4.9651+2n\pi,\text{ where }n\text{ is any}$

integer$\}$

47. $2-\sin2\theta=4\sin2\theta$

$2-\sin2\theta=4\sin2\theta\Rightarrow2=5\sin2\theta\Rightarrow$

$\sin2\theta=\dfrac{2}{5}\Rightarrow\sin2\theta=0.4$

Since $0\le\theta<360°,\ \ 0°\le2\theta<720°.$ In quadrant

I and II, sine is positive.

$\sin2\theta=0.4\Rightarrow2\theta=23.6°,156.4°,383.6°,516.4°$

Thus, $\theta=11.8°,78.2°,191.8°,258.2°.$

Solution set:

$\{11.8°+360°n,\ 78.2°+360°n,\ 191.8°+360°n,$

$258.2°+360°n,\text{where }n\text{ is any integer}\}$ or

$\{11.8°+180°n,\ 78.2°+180°n,$

where n is any integer$\}$

49. $2\cos^2 2\theta=1-\cos2\theta$

$2\cos^2 2\theta+\cos2\theta-1=0$

$(2\cos2\theta-1)(\cos2\theta+1)=0$

$0\le\theta<360°\Rightarrow0°\le2\theta<720°,$ so

$2\cos2\theta-1=0\Rightarrow2\cos2\theta=1\Rightarrow$

$\cos2\theta=\dfrac{1}{2}.$

Thus, $2\theta=60°,300°,420°,660°\Rightarrow$

$\theta=30°,150°,210°,330°$ or

$\cos2\theta+1=0\Rightarrow\cos2\theta=-1$

$2\theta=180°,540°\Rightarrow\theta=90°,270°$

Solution set:

$\{30°+360°n,\ 90°+360°n,$

$150°+360°n,\ 210°+360°n,\ 270°+360°n,$

$330°+360°n,\text{where }n\text{ is any integer}\}$ or

$\{30°+180°n,\ 90°+180°n,\ 150°+180°n,$

where n is any integer$\}$

51. $\sin\dfrac{x}{2}-\cos\dfrac{x}{2}=0\Rightarrow\sin\dfrac{x}{2}=\cos\dfrac{x}{2}\Rightarrow$

$\dfrac{x}{2}=\dfrac{\pi}{4}\Rightarrow x=\dfrac{\pi}{2}$

Solution set: $\left\{\dfrac{\pi}{2}\right\}$

53. $\tan 2x + \sec 2x = 3$

Let $u = 2x$. Then,

$\tan 2x + \sec 2x = 3 \Rightarrow \tan u + \sec u = 3 \Rightarrow$

$\sec u = 3 - \tan u \Rightarrow$

$\sec^2 u = 9 - 6 \tan u + \tan^2 u \Rightarrow$

$1 + \tan^2 u = 9 - 6 \tan u + \tan^2 u \Rightarrow$

$6 \tan u = 8 \Rightarrow \tan u = \dfrac{4}{3} \Rightarrow$

$u \approx 0.92730,\ 0.92730 + \pi,\ 0.92730 + 2\pi \Rightarrow$

$x \approx 0.4636,\ 0.4636 + \dfrac{\pi}{2},\ 0.4636 + \pi$

The solutions were obtained by squaring, so we must check each solution.

$\tan(2 \cdot 0.4636) + \sec(2 \cdot 0.4636) \overset{?}{=} 3$

$3 = 3$ True

Thus, $x = 0.4636$ is a solution.

$\tan\left[2\left(0.4636 + \dfrac{\pi}{2}\right)\right]$

$+ \sec\left[2\left(0.4636 + \dfrac{\pi}{2}\right)\right] \overset{?}{=} 3$

$-0.3 = 3$ False

Thus, $x = 0.4636 + \dfrac{\pi}{2} = 1.2490$ is not a solution.

$\tan\left[2\left(0.4636 + \pi\right)\right] + \sec\left[2\left(0.4636 + \pi\right)\right] \overset{?}{=} 3$

$3 = 3$ True

Thus, $x \approx 0.4636 + \pi \approx 3.6052$ is a solution.
Solution set: $\{0.4636,\ 3.6052\}$

55. The x-intercept method is shown below.

$y_1 = 2\sin 2x - x^3 + 1$ is graphed in the window $[0,\ 2\pi] \times [-4,\ 4]$.

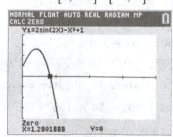

Solution set: $\{1.2802\}$

57. **(a)**

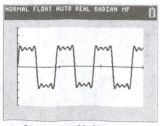

$[0, 0.03] \times [0.005, 0.005]$

$P =$

$$0.003\sin 220\pi t + \dfrac{0.003}{3}\sin 660\pi t$$
$$+ \dfrac{0.003}{5}\sin 1100\pi t + \dfrac{0.003}{7}\sin 1540\pi t$$

(b) The graph is periodic, and the wave has "jagged square" tops and bottoms.

(c) The eardrum is moving outward when $P < 0$.

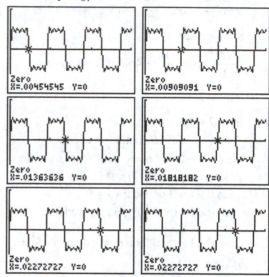

This occurs for the time intervals
$(0.0045, 0.0091)$, $(0.0136, 0.0182)$,
$(0.0227, 0.0273)$.

59. For $x = t$,

$$P(t) = \dfrac{1}{2}\sin[2\pi(220)t] + \dfrac{1}{3}\sin[2\pi(330)t] + \dfrac{1}{4}\sin[2\pi(440)t]$$

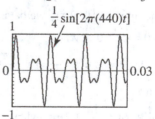

(b) Answers may vary slightly depending on the calculator settings.
$0.0007569,\ 0.009849,\ 0.01894,\ 0.02803$

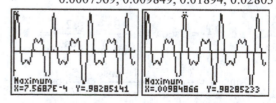

(continued on next page)

(*continued*)

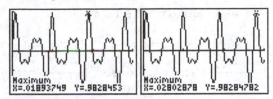

(c) 110 Hz

(d) For $x = t$,

$$P(t) = \sin[2\pi(110)t] + \frac{1}{2}\sin[2\pi(220)t] +$$
$$\frac{1}{3}\sin[2\pi(330)t] + \frac{1}{4}\sin[2\pi(440)t]$$

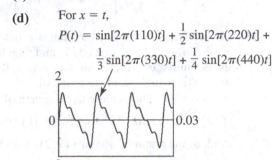

61. (a) $14\sin\left[\dfrac{\pi}{6}(x-4)\right] + 50 = 64$

$$14\sin\left[\frac{\pi}{6}(x-4)\right] = 14$$

$$\sin\left[\frac{\pi}{6}(x-4)\right] = 1$$

$$\frac{\pi}{6}(x-4) = \frac{\pi}{2}$$

$$x - 4 = 3 \Rightarrow x = 7$$

The average monthly temperature is 64°F in the seventh month, July.

(b) $14\sin\left[\dfrac{\pi}{6}(x-4)\right] + 50 = 39$

$$14\sin\left[\frac{\pi}{6}(x-4)\right] = -11$$

$$\sin\left[\frac{\pi}{6}(x-4)\right] = -\frac{11}{14}$$

$$\sin^{-1}\left(-\frac{11}{14}\right) = \frac{\pi}{6}(x-4)$$

$$\frac{6}{\pi}\left[\sin^{-1}\left(-\frac{11}{14}\right)\right] = x - 4$$

$$\frac{6}{\pi}\left[\sin^{-1}\left(-\frac{11}{14}\right)\right] + 4 = x \approx 2.3 \text{ or}$$

$$\frac{6}{\pi}\left[\pi - \sin^{-1}\left(-\frac{11}{14}\right)\right] + 4 = x \approx 11.7$$

The average monthly temperature is 39°F in the second month, February, and in the eleventh month, November.

63. $i = I_{\max}\sin 2\pi ft$

Let $i = 40$, $I_{\max} = 100$, $f = 60$.

$$40 = 100\sin\left[2\pi(60)t\right]$$
$$40 = 100\sin 120\pi t \Rightarrow 0.4 = \sin 120\pi t$$

Using calculator,

$$120\pi t \approx 0.4115168 \Rightarrow t \approx \frac{0.4115168}{120\pi} \Rightarrow$$

$$t \approx 0.0010916 \Rightarrow t \approx 0.001 \text{ sec}$$

65. $i = I_{\max}\sin 2\pi ft$

Let $i = I_{\max}$, $f = 60$.

$$I_{\max} = I_{\max}\sin\left[2\pi(60)t\right] \Rightarrow 1 = \sin 120\pi t \Rightarrow$$

$$120\pi t = \frac{\pi}{2} \Rightarrow 120t = \frac{1}{2} \Rightarrow t = \frac{1}{240} \approx 0.004 \text{ sec}$$

Chapter 6 Quiz
(Sections 6.1–6.3)

1. Domain: $[-1, 1]$; range: $[0, \pi]$

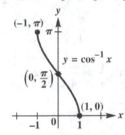

3. (a) $\theta = \arccos 0.92341853 \approx 22.568922°$

(b) $\theta = \cot^{-1}(-1.08886767) \approx 137.431085°$

5. $2\sin\theta - \sqrt{3} = 0 \Rightarrow 2\sin\theta = \sqrt{3} \Rightarrow \sin\theta = \dfrac{\sqrt{3}}{2}$

Over the interval $[0°, 360°)$, the equation $\sin\theta = \dfrac{\sqrt{3}}{2}$ has two solutions, the angles in quadrants I and II that have a reference angle of 60°. These are 60° and 120°.
Solution set: {60°, 120°}

7. $V = \cos 2\pi t, 0 \le t \le \dfrac{1}{2}$

(a) $V = 1, \cos 2\pi t = 1 \Rightarrow 2\pi t = \cos^{-1}1 \Rightarrow$

$$2\pi t = 0 \Rightarrow t = \frac{0}{2\pi} = 0 \text{ sec}$$

(b) $V = 0.30, \cos 2\pi t = 0.30$

$$2\pi t = \cos^{-1}0.30 \Rightarrow$$
$$2\pi t \approx 1.266103673$$
$$t = \frac{1.266103673}{2\pi} = 0.20 \text{ sec}$$

9. $3\cot 2x - \sqrt{3} = 0 \Rightarrow \cot 2x = \dfrac{\sqrt{3}}{3}$

$0 \le x < 2\pi \Rightarrow 0 \le 2x < 4\pi$, so,

$2x = \dfrac{\pi}{3}, \dfrac{4\pi}{3}, \dfrac{7\pi}{3}, \dfrac{10\pi}{3}$ or

$x = \dfrac{\pi}{6}, \dfrac{2\pi}{3}, \dfrac{7\pi}{6}, \dfrac{5\pi}{3}.$

Solution set: $\left\{ \dfrac{\pi}{6}, \dfrac{2\pi}{3}, \dfrac{7\pi}{6}, \dfrac{5\pi}{3} \right\}$

Section 6.4 Equations Involving Inverse Trigonometric Functions

1. C   **3.** C   **5.** A

7. $y = 5\cos x \Rightarrow \dfrac{y}{5} = \cos x \Rightarrow x = \arccos \dfrac{y}{5}$,

$0 \le x \le \pi$

9. $y = 3\tan 2x \Rightarrow \dfrac{y}{3} = \tan 2x \Rightarrow 2x = \arctan \dfrac{y}{3} \Rightarrow$

$x = \dfrac{1}{2}\arctan \dfrac{y}{3}, \ -\dfrac{\pi}{4} < x < \dfrac{\pi}{4}$

11. $y = 6\cos \dfrac{x}{4} \Rightarrow \dfrac{y}{6} = \cos \dfrac{x}{4} \Rightarrow \dfrac{x}{4} = \arccos \dfrac{y}{6} \Rightarrow$

$x = 4\arccos \dfrac{y}{6}, \ 0 \le x \le 4\pi$

13. $y = -2\cos 5x \Rightarrow -\dfrac{y}{2} = \cos 5x \Rightarrow$

$5x = \arccos\left(-\dfrac{y}{2}\right) \Rightarrow x = \dfrac{1}{5}\arccos\left(-\dfrac{y}{2}\right),$

$0 \le x \le \dfrac{\pi}{5}$

15. $y = \sin x - 2 \Rightarrow y + 2 = \sin x \Rightarrow$

$x = \arcsin(y+2), \ -\dfrac{\pi}{2} \le x \le \dfrac{\pi}{2}$

17. $y = -4 + 2\sin x \Rightarrow y + 4 = 2\sin x \Rightarrow$

$\dfrac{y+4}{2} = \sin x \Rightarrow x = \arcsin\left(\dfrac{y+4}{2}\right),$

$-\dfrac{\pi}{2} \le x \le \dfrac{\pi}{2}$

19. $y = \dfrac{1}{2}\cot 3x \Rightarrow 2y = \cot 3x \Rightarrow$

$3x = \text{arccot }2y \Rightarrow x = \dfrac{1}{3}\text{arccot }2y, \ 0 < x < \dfrac{\pi}{3}$

21. $y = \cos(x+3) \Rightarrow x + 3 = \arccos y \Rightarrow$

$x = -3 + \arccos y, \ -3 \le x \le \pi - 3$

23. $y = \sqrt{2} + 3\sec 2x \Rightarrow y - \sqrt{2} = 3\sec 2x \Rightarrow$

$\dfrac{y - \sqrt{2}}{3} = \sec 2x \Rightarrow 2x = \sec^{-1}\left(\dfrac{y - \sqrt{2}}{3}\right) \Rightarrow$

$x = \dfrac{1}{2}\sec^{-1}\left(\dfrac{y - \sqrt{2}}{3}\right), \ 0 \le x \le \dfrac{\pi}{2}, \ x \ne \dfrac{\pi}{4}$

25. The argument of the sine function is x, not $x - 2$. To solve for x, first add 2, and then use the definition of arcsine. Another way to think about this is to think of the graph of $y = \sin x - 2$. This represents the graph of $f(x) = \sin x$, shifted 2 units down. If you think of the graph of $y = \sin(x - 2)$, which represents the graph of $f(x) = \sin x$, shifted 2 units right. The graphs aren't the same, so $\sin x - 2 \ne \sin(x - 2)$.

27. $-4\arcsin x = \pi \Rightarrow \arcsin x = -\dfrac{\pi}{4} \Rightarrow$

$x = \sin\left(-\dfrac{\pi}{4}\right) = -\dfrac{\sqrt{2}}{2}$

Solution set: $\left\{ -\dfrac{\sqrt{2}}{2} \right\}$

29. $\dfrac{4}{3}\cos^{-1}\dfrac{x}{4} = \pi \Rightarrow \cos^{-1}\dfrac{x}{4} = \dfrac{3\pi}{4} \Rightarrow$

$\dfrac{x}{4} = \cos\dfrac{3\pi}{4} \Rightarrow \dfrac{yx}{4} = -\dfrac{\sqrt{2}}{2} \Rightarrow x = -2\sqrt{2}$

Solution set: $\left\{ -2\sqrt{2} \right\}$

31. $2\arccos\left(\dfrac{x}{3} - \dfrac{\pi}{3}\right) = 2\pi \Rightarrow$

$\arccos\left(\dfrac{x}{3} - \dfrac{\pi}{3}\right) = \pi \Rightarrow \dfrac{x}{3} - \dfrac{\pi}{3} = \cos \pi$

$\dfrac{x}{3} - \dfrac{\pi}{3} = -1 \Rightarrow x - \pi = -3 \Rightarrow x = \pi - 3$

Solution set: $\{\pi - 3\}$

33. $\arcsin x = \arctan \dfrac{3}{4}$

Let $\arctan \dfrac{3}{4} = u$, so $\tan u = \dfrac{3}{4}$, u is in

quadrant I. Sketch a triangle and label it. The

hypotenuse is $\sqrt{3^2 + 4^2} = \sqrt{9 + 16} = \sqrt{25} = 5$.

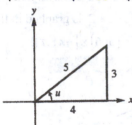

Therefore, $\sin u = \dfrac{3}{r} = \dfrac{3}{5}$. This equation

becomes $\arcsin x = u$, or $x = \sin u$. Thus,

$x = \dfrac{3}{5}$. Solution set: $\left\{ \dfrac{3}{5} \right\}$

35. $\cos^{-1} x = \sin^{-1} \dfrac{3}{5}$

Let $\sin^{-1} \dfrac{3}{5} = u$, so $\sin u = \dfrac{3}{5}$. Sketch a

triangle and label it. The hypotenuse is

$\sqrt{3^2 + 4^2} = \sqrt{9 + 16} = \sqrt{25} = 5$.

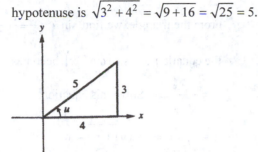

Therefore, $\cos u = \dfrac{4}{5}$. The equation becomes

$\cos^{-1} x = u$, or $x = \cos u$. Thus, $x = \dfrac{4}{5}$.

Solution set: $\left\{ \dfrac{4}{5} \right\}$

37. $\sin^{-1} x - \tan^{-1} 1 = -\dfrac{\pi}{4}$

$\sin^{-1} x - \tan^{-1} 1 = -\dfrac{\pi}{4} \Rightarrow$

$\sin^{-1} x = \tan^{-1} 1 - \dfrac{\pi}{4} \Rightarrow \sin^{-1} x = \dfrac{\pi}{4} - \dfrac{\pi}{4} \Rightarrow$

$\sin^{-1} x = 0 \Rightarrow \sin 0 = x \Rightarrow x = 0$

Solution set: $\{0\}$

39. $\arccos x + 2 \arcsin \dfrac{\sqrt{3}}{2} = \pi$

$\arccos x + 2 \arcsin \dfrac{\sqrt{3}}{2} = \pi \Rightarrow$

$\arccos x = \pi - 2 \arcsin \dfrac{\sqrt{3}}{2} \Rightarrow$

$\arccos x = \pi - 2 \left(\dfrac{\pi}{3} \right)$

$\arccos x = \pi - \dfrac{2\pi}{3} \Rightarrow \arccos x = \dfrac{\pi}{3} \Rightarrow$

$x = \cos \dfrac{\pi}{3} \Rightarrow x = \dfrac{1}{2}$

Solution set: $\left\{ \dfrac{1}{2} \right\}$

41. $\arcsin 2x + \arccos x = \dfrac{\pi}{6}$

$\arcsin 2x + \arccos x = \dfrac{\pi}{6}$

$\arcsin 2x = \dfrac{\pi}{6} - \arccos x$

$2x = \sin \left(\dfrac{\pi}{6} - \arccos x \right)$

Use the identity

$\sin (A - B) = \sin A \cos B - \cos A \sin B$.

$2x = \sin \dfrac{\pi}{6} \cos (\arccos x) - \cos \dfrac{\pi}{6} \sin (\arccos x)$

Let $u = \arccos x$. Thus, $\cos u = x = \dfrac{x}{1}$.

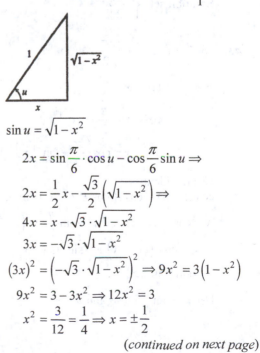

$\sin u = \sqrt{1 - x^2}$

$2x = \sin \dfrac{\pi}{6} \cdot \cos u - \cos \dfrac{\pi}{6} \sin u \Rightarrow$

$2x = \dfrac{1}{2} x - \dfrac{\sqrt{3}}{2} \left(\sqrt{1 - x^2} \right) \Rightarrow$

$4x = x - \sqrt{3} \cdot \sqrt{1 - x^2}$

$3x = -\sqrt{3} \cdot \sqrt{1 - x^2}$

$(3x)^2 = \left(-\sqrt{3} \cdot \sqrt{1 - x^2} \right)^2 \Rightarrow 9x^2 = 3 \left(1 - x^2 \right)$

$9x^2 = 3 - 3x^2 \Rightarrow 12x^2 = 3$

$x^2 = \dfrac{3}{12} = \dfrac{1}{4} \Rightarrow x = \pm \dfrac{1}{2}$

(*continued on next page*)

(*continued*)

Check these proposed solutions because they were found by squaring both side of an equation.

Check $x = \dfrac{1}{2}$.

$$\arcsin 2x + \arccos x = \dfrac{\pi}{6}$$

$$\arcsin\left(2 \cdot \dfrac{1}{2}\right) + \arccos\left(\dfrac{1}{2}\right) = \dfrac{\pi}{6} \ ?$$

$$\dfrac{\pi}{2} + \dfrac{\pi}{3} = \dfrac{\pi}{6} \ ?$$

$$\dfrac{5\pi}{6} = \dfrac{\pi}{6} \quad \text{False}$$

$\dfrac{1}{2}$ is not a solution.

Check $x = -\dfrac{1}{2}$.

$$\arcsin 2x + \arccos x = \dfrac{\pi}{6}$$

$$\arcsin\left(2 \cdot -\dfrac{1}{2}\right) + \arccos\left(-\dfrac{1}{2}\right) = \dfrac{\pi}{6} \ ?$$

$$-\dfrac{\pi}{2} + \dfrac{2\pi}{3} = \dfrac{\pi}{6} \ ?$$

$$\dfrac{\pi}{6} = \dfrac{\pi}{6} \quad \text{True}$$

$-\dfrac{1}{2}$ is a solution.

Solution set: $\left\{-\dfrac{1}{2}\right\}$

43. $\cos^{-1} x + \tan^{-1} x = \dfrac{\pi}{2}$

$$\cos^{-1} x + \tan^{-1} x = \dfrac{\pi}{2}$$

$$\cos^{-1} x = \dfrac{\pi}{2} - \tan^{-1} x$$

$$x = \cos\left(\dfrac{\pi}{2} - \tan^{-1} x\right)$$

Use the identity
$\cos(A - B) = \cos A \cos B + \sin A \sin B.$

$$x = \cos\dfrac{\pi}{2}\cos\left(\tan^{-1} x\right) + \sin\dfrac{\pi}{2}\sin\left(\tan^{-1} x\right)$$

$$x = 0 \cdot \cos\left(\tan^{-1} x\right) + 1 \cdot \sin\left(\tan^{-1} x\right)$$

$$x = \sin\left(\tan^{-1} x\right)$$

Let $u = \tan^{-1} x$. So, $\tan u = x$.

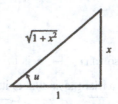

From the triangle, we find $\sin u = \dfrac{x}{\sqrt{1+x^2}}$, so

the equation $x = \sin\left(\tan^{-1} x\right)$ becomes

$$x = \dfrac{x}{\sqrt{1+x^2}}. \text{ Solve this equation.}$$

$$x = \dfrac{x}{\sqrt{1+x^2}} \Rightarrow x\sqrt{1+x^2} = x$$

$$x\sqrt{1+x^2} - x = 0 \Rightarrow x\left(\sqrt{1+x^2} - 1\right) = 0$$

$$x = 0 \text{ or } \sqrt{1+x^2} - 1 = 0 \Rightarrow \sqrt{1+x^2} = 1 \Rightarrow$$

$$1 + x^2 = 1 \Rightarrow x^2 = 0 \Rightarrow x = 0$$

Solution set: $\{0\}$

45. $y_1 = \sin^{-1} x - \cos^{-1} x - \dfrac{\pi}{6}$

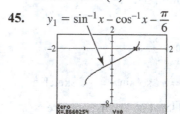

47. The x-intercept method is shown below.

$y_1 = \left(\arctan x\right)^3 - x + 2$ is graphed in the

window $[0, 6] \times [0, 6] \times [-\pi, \pi]$.

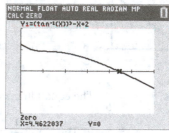

Solution set: $\{4.4622\}$

49. $A = \sqrt{\begin{array}{c}(A_1\cos\phi_1 + A_2\cos\phi_2)^2 \\ + (A_1\sin\phi_1 + A_2\sin\phi_2)^2\end{array}}$ and

$\phi = \arctan\left(\dfrac{A_1\sin\phi_1 + A_2\sin\phi_2}{A_1\cos\phi + A_2\cos\phi_2}\right)$

Make sure your calculator is in radian mode.

(a) Let $A_1 = 0.0012$, $\phi_1 = 0.052$, $A_2 = 0.004$, and $\phi_2 = 0.61$.

$A = \sqrt{\begin{array}{c}(0.0012\cos .052 + 0.004\cos 0.61)^2 \\ + (0.0012\sin 0.052 + 0.004\sin 0.61)^2\end{array}}$

≈ 0.00506

$\phi = \arctan\left(\dfrac{0.0012\sin 0.052 + 0.004\sin 0.61}{0.0012\cos 0.052 + 0.004\cos 0.61}\right)$

≈ 0.484

If $f = 220$, then $P = A\sin(2\pi ft + \phi)$

becomes $P = 0.00506\sin(440\pi t + 0.484)$.

(b) For $x = t$,
$P(t) = 0.00506\sin(440\pi t + 0.484)$
$P_1(t) + P_2(t) = 0.0012\sin(440\pi t + 0.052)$
$+ 0.004\sin(440\pi t + 0.61)$

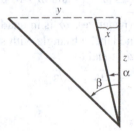

The two graphs are the same.

51. (a) $\tan\alpha = \dfrac{x}{z}$ and $\tan\beta = \dfrac{x+y}{z}$

(b) Because

$\tan\alpha = \dfrac{x}{z} \Rightarrow z\tan\alpha = x \Rightarrow z = \dfrac{x}{\tan\alpha}$

and $\tan\beta = \dfrac{x+y}{z} \Rightarrow z\tan\beta = x+y \Rightarrow$

$z = \dfrac{x+y}{\tan\beta}$, we have $\dfrac{x}{\tan\alpha} = \dfrac{x+y}{\tan\beta}$

(c) $(x+y)\tan\alpha = x\tan\beta \Rightarrow$

$\tan\alpha = \dfrac{x\tan\beta}{x+y} \Rightarrow \alpha = \arctan\left(\dfrac{x\tan\beta}{x+y}\right)$

(d) $x\tan\beta = (x+y)\tan\alpha$

$\tan\beta = \dfrac{(x+y)\tan\alpha}{x}$

$\beta = \arctan\left(\dfrac{(x+y)\tan\alpha}{x}\right)$

53. (a) $E = E_{max}\sin 2\pi ft \Rightarrow \dfrac{E}{E_{max}} = \sin 2\pi ft \Rightarrow$

$2\pi ft = \arcsin\dfrac{E}{E_{max}} \Rightarrow$

$t = \dfrac{1}{2\pi f}\arcsin\dfrac{E}{E_{max}}$

(b) Let $E_{max} = 12$, $E = 5$, and $f = 100$.

$t = \dfrac{1}{2\pi(100)}\arcsin\dfrac{5}{12}$

$= \dfrac{1}{200\pi}\arcsin\dfrac{5}{12} \approx 0.00068\sec$

55. $y = \dfrac{1}{3}\sin\dfrac{4\pi t}{3}$

(a) $3y = \sin\dfrac{4\pi t}{3} \Rightarrow \dfrac{4\pi t}{3} = \arcsin 3y \Rightarrow$

$4\pi t = 3\arcsin 3y \Rightarrow t = \dfrac{3}{4\pi}\arcsin 3y$

(b) If $y = 0.3$ radian,

$t = \dfrac{3}{4\pi}\arcsin 0.9 \Rightarrow t \approx 0.27\sec$.

Chapter 6   Review Exercises

1.

Domain: $[-1, 1]$

Range: $\left[-\dfrac{\pi}{2}, \dfrac{\pi}{2}\right]$

$y = \sin^{-1}x$

Domain: $[-1, 1]$

Range: $[0, \pi]$

$y = \cos^{-1}x$

(continued on next page)

(*continued*)

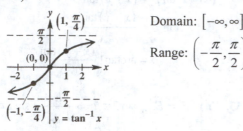

Domain: $[-\infty, \infty]$

Range: $\left(-\dfrac{\pi}{2}, \dfrac{\pi}{2}\right)$

3. False. $\arcsin\left(-\dfrac{1}{2}\right) = -\dfrac{\pi}{6}$, not $\dfrac{11\pi}{6}$.

5. $y = \sin^{-1}\dfrac{\sqrt{2}}{2} \Rightarrow \sin y = \dfrac{\sqrt{2}}{2}$

$-\dfrac{\pi}{2} \leq y \leq \dfrac{\pi}{2}$, so $y = \dfrac{\pi}{4}$.

7. $y = \tan^{-1}\left(-\sqrt{3}\right) \Rightarrow \tan y = -\sqrt{3}$

$-\dfrac{\pi}{2} < y < \dfrac{\pi}{2}$, so $y = -\dfrac{\pi}{3}$.

9. $y = \cos^{-1}\left(-\dfrac{\sqrt{2}}{2}\right) \Rightarrow \cos y = -\dfrac{\sqrt{2}}{2}$

$0 \leq y \leq \pi$, so $y = \dfrac{3\pi}{4}$.

11. $y = \sec^{-1}(-2) \Rightarrow \sec y = -2$

$0 \leq y \leq \pi$ and $y \neq \dfrac{\pi}{2}$, so $y = \dfrac{2\pi}{3}$.

13. $y = \operatorname{arccot}(-1) \Rightarrow \cot y = -1$

$0 < y < \pi$, so $y = \dfrac{3\pi}{4}$.

15. $\theta = \arcsin\left(-\dfrac{\sqrt{3}}{2}\right) \Rightarrow \sin\theta = -\dfrac{\sqrt{3}}{2}$

$-90° \leq \theta \leq 90°$, so $\theta = -60°$.

For Exercises 17–21, be sure that your calculator is in degree mode.

17. $\theta = \arctan 1.7804675 = 60.67924514°$

19. $\theta = \cos^{-1} 0.80396577 \approx 36.4895081°$

21. $\theta = \operatorname{arc\,sec} 3.4723155 \approx 73.26220613°$

23. $\cos\left(\arccos(-1)\right) = \cos\pi = -1$ or

$\cos\left(\arccos(-1)\right) = \cos 180° = -1$

25. $\arccos\left(\cos\dfrac{3\pi}{4}\right) = \arccos\left(-\dfrac{\sqrt{2}}{2}\right) = \dfrac{3\pi}{4}$

27. $\tan^{-1}\left(\tan\dfrac{\pi}{4}\right) = \tan^{-1}\dfrac{\sqrt{2}}{2} = \dfrac{\pi}{4}$

29. $\sin\left(\arccos\dfrac{3}{4}\right)$

Let $\omega = \arccos\dfrac{3}{4}$, so that $\cos\omega = \dfrac{3}{4}$.

Because arccos is defined only in quadrants I and II, and $\dfrac{3}{4}$ is positive, ω is in quadrant I.

Sketch ω and label a triangle with the side opposite ω equal to

$$\sqrt{4^2 - 3^2} = \sqrt{16 - 9} = \sqrt{7}.$$

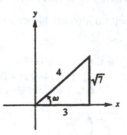

$$\sin\left(\arccos\dfrac{3}{4}\right) = \sin\omega = \dfrac{\sqrt{7}}{4}$$

31. $\cos\left(\csc^{-1}(-2)\right)$

Let $\omega = \csc^{-1}(-2)$, so that $\csc\omega = -2$.

Because $-\dfrac{\pi}{2} \leq \omega \leq \dfrac{\pi}{2}$ and $\omega \neq 0$, and $\csc\omega = -2$ (negative), ω is in quadrant IV. Sketch ω and label a triangle with side adjacent to ω equal to

$$\sqrt{2^2 - (-1)^2} = \sqrt{4 - 1} = \sqrt{3}.$$

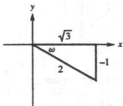

$$\cos\left(\csc^{-1}(-2)\right) = \cos\omega = \dfrac{\sqrt{3}}{2}$$

33. $\tan\left(\arcsin\dfrac{3}{5} + \arccos\dfrac{5}{7}\right)$

Let $\omega_1 = \arcsin\dfrac{3}{5}$, $\omega_2 = \arccos\dfrac{5}{7}$. Sketch

angles ω_1 and ω_2 in quadrant I. The side

adjacent to ω_1 is

$\sqrt{5^2 - 3^2} = \sqrt{25-9} = \sqrt{16} = 4$. The side

opposite ω_2 is

$\sqrt{7^2 - 5^2} = \sqrt{49-25} = \sqrt{24} = 2\sqrt{6}$.

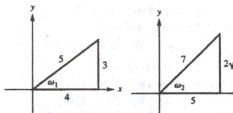

We have $\tan\omega_1 = \dfrac{3}{4}$ and $\tan\omega_2 = \dfrac{2\sqrt{6}}{5}$.

$\tan\left(\arcsin\dfrac{3}{5} + \arccos\dfrac{5}{7}\right)$

$= \tan(\omega_1 + \omega_2) = \dfrac{\tan\omega_1 + \tan\omega_2}{1 - \tan\omega_1 \tan\omega_2}$

$= \dfrac{\dfrac{3}{4} + \dfrac{2\sqrt{6}}{5}}{1 - \left(\dfrac{3}{4}\right)\left(\dfrac{2\sqrt{6}}{5}\right)} = \dfrac{\dfrac{15 + 8\sqrt{6}}{20}}{\dfrac{20 - 6\sqrt{6}}{20}}$

$= \dfrac{15 + 8\sqrt{6}}{20 - 6\sqrt{6}} = \dfrac{15 + 8\sqrt{6}}{20 - 6\sqrt{6}} \cdot \dfrac{20 + 6\sqrt{6}}{20 + 6\sqrt{6}}$

$= \dfrac{588 + 250\sqrt{6}}{184} = \dfrac{294 + 125\sqrt{6}}{92}$

35. $\tan\left(\operatorname{arcsec}\dfrac{\sqrt{u^2 + 1}}{u}\right)$

Let $\theta = \operatorname{arcsec}\dfrac{\sqrt{u^2 + 1}}{u}$, so $\sec\theta = \dfrac{\sqrt{u^2 + 1}}{u}$.

If $u > 0$, $0 < \theta < \dfrac{\pi}{2}$.

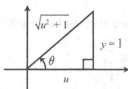

From the Pythagorean theorem,

$y = \sqrt{\left(\sqrt{u^2 + 1}\right)^2 - u^2} = \sqrt{u^2 + 1 - u^2} = \sqrt{1} = 1$.

Therefore $\tan\theta = \dfrac{1}{u}$. Thus,

$\tan\left(\operatorname{arcsec}\dfrac{\sqrt{u^2 + 1}}{u}\right) = \dfrac{1}{u}$.

37. $2\tan x - 1 = 0$

$2\tan x - 1 = 0 \Rightarrow 2\tan x = 1 \Rightarrow \tan x = \dfrac{1}{2}$

Over the interval $[0, 2\pi)$, the equation

$\tan x = \dfrac{1}{2}$ has two solutions. One solution is

in quadrant I and the other is in quadrant III.
Using a calculator, the quadrant I solution is
approximately 0.4636. The quadrant III
solution would be approximately
$0.4636 + \pi \approx 3.6052$.
Solution set: $\{0.4636, 3.6052\}$

39. $\tan x = \cot x$

Use the identity $\cot x = \dfrac{1}{\tan x}$, $\tan x \neq 0$.

$\tan x = \cot x \Rightarrow \tan x = \dfrac{1}{\tan x} \Rightarrow$

$\tan^2 x = 1 \Rightarrow \tan x = \pm 1$

Over the interval $[0, 2\pi)$, the equation
$\tan x = 1$ has two solutions. One solution is in
quadrant I and the other is in quadrant III.

These solutions are $\dfrac{\pi}{4}$ and $\dfrac{5\pi}{4}$. In the same

interval, the equation $\tan x = -1$ has two
solutions. One solution is in quadrant II and
the other is in quadrant IV. These solutions

are $\dfrac{3\pi}{4}$ and $\dfrac{7\pi}{4}$. Solution set:

$\left\{\dfrac{\pi}{4}, \dfrac{3\pi}{4}, \dfrac{5\pi}{4}, \dfrac{7\pi}{4}\right\}$

41. $\tan^2 2x - 1 = 0$

$\tan^2 2x - 1 = 0 \Rightarrow \tan^2 2x = 1 \Rightarrow \tan 2x = \pm 1$

$0 \le x < 2\pi \Rightarrow 0 \le 2x < 4\pi$. Thus,

$2x = \dfrac{\pi}{4}, \dfrac{3\pi}{4}, \dfrac{5\pi}{4}, \dfrac{7\pi}{4}, \dfrac{9\pi}{4}, \dfrac{11\pi}{4}, \dfrac{13\pi}{4}, \dfrac{15\pi}{4}$

implies

$x = \dfrac{\pi}{8}, \dfrac{3\pi}{8}, \dfrac{5\pi}{8}, \dfrac{7\pi}{8}, \dfrac{9\pi}{8}, \dfrac{11\pi}{8}, \dfrac{13\pi}{8}, \dfrac{15\pi}{8}$.

Solution set:

$\left\{\dfrac{\pi}{8}, \dfrac{3\pi}{8}, \dfrac{5\pi}{8}, \dfrac{7\pi}{8}, \dfrac{9\pi}{8}, \dfrac{11\pi}{8}, \dfrac{13\pi}{8}, \dfrac{15\pi}{8}\right\}$

43. $\cos 2x + \cos x = 0$

$\cos 2x + \cos x = 0 \Rightarrow 2\cos^2 x - 1 + \cos x = 0 \Rightarrow$

$2\cos^2 x + \cos x - 1 = 0 \Rightarrow$

$(2\cos x - 1)(\cos x + 1) = 0$

$2\cos x - 1 = 0 \Rightarrow 2\cos x = 1 \Rightarrow \cos x = \dfrac{1}{2}$ or

$\cos x + 1 = 0 \Rightarrow \cos x = -1$

Over the interval $[0, 2\pi)$, the equation

$\cos x = \dfrac{1}{2}$ has two solutions. The angles in

quadrants I and IV that have a reference angle

of $\dfrac{\pi}{3}$ are $\dfrac{\pi}{3}$ and $\dfrac{5\pi}{3}$. In the same interval,

$\cos x = -1$ when the angle is π.

Solution set:

$\left\{ \dfrac{\pi}{3} + 2n\pi,\ \pi + 2n\pi,\ \dfrac{5\pi}{3} + 2n\pi, \right.$

where n is any integer $\Big\}$

45. $\sin^2 \theta + 3\sin \theta + 2 = 0$

$\sin^2 \theta + 3\sin \theta + 2 = 0$

$(\sin \theta + 2)(\sin \theta + 1) = 0$

In the interval $[0°, 360°)$, we have

$\sin \theta + 1 = 0 \Rightarrow \sin \theta = -1 \Rightarrow \theta = 270°$ and

$\sin \theta + 2 = 0 \Rightarrow \sin \theta = -2 < -1 \Rightarrow$ no solution

Solution set: $\{270°\}$

47. $\sin 2\theta = \cos 2\theta + 1$

$\sin 2\theta = \cos 2\theta + 1$

$(\sin 2\theta)^2 = (\cos 2\theta + 1)^2$

$\sin^2 2\theta = \cos^2 2\theta + 2\cos 2\theta + 1$

$1 - \cos^2 2\theta = \cos^2 2\theta + 2\cos 2\theta + 1$

$2\cos^2 2\theta + 2\cos 2\theta = 0$

$\cos^2 2\theta + \cos 2\theta = 0$

$\cos 2\theta (\cos 2\theta + 1) = 0$

$0° \le \theta < 360° \Rightarrow 0° \le 2\theta < 720°.$

$\cos 2\theta = 0 \Rightarrow 2\theta = 90°,\ 270°,\ 450°,\ 630° \Rightarrow$

$\theta = 45°,\ 135°,\ 225°,\ 315°$

$\cos 2\theta + 1 = 0 \Rightarrow \cos 2\theta = -1$

$2\theta = 180°,\ 540° \Rightarrow \theta = 90°,\ 270°$

Possible values for θ are

$\theta = 45°, 90°, 135°, 225°, 270°, 315°.$

All proposed solutions must be checked because the solutions were found by squaring an equation. A value for θ will be a solution if $\sin 2\theta - \cos 2\theta = 1$.

$\theta = 45°, 2\theta = 90° \Rightarrow$

$\sin 90° - \cos 90° = 1 - 0 = 1$

$\theta = 90°, 2\theta = 180° \Rightarrow$

$\sin 180° - \cos 180° = 0 - (-1) = 1$

$\theta = 135°, 2\theta = 270° \Rightarrow$

$\sin 270° - \cos 270° = -1 - 0 \ne 1$

$\theta = 225°, 2\theta = 450° \Rightarrow$

$\sin 450° - \cos 450° = 1 - 0 = 1$

$\theta = 270°, 2\theta = 540° \Rightarrow$

$\sin 540° - \cos 540° = 0 - (-1) = 1$

$\theta = 315°, 2\theta = 630° \Rightarrow$

$\sin 630° - \cos 630° = -1 - 0 \ne 1$

Thus, $\theta = 45°, 90°, 225°, 270°$.

Solution set: $\{45°, 90°, 225°, 270°\}$

49. $3\cos^2 \theta + 2\cos \theta - 1 = 0$

$(3\cos \theta - 1)(\cos \theta + 1) = 0$

In the interval $[0°, 360°)$, we have

$3\cos \theta - 1 = 0 \Rightarrow \cos \theta = \dfrac{1}{3} \Rightarrow$

$\theta \approx 70.5°$ and $289.5°$ (using a calculator)

$\cos \theta + 1 = 0 \Rightarrow \cos \theta = -1 \Rightarrow \theta = 180°$

Solution set: $\{70.5°, 180°, 289.5°\}$

51. $2\sqrt{3} \cos \dfrac{\theta}{2} = -3 \Rightarrow \cos \dfrac{\theta}{2} = -\dfrac{3}{2\sqrt{3}} = -\dfrac{\sqrt{3}}{2} \Rightarrow$

$\dfrac{\theta}{2} = \cos^{-1}\left(-\dfrac{\sqrt{3}}{2} \right) \Rightarrow \theta = 2\cos^{-1}\left(-\dfrac{\sqrt{3}}{2} \right) \Rightarrow$

$\theta = 2(150°) + 2(360°)n$ or

$\theta = 2(210°) + 2(360°)n \Rightarrow$

$\theta = 300° + 720°n$ or $\theta = 420° + 720°n$

Solution set: $\{300° + 720°n,\ 420° + 720°n\}$

53. $\tan \theta - \sec \theta = 1 \Rightarrow \tan \theta = 1 + \sec \theta \Rightarrow$

$\tan^2 \theta = 1 + 2\sec \theta + \sec^2 \theta \Rightarrow$

$\tan^2 \theta = 1 + 2\sec \theta + 1 + \tan^2 \theta \Rightarrow$

$-2 = 2\sec \theta \Rightarrow \sec \theta = -1 \Rightarrow \cos \theta = -1 \Rightarrow$

$\theta = 180°$

Check the proposed solution because it was obtained by squaring.

$\tan 180° - \sec 180° \overset{?}{=} 1$

$0 - (-1) \overset{?}{=} 1$

$1 = 1$ True

Solution set: $\{180° + 360°n\}$

55. $\dfrac{4}{3}\arctan\dfrac{x}{2}=\pi\Rightarrow\arctan\dfrac{x}{2}=\dfrac{3\pi}{4}$

But, by definition, the range of arctan is

$\left(-\dfrac{\pi}{2},\dfrac{\pi}{2}\right)$. So, this equation has no solution.

Solution set: $\varnothing$

57. $\arccos x+\arctan 1=\dfrac{11\pi}{12}$

$\arccos x+\arctan 1=\dfrac{11\pi}{12}$

$\arccos x=\dfrac{11\pi}{12}-\arctan 1$

$\arccos x=\dfrac{11\pi}{12}-\dfrac{\pi}{4}=\dfrac{8\pi}{12}=\dfrac{2\pi}{3}$

$\cos\dfrac{2\pi}{3}=x\Rightarrow x=-\dfrac{1}{2}$

Solution set: $\left\{-\dfrac{1}{2}\right\}$

59. $y=\dfrac{1}{2}\sin x\Rightarrow 2y=\sin x\Rightarrow x=\sin^{-1}(2y)$,

$-\dfrac{\pi}{2}\le x\le\dfrac{\pi}{2}$

61. $y=\dfrac{1}{2}\tan(3x+2)\Rightarrow 2y=\tan(3x+2)\Rightarrow$

$3x+2=\arctan 2y\Rightarrow 3x=\arctan 2y-2\Rightarrow$

$x=\left(\dfrac{1}{3}\arctan 2y\right)-\dfrac{2}{3},\ -\dfrac{2}{3}-\dfrac{\pi}{6}<x<-\dfrac{2}{3}+\dfrac{\pi}{6}$

63. (a) Let α be the angle to the left of θ.

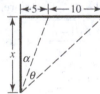

Thus, we have

$\tan(\alpha+\theta)=\dfrac{5+10}{x}$

$\alpha+\theta=\tan^{-1}\left(\dfrac{15}{x}\right)$

$\theta=\tan^{-1}\left(\dfrac{15}{x}\right)-\alpha$

$\theta=\tan^{-1}\left(\dfrac{15}{x}\right)-\tan^{-1}\left(\dfrac{5}{x}\right)$

(b) The maximum occurs at approximately 8.66026 ft. There may be a discrepancy in the final digits.

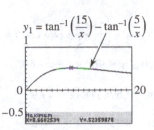

65. If $\theta_1>48.8°$, then $\theta_2>90°$ and the light beam is completely underwater.

Chapter 6   Test

1.

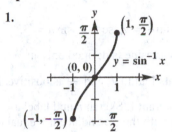

Domain: $[-1,1]$; range: $\left[-\dfrac{\pi}{2},\dfrac{\pi}{2}\right]$

2. (a) $y=\arccos\left(-\dfrac{1}{2}\right)\Rightarrow\cos y=-\dfrac{1}{2}$

$0\le y\le\pi$, so $y=\dfrac{2\pi}{3}$.

(b) $y=\sin^{-1}\left(-\dfrac{\sqrt{3}}{2}\right)\Rightarrow\sin y=-\dfrac{\sqrt{3}}{2}$

$-\dfrac{\pi}{2}\le y\le\dfrac{\pi}{2}$, so $y=-\dfrac{\pi}{3}$.

(c) $y=\tan^{-1}0\Rightarrow\tan y=0$

$-\dfrac{\pi}{2}<y<\dfrac{\pi}{2}$, so $y=0$.

(d) $y=\text{arcsec}(-2)\Rightarrow\sec y=-2$

$0\le y\le\pi$ and $y\ne\dfrac{\pi}{2}$, so $y=\dfrac{2\pi}{3}$.

3. (a) $\theta=\arccos\dfrac{\sqrt{3}}{2}\Rightarrow\cos\theta=\dfrac{\sqrt{3}}{2}$

$0\le y\le 180°$, so $y=30°$.

(b) $\theta=\tan^{-1}(-1)\Rightarrow\tan\theta=-1$

$-90°<y<90°$, so $y=-45°$.

(c) $\theta=\cot^{-1}(-1)\Rightarrow\cot\theta=-1$

$0°<y<180°$, so $y=135°$.

(d) $\theta = \csc^{-1}\left(-\dfrac{2\sqrt{3}}{3}\right) \Rightarrow \csc\theta = -\dfrac{2\sqrt{3}}{3}$

$-90 \le y \le 90°,$ so $\theta = -60°.$

4. (a) $\sin^{-1} 0.69431882 \approx 43.97°$

(b) $\sec^{-1} 1.0840880 \approx 22.72°$

(c) $\cot^{-1}(-0.7125586) \approx 125.47°$

5. (a) $\cos\left(\arcsin\dfrac{2}{3}\right)$

Let $\arcsin\dfrac{2}{3} = u$, so that $\sin u = \dfrac{2}{3}$.
Because arcsine is defined only in
quadrants I and IV, and $\dfrac{2}{3}$ is positive, u
is in quadrant I. Sketch u and label a
triangle with the side adjacent u equal to
$\sqrt{3^2 - 2^2} = \sqrt{9-4} = \sqrt{5}.$

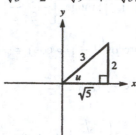

$\cos\left(\arcsin\dfrac{2}{3}\right) = \cos u = \dfrac{\sqrt{5}}{3}$

(b) $\sin\left(2\cos^{-1}\dfrac{1}{3}\right)$

Let $\theta = \cos^{-1}\dfrac{1}{3}$, so that $\cos\theta = \dfrac{1}{3}$.
Because arccosine is defined only in
quadrants I and II, and $\dfrac{1}{3}$ is positive, θ
is in quadrant I. Sketch θ and label a
triangle with the side opposite to θ equal
to $\theta = \sqrt{3^2 - (-1)^2} = \sqrt{9-1} = \sqrt{8} = 2\sqrt{2}.$

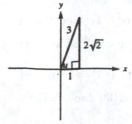

Thus, $\sin\theta = \dfrac{2\sqrt{2}}{3}$ and

$\sin\left(2\cos^{-1}\dfrac{1}{3}\right) = \sin 2\theta.$

$\sin\left(2\cos^{-1}\dfrac{1}{3}\right) = \sin 2\theta = 2\sin\theta\cos\theta$

$= 2\left(\dfrac{2\sqrt{2}}{3}\right)\left(\dfrac{1}{3}\right) = \dfrac{4\sqrt{2}}{9}$

6. There is no value of θ for which $\sin\theta = 3$
because $-1 \le \sin\theta \le 1$. Thus, $\sin^{-1} 3$ is not
defined.

7. $\arcsin\left(\sin\dfrac{5\pi}{6}\right) = \arcsin\left(\dfrac{1}{2}\right) = \dfrac{\pi}{6} \ne \dfrac{5\pi}{6}$

8. $\tan(\arcsin u)$

Let $\theta = \arcsin u$, so $\sin\theta = u = \dfrac{u}{1}$. If $u > 0$,

$0 < \theta < \dfrac{\pi}{2}.$

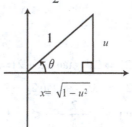

From the Pythagorean theorem,
$x = \sqrt{1^2 - u^2} = \sqrt{1 - u^2}$. Therefore,

$\tan\theta = \dfrac{u}{\sqrt{1-u^2}} = \dfrac{u\sqrt{1-u^2}}{1-u^2}.$ Thus,

$\tan(\arcsin u) = \dfrac{u\sqrt{1-u^2}}{1-u^2}.$

9. $-3\sec\theta + 2\sqrt{3} = 0 \Rightarrow \sec\theta = \dfrac{2\sqrt{3}}{3}$

Over the interval $[0, 360°)$, the equation

$\sec x = \dfrac{2\sqrt{3}}{3}$ has two solutions. One solution
is in quadrant I and the other is in quadrant
IV. These solutions are 30° and 330°.
Solution set: {30°, 330°}

10. $\sin^2\theta = \cos^2\theta + 1$

$\sin^2\theta = 1 - \sin^2\theta + 1$

$2\sin^2\theta = 2 \Rightarrow \sin^2\theta = 1 \Rightarrow \sin\theta = \pm 1$

Over the interval $[0, 360°)$, the equation $\sin\theta = 1$ has one solution, 90°. Over the interval $[0, 2\pi)$, the equation $\sin\theta = -1$ has one solution, 270°. Solution set: $\{90°, 270°\}$

11. $\csc^2\theta - 2\cot\theta = 4$

$1 + \cot^2\theta - 2\cot\theta = 4$

$\cot^2\theta - 2\cot\theta - 3 = 0$

$(\cot\theta - 3)(\cot\theta + 1) = 0 \Rightarrow$

$\cot\theta = 3$ or $\cot\theta = -1$

Over the interval $[0, 360°)$, the equation $\cot\theta = 3$ has two solutions, 18.4° and 198.4° (found using a calculator.) Over the interval $[0, 2\pi)$, the equation $\cot\theta = -1$ has two solutions, 135° and 315°.
Solution set: $\{18.4°, 135°, 198.4°, 315°\}$

12. $\cos x = \cos 2x$

$\cos x = 2\cos^2 x - 1$

$2\cos^2 x - \cos x - 1 = 0$

$(2\cos x + 1)(\cos x - 1) = 0 \Rightarrow$

$\cos x = -\dfrac{1}{2}$ or $\cos x = 1$

Over the interval $[0, 2\pi)$, the equation

$\cos x = -\dfrac{1}{2}$ has two solutions, $\dfrac{2\pi}{3}$ and $\dfrac{4\pi}{3}$.

Over the interval $[0, 2\pi)$, the equation

$\cos x = 1$ has one solution, 0.

Solution set: $\left\{0, \dfrac{2\pi}{3}, \dfrac{4\pi}{3}\right\}$

13. $\sqrt{2}\cos 3x - 1 = 0 \Rightarrow \cos 3x = \dfrac{1}{\sqrt{2}} = \dfrac{\sqrt{2}}{2}$

$0 \le x < 2\pi \Rightarrow 0 \le 3x < 6\pi.$ Thus

$3x = \dfrac{\pi}{4}, \dfrac{7\pi}{4}, \dfrac{9\pi}{4}, \dfrac{15\pi}{4}, \dfrac{17\pi}{4}, \dfrac{23\pi}{4} \Rightarrow$

$x = \dfrac{\pi}{12}, \dfrac{7\pi}{12}, \dfrac{3\pi}{4}, \dfrac{5\pi}{4}, \dfrac{17\pi}{12}, \dfrac{23\pi}{12}$

Solution set: $\left\{\dfrac{\pi}{12}, \dfrac{7\pi}{12}, \dfrac{3\pi}{4}, \dfrac{5\pi}{4}, \dfrac{17\pi}{12}, \dfrac{23\pi}{12}\right\}$

14. $\sin x \cos x = \dfrac{1}{3} \Rightarrow 2\sin x \cos x = \dfrac{2}{3} \Rightarrow$

$\sin 2x = \dfrac{2}{3}$

$0 \le x < 2\pi \Rightarrow 0 \le 2x < 4\pi.$ Use a calculator to find $2x$.

$2x \approx 0.72672, 2.4118, 7.0129, 8.69505 \Rightarrow$

$x \approx 0.3649, 1.2059, 3.5065, 4.3475$

Solution set: $\{0.3649, 1.2059, 3.5065, 4.3475\}$

15. $\sin^2\theta = -\cos 2\theta$

$\sin^2\theta = -\left(1 - 2\sin^2\theta\right)$

$\sin^2\theta = 1 \Rightarrow \sin\theta = \pm 1$

Over the interval $[0, 360°)$, the equation $\sin\theta = 1$ has one solution, 90°. Over the interval $[0, 360°)$, the equation $\sin\theta = -1$ has one solution, 270°. Because 270° = 90° + 180°, the solution set is $\{90° + 180°n$, where n is any integer$\}$.

16. $2\sqrt{3}\sin\dfrac{x}{2} = 3 \Rightarrow \sin\dfrac{x}{2} = \dfrac{3}{2\sqrt{3}} \Rightarrow$

$\dfrac{x}{2} = \sin^{-1}\dfrac{3}{2\sqrt{3}} \Rightarrow \dfrac{x}{2} = \sin^{-1}\dfrac{\sqrt{3}}{2}$

Since $0 \le x < 2\pi \Rightarrow 0° \le \dfrac{x}{2} < \pi$, we have

$\dfrac{x}{2} = \dfrac{\pi}{3}, \dfrac{2\pi}{3} \Rightarrow x = \dfrac{2\pi}{3}, \dfrac{4\pi}{3}$

Solution set:

$\left\{\dfrac{2\pi}{3} + 2\pi n, \dfrac{4\pi}{3} + 2\pi n, \right.$

where n is any integer$\Big\}$

17. $\csc x - \cot x = 1 \Rightarrow \csc x = 1 + \cot x \Rightarrow$

$\csc^2 x = 1 + 2\cot x + \cot^2 x$

$\csc^2 x = 2\cot x + \csc^2 x \Rightarrow 0 = 2\cot x \Rightarrow$

$\cot x = 0 \Rightarrow x = \dfrac{\pi}{2}, \dfrac{3\pi}{2}$

Because the solutions were obtained by squaring, we must check them.

$\csc\dfrac{\pi}{2} - \cot\dfrac{\pi}{2} \overset{?}{=} 1$

$1 - 0 \overset{?}{=} 1$

$1 = 1$ True

$\csc\dfrac{3\pi}{2} - \cot\dfrac{3\pi}{2} \overset{?}{=} 1$

$-1 - 0 \overset{?}{=} 1$

$-1 = 1$ False

Solution set:

$\left\{\dfrac{\pi}{2} + 2n\pi, \text{ where } n \text{ is any integer}\right\}$

18. (a) $y = \cos 3x \Rightarrow 3x = \arccos y \Rightarrow$

$x = \dfrac{1}{3}\arccos y,\ 0 \le x \le \dfrac{\pi}{3}$

(b) $y = 4 + 3\cot x \Rightarrow y - 4 = 3\cot x \Rightarrow$

$\dfrac{y-4}{3} = \cot x \Rightarrow x = \cot^{-1}\left(\dfrac{y-4}{3}\right),$

$0 < x < \pi$

19. (a) $\arcsin x = \arctan \dfrac{4}{3}$

Let $\omega = \arctan \dfrac{4}{3}$. Then $\tan \omega = \dfrac{4}{3}$.

Sketch ω in quadrant I and label a
triangle with the hypotenuse equal to

$\sqrt{4^2 + 3^2} = \sqrt{16+9} = \sqrt{25} = 5.$

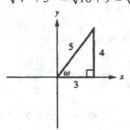

Thus, we have

$\arcsin x = \arctan \dfrac{4}{3} \Rightarrow \arcsin x = \omega \Rightarrow$

$x = \sin \omega = \dfrac{4}{5}$

Solution set: $\left\{\dfrac{4}{5}\right\}$

(b) $\operatorname{arccot} x + 2\arcsin \dfrac{\sqrt{3}}{2} = \pi \Rightarrow$

$\operatorname{arccot} x + 2\left(\dfrac{\pi}{3}\right) = \pi \Rightarrow \operatorname{arccot} x = \dfrac{\pi}{3} \Rightarrow$

$x = \cot \dfrac{\pi}{3} = \dfrac{1}{\tan \frac{\pi}{3}} = \dfrac{1}{\sqrt{3}} = \dfrac{\sqrt{3}}{3}$

Solution set: $\left\{\dfrac{\sqrt{3}}{3}\right\}$

20.

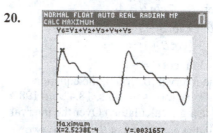

P first reaches its maximum at
approximately 2.5×10^{-4}. The maximum is
approximately 0.003166.

Chapter 7

Applications of Trigonometry and Vectors

Section 7.1 Oblique Triangles and the Law of Sines

1. A triangle that is not a right triangle is an **oblique** triangle.

3. If we know three **angles** of a triangle, we cannot find a unique solution for the triangle.

5. An alternative form of the law of sines is
$$\frac{\sin A}{a} = \frac{\sin B}{b} = \frac{\sin C}{c}.$$

7. Yes, the law of sines may be used.

9. No, there is insufficient information to use the law of sines.

11. The measure of angle C is
$$180° - (60° + 75°) = 180° - 135° = 45°.$$

$$\frac{a}{\sin A} = \frac{c}{\sin C} \Rightarrow \frac{a}{\sin 60°} = \frac{\sqrt{2}}{\sin 45°} \Rightarrow$$

$$a = \frac{\sqrt{2}\sin 60°}{\sin 45°} = \frac{\sqrt{2} \cdot \frac{\sqrt{3}}{2}}{\frac{\sqrt{2}}{2}} = \sqrt{2} \cdot \frac{\sqrt{3}}{2} \cdot \frac{2}{\sqrt{2}}$$

$$= \sqrt{3}$$

13. $A = 37°$, $B = 48°$, $c = 18$ m
$$C = 180° - A - B \Rightarrow$$
$$C = 180° - 37° - 48° = 95°$$
$$\frac{b}{\sin B} = \frac{c}{\sin C} \Rightarrow \frac{b}{\sin 48°} = \frac{18}{\sin 95°} \Rightarrow$$
$$b = \frac{18\sin 48°}{\sin 95°} \approx 13 \text{ m}$$
$$\frac{a}{\sin A} = \frac{c}{\sin C} \Rightarrow \frac{a}{\sin 37°} = \frac{18}{\sin 95°} \Rightarrow$$
$$a = \frac{18\sin 37°}{\sin 95°} \approx 11 \text{ m}$$

15. $A = 27.2°$, $C = 115.5°$, $c = 76.0$ ft
$$B = 180° - A - C \Rightarrow$$
$$B = 180° - 27.2° - 115.5° = 37.3°$$
$$\frac{a}{\sin A} = \frac{c}{\sin C} \Rightarrow \frac{a}{\sin 27.2°} = \frac{76.0}{\sin 115.5°} \Rightarrow$$
$$a = \frac{76.0\sin 27.2°}{\sin 115.5°} \approx 38.5 \text{ ft}$$

$$\frac{b}{\sin B} = \frac{c}{\sin C} \Rightarrow \frac{b}{\sin 37.3°} = \frac{76.0}{\sin 115.5°} \Rightarrow$$
$$b = \frac{76.0\sin 37.3°}{\sin 115.5°} \approx 51.0 \text{ ft}$$

17. $A = 68.41°$, $B = 54.23°$, $a = 12.75$ ft
$$C = 180° - A - B \Rightarrow$$
$$C = 180° - 68.41° - 54.23° = 57.36°$$
$$\frac{a}{\sin A} = \frac{b}{\sin B} \Rightarrow \frac{12.75}{\sin 68.41°} = \frac{b}{\sin 54.23°} \Rightarrow$$
$$b = \frac{12.75\sin 54.23°}{\sin 68.41°} \approx 11.13 \text{ ft}$$
$$\frac{a}{\sin A} = \frac{c}{\sin C} \Rightarrow \frac{12.75}{\sin 68.41°} = \frac{c}{\sin 57.36°} \Rightarrow$$
$$c = \frac{12.75\sin 57.36°}{\sin 68.41°} \approx 11.55 \text{ ft}$$

19. $A = 87.2°$, $b = 75.9$ yd, $C = 74.3°$
$$B = 180° - A - C \Rightarrow$$
$$B = 180° - 87.2° - 74.3° = 18.5°$$
$$\frac{a}{\sin A} = \frac{b}{\sin B} \Rightarrow \frac{a}{\sin 87.2°} = \frac{75.9}{\sin 18.5°} \Rightarrow$$
$$a = \frac{75.9\sin 87.2°}{\sin 18.5°} \approx 239 \text{ yd}$$
$$\frac{b}{\sin B} = \frac{c}{\sin C} \Rightarrow \frac{75.9}{\sin 18.5°} = \frac{c}{\sin 74.3°} \Rightarrow$$
$$c = \frac{75.9\sin 74.3°}{\sin 18.5°} \approx 230 \text{ yd}$$

21. $B = 20°50'$, $AC = 132$ ft, $C = 103°10'$
$$A = 180° - B - C$$
$$A = 180° - 20°50' - 103°10' \Rightarrow A = 56°00'$$
$$\frac{AC}{\sin B} = \frac{AB}{\sin C} \Rightarrow \frac{132}{\sin 20°50'} = \frac{AB}{\sin 103°10'} \Rightarrow$$
$$AB = \frac{132\sin 103°10'}{\sin 20°50'} \approx 361 \text{ ft}$$
$$\frac{BC}{\sin A} = \frac{AC}{\sin B} \Rightarrow \frac{BC}{\sin 56°00'} = \frac{132}{\sin 20°50'} \Rightarrow$$
$$BC = \frac{132\sin 56°00'}{\sin 20°50'} \approx 308 \text{ ft}$$

23. $A = 39.70°, C = 30.35°, b = 39.74$ m

$B = 180° - A - C \Rightarrow$
$B = 180° - 39.70° - 30.35° \Rightarrow$
$B = 109.95° \approx 110.0°$ (rounded)

$$\frac{a}{\sin A} = \frac{b}{\sin B} \Rightarrow \frac{a}{\sin 39.70°} = \frac{39.74}{\sin 109.95°} \Rightarrow$$
$$a = \frac{39.74 \sin 39.70°}{\sin 109.95°} \approx 27.01 \text{ m}$$

$$\frac{b}{\sin B} = \frac{c}{\sin C} \Rightarrow \frac{39.74}{\sin 109.95°} = \frac{c}{\sin 30.35°} \Rightarrow$$
$$c = \frac{39.74 \sin 30.35°}{\sin 109.95°} \approx 21.36 \text{ m}$$

25. $B = 42.88°, C = 102.40°, b = 3974$ ft

$A = 180° - B - C \Rightarrow$
$A = 180° - 42.88° - 102.40° = 34.72°$

$$\frac{a}{\sin A} = \frac{b}{\sin B} \Rightarrow \frac{a}{\sin 34.72°} = \frac{3974}{\sin 42.88°} \Rightarrow$$
$$a = \frac{3974 \sin 34.72°}{\sin 42.88°} \approx 3326 \text{ ft}$$

$$\frac{b}{\sin B} = \frac{c}{\sin C} \Rightarrow \frac{3974}{\sin 42.88°} = \frac{c}{\sin 102.40°} \Rightarrow$$
$$c = \frac{3974 \sin 102.40°}{\sin 42.88°} \approx 5704 \text{ ft}$$

27. $A = 39°54', a = 268.7$ m, $B = 42°32'$

$C = 180° - A - B \Rightarrow$
$C = 180° - 39°54' - 42°32' = 97°34'$

$$\frac{a}{\sin A} = \frac{b}{\sin B} \Rightarrow \frac{268.7}{\sin 39°54'} = \frac{b}{\sin 42°32'} \Rightarrow$$
$$b = \frac{268.7 \sin 42°32'}{\sin 39°54'} \approx 283.2 \text{ m}$$

$$\frac{a}{\sin A} = \frac{c}{\sin C} \Rightarrow \frac{268.7}{\sin 39°54'} = \frac{c}{\sin 97°54'} \Rightarrow$$
$$c = \frac{268.7 \sin 97°54'}{\sin 39°54'} \approx 415.2 \text{ m}$$

29. Answers may vary. Sample answer: To use the law of sines, we must know an angle measure, the length of the side opposite it, and at least one other angle measure or side length.

31. Answers may vary. Sample answer: If two angles and a side are given, the third angle can be determined using the angle sum formula. Then, the ASA congruence axiom can be applied. This triangle is uniquely determined because there is only one possible triangle that meets these initial conditions.

33. $B = 112°10'; C = 15°20'; BC = 354$ m

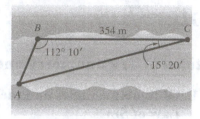

$A = 180° - B - C$
$A = 180° - 112°10' - 15°20'$
$\quad = 179°60' - 127°30' = 52°30'$

$$\frac{BC}{\sin A} = \frac{AB}{\sin C}$$
$$\frac{354}{\sin 52°30'} = \frac{AB}{\sin 15°20'}$$
$$AB = \frac{354 \sin 15°20'}{\sin 52°30'} \approx 118 \text{ m}$$

35. Let d = the distance the ship traveled between the two observations; L = the location of the lighthouse.
$L = 180° - 38.8° - 44.2°$
$\quad = 97.0°$

$$\frac{d}{\sin 97°} = \frac{12.5}{\sin 44.2°}$$
$$d = \frac{12.5 \sin 97°}{\sin 44.2°}$$
$$\approx 17.8 \text{ km}$$

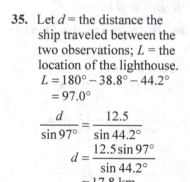
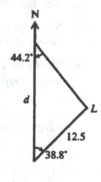

37. Let x = the distance to the lighthouse at bearing N 37° E; y = the distance to the lighthouse at bearing N 25° E.
$\theta = 180° - 37° = 143°$
$\alpha = 180° - \theta - 25° = 180° - 143° - 25° = 12°$

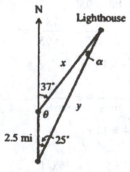

$$\frac{2.5}{\sin \alpha} = \frac{x}{\sin 25°} \Rightarrow x = \frac{2.5 \sin 25°}{\sin 12°} \approx 5.1 \text{ mi}$$

$$\frac{2.5}{\sin \alpha} = \frac{y}{\sin \theta} \Rightarrow \frac{2.5}{\sin 12°} = \frac{y}{\sin 143°} \Rightarrow$$
$$y = \frac{2.5 \sin 143°}{\sin 12°} \approx 7.2 \text{ mi}$$

39. Let A = the location of the balloon;
B = the location of the farther town;
C = the location of the closer town.
Angle $ABC = 31°$ and angle $ACB = 35°$
because the angles of depression are alternate
interior angles with the angles of the triangle.

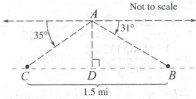
Not to scale

Angle $BAC = 180° - 31° - 35° = 114°$

$$\frac{1.5}{\sin BAC} = \frac{AB}{\sin ACB} \Rightarrow \frac{1.5}{\sin 114°} = \frac{AB}{\sin 35°} \Rightarrow$$

$$AB = \frac{1.5\sin 35°}{\sin 114°} \approx 0.94178636$$

$$\sin ABC = \frac{AD}{AB} \Rightarrow \sin 31° = \frac{AD}{0.94178636} \Rightarrow$$

$$AD = 0.94178636 \cdot \sin 31° \approx 0.49$$

The balloon is 0.49 mi above the ground.

41. We cannot find θ directly because the length
of the side opposite angle θ is not given.
Redraw the triangle shown in the figure to the
right and label the third angle as α.

$$\frac{\sin\alpha}{1.6+2.7} = \frac{\sin 38°}{1.6+3.6}$$

$$\frac{\sin\alpha}{4.3} = \frac{\sin 38°}{5.2}$$

$$\sin\alpha = \frac{4.3\sin 38°}{5.2} \approx 0.50910468$$

$$\alpha \approx \sin^{-1}(0.50910468) \approx 31°$$

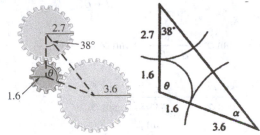

Thus,
$\theta = 180° - 38° - \alpha \approx 180° - 38° - 31° = 111°$

43. Angle C is equal to the difference between the
angles of elevation.
$C = B - A = 52.7430° - 52.6997° = 0.0433°$
The distance BC to the moon can be
determined using the law of sines.

$$\frac{BC}{\sin A} = \frac{AB}{\sin C}$$

$$\frac{BC}{\sin 52.6997°} = \frac{398}{\sin 0.0433°}$$

$$BC = \frac{398\sin 52.6997°}{\sin 0.0433°}$$
$$BC \approx 418,930 \text{ km}$$

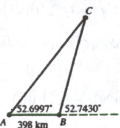

To find AC, we have

$$\frac{AC}{\sin B} = \frac{AB}{\sin C}$$

$$\frac{AC}{\sin(180° - 52.7430°)} = \frac{398}{\sin 0.0433°}$$

$$\frac{AC}{\sin 127.2570°} = \frac{398}{\sin 0.0433°}$$

$$AC = \frac{398\sin 127.2570°}{\sin 0.0433°}$$
$$\approx 419,171 \text{ km}$$

In either case the distance is approximately
419,000 km compared to the actual value of
406,000 km.

45. We need to determine the length of CB, which
is the sum of CD and DB.

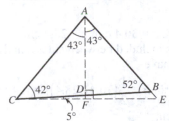

Triangle ACE is isosceles, so

$$m\angle ACE = m\angle AEC = \frac{180° - 86°}{2} = 47°. \text{ It}$$

follows that $m\angle ACD = 47° - 5° = 42°$.
$m\angle CAD = m\angle DAB = 43°$ because the
photograph was taken with no tilt and AD is
the bisector of $\angle CAB$. The length of AD is
3500 ft, and
$m\angle ABC = 180° - (86° + 42°) = 52°$. Using the
law of sines, we have

$$CD = \frac{3500\sin 43°}{\sin 42°} \approx 3567.3 \text{ ft and}$$

$$DB = \frac{3500\sin 43°}{\sin 52°} \approx 3029.1 \text{ ft. Thus,}$$

$$CB = CD + DB \approx 3567.3 + 3029.1$$
$$\approx 6596.4 \approx 6600 \text{ ft}$$

47. To find the area of the triangle, use $A = \frac{1}{2}bh$, with $b = 1$ and $h = \sqrt{3}$. $A = \frac{1}{2}(1)(\sqrt{3}) = \frac{\sqrt{3}}{2}$

Now use $A = \frac{1}{2}ab\sin C$, with $a = \sqrt{3}$, $b = 1$, and $C = 90°$.

$A = \frac{1}{2}(\sqrt{3})(1)\sin 90° = \frac{1}{2}(\sqrt{3})(1)(1) = \frac{\sqrt{3}}{2}$

49. To find the area of the triangle, use $A = \frac{1}{2}bh$, with $b = 1$ and $h = \sqrt{2}$. $A = \frac{1}{2}(1)(\sqrt{2}) = \frac{\sqrt{2}}{2}$

Now use $A = \frac{1}{2}ab\sin C$, with $a = 2$, $b = 1$, and $C = 45°$.

$A = \frac{1}{2}(2)(1)\sin 45° = \frac{1}{2}(2)(1)\left(\frac{\sqrt{2}}{2}\right) = \frac{\sqrt{2}}{2}$

51. $A = 42.5°$, $b = 13.6$ m, $c = 10.1$ m
Angle A is included between sides b and c.
Thus, we have

$A = \frac{1}{2}bc\sin A = \frac{1}{2}(13.6)(10.1)\sin 42.5°$
≈ 46.4 m^2

53. $B = 124.5°$, $a = 30.4$ cm, $c = 28.4$ cm
Angle B is included between sides a and c.
Thus, we have

$A = \frac{1}{2}ac\sin B = \frac{1}{2}(30.4)(28.4)\sin 124.5°$
≈ 356 cm^2

55. $A = 56.80°$, $b = 32.67$ in., $c = 52.89$ in.
Angle A is included between sides b and c.
Thus, we have

$A = \frac{1}{2}bc\sin A = \frac{1}{2}(32.67)(52.89)\sin 56.80°$
≈ 722.9 in.^{2}

57. $A = 30.50°$, $b = 13.00$ cm, $C = 112.60°$
In order to use the area formula, we need to find either a or c.
$B = 180° - A - C \Rightarrow$
$B = 180° - 30.50° - 112.60° = 36.90°$
Finding a:

$\frac{a}{\sin A} = \frac{b}{\sin B} \Rightarrow \frac{a}{\sin 30.5°} = \frac{13.00}{\sin 36.90°} \Rightarrow$

$a = \frac{13.00\sin 30.5°}{\sin 36.90°} \approx 10.9890$ cm

$A = \frac{1}{2}ab\sin C$
$= \frac{1}{2}(10.9890)(13.00)\sin 112.60°$
≈ 65.94 cm^2

Finding c:

$\frac{b}{\sin B} = \frac{c}{\sin C} \Rightarrow \frac{13.00}{\sin 36.9°} = \frac{c}{\sin 112.6°} \Rightarrow$

$c = \frac{13.00\sin 112.6°}{\sin 36.9°} \approx 19.9889$ cm

$A = \frac{1}{2}bc\sin A = \frac{1}{2}(19.9889)(13.00)\sin 30.5°$
≈ 65.94 cm^2

59. $A = \frac{1}{2}ab\sin C$

$= \frac{1}{2}(16.1)(15.2)\sin 125° \approx 100$ m^2

61. $\frac{a}{\sin A} = \frac{b}{\sin B} = \frac{c}{\sin C} = 2r$ and $r = \frac{1}{2}$
(because the diameter is 1), so we have

$\frac{a}{\sin A} = \frac{b}{\sin B} = \frac{c}{\sin C} = 2\left(\frac{1}{2}\right) = 1$

Then, $a = \sin A$, $b = \sin B$, and $c = \sin C$.

63. Triangles ACD and BCD are right triangles, so

we have $\tan \alpha = \frac{x}{d + BC}$ and $\tan \beta = \frac{x}{BC}$.

$\tan \beta = \frac{x}{BC} \Rightarrow BC = \frac{x}{\tan \beta}$, so we can

substitute into $\tan \alpha = \frac{x}{d + BC}$ and solve for

x.

$\tan \alpha = \frac{x}{d + BC} \Rightarrow \tan \alpha = \frac{x}{d + \frac{x}{\tan \beta}} \Rightarrow$

$\frac{\sin \alpha}{\cos \alpha} = \frac{x}{d + \frac{x\cos \beta}{\sin \beta}} \cdot \frac{\sin \beta}{\sin \beta}$

$\frac{\sin \alpha}{\cos \alpha} = \frac{x\sin \beta}{d\sin \beta + x\cos \beta}$

$\sin \alpha(d\sin \beta + x\cos \beta) = x\sin \beta(\cos \alpha)$
$d\sin \alpha \sin \beta + x\sin \alpha \cos \beta = x\cos \alpha \sin \beta$

(continued on next page)

(continued)

$$d \sin \alpha \sin \beta = x \cos \alpha \sin \beta - x \sin \alpha \cos \beta$$
$$= -x \left(\sin \alpha \cos \beta - \cos \alpha \sin \beta \right)$$
$$= -x \sin \left(a - \beta \right) = x \sin \left[-(\alpha - \beta) \right]$$

$$d \sin \alpha \sin \beta = x \sin \left(\beta - \alpha \right)$$
$$\frac{d \sin \alpha \sin \beta}{\sin \left(\beta - \alpha \right)} = x$$

Section 7.2 The Ambiguous Case of the Law of Sines

1. A. These values lead to two possible measures for angle B.

$$\frac{\sin A}{a} = \frac{\sin B}{b} \Rightarrow \frac{\sin 50°}{19} = \frac{\sin B}{21} \Rightarrow$$

$$\sin B = \frac{0.7660 \cdot 21}{19} \approx 0.8467 \Rightarrow$$

$B \approx 57.9°$ or $B \approx 122.1°$

Choices B, C, and D all lead to unique triangles.

3. The vertical distance from the point $(3, 4)$ to the x-axis is 4.

(a) If L is more than 4, two triangles can be drawn. But h must be less than 5 for both triangles to be on the positive x-axis. So, $4 < L < 5$.

(b) If $L = 4$, then exactly one triangle is possible. If $L > 5$, then only one triangle is possible on the positive x-axis.

(c) If $L < 4$, then no triangle is possible, because the side of length L would not reach the x-axis.

5. $a = 50$, $b = 26$, $A = 95°$

$$\frac{\sin A}{a} = \frac{\sin B}{b} \Rightarrow \frac{\sin 95°}{50} = \frac{\sin B}{26} \Rightarrow$$

$$\sin B = \frac{26 \sin 95°}{50} \approx 0.51802124 \Rightarrow B \approx 31.2°$$

Another possible value for B is $180° - 21.2° = 148.8°$, but this is too large to be in a triangle that also has a 95° angle. Therefore, only one triangle is possible.

7. $a = 31$, $b = 26$, $B = 48°$

$$\frac{\sin A}{a} = \frac{\sin B}{b} \Rightarrow \frac{\sin A}{31} = \frac{\sin 48°}{26} \Rightarrow$$

$$\sin A = \frac{31 \sin 48°}{26} \approx 0.88605729 \Rightarrow A \approx 62.4°$$

Another possible value for A is $180° - 62.4° = 117.6°$. Therefore, two triangles are possible.

9. $c = 50$, $b = 61$, $C = 58°$

$$\frac{\sin C}{c} = \frac{\sin B}{b} \Rightarrow \frac{\sin 58°}{50} = \frac{\sin B}{61} \Rightarrow$$

$$\sin B = \frac{61 \sin 58°}{50} \approx 1.03461868$$

Because $\sin B > 1$ is impossible, no triangle is possible for the given parts.

11. $a = \sqrt{6}$, $b = 2$, $A = 60°$

$$\frac{\sin B}{b} = \frac{\sin A}{a} \Rightarrow \frac{\sin B}{2} = \frac{\sin 60°}{\sqrt{6}} \Rightarrow$$

$$\sin B = \frac{2 \sin 60°}{\sqrt{6}} = \frac{2 \cdot \frac{\sqrt{3}}{2}}{\sqrt{6}} = \frac{\sqrt{3}}{\sqrt{6}} = \sqrt{\frac{3}{6}}$$

$$= \sqrt{\frac{1}{2}} = \frac{1}{\sqrt{2}} \cdot \frac{\sqrt{2}}{2} = \frac{\sqrt{2}}{2}$$

$$\sin B = \frac{\sqrt{2}}{2} \Rightarrow B = 45°$$

There is another angle between 0° and 180° whose sine is $\frac{\sqrt{2}}{2}$: $180° - 45° = 135°$.

However, this is too large because $A = 60°$ and $60° + 135° = 195° > 180°$, so there is only one solution, $B = 45°$.

13. $A = 29.7°$, $b = 41.5$ ft, $a = 27.2$ ft

$$\frac{\sin B}{b} = \frac{\sin A}{a} \Rightarrow \frac{\sin B}{41.5} = \frac{\sin 29.7°}{27.2} \Rightarrow$$

$$\sin B = \frac{41.5 \sin 29.7°}{27.2} \approx 0.75593878$$

There are two angles B between 0° and 180° that satisfy the condition. Because $\sin B \approx 0.75593878$, to the nearest tenth value of B is $B_1 = 49.1°$. Supplementary angles have the same sine value, so another possible value of B is $B_2 = 180° - 49.1° = 130.9°$. This is a valid angle measure for this triangle because
$A + B_2 = 29.7° + 130.9° = 160.6° < 180°$.

Solving separately for angles C_1 and C_2 we have the following.

$$C_1 = 180° - A - B_1$$
$$= 180° - 29.7° - 49.1° = 101.2°$$
$$C_2 = 180° - A - B_2$$
$$= 180° - 29.7° - 130.9° = 19.4°$$

15. $C = 41°20'$, $b = 25.9$ m, $c = 38.4$ m

$$\frac{\sin B}{b} = \frac{\sin C}{c} \Rightarrow \frac{\sin B}{25.9} = \frac{\sin 41°20'}{38.4} \Rightarrow$$

$$\sin B = \frac{25.9 \sin 41°20'}{38.4} \approx 0.44545209$$

There are two angles B between $0°$ and $180°$ that satisfy the condition. Because $\sin B \approx 0.44545209$, $B_1 \approx 26.5° = 26°30'$. Supplementary angles have the same sine value, so another possible value of B is $B_2 = 180° - 26°30' = 153°30'$. This is not a valid angle measure for this triangle because $C + B_2 = 41°20' + 153°30' = 194°30' > 180°$.

Thus, $A = 180° - 26°30' - 41°20' = 112°10'$.

17. $B = 74.3°$, $a = 859$ m, $b = 783$ m

$$\frac{\sin A}{a} = \frac{\sin B}{b} \Rightarrow \frac{\sin A}{859} = \frac{\sin 74.3°}{783} \Rightarrow$$

$$\sin A = \frac{859 \sin 74.3°}{783} \approx 1.0561331$$

Because $\sin A > 1$ is impossible, no such triangle exists.

19. $A = 142.13°$, $b = 5.432$ ft, $a = 7.297$ ft

$$\frac{\sin B}{b} = \frac{\sin A}{a} \Rightarrow \sin B = \frac{b \sin A}{a} \Rightarrow$$

$$\sin B = \frac{5.432 \sin 142.13°}{7.297} \approx 0.45697580 \Rightarrow$$

$$B \approx 27.19°$$

Because angle A is obtuse, angle B must be acute, so this is the only possible value for B and there is one triangle with the given measurements.

$$C = 180° - A - B$$
$$= 180° - 142.13° - 27.19° = 10.68°$$

Thus, $B \approx 27.19°$ and $C \approx 10.68°$.

21. $A = 42.5°$, $a = 15.6$ ft, $b = 8.14$ ft

$$\frac{\sin B}{b} = \frac{\sin A}{a} \Rightarrow \frac{\sin B}{8.14} = \frac{\sin 42.5°}{15.6} \Rightarrow$$

$$\sin B = \frac{8.14 \sin 42.5°}{15.6} \approx 0.35251951$$

There are two angles B between $0°$ and $180°$ that satisfy the condition. Because $\sin B \approx 0.35251951$, to the nearest tenth value of B is $B_1 = 20.6°$. Supplementary angles have the same sine value, so another possible value of B is $B_2 = 180° - 20.6° = 159.4°$.

This is not a valid angle measure for this triangle because $A + B_2 = 42.5° + 159.4° = 201.9° > 180°$. Thus, $C = 180° - 42.5° - 20.6° = 116.9°$. Solving for c, we have the following.

$$\frac{c}{\sin C} = \frac{a}{\sin A} \Rightarrow \frac{c}{\sin 116.9°} = \frac{15.6}{\sin 42.5°} \Rightarrow$$

$$c = \frac{15.6 \sin 116.9°}{\sin 42.5°} \approx 20.6 \text{ ft}$$

23. $B = 72.2°$, $b = 78.3$ m, $c = 145$ m

$$\frac{\sin C}{c} = \frac{\sin B}{b} \Rightarrow \frac{\sin C}{145} = \frac{\sin 72.2°}{78.3} \Rightarrow$$

$$\sin C = \frac{145 \sin 72.2°}{78.3} \approx 1.7632026$$

Because $\sin C > 1$ is impossible, no such triangle exists.

25. $A = 38°40'$, $a = 9.72$ m, $b = 11.8$ m

$$\frac{\sin B}{b} = \frac{\sin A}{a} \Rightarrow \frac{\sin B}{11.8} = \frac{\sin 38°40'}{9.72} \Rightarrow$$

$$\sin B = \frac{11.8 \sin 38°40'}{9.72} \approx 0.75848811$$

There are two angles B between $0°$ and $180°$ that satisfy the condition. Because $\sin B \approx 0.75848811$, to the nearest tenth value of B is $B_1 = 49.3° \approx 49°20'$. Supplementary angles have the same sine value, so another possible value of B is

$$B_2 = 180° - 49°20'$$
$$= 179°60' - 49°20' = 130°40'$$

This is a valid angle measure for this triangle because

$$A + B_2 = 38°40' + 130°40' = 169°20' < 180°.$$

Solving separately for triangles AB_1C_1 and AB_2C_2 we have the following.

AB_1C_1 :

$$C_1 = 180° - A - B_1 = 180° - 38°40' - 49°20'$$
$$= 180° - 88°00' = 92°00'$$

$$\frac{c_1}{\sin C_1} = \frac{a}{\sin A} \Rightarrow \frac{c_1}{\sin 92°00'} = \frac{9.72}{\sin 38°40'} \Rightarrow$$

$$c_1 = \frac{9.72 \sin 92°00'}{\sin 38°40'} \approx 15.5 \text{ m}$$

AB_2C_2:

$$C_2 = 180° - A - B_2$$
$$= 180° - 38°40' - 130°40' = 10°40'$$

$$\frac{c_2}{\sin C_2} = \frac{a}{\sin A} \Rightarrow \frac{c_2}{\sin 10°40'} = \frac{9.72}{\sin 38°40'} \Rightarrow$$

$$c_2 = \frac{9.72 \sin 10°40'}{\sin 38°40'} \approx 2.88 \text{ m}$$

27. $A = 96.80°$, $b = 3.589$ ft, $a = 5.818$ ft

$$\frac{\sin B}{b} = \frac{\sin A}{a} \Rightarrow \frac{\sin B}{3.589} = \frac{\sin 96.80°}{5.818} \Rightarrow$$

$$\sin B = \frac{3.589 \sin 96.80°}{5.818} \approx 0.61253922$$

There are two angles B between $0°$ and $180°$ that satisfy the condition. Because $\sin B \approx 0.61253922$, $B_1 \approx 37.77°$. Supplementary angles have the same sine value, so another possible value of B is $B_2 = 180° - 37.77° = 142.23°$. This is not a valid angle measure for this triangle because $A + B_2 = 96.80° + 142.23° = 239.03° > 180°$. Thus $C = 180° - 96.80° - 37.77° = 45.43°$. Solving for c, we have the following.

$$\frac{c}{\sin C} = \frac{a}{\sin A} \Rightarrow \frac{c}{\sin 45.43°} = \frac{5.8518}{\sin 96.80°} \Rightarrow$$

$$c = \frac{5.8518 \sin 45.43°}{\sin 96.80°} \approx 4.174 \text{ ft}$$

29. $B = 39.68°$, $a = 29.81$ m, $b = 23.76$ m

$$\frac{\sin A}{a} = \frac{\sin B}{b} \Rightarrow \frac{\sin A}{29.81} = \frac{\sin 39.68°}{23.76} \Rightarrow$$

$$\sin A = \frac{29.81 \sin 39.68°}{23.76} \approx 0.80108002$$

There are two angles A between $0°$ and $180°$ that satisfy the condition.
Because $\sin A \approx 0.80108002$, to the nearest hundredth, the value of A is $A_1 = 53.23°$. Supplementary angles have the same sine value, so another possible value of A is $A_2 = 180° - 53.23° = 126.77°$. This is a valid angle measure for this triangle because $B + A_2 = 39.68° + 126.77° = 166.45° < 180°$. Solving separately for triangles A_1BC_1 and A_2BC_2 we have the following.

A_1BC_1:

$$C_1 = 180° - A_1 - B$$
$$= 180° - 53.23° - 39.68° = 87.09°$$

$$\frac{c_1}{\sin C_1} = \frac{b}{\sin B} \Rightarrow \frac{c_1}{\sin 87.09°} = \frac{23.76}{\sin 39.68°} \Rightarrow$$

$$c_1 = \frac{23.76 \sin 87.09°}{\sin 39.68°} \approx 37.16 \text{ m}$$

A_2BC_2:

$$C_2 = 180° - A_2 - B$$
$$= 180° - 126.77° - 39.68° = 13.55°$$

$$\frac{c_2}{\sin C_2} = \frac{b}{\sin B} \Rightarrow \frac{c_2}{\sin 13.55°} = \frac{23.76}{\sin 39.68°} \Rightarrow$$

$$c_2 = \frac{23.76 \sin 13.55°}{\sin 39.68°} \approx 8.719 \text{ m}$$

31. $a = \sqrt{5}$, $c = 2\sqrt{5}$, $A = 30°$

$$\frac{\sin C}{c} = \frac{\sin A}{a} \Rightarrow \frac{\sin C}{2\sqrt{5}} = \frac{\sin 30°}{\sqrt{5}} \Rightarrow$$

$$\sin C = \frac{2\sqrt{5} \sin 30°}{\sqrt{5}} = \frac{2\sqrt{5} \cdot \frac{1}{2}}{\sqrt{5}} = \frac{\sqrt{5}}{\sqrt{5}} = 1 \Rightarrow$$

$$C = 90°$$

This is a right triangle.

33. Answers will vary. Sample answer: Angle A is an obtuse angle, and because there can be only one obtuse angle in a triangle, the longest side, a, will be opposite this angle. However, we are given that $b > a$, so no triangle exists.

35. $Y = 43°30'$, $Z = 95°30'$, $XY = 960$ m

We are looking for XZ.

$$\frac{XZ}{\sin Y} = \frac{XY}{\sin Z} \Rightarrow \frac{XZ}{\sin 43°30'} = \frac{960}{\sin 95°30'} \Rightarrow$$

$$XZ = \frac{960 \sin 43°30'}{\sin 95°30'} \approx 664 \text{ m}$$

37. The height of the building is CD.

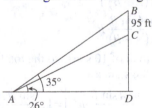

In right triangle ABD, we have $m\angle B = 90° - 35° = 55°$. In triangle ABC, we have $m\angle CAB = 35° - 26° = 9°$. This gives

$$\frac{BC}{\sin \angle CAB} = \frac{AC}{\sin \angle B} \Rightarrow \frac{95}{\sin 9°} = \frac{AC}{\sin 55°} \Rightarrow$$

$$AC = \frac{95 \sin 55°}{\sin 9°} \approx 497.5$$

In triangle ACD,

$$\sin \angle CAD = \frac{CD}{AC} \Rightarrow \sin 26° = \frac{CD}{497.5} \Rightarrow$$

$$CD = 497.5 \sin 26° \approx 218 \text{ ft}$$

39. Prove that $\dfrac{a+b}{b} = \dfrac{\sin A + \sin B}{\sin B}$.

Start with the law of sines.

$$\dfrac{a}{\sin A} = \dfrac{b}{\sin B} \Rightarrow a = \dfrac{b\sin A}{\sin B}$$

Substitute for a in the expression $\dfrac{a+b}{b}$.

$$\dfrac{a+b}{b} = \dfrac{\dfrac{b\sin A}{\sin B}+b}{b} = \dfrac{\dfrac{b\sin A}{\sin B}+b}{b}\cdot\dfrac{\sin B}{\sin B}$$

$$= \dfrac{b\sin A + b\sin B}{b\sin B} = \dfrac{b(\sin A + \sin B)}{\sin B}$$

$$= \dfrac{\sin A + \sin B}{\sin B}$$

41. We can use the area formula $\mathcal{A} = \dfrac{1}{2}rR\sin B$

for this triangle. By the law of sines, we have

$$\dfrac{r}{\sin A} = \dfrac{R}{\sin C} \Rightarrow r = \dfrac{R\sin A}{\sin C}$$

$$\sin C = \sin\left[180^\circ - (A+B)\right] = \sin(A+B), \text{ so}$$

we have $r = \dfrac{R\sin A}{\sin C} \Rightarrow r = \dfrac{R\sin A}{\sin(A+B)}$

By substituting into the area formula, we have

$$\mathcal{A} = \dfrac{1}{2}rR\sin B \Rightarrow$$

$$\mathcal{A} = \dfrac{1}{2}\left[\dfrac{R\sin A}{\sin(A+B)}\right]R\sin B \Rightarrow$$

$$\mathcal{A} = \dfrac{1}{2}\cdot\dfrac{\sin A\sin B}{\sin(A+B)}R^2$$

There are a total of 10 stars, so the total area covered by the stars is

$$\mathcal{A} = 10\left[\dfrac{1}{2}\cdot\dfrac{\sin A\sin B}{\sin(A+B)}R^2\right] = \left[5\dfrac{\sin A\sin B}{\sin(A+B)}\right]R^2$$

43. **(a)** $11.4 \text{ in.}\cdot\dfrac{10}{13}\text{ in.} \approx 8.77 \text{ in.}^2$

(b) $\mathcal{A} = 50\left[5\dfrac{\sin 18^\circ\sin 36^\circ}{\sin(18^\circ + 36^\circ)}\right]\cdot 0.308^2$

$\approx 5.32 \text{ in.}^2$

Section 7.3 The Law of Cosines

1. a, b, and C

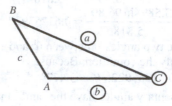

(a) SAS

(b) law of cosines

3. a, b, and A

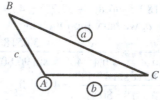

(a) SSA

(b) law of sines

5. A, B, and c

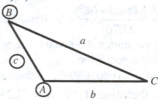

(a) ASA

(b) law of sines

7. a, b, and c

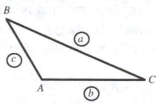

(a) SSS

(b) law of cosines

9. $a^2 = 1^2 + \left(4\sqrt{2}\right)^2 - 2(1)\left(4\sqrt{2}\right)\cos 45^\circ$

$$= 1 + 32 - 8\sqrt{2}\left(\dfrac{\sqrt{2}}{2}\right) = 33 - 8 = 25$$

$$a = \sqrt{25} = 5$$

11. $\cos\theta = \dfrac{b^2 + c^2 - a^2}{2bc} = \dfrac{3^2 + 5^2 - 7^2}{2(3)(5)}$

$$= \dfrac{9 + 25 - 49}{30} = -\dfrac{1}{2} \Rightarrow \theta = 120^\circ$$

13. $A = 121°$, $b = 5$, $c = 3$

Start by finding a with the law of cosines.

$a^2 = b^2 + c^2 - 2bc \cos A \Rightarrow$

$a^2 = 5^2 + 3^2 - 2(5)(3)\cos 121° \approx 49.5 \Rightarrow$

$a \approx 7.04 \approx 7.0$

C must be smaller than B because it is opposite the shorter of the two sides b and c. Therefore, C cannot be obtuse.

$\dfrac{\sin A}{a} = \dfrac{\sin C}{C} \Rightarrow \dfrac{\sin 121°}{7.04} = \dfrac{\sin C}{3} \Rightarrow$

$\sin C = \dfrac{3\sin 121°}{7.04} \approx 0.36527016 \Rightarrow C \approx 21.4°$

Thus, $B = 180° - 121° - 21.4° = 37.6°$.

15. $a = 12$, $b = 10$, $c = 10$

We can use the law of cosines to solve for any angle of the triangle. Because b and c have the same measure, this is an isosceles triangle and angles B and C also have the same measure. If we solve for B, we obtain

$b^2 = a^2 + c^2 - 2ac \cos B$

$\cos B = \dfrac{a^2 + c^2 - b^2}{2ac}$

$\cos B = \dfrac{12^2 + 10^2 - 10^2}{2(12)(10)} = \dfrac{144}{240} = \dfrac{3}{5} \Rightarrow$

$B \approx 53.13° \approx 53.1°$

Therefore, $C = B \approx 53.13° \approx 53.1°$ and $A = 180° - 2(53.13°) = 37.74° \approx 37.7°$.

If we solve for A directly, however, we obtain

$a^2 = b^2 + c^2 - 2bc \cos A$

$\cos A = \dfrac{b^2 + c^2 - a^2}{2bc}$

$\cos A = \dfrac{10^2 + 10^2 - 12^2}{2(10)(10)} = \dfrac{56}{200} = \dfrac{7}{5} \Rightarrow$

$A \approx 73.7°$

The angles may not sum to 180° due to rounding.

17. $B = 55°$, $a = 90$, $c = 100$

Start by finding b with the law of cosines.

$b^2 = a^2 + c^2 - 2ac \cos B$

$b^2 = 90^2 + 100^2 - 2(90)(100)\cos 55° \approx 7775.6$

$b \approx 88.18$

(will be rounded as 88.2) Of the remaining angles A and C, A must be smaller because it is opposite the shorter of the two sides a and c. Therefore, A cannot be obtuse.

$\dfrac{\sin A}{a} = \dfrac{\sin B}{b} \Rightarrow \dfrac{\sin A}{90} = \dfrac{\sin 55°}{88.18} \Rightarrow$

$\sin A = \dfrac{90\sin 55°}{88.18} \approx 0.83605902 \Rightarrow A \approx 56.7°$

Thus, $C = 180° - 55° - 56.7° = 68.3°$.

19. $A = 41.4°$, $b = 2.78$ yd, $c = 3.92$ yd

First find a.

$a^2 = b^2 + c^2 - 2bc \cos A$

$a^2 = 2.78^2 + 3.92^2 - 2(2.78)(3.92)\cos 41.4°$

$\approx 6.7460 \Rightarrow a \approx 2.60$ yd

Find B next, because angle B is smaller than angle C (because $b < c$), and thus angle B must be acute.

$\dfrac{\sin B}{b} = \dfrac{\sin A}{a} \Rightarrow \dfrac{\sin B}{2.78} = \dfrac{\sin 41.4°}{2.597} \Rightarrow$

$\sin B = \dfrac{2.78\sin 41.4°}{2.597} \approx 0.707091182 \Rightarrow$

$B \approx 45.1°$

Finally, $C = 180° - 41.4° - 45.1° = 93.5°$.

21. $C = 45.6°$, $b = 8.94$ m, $a = 7.23$ m

First find c.

$c^2 = a^2 + b^2 - 2ab \cos C \Rightarrow$

$c^2 = 7.23^2 + 8.94^2 - 2(7.23)(8.94)\cos 45.6°$

$\approx 41.7493 \Rightarrow c \approx 6.46$ m

Find A next, because angle A is smaller than angle B (because $a < b$), and thus angle A must be acute.

$\dfrac{\sin A}{a} = \dfrac{\sin C}{c} \Rightarrow \dfrac{\sin A}{7.23} = \dfrac{\sin 45.6°}{6.461} \Rightarrow$

$\sin A = \dfrac{7.23\sin 45.6°}{6.461} \approx 0.79951052 \Rightarrow$

$A \approx 53.1°$

Finally, $B = 180° - 53.1° - 45.6° = 81.3°$.

23. $a = 9.3$ cm, $b = 5.7$ cm, $c = 8.2$ cm

We can use the law of cosines to solve for any of angle of the triangle. We solve for A, the largest angle. We will know that A is obtuse if $\cos A < 0$.

$a^2 = b^2 + c^2 - 2bc \cos A \Rightarrow$

$\cos A = \dfrac{5.7^2 + 8.2^2 - 9.3^2}{2(5.7)(8.2)} \approx 0.14163457 \Rightarrow$

$A \approx 82°$

Find B next, because angle B is smaller than angle C (because $b < c$), and thus angle B must be acute.

$\dfrac{\sin B}{b} = \dfrac{\sin A}{a} \Rightarrow \dfrac{\sin B}{5.7} = \dfrac{\sin 82°}{9.3} \Rightarrow$

$\sin B = \dfrac{5.7\sin 82°}{9.3} \approx 0.60693849 \Rightarrow B \approx 37°$

Thus, $C = 180° - 82° - 37° = 61°$.

25. $a = 42.9$ m, $b = 37.6$ m, $c = 62.7$ m

Angle C is the largest, so find it first.

$$c^2 = a^2 + b^2 - 2ab\cos C \Rightarrow$$

$$\cos C = \frac{42.9^2 + 37.6^2 - 62.7^2}{2(42.9)(37.6)}$$

$$\approx -0.20988940 \Rightarrow$$

$$C \approx 102.1° \approx 102°10'$$

Find B next, because angle B is smaller than angle A (because $b < a$), and thus angle B must be acute.

$$\frac{\sin B}{b} = \frac{\sin C}{c} \Rightarrow \frac{\sin B}{37.6} = \frac{\sin 102.1°}{62.7} \Rightarrow$$

$$\sin B = \frac{37.6\sin 102.1°}{62.7} \approx 0.58635805 \Rightarrow$$

$$B \approx 35.9° \approx 35°50'$$

Thus,

$$A = 180° - 35°50' - 102°10'$$

$$= 180° - 138° = 42°00'$$

27. $a = 965$ ft, $b = 876$ ft, $c = 1240$ ft

Angle C is the largest, so find it first.

$$c^2 = a^2 + b^2 - 2ab\cos C \Rightarrow$$

$$\cos C = \frac{965^2 + 876^2 - 1240^2}{2(965)(876)} \approx 0.09522855 \Rightarrow$$

$$C \approx 84.5° \text{ or } 84°30'$$

Find B next, because angle B is smaller than angle A (because $b < a$), and thus angle B must be acute.

$$\frac{\sin B}{b} = \frac{\sin C}{c} \Rightarrow \frac{\sin B}{876} = \frac{\sin 84°30'}{1240} \Rightarrow$$

$$\sin B = \frac{876\sin 84°30'}{1240} \approx 0.70319925 \Rightarrow$$

$$B \approx 44.7° \text{ or } 44°40'$$

Thus,

$$A = 180° - 44°40' - 84°30'$$

$$= 179°60' - 129°10' = 50°50'$$

29. $A = 80°40'$ $b = 143$ cm, $c = 89.6$ cm

First find a.

$$a^2 = b^2 + c^2 - 2bc\cos A \Rightarrow$$

$$a^2 = 143^2 + 89.6^2 - 2(143)(89.6)\cos 80°40'$$

$$\approx 24,321.25 \Rightarrow a \approx 156 \text{ cm}$$

Find C next, because angle C is smaller than angle B (because $c < a$), and thus angle C must be acute.

$$\frac{\sin C}{c} = \frac{\sin A}{a} \Rightarrow \frac{\sin C}{89.6} = \frac{\sin 80°40'}{156.0} \Rightarrow$$

$$\sin C = \frac{89.6\sin 80°40'}{156.0} \approx 0.56675534 \Rightarrow$$

$$C \approx 34.5° = 34°30'$$

Finally,

$$B = 180° - 80°40' - 34°30'$$

$$= 179°60' - 115°10' = 64°50'$$

31. $B = 74.8°$, $a = 8.92$ in., $c = 6.43$ in.

First find b.

$$b^2 = a^2 + c^2 - 2ac\cos B \Rightarrow$$

$$b^2 = 8.92^2 + 6.43^2$$

$$- 2(8.92)(6.43)\cos 74.8°$$

$$\approx 90.8353 \Rightarrow b \approx 9.53 \text{ in.}$$

Find C next, because angle C is smaller than angle A (because $c < a$), and thus angle C must be acute.

$$\frac{\sin C}{c} = \frac{\sin B}{b} \Rightarrow \frac{\sin C}{6.43} = \frac{\sin 74.8°}{9.53} \Rightarrow$$

$$\sin C = \frac{6.43\sin 74.8°}{9.53} \approx 0.6540561811 \Rightarrow$$

$$C \approx 40.6°$$

Thus, $A = 180° - 74.8° - 40.6° = 64.6°$.

33. $A = 112.8°$, $b = 6.28$ m, $c = 12.2$ m

First find a.

$$a^2 = b^2 + c^2 - 2bc\cos A \Rightarrow$$

$$a^2 = 6.28^2 + 12.2^2 - 2(6.28)(12.2)\cos 112.8°$$

$$\approx 247.658 \Rightarrow a \approx 15.7 \text{ m}$$

Find B next, because angle B is smaller than angle C (because $b < c$), and thus angle B must be acute.

$$\frac{\sin B}{b} = \frac{\sin A}{a} \Rightarrow \frac{\sin B}{6.28} = \frac{\sin 112.8°}{15.74} \Rightarrow$$

$$\sin B = \frac{6.28\sin 112.8°}{15.74} \approx 0.36780817 \Rightarrow$$

$$B \approx 21.6°$$

Finally, $C = 180° - 112.8° - 21.6° = 45.6°$.

35. $a = 3.0$ ft, $b = 5.0$ ft, $c = 6.0$ ft

Angle C is the largest, so find it first.

$$c^2 = a^2 + b^2 - 2ab\cos C \Rightarrow$$

$$\cos C = \frac{3.0^2 + 5.0^2 - 6.0^2}{2(3.0)(5.0)} = -\frac{2}{30} = -\frac{1}{15}$$

$$\approx -0.06666667 \Rightarrow C \approx 94°$$

Find A next, because angle A is smaller than angle B (because $a < b$), and thus angle A must be acute.

$$\frac{\sin A}{a} = \frac{\sin C}{c} \Rightarrow \frac{\sin A}{3} = \frac{\sin 94°}{6} \Rightarrow$$

$$\sin A = \frac{3\sin 94°}{6} \approx 0.49878203 \Rightarrow A \approx 30°$$

Thus, $B = 180° - 30° - 94° = 56°$.

37. There are three ways to apply the law of cosines when $a = 3$, $b = 4$, and $c = 10$. Solving for A:

$$a^2 = b^2 + c^2 - 2bc \cos A \Rightarrow$$

$$\cos A = \frac{4^2 + 10^2 - 3^2}{2(4)(10)} = \frac{107}{80} = 1.3375$$

Solving for B:

$$b^2 = a^2 + c^2 - 2ac \cos B \Rightarrow$$

$$\cos B = \frac{3^2 + 10^2 - 4^2}{2(3)(10)} = \frac{93}{60} = \frac{31}{20} = 1.55$$

Solving for C:

$$c^2 = a^2 + b^2 - 2ab \cos C \Rightarrow$$

$$\cos C = \frac{3^2 + 4^2 - 10^2}{2(3)(4)} = \frac{-75}{24} = -\frac{25}{8} = -3.125$$

The cosine of any angle of a triangle must be between -1 and 1, so a triangle cannot have sides 3, 4, and 10.

39. A and B are on opposite sides of False River. We must find AB, or c, in the following triangle.

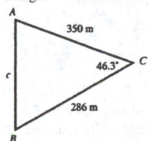

$$c^2 = a^2 + b^2 - 2ab \cos C$$

$$c^2 = 286^2 + 350^2 - 2(286)(350) \cos 46.3°$$

$$c^2 \approx 65,981.3 \Rightarrow c \approx 257$$

The length of AB is 257 m.

41. Using the law of cosines we can solve for the measure of angle A.

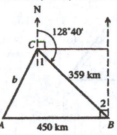

$$\cos A = \frac{25.9^2 + 32.5^2 - 57.8^2}{2(25.9)(32.5)}$$

$$\approx -0.95858628 \Rightarrow A \approx 163.5°$$

43. Find AC, or b, in the following triangle.

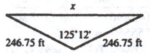

$$m\angle 1 = 180° - 128°40' = 51°20'$$

Angles 1 and 2 are alternate interior angles formed when parallel lines (the north lines) are cut by a transversal, line BC, so

$$m\angle 2 = m\angle 1 = 51°20'.$$

$$m\angle ABC = 90° - m\angle 2 = 90° - 51°20' = 38°40'$$

$$b^2 = a^2 + c^2 - 2ac \cos B \Rightarrow$$

$$b^2 = 359^2 + 450^2 - 2(359)(450) \cos 38°40'$$

$$\approx 79,106 \Rightarrow b \approx 281 \text{ km}$$

C is about 281 km from A.

45. Let $x =$ the distance between the ends of the two equal sides.

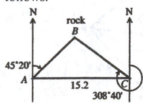

Use the law of cosines to find x.

$$x^2 = 246.75^2 + 246.75^2$$

$$- 2(246.75)(246.75) \cos 125°12'$$

$$\approx 191,963.937 \Rightarrow x \approx 438.14$$

The distance between the ends of the two equal sides is 438.14 feet.

47. Sketch a triangle showing the situation as follows.

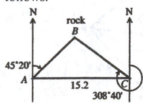

$$m\angle A = 90° - 45°20' = 44°40'$$

$$m\angle C = 308°40' - 270° = 38°40'$$

$$m\angle B = 180° - A - C = 180° - 44°40' - 38°40'$$

$$= 96°40'$$

We have only one side of a triangle, so use the law of sines to find $BC = a$.

$$\frac{a}{\sin A} = \frac{b}{\sin B} \Rightarrow \frac{a}{\sin 44° \ 40'} = \frac{15.2}{\sin 96° \ 40'} \Rightarrow$$

$$a = \frac{15.2 \sin 44° \ 40'}{\sin 96° \ 40'} \approx 10.8$$

The distance between the ship and the rock is about 10.8 miles.

49. Use the law of cosines to find the angle, θ.

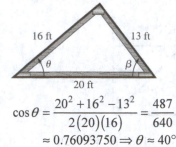

$$\cos \theta = \frac{20^2 + 16^2 - 13^2}{2(20)(16)} = \frac{487}{640}$$
$$\approx 0.76093750 \Rightarrow \theta \approx 40°$$

51. Let A = the angle between the beam and the 45-ft cable. Let B = the angle between the beam and the 60-ft cable.

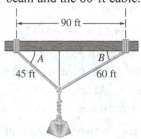

$$\cos A = \frac{45^2 + 90^2 - 60^2}{2(45)(90)} = \frac{6525}{8100} = \frac{29}{36}$$
$$\approx 0.80555556 \Rightarrow A \approx 36°$$
$$\cos B = \frac{90^2 + 60^2 - 45^2}{2(90)(60)} = \frac{9675}{10,800} = \frac{43}{48}$$
$$\approx 0.89583333 \Rightarrow B \approx 26°$$

53. Let A = home plate; B = first base; C = second base; D = third base; P = pitcher's rubber. Draw AC through P, draw PB and PD.

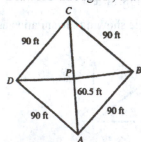

In triangle ABC, $m\angle B = 90°$, and $m\angle A = m\angle C = 45°$.

$$AC = \sqrt{90^2 + 90^2} = \sqrt{2 \cdot 90^2} = 90\sqrt{2} \text{ and}$$
$$PC = 90\sqrt{2} - 60.5 \approx 66.8 \text{ ft}$$
In triangle APB, $m\angle A = 45°$.
$$PB^2 = AP^2 + AB^2 - 2(AP)(AB)\cos A$$
$$PB^2 = 60.5^2 + 90^2 - 2(60.5)(90)\cos 45°$$
$$PB^2 \approx 4059.86 \Rightarrow PB \approx 63.7 \text{ ft}$$

Triangles APB and APD are congruent, so $PB = PD = 63.7$ ft.
The distance to second base is 66.8 ft and the distance to both first and third base is 63.7 ft.

55. Find the distance of the ship from point A.

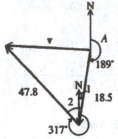

$$m\angle 1 = 189° - 180° = 9°$$
$$m\angle 2 = 360° - 317° = 43°$$
$$m\angle 1 + m\angle 2 = 9° + 43° = 52°$$
Use the law of cosines to find v.
$$v^2 = 47.8^2 + 18.5^2 - 2(47.8)(18.5)\cos 52°$$
$$\approx 1538.23 \Rightarrow v \approx 39.2 \text{ km}$$

57. Let c = the length of the property line that cannot be directly measured.

Not to scale

Using the law of cosines, we have
$$c^2 = 14.0^2 + 13.0^2 - 2(14.0)(13.0)\cos 70°$$
$$\approx 240.5 \Rightarrow c \approx 15.5 \text{ ft}$$
(rounded to three significant digits)
The length of the property line is approximately $18.0 + 15.5 + 14.0 = 47.5$ feet

59. Let c = the length of the tunnel.

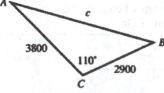

Use the law of cosines to find c.
$$c^2 = 3800^2 + 2900^2 - 2(3800)(2900)\cos 110°$$
$$\approx 30,388,124 \Rightarrow c \approx 5512.5$$
The tunnel is 5500 meters long. (rounded to two significant digits)

61. Let a be the length of the segment from $(0, 0)$ to $(6, 8)$. Use the distance formula.

$$a = \sqrt{(6-0)^2 + (8-0)^2} = \sqrt{6^2 + 8^2}$$
$$= \sqrt{36 + 64} = \sqrt{100} = 10$$

Let b be the length of the segment from $(0, 0)$ to $(4, 3)$.

$$b = \sqrt{(4-0)^2 + (3-0)^2} = \sqrt{4^2 + 3^2}$$
$$= \sqrt{16 + 9} = \sqrt{25} = 5$$

Let c be the length of the segment from $(4, 3)$ to $(6, 8)$.

$$c = \sqrt{(6-4)^2 + (8-3)^2} = \sqrt{2^2 + 5^2}$$
$$= \sqrt{4 + 25} = \sqrt{29}$$

$$\cos\theta = \frac{a^2 + b^2 - c^2}{2ab} \Rightarrow$$

$$\cos\theta = \frac{10^2 + 5^2 - \left(\sqrt{29}\right)^2}{2(10)(5)} = \frac{100 + 25 - 29}{100}$$

$$= 0.96 \Rightarrow \theta \approx 16.26°$$

63. $\mathcal{A} = \dfrac{1}{2}bh \Rightarrow$

$$\mathcal{A} = \frac{1}{2}(16)\left(3\sqrt{3}\right) = 24\sqrt{3} \approx 41.57.$$

To use Heron's Formula, first find the semiperimeter,

$$s = \frac{1}{2}(a+b+c) = \frac{1}{2}(6+14+16) = \frac{1}{2}\cdot 36 = 18.$$

Now find the area of the triangle.

$$\mathcal{A} = \sqrt{s(s-a)(s-b)(s-c)}$$
$$= \sqrt{18(18-6)(18-14)(18-16)}$$
$$= \sqrt{18(12)(4)(2)} = \sqrt{1728} = 24\sqrt{33} \approx 41.57$$

Both formulas give the same area.

65. $a = 12$ m, $b = 16$ m, $c = 25$ m

$$s = \frac{1}{2}(a+b+c) = \frac{1}{2}(12+16+25)$$
$$= \frac{1}{2}\cdot 53 = 26.5$$

$$\mathcal{A} = \sqrt{s(s-a)(s-b)(s-c)}$$
$$= \sqrt{26.5(26.5-12)(26.5-16)(26.5-25)}$$
$$= \sqrt{26.5(14.5)(10.5)(1.5)} \approx 78 \text{ m}^2$$

(rounded to two significant digits)

67. $a = 154$ cm, $b = 179$ cm, $c = 183$ cm

$$s = \frac{1}{2}(a+b+c) = \frac{1}{2}(154+179+183)$$
$$= \frac{1}{2}\cdot 516 = 258$$

$$\mathcal{A} = \sqrt{s(s-a)(s-b)(s-c)}$$
$$= \sqrt{258(258-154)(258-179)(258-183)}$$
$$= \sqrt{258(104)(79)(75)} \approx 12,600 \text{ cm}^2$$

(rounded to three significant digits)

69. $a = 76.3$ ft, $b = 109$ ft, $c = 98.8$ ft

$$s = \frac{1}{2}(a+b+c) = \frac{1}{2}(76.3+109+98.8)$$
$$= \frac{1}{2}\cdot 284.1 = 142.05$$

$$\mathcal{A} = \sqrt{s(s-a)(s-b)(s-c)}$$
$$= \sqrt{\begin{array}{c}142.05(142.05-76.3)(142.05-109)\cdot\\(142.05-98.8)\end{array}}$$
$$= \sqrt{142.05(65.75)(33.05)(43.25)} \approx 3650 \text{ ft}^2$$

(rounded to three significant digits)

71. Perimeter: $9 + 10 + 17 = 36$ feet, so the semi-perimeter is $\dfrac{1}{2}\cdot 36 = 18$ feet.

Use Heron's Formula to find the area.

$$\mathcal{A} = \sqrt{s(s-a)(s-b)(s-c)}$$
$$= \sqrt{18(18-9)(18-10)(18-17)}$$
$$= \sqrt{18(9)(8)(1)} = \sqrt{1296} = 36 \text{ ft}$$

The perimeter and area both equal 36, so the triangle is a *perfect triangle*.

73. Find the area of the Bermuda Triangle using Heron's Formula.

$$s = \frac{1}{2}(a+b+c) = \frac{1}{2}(850+925+1300)$$
$$= \frac{1}{2}\cdot 3075 = 1537.5$$

$$\mathcal{A} = \sqrt{s(s-a)(s-b)(s-c)}$$
$$= \sqrt{\begin{array}{c}1537.5(1537.5-850)\cdot\\(1537.5-925)(1537.5-1300)\end{array}}$$
$$= \sqrt{1537.5(687.5)(612.5)(237.5)}$$
$$\approx 392,128.82$$

The area of the Bermuda Triangle is about $390,000 \text{ mi}^2$.

75. **(a)** Using the law of sines, we have

$$\frac{\sin C}{c} = \frac{\sin A}{a} \Rightarrow \frac{\sin C}{15} = \frac{\sin 60°}{13} \Rightarrow \sin C = \frac{15 \sin 60°}{13} = \frac{15}{13} \cdot \frac{\sqrt{3}}{2} \approx 0.99926008$$

There are two angles C between $0°$ and $180°$ that satisfy the condition. Because $\sin C \approx 0.99926008$, to the nearest tenth value of C is $C_1 = 87.8°$. Supplementary angles have the same sine value, so another possible value of C is $B_2 = 180° - 87.8° = 92.2°$.

(b) By the law of cosines, we have

$$\cos C = \frac{a^2 + b^2 - c^2}{2ab} = \frac{13^2 + 7^2 - 15^2}{2(13)(7)} = \frac{-7}{182} = -\frac{1}{26} \approx -0.03846154 \Rightarrow C \approx 92.2°$$

(c) With the law of cosines, we are required to find the inverse cosine of a negative number; therefore; we know angle C is greater than $90°$.

77.

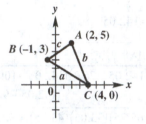

$$a = \sqrt{(-1-4)^2 + (3-0)^2} = \sqrt{25 + 9} = \sqrt{34}$$

$$b = \sqrt{(2-4)^2 + (5-0)^2} = \sqrt{4 + 25} = \sqrt{29}$$

$$c = \sqrt{(-1-2)^2 + (3-5)^2} = \sqrt{9 + 4} = \sqrt{13}$$

79. First find the semiperimeter. $s = \frac{1}{2}(a + b + c) = \frac{1}{2}\left(\sqrt{34} + \sqrt{29} + \sqrt{13}\right)$

Using Heron's formula, we have

$$\mathcal{A} = \sqrt{s(s-a)(s-b)(s-c)}$$

$$= \sqrt{\frac{1}{2}\left(\sqrt{34} + \sqrt{29} + \sqrt{13}\right)\left(\frac{1}{2}\left(\sqrt{34} + \sqrt{29} + \sqrt{13}\right) - \sqrt{34}\right)\left(\frac{1}{2}\left(\sqrt{34} + \sqrt{29} + \sqrt{13}\right) - \sqrt{29}\right)\left(\frac{1}{2}\left(\sqrt{34} + \sqrt{29} + \sqrt{13}\right) - \sqrt{13}\right)}$$

$$= 9.5 \text{ sq units (found using a calculator)}$$

Chapter 7   Quiz
(Sections 7.1−7.3)

1. Using the law of sines, we have

$$\frac{\sin B}{b} = \frac{\sin C}{c} \Rightarrow \frac{\sin 30.6°}{7.42} = \frac{\sin C}{4.54} \Rightarrow \sin C = \frac{4.54 \sin 30.6°}{7.42} \approx 0.311462 \Rightarrow C \approx 18.1°$$

$$A = 180° - B - C = 180° - 30.6° - 18.1° = 131.3° \approx 131° \text{ (rounded to three significant digits)}$$

3. Using the law of cosines, we have

$$c^2 = a^2 + b^2 - 2ab \cos C$$

$$21.2^2 = 28.4^2 + 16.9^2 - 2 \cdot 28.4 \cdot 16.9 \cos C$$

$$-642.73 = -959.92 \cos C$$

$$\frac{642.73}{959.92} = \cos C \Rightarrow C \approx 48.0° \text{ (rounded to three significant digits)}$$

5. First find the semiperimeter:

$$s = \frac{1}{2}(19.5 + 21.0 + 22.5) = 31.5$$

Using Heron's formula, we have

$$\begin{aligned}\mathcal{A} &= \sqrt{s(s-a)(s-b)(s-c)} \\ &= \sqrt{31.5(31.5-19.5)(31.5-21.0)(31.5-22.5)} \\ &= \sqrt{31.5(12)(10.5)(9)} \\ &= \sqrt{35,721} = 189 \text{ km}^2 \end{aligned}$$

7. $\angle C = 180° - 111° - 41° = 28°$

Using the law of sines, we have

$$\frac{a}{\sin A} = \frac{c}{\sin C} \Rightarrow \frac{a}{\sin 111°} = \frac{326}{\sin 28°} \Rightarrow$$

$$a = \frac{326 \sin 111°}{\sin 28°} \approx 648$$

$$\frac{b}{\sin B} = \frac{c}{\sin C} \Rightarrow \frac{b}{\sin 41°} = \frac{326}{\sin 28°} \Rightarrow$$

$$b = \frac{326 \sin 41°}{\sin 28°} \approx 456$$

Note that both a and b have been rounded to three significant digits.

9. $AB = 22.47928$ mi, $AC = 28.14276$ mi, $A = 58.56989°$

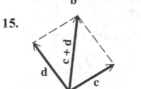

This is SAS, so use the law of cosines.

$$BC^2 = AC^2 + AB^2 - 2(AC)(AB)\cos A$$

$$BC^2 = 28.14276^2 + 22.47928^2$$
$$\qquad - 2(28.14276)(22.47928)\cos 58.56989°$$

$$BC^2 \approx 637.55393$$
$$BC \approx 25.24983$$

BC is approximately 25.24983 mi.
(rounded to seven significant digits)

Section 7.4 Geometrically Defined Vectors and Applications

1. Equal vectors have the same magnitude and direction. Equal vectors are **m** and **p**; **n** and **r**.

3. One vector is a positive scalar multiple of another if the two vectors point in the same direction; they may have different magnitudes.
m = 1**p**; **m** = 2**t**; **n** = 1**r**; **p** = 2**t** or

$$\mathbf{p} = 1\mathbf{m}; \ \mathbf{t} = \frac{1}{2}\mathbf{m}; \ \mathbf{r} = 1\mathbf{n}; \ \mathbf{t} = \frac{1}{2}\mathbf{p}$$

5.

7.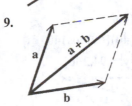

9.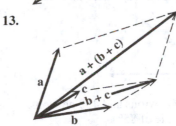

11.

13.

15.

17. **a** + (**b** + **c**) = (**a** + **b**) + **c**

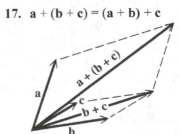

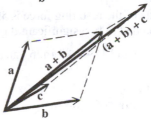

Yes, vector addition is associative.

19. $|\mathbf{u}| = 12, |\mathbf{v}| = 20, \theta = 27°$

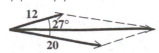

21. $|\mathbf{u}| = 20, |\mathbf{v}| = 30, \theta = 30°$

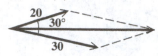

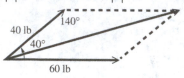

23. $\alpha = 180° - 40° = 140°$

$|\mathbf{v}|^2 = 40^2 + 60^2 - 2(40)(60)\cos 140°$

$|\mathbf{v}|^2 \approx 8877.0133 \Rightarrow |\mathbf{v}| \approx 94.2$ lb

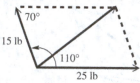

25. $\alpha = 180° - 110° = 70°$

$|\mathbf{v}|^2 = 15^2 + 25^2 - 2(15)(25)\cos 70°$

$|\mathbf{v}|^2 \approx 593.48489 \Rightarrow |\mathbf{v}| \approx 24.4$ lb

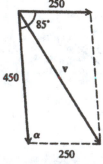

27. Forces of 250 newtons and 450 newtons, forming an angle of 85°

$\alpha = 180° - 85° = 95°$

$|\mathbf{v}|^2 = 250^2 + 450^2 - 2(250)(450)\cos 95°$

$|\mathbf{v}|^2 \approx 284,610.04 \Rightarrow |\mathbf{v}| \approx 533.5$

The magnitude of the resulting force is 530 newtons. (rounded to two significant digits)

29. Forces of 116 lb and 139 lb, forming an angle of 140° 50′

$\alpha = 180° - 140° 50′$
$\quad = 179° 60′ - 140° 50′ = 39° 10′$

$|\mathbf{v}|^2 = 139^2 + 116^2 - 2(139)(116)\cos 39° 10′$

$|\mathbf{v}|^2 \approx 7774.7359 \Rightarrow |\mathbf{v}| \approx 88.174$

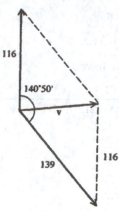

The magnitude of the resulting force is 88.2 lb. (rounded to three significant digits)

31. Find the direction and magnitude of the equilibrant.

$A = 180° - 28.2° = 151.8°$, so we can use the law of cosines to find the magnitude of the resultant, **v**.

$|\mathbf{v}|^2 = 1240^2 + 1480^2 - 2(1240)(1480)\cos 151.8°$

$\approx 6962736.2 \Rightarrow |\mathbf{v}| \approx 2639$ lb

(will be rounded as 2640)
Use the law of sines to find α.

$\dfrac{\sin \alpha}{1240} = \dfrac{\sin 151.8°}{2639}$

$\sin \alpha = \dfrac{1240 \sin 151.8°}{2639} \approx 0.22203977$

$\alpha \approx 12.8°$

Thus, we have 2640 lb at an angle of $\theta = 180° - 12.8° = 167.2°$ with the 1480-lb force.

33. Let α = the angle between the forces. To find α, use the law of cosines to find θ.

$786^2 = 692^2 + 423^2 - 2(692)(423)\cos \theta$

$\cos \theta = \dfrac{692^2 + 423^2 - 786^2}{2(692)(423)} \approx 0.06832049$

$\theta \approx 86.1°$

Thus, $\alpha = 180° - 86.1° = 93.9°$.

35. Use the parallelogram rule. In the figure, **x** represents the second force and **v** is the resultant.

$\alpha = 180° - 78°50' = 101°10'$ and
$\beta = 78°50' - 41°10' = 37°40'$

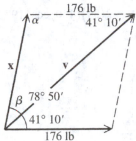

Using the law of sines, we have

$$\frac{|\mathbf{x}|}{\sin 41°10'} = \frac{176}{\sin 37°40'} \Rightarrow$$

$$|\mathbf{x}| = \frac{176\sin 41°10'}{\sin 37°40'} \approx 190$$

$$\frac{|\mathbf{v}|}{\sin \alpha} = \frac{176}{\sin 37°40'} \Rightarrow$$

$$|\mathbf{v}| = \frac{176\sin 101°10'}{\sin 37°40'} \approx 283$$

Thus, the magnitude of the second force is about 190 lb and the magnitude of the resultant is about 283 lb.

37. Let θ = the angle that the hill makes with the horizontal.

The 80-lb downward force has a 25-lb component parallel to the hill. The two right triangles are similar and have congruent angles.

$$\sin \theta = \frac{25}{80} = \frac{5}{16} = 0.3125 \Rightarrow \theta \approx 18°$$

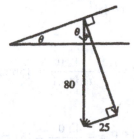

39. Find the force needed to hold a 60-ton monolith along the causeway.
The force needed to pull 60 tons is equal to the magnitude of **x**, the component parallel to the causeway. The length of the causeway is not relevant in this problem.

$$\sin 2.3° = \frac{|\mathbf{x}|}{60} \Rightarrow |\mathbf{x}| = 60\sin 2.3° \approx 2.4 \text{ tons}$$

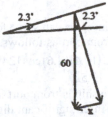

The force needed is 2.4 tons.

41. As in Example 4 on page 329 of the text, angle B equals angle θ, and here the magnitude of vector **BA** represents the weight of the stump grinder. The vector **AC** equals vector **BE**, which represents the force required to hold the stump grinder on the incline. Thus, we have

$$\sin B = \frac{18.0}{60.0} = \frac{3.0}{10.0} = 0.3 \Rightarrow B \approx 17.5°$$

43. Let **r** = the vertical component of the person exerting a 114-lb force;
s = the vertical component of the person exerting a 150-lb force.
The weight of the box is the sum of the magnitudes of the vertical components of the two vectors representing the forces exerted by the two people.

$$|\mathbf{r}| = 114\sin 54.9° \approx 93.27 \text{ and}$$

$$|\mathbf{s}| = 150\sin 62.4° \approx 132.93$$

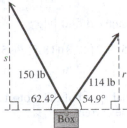

Thus, the weight of box is
$$|\mathbf{r}| + |\mathbf{s}| \approx 93.27 + 132.93 = 226.2 \approx 226 \text{ lb.}$$

45. Refer to the diagram. In order for the ship to turn due east, the ship must turn the measure of angle CAB, which is $90° - 34° = 56°$.
Angle DAB is therefore $180° - 56° = 124°$.

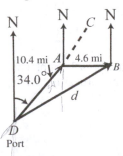

(continued on next page)

(*continued*)

Using the law of cosines, we can solve for the distance the ship is from port as follows.

$$d^2 = 10.4^2 + 4.6^2 - 2(10.4)(4.6)\cos 124°$$
$$\approx 182.824 \Rightarrow d \approx 13.52$$

Thus, the distance the ship is from port is 13.5 mi. (rounded to three significant digits) To find the bearing, we first seek the measure of angle ADB, which we will refer to as angle D. Using the law of cosines we have

$$\cos D = \frac{13.52^2 + 10.4^2 - 4.6^2}{2(13.52)(10.4)} \approx 0.95937073 \Rightarrow$$

$$D \approx 16.4°$$

Thus, the bearing is
$$34.0° + D = 34.0° + 16.4° = 50.4°.$$

47. Find the distance of the ship from point A.
Angle 1 = 189° − 180° = 9°
Angle 2 = 360° − 317° = 43°
Angle 1 + Angle 2 = 9° + 43° = 52°

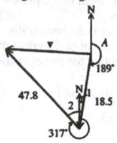

Use the law of cosines to find $|\mathbf{v}|$.

$$|\mathbf{v}|^2 = 47.8^2 + 18.5^2 - 2(47.8)(18.5)\cos 52°$$
$$\approx 1538.23 \Rightarrow |\mathbf{v}| \approx 39.2 \text{ km}$$

49. Let x = be the actual speed of the motorboat; y = the speed of the current.

$$\sin 10° = \frac{y}{20.0} \Rightarrow y = 20.0\sin 10° \approx 3.5$$

$$\cos 10° = \frac{x}{20.0} \Rightarrow x = 20.0\cos 10° \approx 19.7$$

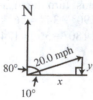

The speed of the current is 3.5 mph and the actual speed of the motorboat is 19.7 mph.

51. Let $\mathbf{v}$ = the ground speed vector. Find the bearing and ground speed of the plane.
Angle A = 233° − 114° = 119°
Use the law of cosines to find $|\mathbf{v}|$.

$$|\mathbf{v}|^2 = 39^2 + 450^2 - 2(39)(450)\cos 119°$$
$$|\mathbf{v}|^2 \approx 221,037.82$$
$$|\mathbf{v}| \approx 470.1$$

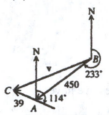

The ground speed is 470 mph. (rounded to two significant digits)
Use the law of sines to find angle B.

$$\frac{\sin B}{39} = \frac{\sin 119°}{470.1} \Rightarrow$$

$$\sin B = \frac{39\sin 119°}{470.1} \approx 0.07255939 \Rightarrow B \approx 4°$$

Thus, the bearing is
$$B + 233° = 4° + 233° = 237°.$$

53. Let $|\mathbf{x}|$ = the airspeed and $|\mathbf{d}|$ = the ground speed.

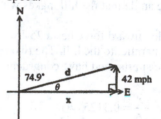

$$\theta = 90° - 74.9° = 15.1°$$

$$\frac{|\mathbf{x}|}{42.0} = \cot 15.1° \Rightarrow$$

$$|\mathbf{x}| = 42.0\cot 15.1° = \frac{42.0}{\tan 15.1°} \approx 156 \text{ mph}$$

$$\frac{|\mathbf{d}|}{42} = \csc 15.1° \Rightarrow$$

$$|\mathbf{d}| = 42.0\csc 15.1° = \frac{42.0}{\sin 15.1°} \approx 161 \text{ mph}$$

55. Let **c** = the ground speed vector.

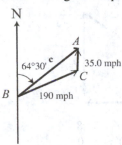

By alternate interior angles, angle $A = 64°30'$.
Use the law of sines to find B.

$$\frac{\sin B}{35.0} = \frac{\sin A}{190.0} \Rightarrow$$

$$\sin B = \frac{35.0 \sin 64°30'}{190.0} \approx 0.16626571 \Rightarrow$$

$$B \approx 9.57° \approx 9°30'$$

Thus, the bearing is
$64°30' + B = 64°30' + 9°30' = 74°00'$.
Because $C = 180° - A - B$
$= 180° - 64.50° - 9.57° = 105.93°$, we use the
law of sines to find the ground speed.

$$\frac{|\mathbf{c}|}{\sin C} = \frac{35.0}{\sin B} \Rightarrow$$

$$|\mathbf{c}| = \frac{35.0 \sin 105.93°}{\sin 9.57°} \approx 202 \text{ mph}$$

The bearing is $74°00'$; the ground speed is
202 mph.

57. Let **v** = the airspeed vector

The ground speed is $\dfrac{400 \text{ mi}}{2.5 \text{ hr}} = 160$ mph.

angle $BAC = 328° - 180° = 148°$
Using the law of cosines to find $|\mathbf{v}|$, we have

$$|\mathbf{v}|^2 = 11^2 + 160^2 - 2(11)(160)\cos 148°$$

$$|\mathbf{v}|^2 \approx 28,706.1 \Rightarrow |\mathbf{v}| \approx 169.4$$

The airspeed must be 170 mph. (rounded to
two significant digits)

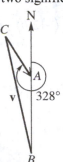

Use the law of sines to find B.

$$\frac{\sin B}{11} = \frac{\sin 148°}{169.4} \Rightarrow \sin B = \frac{11 \sin 148°}{169.4} \Rightarrow$$

$$\sin B \approx 0.03441034 \Rightarrow B \approx 2.0°$$

The bearing must be approximately
$360° - 2.0° = 358°$.

59. Find the ground speed and resulting bearing.
Angle $A = 245° - 174° = 71°$
Use the law of cosines to find $|\mathbf{v}|$.

$$|\mathbf{v}|^2 = 30^2 + 240^2 - 2(30)(240)\cos 71°$$

$$|\mathbf{v}|^2 \approx 53,811.8 \Rightarrow |\mathbf{v}| \approx 232.1$$

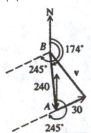

The ground speed is 230 km per hr. (rounded
to two significant digits)
Use the law of sines to find angle B.

$$\frac{\sin B}{30} = \frac{\sin 71°}{230} \Rightarrow$$

$$\sin B = \frac{30 \sin 71°}{230} \approx 0.12332851 \Rightarrow B \approx 7°$$

Thus, the bearing is
$174° - b = 174° - 7° = 167°$.

Section 7.5   Algebraically Defined Vectors and the Dot Product

1. The magnitude of vector **u** is $\underline{2}$.

3. The horizontal component, a, of vector **v** is
$\underline{\dfrac{\sqrt{2}}{2}}$.

5. The sum of the vectors $\mathbf{u} = \langle -3, 5 \rangle$ and
$\mathbf{v} = \langle 7, 4 \rangle$ is $\underline{\langle 4, 9 \rangle}$.

7. The formula for the dot product of the two
vectors $\mathbf{u} = \langle a, b \rangle$ and $\mathbf{v} = \langle c, d \rangle$ is
$\mathbf{u} \cdot \mathbf{v} = \underline{ac + bd}$.

9. Magnitude:

$$\sqrt{15^2 + (-8)^2} = \sqrt{225 + 64} = \sqrt{289} = 17$$

Angle:

$$\tan\theta' = \frac{b}{a} \Rightarrow \tan\theta' = \frac{-8}{15} \Rightarrow$$

$$\theta' = \tan^{-1}\left(-\frac{8}{15}\right) \approx -28.1° \Rightarrow$$

$$\theta = -28.1° + 360° = 331.9°$$
(θ lies in quadrant IV)

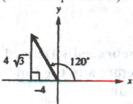

11. Magnitude:

$$\sqrt{(-4)^2 + (4\sqrt{3})^2} = \sqrt{16 + 48} = \sqrt{64} = 8$$

Angle:

$$\tan\theta' = \frac{b}{a} \Rightarrow \tan\theta' = \frac{4\sqrt{3}}{-4} \Rightarrow$$

$$\theta' = \tan^{-1}\left(-\sqrt{3}\right) = -60° \Rightarrow$$

$$\theta = -60° + 180° = 120°$$
(θ lies in quadrant II)

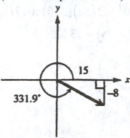

In Exercises 13−17, **x** is the horizontal component of **v**, and **y** is the vertical component of **v**. Thus, $|\mathbf{x}|$ is the magnitude of **x** and $|\mathbf{y}|$ is the magnitude of **y**.

13. $\theta = 20°$, $|\mathbf{v}| = 50$

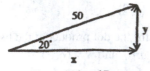

$$\mathbf{x} = 50\cos 20° \approx 47$$
$$\mathbf{y} = 50\sin 20° \approx 17$$

15. $\theta = 35°50'$, $|\mathbf{v}| = 47.8$

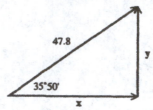

$$\mathbf{x} = 47.8\cos 35°50' \approx 38.8$$
$$\mathbf{y} = 47.8\sin 35°50' \approx 28.0$$

17. $\theta = 128.5°$, $|\mathbf{v}| = 198$

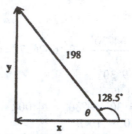

$$\mathbf{x} = 198\cos 128.5° \approx -123$$
$$\mathbf{y} = 198\sin 128.5° \approx 155$$

19. $\mathbf{u} = \langle a, b \rangle = \langle 5\cos(30°), 5\sin(30°)\rangle$

$$= \left\langle \frac{5\sqrt{3}}{2}, \frac{5}{2} \right\rangle$$

21. $\mathbf{v} = \langle a, b \rangle = \langle 4\cos(410°), 4\sin(140°)\rangle$

$$\approx \langle -3.0642, 2.5712 \rangle$$

23. $\mathbf{v} = \langle a, b \rangle = \langle 5\cos(-35°), 5\sin(-35°)\rangle$

$$\approx \langle 4.0958, -2.8679 \rangle$$

25. From the figure, $\mathbf{u} = \langle -8, 8 \rangle$ and $\mathbf{v} = \langle 4, 8 \rangle$.

   (a) $\mathbf{u} + \mathbf{v} = \langle -8, 8 \rangle + \langle 4, 8 \rangle$
   $$= \langle -8 + 4, 8 + 8 \rangle = \langle -4, 16 \rangle$$

   (b) $\mathbf{u} - \mathbf{v} = \langle -8, 8 \rangle - \langle 4, 8 \rangle$
   $$= \langle -8 - 4, 8 - 8 \rangle = \langle -12, 0 \rangle$$

   (c) $-\mathbf{u} = -\langle -8, 8 \rangle = \langle 8, -8 \rangle$

27. From the figure, $\mathbf{u} = \langle 4, 8 \rangle$ and $\mathbf{v} = \langle 4, -8 \rangle$.

   (a) $\mathbf{u} + \mathbf{v} = \langle 4, 8 \rangle + \langle 4, -8 \rangle$
   $$= \langle 4 + 4, 8 - 8 \rangle = \langle 8, 0 \rangle$$

   (b) $\mathbf{u} - \mathbf{v} = \langle 4, 8 \rangle - \langle 4, -8 \rangle$
   $$= \langle 4 - 4, 8 - (-8) \rangle = \langle 0, 16 \rangle$$

   (c) $-\mathbf{u} = -\langle 4, 8 \rangle = \langle -4, -8 \rangle$

29. From the figure, $\mathbf{u} = \langle -8, 4 \rangle$ and $\mathbf{v} = \langle 8, 8 \rangle$.

(a) $\mathbf{u} + \mathbf{v} = \langle -8, 4 \rangle + \langle 8, 8 \rangle$
$= \langle -8 + 8, 4 + 8 \rangle = \langle 0, 12 \rangle$

(b) $\mathbf{u} - \mathbf{v} = \langle -8, 4 \rangle - \langle 8, 8 \rangle$
$= \langle -8 - 8, 4 - 8 \rangle = \langle -16, -4 \rangle$

(c) $-\mathbf{u} = -\langle -8, 4 \rangle = \langle 8, -4 \rangle$

31. (a) $2\mathbf{u} = 2(2\mathbf{i}) = 4\mathbf{i}$

(b) $2\mathbf{u} + 3\mathbf{v} = 2(2\mathbf{i}) + 3(\mathbf{i} + \mathbf{j})$
$= 4\mathbf{i} + 3\mathbf{i} + 3\mathbf{j} = 7\mathbf{i} + 3\mathbf{j}$

(c) $\mathbf{v} - 3\mathbf{u} = \mathbf{i} + \mathbf{j} - 3(2\mathbf{i}) = \mathbf{i} + \mathbf{j} - 6\mathbf{i} = -5\mathbf{i} + \mathbf{j}$

33. (a) $2\mathbf{u} = 2\langle -1, 2 \rangle = \langle -2, 4 \rangle$

(b) $2\mathbf{u} + 3\mathbf{v} = 2\langle -1, 2 \rangle + 3\langle 3, 0 \rangle$
$= \langle -2, 4 \rangle + \langle 9, 0 \rangle$
$= \langle -2 + 9, 4 + 0 \rangle = \langle 7, 4 \rangle$

(c) $\mathbf{v} - 3\mathbf{u} = \langle 3, 0 \rangle - 3\langle -1, 2 \rangle = \langle 3, 0 \rangle - \langle -3, 6 \rangle$
$= \langle 3 - (-3), 0 - 6 \rangle = \langle 6, -6 \rangle$

For Exercises 35–41, $\mathbf{u} = \langle -2, 5 \rangle$ and $\mathbf{v} = \langle 4, 3 \rangle$.

35. $\mathbf{u} - \mathbf{v} = \langle -2, 5 \rangle - \langle 4, 3 \rangle$
$= \langle -2 - 4, 5 - 3 \rangle = \langle -6, 2 \rangle$

37. $-4\mathbf{u} = -4\langle -2, 5 \rangle = \langle -4(-2), -4(5) \rangle = \langle 8, -20 \rangle$

39. $3\mathbf{u} - 6\mathbf{v} = 3\langle -2, 5 \rangle - 6\langle 4, 3 \rangle$
$= \langle -6, 15 \rangle - \langle 24, 18 \rangle$
$= \langle -6 - 24, 15 - 18 \rangle = \langle -30, -3 \rangle$

41. $\mathbf{u} + \mathbf{v} - 3\mathbf{u} = \langle -2, 5 \rangle + \langle 4, 3 \rangle - 3\langle -2, 5 \rangle$
$= \langle -2, 5 \rangle + \langle 4, 3 \rangle - \langle 3(-2), 3(5) \rangle$
$= \langle -2, 5 \rangle + \langle 4, 3 \rangle - \langle -6, 15 \rangle$
$= \langle -2 + 4, 5 + 3 \rangle - \langle -6, 15 \rangle$
$= \langle 2, 8 \rangle - \langle -6, 15 \rangle$
$= \langle 2 - (-6), 8 - 15 \rangle = \langle 8, -7 \rangle$

43. $\langle -5, 8 \rangle = -5\mathbf{i} + 8\mathbf{j}$

45. $\langle 2, 0 \rangle = 2\mathbf{i} + 0\mathbf{j} = 2\mathbf{i}$

47. $\langle 6, -1 \rangle \cdot \langle 2, 5 \rangle = 6(2) + (-1)(5) = 12 - 5 = 7$

49. $\langle 5, 2 \rangle \cdot \langle -4, 10 \rangle = 5(-4) + 2(10) = -20 + 20 = 0$

51. $4\mathbf{i} = \langle 4, 0 \rangle; 5\mathbf{i} - 9\mathbf{j} = \langle 5, -9 \rangle$
$\langle 4, 0 \rangle \cdot \langle 5, -9 \rangle = 4(5) + 0(-9) = 20 - 0 = 20$

53. $\langle 2, 1 \rangle \cdot \langle -3, 1 \rangle$

$\cos\theta = \dfrac{\langle 2, 1 \rangle \cdot \langle -3, 1 \rangle}{\sqrt{2^2 + 1^2} \cdot \sqrt{(-3)^2 + 1^2}} = \dfrac{-6 + 1}{\sqrt{5} \cdot \sqrt{10}}$

$= \dfrac{-5}{5\sqrt{2}} = \dfrac{-1}{\sqrt{2}} = -\dfrac{\sqrt{2}}{2} \Rightarrow \theta = 135°$

55. $\langle 1, 2 \rangle \cdot \langle -6, 3 \rangle$

$\cos\theta = \dfrac{\langle 1, 2 \rangle \cdot \langle -6, 3 \rangle}{\sqrt{1^2 + 2^2} \cdot \sqrt{(-6)^2 + 3^2}} = \dfrac{-6 + 6}{\sqrt{5}\sqrt{45}}$

$= \dfrac{0}{15} = 0 \Rightarrow \theta = 90°$

57. First write the given vectors in component form: $3\mathbf{i} + 4\mathbf{j} = \langle 3, 4 \rangle$ and $\mathbf{j} = \langle 0, 1 \rangle$

$\cos\theta = \dfrac{\langle 3, 4 \rangle \cdot \langle 0, 1 \rangle}{|\langle 3, 4 \rangle||\langle 0, 1 \rangle|} = \dfrac{\langle 3, 4 \rangle \cdot \langle 0, 1 \rangle}{\sqrt{3^2 + 4^2} \cdot \sqrt{0^2 + 1^2}}$

$= \dfrac{0 + 4}{\sqrt{25} \cdot \sqrt{1}} = \dfrac{4}{5 \cdot 1} = \dfrac{4}{5} = 0.8 \Rightarrow$

$\theta = \cos^{-1} 0.8 \approx 36.87°$

For Exercises 59–61, $\mathbf{u} = \langle -2, 1 \rangle$, $\mathbf{v} = \langle 3, 4 \rangle$, and $\mathbf{w} = \langle -5, 12 \rangle$.

59. $(3\mathbf{u}) \cdot \mathbf{v} = (3\langle -2, 1 \rangle) \cdot \langle 3, 4 \rangle$
$= \langle -6, 3 \rangle \cdot \langle 3, 4 \rangle = -18 + 12 = -6$

61. $\mathbf{u} \cdot \mathbf{v} - \mathbf{u} \cdot \mathbf{w} = \langle -2, 1 \rangle \cdot \langle 3, 4 \rangle - \langle -2, 1 \rangle \cdot \langle -5, 12 \rangle$
$= (-6 + 4) - (10 + 12)$
$= -2 - 22 = -24$

63. $\langle 1, 2 \rangle \cdot \langle -6, 3 \rangle = -6 + 6 = 0$, so the vectors are orthogonal.

65. $\langle 1, 0 \rangle \cdot \langle \sqrt{2}, 0 \rangle = \sqrt{2} + 0 = \sqrt{2} \neq 0$, so the vectors are not orthogonal

67. $\sqrt{5}\mathbf{i} - 2\mathbf{j} = \langle \sqrt{5}, -2 \rangle; -5\mathbf{i} + 2\sqrt{5}\mathbf{j} = \langle -5, 2\sqrt{5} \rangle$

$\langle \sqrt{5}, -2 \rangle \cdot \langle -5, 2\sqrt{5} \rangle = -5\sqrt{5} - 4\sqrt{5}$

$= -9\sqrt{5} \neq 0$, so the vectors are not orthogonal.

69. $\mathbf{R} = \mathbf{i} - 2\mathbf{j}$ and $\mathbf{A} = 0.5\mathbf{i} + \mathbf{j}$

(a) Write the given vector in component form. $\mathbf{R} = \mathbf{i} - 2\mathbf{j} = \langle 1, -2 \rangle$ and

$\mathbf{A} = 0.5\mathbf{i} + \mathbf{j} = \langle 0.5, 1 \rangle$

$|\mathbf{R}| = \sqrt{1^2 + (-2)^2} = \sqrt{1+4} = \sqrt{5} \approx 2.2$

and $|\mathbf{A}| = \sqrt{0.5^2 + 1^2} = \sqrt{0.25 + 1} \approx 1.1$

About 2.2 in. of rain fell. The area of the opening of the rain gauge is about 1.1 in.^{2}.

(b) $V = |\mathbf{R} \cdot \mathbf{A}| = |\langle 1, -2 \rangle \cdot \langle 0.5, 1 \rangle|$

$= |0.5 + (-2)| = |-1.5| = 1.5$

The volume of rain was 1.5 in.^{3}.

71. Draw a line parallel to the x-axis and the vector $\mathbf{u} + \mathbf{v}$ (shown as a dashed line) Because $\theta_1 = 110°$, its supplementary angle is 70°. Further, because $\theta_2 = 260°$, the angle α is $260° - 180° = 80°$. Then the angle CBA becomes $180 - (80 + 70) = 180 - 150 = 30°$.

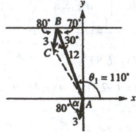

Using the law of cosines, the magnitude of $\mathbf{u} + \mathbf{v}$ is found as follows:

$|\mathbf{u} + \mathbf{v}|^2 = a^2 + c^2 - 2ac \cos B$

$|\mathbf{u} + \mathbf{v}|^2 = 3^2 + 12^2 - 2(3)(12) \cos 30°$

$= 9 + 144 - 72 \cdot \dfrac{\sqrt{3}}{2} = 153 - 36\sqrt{3}$

≈ 90.646171

Thus, $|\mathbf{u} + \mathbf{v}| \approx 9.5208$.

Using the law of sines, we have

$\dfrac{\sin A}{a} = \dfrac{\sin B}{b} \Rightarrow \dfrac{\sin A}{3} = \dfrac{\sin 30°}{9.5208} \Rightarrow$

$\sin A = \dfrac{3 \sin 30°}{9.5208} = \dfrac{3 \cdot \frac{1}{2}}{9.5208} \approx 0.15754979 \Rightarrow$

$A \approx 9.0647°$

The direction angle of $\mathbf{u} + \mathbf{v}$ is $110° + 9.0647° = 119.0647°$.

73. $c = 3 \cos 260° \approx -0.52094453$ and $d = 3 \sin 260° \approx -2.95442326$, so $\langle c, d \rangle \approx \langle -0.5209, -2.9544 \rangle$.

75. Magnitude:

$\sqrt{(-4.62518625)^2 + 8.32188819^2} \approx 9.5208$

Angle:

$\tan \theta' = \dfrac{8.32188819}{-4.625186258} \Rightarrow \theta' \approx -60.9353° \Rightarrow$

$\theta = -60.9353° + 180° = 119.0647°$

(θ lies in quadrant II)

Summary Exercises on Applications of Trigonometry and Vectors

1. Consider the diagram below.

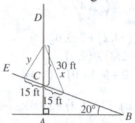

If we extend the flagpole, a right triangle CAB is formed. Thus, the measure of angle BCA is $90° - 20° = 70°$. Angles DCB and BCA are supplementary, so Sthe measure of angle DCB is $180° - 70° = 110°$. We can now use the law of cosines to find the measure of the support wire on the right, x.

$x^2 = 30^2 + 15^2 - 2(30)(15) \cos 110°$

$\approx 1432.818 \Rightarrow x \approx 37.85 \approx 38$ ft

Now, to find the length of the support wire on the left, we have different ways to find it. One way would be to use the approximation for x and use the law of cosines. To avoid using the approximate value, we will find y with the same method as for x. Angles DCB and DCE are supplementary, soS the measure of angle DCE is $180° - 70° = 110°$. We can now use the law of cosines to find the measure of the support wire on the left, y.

$y^2 = 30^2 + 15^2 - 2(30)(15) \cos 70°$

$\approx 817.182 \Rightarrow x \approx 28.59 \approx 29$ ft

The lengths of the two wires are about 29 ft and 38 ft.

3. Let c be the distance between the two lighthouses. Refer to the figure on the next page.

Angles DAC and CAB form a line, so angle CAB is the supplement of angle DAC. Thus, the measure of angle CAB is the following.

$180° - 129°43' = 179°60' - 129°43' = 50°17'$

The angles of a triangle must add up to $180°$, so the measure of angle ACB is

$180° - 39°43' - 50°17' = 180° - 90° = 90°$

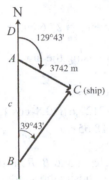

Thus, we have the following.

$\cos 50°17' = \dfrac{3742}{c} \Rightarrow c = \dfrac{3742}{\cos 50°17'} \approx 5856$

The two lighthouses are 5856 m apart.

5. Let $\mathbf{x}$ be the horizontal force.

$\tan 40° = \dfrac{|\mathbf{x}|}{50} \Rightarrow |\mathbf{x}| = 50 \tan 40° \approx 42 \text{ lb}$

7. $\mathbf{v} = 6\mathbf{i} + 8\mathbf{j} = \langle 6, 8 \rangle$

(a) The speed of the wind is

$|\mathbf{v}| = \sqrt{6^2 + 8^2} = \sqrt{36 + 64}$
$= \sqrt{100} = 10 \text{ mph}$

(b) $3\mathbf{v} = \langle 3 \cdot 6, 3 \cdot 8 \rangle = \langle 18, 24 \rangle = 18\mathbf{i} + 24\mathbf{j}$;

$|3\mathbf{v}| = \sqrt{18^2 + 24^2} = \sqrt{324 + 576} = 30$

This represents a 30 mph wind in the direction of $\mathbf{v}$.

(c) $\mathbf{u}$ represents a southeast wind of

$|\mathbf{u}| = \sqrt{(-8)^2 + 8^2} = \sqrt{64 + 64}$
$= \sqrt{128} = 8\sqrt{2} \approx 11.3 \text{ mph}$

9.

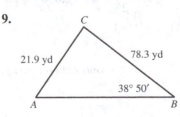

Using the law of sines, we have

$\dfrac{\sin A}{BC} = \dfrac{\sin B}{AC} \Rightarrow \dfrac{\sin A}{78.3} = \dfrac{\sin 38°50'}{21.9} \Rightarrow$

$\sin A = \dfrac{78.3 \sin 38°50'}{21.9} \approx 2.24$

Because $-1 \le \sin A \le 1$, the triangle cannot exist.

Chapter 7   Review Exercises

1. Find b, given $C = 74.2°$, $c = 96.3$ m, $B = 39.5°$.
Use the law of sines to find b.

$\dfrac{b}{\sin B} = \dfrac{c}{\sin C} \Rightarrow \dfrac{b}{\sin 39.5°} = \dfrac{96.3}{\sin 74.2°} \Rightarrow$

$b = \dfrac{96.3 \sin 39.5°}{\sin 74.2°} \approx 63.7 \text{ m}$

3. Find B, given $C = 51.3°$, $c = 68.3$ m, $b = 58.2$ m.
Use the law of sines to find B.

$\dfrac{\sin B}{b} = \dfrac{\sin C}{c} \Rightarrow \dfrac{\sin B}{58.2} = \dfrac{\sin 51.3°}{68.3} \Rightarrow$

$\sin B = \dfrac{58.2 \sin 51.3°}{68.3} \approx 0.66502269$

There are two angles B between $0°$ and $180°$ that satisfy the condition. Because $\sin B \approx 0.66502269$, to the nearest tenth, th value of B is $B_1 = 41.7°$. Supplementary angles have the same sine value, so another possible value of B is $B_2 = 180° - 41.7° = 138.3°$. This is not a valid angle measure for this triangle because $C + B_2 = 51.3° + 138.3° = 189.6° > 180°$.

Thus, $B = 41.7°$.

5. Find A, given $B = 39°50'$, $b = 268$ m, $a = 340$ m.
Use the law of sines to find A.

$\dfrac{\sin A}{a} = \dfrac{\sin B}{b} \Rightarrow \dfrac{\sin A}{340} = \dfrac{\sin 39°50'}{268} \Rightarrow$

$\sin A = \dfrac{340 \sin 39°50'}{268} \approx 0.81264638$

There are two angles A between $0°$ and $180°$ that satisfy the condition. Because $\sin A \approx 0.81264638$, to the nearest tenth value of A is $A_1 = 54.4° \approx 54°20'$. Supplementary angles have the same sine value, so another possible value of A is $A_2 = 180° - 54°20'$
$= 179°60' - 54°20' = 125°40'$. This is a valid angle measure for this triangle because $B + A_2 = 39°50' + 125°40' = 165°30' < 180°$.

$A = 54°20'$ or $A = 125°40'$

7. No. If two angles of a triangle are given, then the third angle is known because the sum of the measures of the three angles is 180°. One side is given, so there will only be one triangle that will satisfy the conditions.

9. $a = 10$, $B = 30°$

(a) The value of b that forms a right triangle would yield exactly one value for A. That is, $b = 10 \sin 30° = 5$. Also, any value of b greater than or equal to 10 would yield a unique value for A.

(b) Any value of b between 5 and 10, would yield two possible values for A.

(c) If b is less than 5, then no value for A is possible.

11. Find A, given $a = 86.14$ in., $b = 253.2$ in., $c = 241.9$ in.

Use the law of cosines to find A.

$a^2 = b^2 + c^2 - 2bc \cos A \Rightarrow$

$\cos A = \dfrac{b^2 + c^2 - a^2}{2bc}$

$= \dfrac{253.2^2 + 241.9^2 - 86.14^2}{2(253.2)(241.9)}$

≈ 0.94046923

Thus, $A \approx 19.87°$ or $19°52'$.

13. Find a, given $A = 51°20'$, $c = 68.3$ m, $b = 58.2$ m.

Use the law of cosines to find a.

$a^2 = b^2 + c^2 - 2bc \cos A \Rightarrow$

$a^2 = 58.2^2 + 68.3^2 - 2(58.2)(68.3) \cos 51°20'$

$\approx 3084.99 \Rightarrow a \approx 55.5$ m

15. Find a, given $A = 60°$, $b = 5.0$ cm, $c = 21$ cm.

Use the law of cosines to find a.

$a^2 = b^2 + c^2 - 2bc \cos A \Rightarrow$

$a^2 = 5.0^2 + 21^2 - 2(5.0)(21) \cos 60° = 361 \Rightarrow$

$a = 19$ cm

17. Solve the triangle, given $A = 25.2°$, $a = 6.92$ yd, $b = 4.82$ yd.

$\dfrac{\sin B}{b} = \dfrac{\sin A}{a} \Rightarrow \sin B = \dfrac{b \sin A}{a} \Rightarrow$

$\sin B = \dfrac{4.82 \sin 25.2°}{6.92} \approx 0.29656881$

There are two angles B between 0° and 180° that satisfy the condition. because $\sin B \approx 0.29656881$, to the nearest tenth value of B is $B_1 = 17.3°$. Supplementary angles have the same sine value, so another possible value of B is $B_2 = 180° - 17.3° = 162.7°$.

This is not a valid angle measure for this triangle because

$A + B_2 = 25.2° + 162.7° = 187.9° > 180°$.

$C = 180° - A - B \Rightarrow$

$C = 180° - 25.2° - 17.3° \Rightarrow C = 137.5°$

Use the law of sines to find c.

$\dfrac{c}{\sin C} = \dfrac{a}{\sin A} \Rightarrow \dfrac{c}{\sin 137.5°} = \dfrac{6.92}{\sin 25.2°} \Rightarrow$

$c = \dfrac{6.92 \sin 137.5°}{\sin 25.2°} \approx 11.0$ yd

19. Solve the triangle, given $a = 27.6$ cm, $b = 19.8$ cm, $C = 42°\,30'$.

This is a SAS case, so using the law of cosines.

$c^2 = a^2 + b^2 - 2ab \cos C \Rightarrow$

$c^2 = 27.6^2 + 19.8^2 - 2(27.6)(19.8) \cos 42°\,30'$

$\approx 347.985 \Rightarrow c \approx 18.65$ cm

(will be rounded as 18.7)

Of the remaining angles A and B, B must be smaller because it is opposite the shorter of the two sides a and b. Therefore, B cannot be obtuse.

$\dfrac{\sin B}{b} = \dfrac{\sin C}{c} \Rightarrow \dfrac{\sin B}{19.8} = \dfrac{\sin 42°30'}{18.65} \Rightarrow$

$\sin B = \dfrac{19.8 \sin 42°30'}{18.65} \approx 0.717124859 \Rightarrow$

$B \approx 45.8° \approx 45°50'$

Thus,

$A = 180° - B - C = 180° - 45°\,50' - 42°\,30'$

$= 179°60' - 88°20' = 91°\,40'$

21. Given $b = 840.6$ m, $c = 715.9$ m, $A = 149.3°$, find the area.

Angle A is included between sides b and c. Thus, we have

$\mathscr{A} = \dfrac{1}{2} bc \sin A = \dfrac{1}{2}(840.6)(715.9) \sin 149.3°$

$\approx 153,600$ m^2

(rounded to four significant digits)

23. Given $a = 0.913$ km, $b = 0.816$ km, $c = 0.582$ km, find the area.

Use Heron's formula to find the area.

$s = \dfrac{1}{2}(a + b + c) = \dfrac{1}{2}(0.913 + 0.816 + 0.582)$

$= \dfrac{1}{2} \cdot 2.311 = 1.1555$

(continued on next page)

(continued)

$$A = \sqrt{s(s-a)(s-b)(s-c)}$$
$$= \sqrt{\begin{array}{c}1.1555(1.1555-0.913)\cdot \\ (1.1555-0.816)(1.1555-0.582)\end{array}}$$
$$= \sqrt{1.1555(0.2425)(0.3395)(0.5735)}$$
$$\approx 0.234 \text{ km}^2$$
(rounded to three significant digits)

25. $B = 58.4°$ and $C = 27.9°$, so
$A = 180° - B - C = 180° - 58.4° - 27.9° = 93.7°$.
Using the law of sines, we have
$$\frac{AB}{\sin C} = \frac{125}{\sin A} \Rightarrow \frac{AB}{\sin 27.9°} = \frac{125}{\sin 93.7°} \Rightarrow$$
$$AB = \frac{125 \sin 27.9°}{\sin 93.7°} \approx 58.61$$
The canyon is 58.6 feet across. (rounded to three significant digits)

27. Let $AC = x =$ the height of the tree.

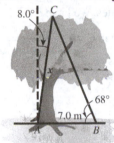

Angle $A = 90° - 8.0° = 82°$
Angle $C = 180° - B - A = 30°$
Use the law of sines to find $AC = b$.
$$\frac{b}{\sin B} = \frac{c}{\sin C}$$
$$\frac{b}{\sin 68°} = \frac{7.0}{\sin 30°}$$
$$b = \frac{7.0 \sin 68°}{\sin 30°}$$
$$b \approx 12.98$$
The tree is 13 meters tall (rounded to two significant digits).

29. Let $h =$ the height of tree.
$\theta = 27.2° - 14.3° = 12.9°$
$\alpha = 90° - 27.2° = 62.8°$
$$\frac{h}{\sin \theta} = \frac{212}{\sin \alpha}$$
$$\frac{h}{\sin 12.9°} = \frac{212}{\sin 62.8°}$$
$$h = \frac{212 \sin 12.9°}{\sin 62.8°} \approx 53.21$$

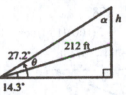

The height of the tree is 53.2 ft. (rounded to three significant digits)

31. Let $x =$ the distance between the boats.
In 3 hours the first boat travels $3(36.2) =$ 108.6 km and the second travels $3(45.6) =$ 136.8 km.
Use the law of cosines to find x.
$$x^2 = 108.6^2 + 136.8^2$$
$$- 2(108.6)(136.8)\cos 54°10'$$
$$\approx 13,113.359 \Rightarrow x \approx 115 \text{ km}$$
They are 115 km apart.

33. Use the distance formula to find the distances between the points.
Distance between $(-8, 6)$ and $(0, 0)$:
$$\sqrt{(-8-0)^2 + (6-0)^2} = \sqrt{(-8)^2 + 6^2}$$
$$= \sqrt{64 + 36} = \sqrt{100} = 10$$
Distance between $(-8, 6)$ and $(3, 4)$:
$$\sqrt{(-8-3)^2 + (6-4)^2} = \sqrt{(-11)^2 + 2^2}$$
$$= \sqrt{121 + 4} = \sqrt{125}$$
$$= 5\sqrt{5} \approx 11.18$$
Distance between $(3, 4)$ and $(0, 0)$:
$$\sqrt{(3-0)^2 + (4-0)^2} = \sqrt{3^2 + 4^2}$$
$$= \sqrt{9 + 16} = \sqrt{25} = 5$$
$$s \approx \frac{1}{2}(10 + 11.18 + 5) = \frac{1}{2} \cdot 26.18 = 13.09$$
$$A = \sqrt{s(s-a)(s-b)(s-c)}$$
$$= \sqrt{\begin{array}{c}13.09(13.09-10)\cdot \\ (13.09-11.18)(13.09-5)\end{array}}$$
$$= \sqrt{13.09(3.09)(1.91)(8.09)}$$
$$\approx 25 \text{ sq units (rounded to two significant digits)}$$

35. $\mathbf{a} - \mathbf{b}$

37. $\alpha = 180° - 52° = 128°$

$|\mathbf{v}|^2 = 100^2 + 130^2 - 2(100)(130)\cos 128°$

$|\mathbf{v}|^2 \approx 42907.2 \Rightarrow |\mathbf{v}| \approx 207$ lb

39. $|\mathbf{v}| = 964,\ \theta = 154°20'$

horizontal:

$x = |\mathbf{v}|\cos\theta = 964\cos 154°\ 20' \approx -869$

vertical: $y = |\mathbf{v}|\sin\theta = 964\sin 154°\ 20' \approx 418$

41. $\mathbf{u} = \langle -9, 12 \rangle$

magnitude:

$|\mathbf{u}| = \sqrt{(-9)^2 + 12^2} = \sqrt{81 + 144} = \sqrt{225} = 15$

Angle: $\tan\theta' = \dfrac{b}{a} \Rightarrow \tan\theta' = \dfrac{12}{-9} \Rightarrow$

$\theta' = \tan^{-1}\left(-\dfrac{4}{3}\right) \approx -53.1° \Rightarrow$

$\theta = -53.1° + 180° = 126.9°$

(θ lies in quadrant II)

43. $\mathbf{v} = 2\mathbf{i} - \mathbf{j},\ \ \mathbf{u} = -3\mathbf{i} + 2\mathbf{j}$

First write the given vectors in component form.

$\mathbf{v} = 2\mathbf{i} - \mathbf{j} = \langle 2, -1 \rangle$ and $\mathbf{u} = -3\mathbf{i} + 2\mathbf{j} = \langle -3, 2 \rangle$

(a) $2\mathbf{v} + \mathbf{u} = 2\langle 2, -1 \rangle + \langle -3, 2 \rangle$

$= \langle 2 \cdot 2, 2(-1) \rangle + \langle -3, 2 \rangle$

$= \langle 4, -2 \rangle + \langle -3, 2 \rangle$

$= \langle 4 + (-3), -2 + 2 \rangle = \langle 1, 0 \rangle = \mathbf{i}$

(b) $2\mathbf{v} = 2\langle 2, -1 \rangle = \langle 2 \cdot 2, 2(-1) \rangle$

$= \langle 4, -2 \rangle = 4\mathbf{i} - 2\mathbf{j}$

(c) $\mathbf{v} - 3\mathbf{u} = \langle 2, -1 \rangle - 3\langle -3, 2 \rangle$

$= \langle 2, -1 \rangle - \langle 3(-3), 3 \cdot 2 \rangle$

$= \langle 2, -1 \rangle - \langle -9, 6 \rangle$

$= \langle 2 - (-9), -1 - 6 \rangle$

$= \langle 11, -7 \rangle = 11\mathbf{i} - 7\mathbf{j}$

45. $\mathbf{u} = \langle 5, -3 \rangle,\ \mathbf{v} = \langle 3, 5 \rangle$

$\mathbf{u} \cdot \mathbf{v} = \langle 5, -3 \rangle \cdot \langle 3, 5 \rangle = 5(3) + (-3) \cdot 5$

$= 15 - 15 = 0$

$\cos\theta = \dfrac{\mathbf{u} \cdot \mathbf{v}}{|\mathbf{u}||\mathbf{v}|} \Rightarrow \cos\theta = \dfrac{0}{|\langle 5, -3 \rangle||\langle 3, 5 \rangle|} = 0 \Rightarrow$

$\theta = \cos^{-1} 0 = 90°$

The vectors are orthogonal.

47. Let $|\mathbf{x}|$ be the resultant force.

$\theta = 180° - 15° - 10° = 155°$

$|\mathbf{x}|^2 = 12^2 + 18^2 - 2(12)(18)\cos 155°$

$|\mathbf{x}|^2 \approx 859.5 \Rightarrow |\mathbf{x}| \approx 29$

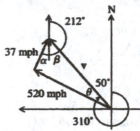

The magnitude of the resultant force on Jessie and the sled is 29 lb.

49. Let $\mathbf{v}$ = the ground speed vector.

$\alpha = 212° - 180° = 32°$ and $\beta = 50°$ because they are alternate interior angles. Angle opposite to 520 is $\alpha + \beta = 82°$.

Using the law of sines, we have

$\dfrac{\sin\theta}{37} = \dfrac{\sin 82°}{520} \Rightarrow \sin\theta = \dfrac{37\sin 82°}{520} \Rightarrow$

$\sin\theta \approx 0.07046138 \Rightarrow \theta \approx 4°$

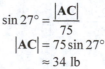

Thus, the bearing is $360° - 50° - \theta = 306°$. The angle opposite $\mathbf{v}$ is $180° - 82° - 4° = 94°$. Using the laws of sines, we have

$\dfrac{|\mathbf{v}|}{\sin 94°} = \dfrac{520}{\sin 82°} \Rightarrow$

$|\mathbf{v}| = \dfrac{520\sin 94°}{\sin 82°} \approx 524$ mph

The pilot should fly on a bearing of 306°. Her actual speed is 524 mph.

51. Refer to Example 3 in section 7.4. The magnitude of vector **AC** gives the magnitude of the required force.

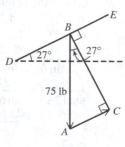

$\sin 27° = \dfrac{|AC|}{75}$

$|AC| = 75\sin 27°$

≈ 34 lb

53. $A = 30°, B = 60°, C = 90°, a = 7, b = 7\sqrt{3}, c = 14$

Newton's formula: $\dfrac{a+b}{c} = \dfrac{\cos\frac{1}{2}(A-B)}{\sin\frac{1}{2}C}$

$\dfrac{7+7\sqrt{3}}{14} \overset{?}{=} \dfrac{\cos\frac{1}{2}(30° - 60°)}{\sin\frac{1}{2}(90°)}$

$\dfrac{1+\sqrt{3}}{2} \overset{?}{=} \dfrac{\cos(-15°)}{\sin 45°} \Rightarrow \dfrac{1+\sqrt{3}}{2} \overset{?}{=} \dfrac{\cos(15°)}{\sin 45°}$

Using the half-angle formula, we have

$\cos 15° = \cos\left(\tfrac{1}{2}\cdot 30°\right) = \sqrt{\dfrac{1+\cos 30°}{2}}$

$= \sqrt{\dfrac{1+\frac{\sqrt{3}}{2}}{2}} = \sqrt{\dfrac{2+\sqrt{3}}{4}} = \dfrac{\sqrt{2+\sqrt{3}}}{2}$

Continuing, we have

$\dfrac{1+\sqrt{3}}{2} \overset{?}{=} \dfrac{\cos(15°)}{\sin 45°} \Rightarrow \dfrac{1+\sqrt{3}}{2} \overset{?}{=} \dfrac{\frac{\sqrt{2+\sqrt{3}}}{2}}{\frac{\sqrt{2}}{2}} \Rightarrow$

$\dfrac{1+\sqrt{3}}{2} \overset{?}{=} \dfrac{\sqrt{2+\sqrt{3}}}{\sqrt{2}} \Rightarrow \dfrac{\sqrt{(1+\sqrt{3})^2}}{\sqrt{2^2}} \overset{?}{=} \dfrac{\sqrt{2+\sqrt{3}}}{\sqrt{2}} \Rightarrow$

$\sqrt{\dfrac{(1+\sqrt{3})^2}{2^2}} \overset{?}{=} \sqrt{\dfrac{2+\sqrt{3}}{2}} \Rightarrow$

$\sqrt{\dfrac{1+2\sqrt{3}+3}{4}} \overset{?}{=} \sqrt{\dfrac{2+\sqrt{3}}{2}} \Rightarrow$

$\sqrt{\dfrac{4+2\sqrt{3}}{4}} \overset{?}{=} \sqrt{\dfrac{2+\sqrt{3}}{2}} \Rightarrow \sqrt{\dfrac{2+\sqrt{3}}{2}} = \sqrt{\dfrac{2+\sqrt{3}}{2}}$

55. Let $a = 2$, $b = 2\sqrt{3}$, $A = 30°$, $B = 60°$.

Verify $\dfrac{\tan\frac{1}{2}(A-B)}{\tan\frac{1}{2}(A+B)} = \dfrac{a-b}{a+b}$.

$\dfrac{\tan\frac{1}{2}(A-B)}{\tan\frac{1}{2}(A+B)} = \dfrac{\tan\frac{1}{2}(30°-60°)}{\tan\frac{1}{2}(30°+60°)}$

$= \dfrac{\tan(-15°)}{\tan 45°} \approx -0.26794919$

$\dfrac{a-b}{a+b} = \dfrac{2-2\sqrt{3}}{2+2\sqrt{3}} \cdot \dfrac{2-2\sqrt{3}}{2-2\sqrt{3}} = \dfrac{4-8\sqrt{3}+12}{4-12}$

$= \dfrac{16-8\sqrt{3}}{-8} = -2+\sqrt{3} \approx -0.26794919$

Thus, $\dfrac{\tan\frac{1}{2}(A-B)}{\tan\frac{1}{2}(A+B)} = \dfrac{a-b}{a+b}$ using the given

values of a, b, A, and B.

Chapter 7 Test

1. Find C, given $A = 25.2°$, $a = 6.92$ yd, $b = 4.82$ yd.
Use the law of sines to first find the measure of angle B.

$\dfrac{\sin 25.2°}{6.92} = \dfrac{\sin B}{4.82} \Rightarrow \sin B = \dfrac{4.82\sin 25.2°}{6.92} \Rightarrow$

$B = \sin^{-1}\left(\dfrac{4.82\sin 25.2°}{6.92}\right) \approx 17.3°$

Use the fact that the angles of a triangle sum to $180°$ to find the measure of angle C.
$C = 180° - A - B = 180° - 25.2° - 17.3° = 137.5°$

2. Find c, given $C = 118°$, $b = 131$ km, $a = 75.0$ km.
Using the law of cosines to find the length of c.

$c^2 = a^2 + b^2 - 2ab\cos C \Rightarrow c^2$
$= 75.0^2 + 131^2 - 2(75.0)(131)\cos 118°$
$\approx 32011.12 \Rightarrow c \approx 178.9$ km
c is approximately 179 km. (rounded to two significant digits)

3. Find B, given $a = 17.3$ ft, $b = 22.6$ ft, $c = 29.8$ ft.
Using the law of cosines, find the measure of angle B.

$b^2 = a^2 + c^2 - 2ac\cos B \Rightarrow$

$\cos B = \dfrac{a^2 + c^2 - b^2}{2ac} = \dfrac{17.3^2 + 29.8^2 - 22.6^2}{2(17.3)(29.8)}$

$\approx 0.65617605 \Rightarrow B \approx 49.0°$
B is approximately 49.0°.

4. $a = 14$, $b = 30$, $c = 40$
We can use Heron's formula to find the area.

$s = \tfrac{1}{2}(a+b+c) = \tfrac{1}{2}(14+30+40) = 42$

$\mathcal{A} = \sqrt{s(s-a)(s-b)(s-c)}$
$= \sqrt{42(42-14)(42-30)(42-40)}$
$= \sqrt{42\cdot 28\cdot 12\cdot 2} = \sqrt{28,224} = 168$ sq units

5. This is SAS, so we can use the formula
$\mathcal{A} = \tfrac{1}{2}zy\sin X$.

$\mathcal{A} = \dfrac{1}{2}\cdot 6\cdot 12\sin 30° = \dfrac{1}{2}\cdot 6\cdot 12\cdot \dfrac{1}{2} = 18$ sq units

6. B is obtuse, so b must be the longest side of the triangle.
(a) $b > 10$

(b) none

(c) $b \le 10$

7. $A = 60°, b = 30$ m, $c = 45$ m

This is SAS, so use the law of cosines to find

a: $a^2 = b^2 + c^2 - 2bc \cos A \Rightarrow$

$a^2 = 30^2 + 45^2 - 2 \cdot 30 \cdot 45 \cos 60° = 1575 \Rightarrow$

$a = 15\sqrt{7} \approx 40$ m

Now use the law of sines to find B:

$\dfrac{\sin B}{b} = \dfrac{\sin A}{a} \Rightarrow \dfrac{\sin B}{30} = \dfrac{\sin 60°}{15\sqrt{7}} \Rightarrow$

$\sin B = \dfrac{30 \sin 60°}{15\sqrt{7}} \Rightarrow B \approx 41°$

$C = 180° - A - B = 180° - 60° - 41° = 79°$

8. $b = 1075$ in., $c = 785$ in., $C = 38°30'$

We can use the law of sines.

$\dfrac{\sin B}{b} = \dfrac{\sin C}{c} \Rightarrow \dfrac{\sin B}{1075} = \dfrac{\sin 38°30'}{785} \Rightarrow$

$\sin B = \dfrac{1075 \sin 38°30'}{785} \Rightarrow$

$B_1 \approx 58.5° = 58°30'$ or

$B_2 = 180° - 58°30' = 121°30'$

Solving separately for triangles

$A_1 B_1 C$ and $A_2 B_2 C$, we have the following.

$A_1 B_1 C$:

$A_1 = 180° - B_1 - C = 180° - 58°30' - 38°30'$

$\quad = 83°00'$

$\dfrac{a_1}{\sin A_1} = \dfrac{b}{\sin B_1} \Rightarrow \dfrac{a_1}{\sin 83°} = \dfrac{1075}{\sin 58°30'} \Rightarrow$

$a_1 = \dfrac{1075 \sin 83°}{\sin 58°30'} \approx 1251 \approx 1250$ in. (rounded

to three significant digits)

$A_2 B_2 C$:

$A_2 = 180° - B_2 - C = 180° - 121°30' - 38°30'$

$\quad = 20°00'$

$\dfrac{a_2}{\sin A_2} = \dfrac{b}{\sin B_2} \Rightarrow \dfrac{a_2}{\sin 20°} = \dfrac{1075}{\sin 121°30'} \Rightarrow$

$a_2 = \dfrac{1075 \sin 20°}{\sin 121°30'} \approx 431$ in. (rounded to

three significant digits)

9. magnitude:

$|\mathbf{v}| = \sqrt{(-6)^2 + 8^2} = \sqrt{36 + 64} = \sqrt{100} = 10$

angle:

$\tan \theta' = \dfrac{y}{x} \Rightarrow$

$\tan \theta' = \dfrac{8}{-6} = -\dfrac{4}{3} \approx -1.33333333 \Rightarrow$

$\theta' \approx -53.1° \Rightarrow \theta = -53.1° + 180° = 126.9°$

(θ lies in quadrant II)

The magnitude $|\mathbf{v}|$ is 10 and $\theta = 126.9°$.

10.

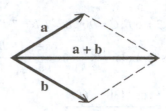

11. $\mathbf{u} = \langle -1, 3 \rangle, \mathbf{v} = \langle 2, -6 \rangle$

(a) $\mathbf{u} + \mathbf{v} = \langle -1, 3 \rangle + \langle 2, -6 \rangle$

$\quad = \langle -1 + 2, 3 + (-6) \rangle = \langle 1, -3 \rangle$

(b) $-3\mathbf{v} = -3\langle 2, -6 \rangle = \langle -3 \cdot 2, -3(-6) \rangle$

$\quad = \langle -6, 18 \rangle$

(c) $\mathbf{u} \cdot \mathbf{v} = \langle -1, 3 \rangle \cdot \langle 2, -6 \rangle = -1(2) + 3(-6)$

$\quad = -2 - 18 = -20$

(d) $|\mathbf{u}| = \sqrt{(-1)^2 + 3^2} = \sqrt{1 + 9} = \sqrt{10}$

12. $\mathbf{u} = \langle 4, 3 \rangle, \mathbf{v} = \langle 1, 5 \rangle$

$\mathbf{u} \cdot \mathbf{v} = \langle 4, 3 \rangle \cdot \langle 1, 5 \rangle = 4(1) + 3(5) = 19$

$|\mathbf{u}| = \sqrt{4^2 + 3^2} = \sqrt{25} = 5$

$|\mathbf{v}| = \sqrt{1^2 + 5^2} = \sqrt{26}$

$\cos \theta = \dfrac{\mathbf{u} \cdot \mathbf{v}}{|\mathbf{u}||\mathbf{v}|} \Rightarrow \cos \theta = \dfrac{19}{5\sqrt{26}} \Rightarrow$

$\theta = \cos^{-1}\left(\dfrac{19}{5\sqrt{26}}\right) \approx 41.8°$

13. $\mathbf{u} = \langle -4, 7 \rangle, \mathbf{v} = \langle -14, -8 \rangle$

$\mathbf{u} \cdot \mathbf{v} = \langle -4, 7 \rangle \cdot \langle -14, -8 \rangle = -4(-14) + 7(-8)$

$\quad = 56 - 56 = 0$

The dot product is 0, so the vectors are orthogonal.

14. Given $A = 24° 50', B = 47° 20'$ and

$AB = 8.4$ mi, first find the measure of angle C.

$C = 180° - 47° 20' - 24° 50'$

$\quad = 179°60' - 72° 10' = 107° 50'$

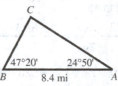

Use this information and the law of sines to find AC.

$\dfrac{AC}{\sin 47° 20'} = \dfrac{8.4}{\sin 107° 50'} \Rightarrow$

$AC = \dfrac{8.4 \sin 47° 20'}{\sin 107°50'} \approx 6.49$ mi

(continued on next page)

(*continued*)

Drop a perpendicular line from C to segment AB.

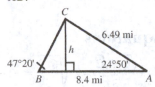

Thus, $\sin 24°50' = \dfrac{h}{6.49} \Rightarrow$

$h \approx 6.49 \sin 24°50' \approx 2.7$ mi.

The balloon is 2.7 miles off the ground.

15. horizontal:

$x = |\mathbf{v}| \cos\theta = 569 \cos 127.5° \approx -346$ and

vertical: $y = |\mathbf{v}| \sin\theta = 569 \sin 127.5° \approx 451$

The vector is $\langle -346, 451 \rangle$.

16.

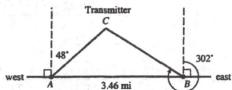

The bearing is 48° from A, so angle A in ABC must be $90° - 48° = 42°$. The bearing is 302° from B, so angle B in ABC must be $302° - 270° = 32°$. The angles of a triangle sum to 180°, so
$C = 180° - A - B = 180° - 42° - 32° = 106°$.
Using the law of sines, we have

$$\frac{b}{\sin B} = \frac{c}{\sin C} \Rightarrow \frac{b}{\sin 32°} = \frac{3.46}{\sin 106°} \Rightarrow$$

$$b = \frac{3.46 \sin 32°}{\sin 106°} \approx 1.91 \text{ mi}$$

The distance from A to the transmitter is 1.91 miles. (rounded to two significant digits)

17.

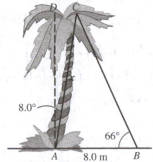

$m\angle DAC = 8.0°$, so
$m\angle CAB = 90° - 8.0° = 82.0°$. $m\angle B = 66°$, so
$m\angle C = 180° - 82° - 66° = 32°$. Now use the law of sines to find AC:

$$\frac{AC}{\sin B} = \frac{AB}{\sin C} \Rightarrow \frac{AC}{\sin 66°} = \frac{8.0}{\sin 32°} \Rightarrow$$

$$AC = \frac{8.0 \sin 66°}{\sin 32°} \approx 13.8 \approx 14 \text{ m}$$

18.

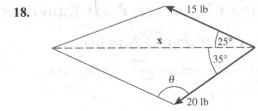

Let $|\mathbf{x}|$ be the equilibrant force.
$\theta = 180° - 35° - 25° = 120°$
Using the law of cosines, we have
$|\mathbf{x}|^2 = 15^2 + 20^2 - 2 \cdot 15 \cdot 20 \cos 120° \Rightarrow$
$|\mathbf{x}|^2 = 925 \Rightarrow \mathbf{x} \approx 30.4 \approx 30 \text{ lb}$

19. $\mathbf{AX}$ is the airspeed vector. The plane is flying 630 miles due north in 3 hours, so the ground speed is 210 mph. The measure of angle ACX is $180° - 42° = 138°$.

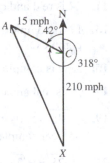

$$\left|\overrightarrow{\mathbf{AX}}\right|^2 = 15^2 + 210^2 - 2(15)(210)\cos 138°$$
$$= 49006.8124 \Rightarrow$$
$$\left|\overrightarrow{\mathbf{AX}}\right| \approx 221.3748 \approx 220 \text{ mph (rounded to two}$$

significant digits)
Using the law of sines to find the measure of angle X, we have

$$\frac{\sin X}{15} = \frac{\sin 138°}{220} \Rightarrow \sin X = \frac{15 \sin 138°}{220} \Rightarrow$$

$$X = \sin^{-1}\left(\frac{15 \sin 138°}{220}\right) \approx 2.6°$$

The plane's bearing is $360° - 2.6° = 357.4° \approx 357°$.

20. $\left|\overrightarrow{AC}\right| = 16.0 \text{ lb}$

$\left|\overrightarrow{BA}\right| = 50.0 \text{ lb}$

$\sin\theta = \dfrac{16.0}{50.0}$

$\theta = \sin^{-1}\left(\dfrac{16.0}{50.0}\right)$

$\approx 18.7°$

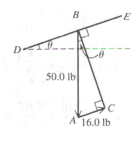

Chapter 8

Complex Numbers, Polar Equations, and Parametric Equations

Section 8.1 Complex Numbers

1. By definition, $i = \sqrt{-1}$, and therefore,

   $i^2 = \underline{-1}$.

3. In terms of i, $\sqrt{-100} = \underline{10i}$.

5. True

7. True

9. False Every real number is a complex number.

11. -4 is real and complex.

13. $13i$ is complex, pure imaginary and nonreal complex.

15. $5 + i$ is complex and nonreal complex.

17. π is real and complex.

19. $\sqrt{-25} = 5i$ is complex, pure imaginary and nonreal complex.

21. $\sqrt{-25} = i\sqrt{25} = 5i$

23. $\sqrt{-10} = i\sqrt{10}$

25. $\sqrt{-288} = i\sqrt{288} = i\sqrt{144 \cdot 2} = 12i\sqrt{2}$

27. $-\sqrt{-18} = -i\sqrt{18} = -i\sqrt{9 \cdot 2} = -3i\sqrt{2}$

29. $\sqrt{-13} \cdot \sqrt{-13} = i\sqrt{13} \cdot i\sqrt{13}$
    $= i^2 \left(\sqrt{13}\right)^2 = -1 \cdot 13 = -13$

31. $\sqrt{-3} \cdot \sqrt{-8} = i\sqrt{3} \cdot i\sqrt{8} = i^2\sqrt{3 \cdot 8}$
    $= -1 \cdot \sqrt{24} = -\sqrt{4 \cdot 6} = -2\sqrt{6}$

33. $\dfrac{\sqrt{-30}}{\sqrt{-10}} = \dfrac{i\sqrt{30}}{i\sqrt{10}} = \sqrt{\dfrac{30}{10}} = \sqrt{3}$

35. $\dfrac{\sqrt{-24}}{\sqrt{8}} = \dfrac{i\sqrt{24}}{\sqrt{8}} = i\sqrt{\dfrac{24}{8}} = i\sqrt{3}$

37. $\dfrac{\sqrt{-10}}{\sqrt{-40}} = \dfrac{i\sqrt{10}}{i\sqrt{40}} = \sqrt{\dfrac{10}{40}} = \sqrt{\dfrac{1}{4}} = \dfrac{1}{2}$

39. $\dfrac{\sqrt{-6} \cdot \sqrt{-2}}{\sqrt{3}} = \dfrac{i\sqrt{6} \cdot i\sqrt{2}}{\sqrt{3}} = i^2\sqrt{\dfrac{6 \cdot 2}{3}} = -1 \cdot \sqrt{\dfrac{12}{3}}$
    $= -\sqrt{4} = -2$

41. $\dfrac{-6 - \sqrt{-24}}{2} = \dfrac{-6 - \sqrt{-4 \cdot 6}}{2} = \dfrac{-6 - 2i\sqrt{6}}{2}$
    $= \dfrac{2\left(-3 - i\sqrt{6}\right)}{2} = -3 - i\sqrt{6}$

43. $\dfrac{10 + \sqrt{-200}}{5} = \dfrac{10 + \sqrt{-100 \cdot 2}}{5}$
    $= \dfrac{10 + 10i\sqrt{2}}{5} = \dfrac{5\left(2 + 2i\sqrt{2}\right)}{5}$
    $= 2 + 2i\sqrt{2}$

45. $\dfrac{-3 + \sqrt{-18}}{24} = \dfrac{-3 + \sqrt{-9 \cdot 2}}{24} = \dfrac{-3 + 3i\sqrt{2}}{24}$
    $= \dfrac{3\left(-1 + i\sqrt{2}\right)}{24} = \dfrac{-1 + i\sqrt{2}}{8}$
    $= -\dfrac{1}{8} + \dfrac{\sqrt{2}}{8}i$

47. $(3 + 2i) + (9 - 3i) = (3 + 9) + \left[2 + (-3)\right]i$
    $= 12 + (-1)i = 12 - i$

49. $(-2 + 4i) - (-4 + 4i)$
    $= \left[-2 - (-4)\right] + (4 - 4)i$
    $= 2 + 0i = 2$

51. $(2 - 5i) - (3 + 4i) - (-1 - 9i)$
    $= \left[2 - 3 - (-1)\right] + \left[-5 - 4 - (-9)\right]i = 0$

53. $-i\sqrt{2} - 2 - \left(6 - 4i\sqrt{2}\right) - \left(5 - i\sqrt{2}\right)$
    $= (-2 - 6 - 5) + \left[-1 - (-4) - (-1)\right]i\sqrt{2}$
    $= -13 + 4i\sqrt{2}$

55. $(2 + i)(3 - 2i)$
    $= 2(3) + 2(-2i) + i(3) + i(-2i)$
    $= 6 - 4i + 3i - 2i^2 = 6 - i - 2(-1)$
    $= 6 - i + 2 = 8 - i$

57. $(2+4i)(-1+3i)$

$= 2(-1)+2(3i)+4i(-1)+4i(3i)$

$= -2+6i-4i+12i^2 = -2+2i+12(-1)$

$= -2+2i-12 = -14+2i$

59. $(3-2i)^2 = 3^2 - 2(3)(2i)+(2i)^2$

$= 9-12i-4 = 5-12i$

61. $(3+i)(3-i) = 3^2 - i^2 = 9-(-1) = 10$

63. $(-2-3i)(-2+3i) = (-2)^2 - (3i)^2 = 4-9i^2$

$= 4-9(-1) = 13$

65. $(\sqrt{6}+i)(\sqrt{6}-i) = (\sqrt{6})^2 - i^2$

$= 6-(-1) = 6+1 = 7$

67. $i(3-4i)(3+4i) = i\left[(3-4i)(3+4i)\right]$

$= i\left[3^2 - (4i)^2\right] = i\left[9-16i^2\right]$

$= i\left[9-16(-1)\right]$

$= i(9+16) = 25i$

69. $3i(2-i)^2 = 3i\left(2^2 - 2(2i)+i^2\right)$

$= 3i(4-4i-1) = 3i(3-4i)$

$= 9i-12i^2 = 9i-12(-1)$

$= 12+9i$

71. $(2+i)(2-i)(4+3i) = \left[(2+i)(2-i)\right](4+3i)$

$= \left[2^2 - i^2\right](4+3i)$

$= \left[4-(-1)\right](4+3i)$

$= 5(4+3i) = 20+15i$

73. $\dfrac{6+2i}{1+2i} = \dfrac{(6+2i)(1-2i)}{(1+2i)(1-2i)}$

$= \dfrac{6-12i+2i-4i^2}{1^2 - (2i)^2} = \dfrac{6-10i-4(-1)}{1-4i^2}$

$= \dfrac{6-10i+4}{1-4(-1)} = \dfrac{10-10i}{1+4} = \dfrac{10-10i}{5}$

$= \dfrac{10}{5} - \dfrac{10}{5}i = 2-2i$

75. $\dfrac{2-i}{2+i} = \dfrac{(2-i)(2-i)}{(2+i)(2-i)} = \dfrac{2^2 - 2(2i)+i^2}{2^2 - i^2}$

$= \dfrac{4-4i+(-1)}{4-(-1)} = \dfrac{3-4i}{5} = \dfrac{3}{5} - \dfrac{4}{5}i$

77. $\dfrac{1-3i}{1+i} = \dfrac{(1-3i)(1-i)}{(1+i)(1-i)} = \dfrac{1-i-3i+3i^2}{1^2 - i^2}$

$= \dfrac{1-4i+3(-1)}{1-(-1)} = \dfrac{1-4i-3}{2}$

$= \dfrac{-2-4i}{2} = \dfrac{-2}{2} - \dfrac{4}{2}i = -1-2i$

79. $\dfrac{-5}{i} = \dfrac{-5(-i)}{i(-i)} = \dfrac{5i}{-i^2} = \dfrac{5i}{-(-1)} = \dfrac{5i}{1}$

$= 5i$ or $0+5i$

81. $\dfrac{8}{-i} = \dfrac{8 \cdot i}{-i \cdot i} = \dfrac{8i}{-i^2} = \dfrac{8i}{-(-1)} = \dfrac{8i}{1} = 8i$ or $0+8i$

83. $\dfrac{2}{3i} = \dfrac{2(-3i)}{3i \cdot (-3i)} = \dfrac{-6i}{-9i^2} = \dfrac{-6i}{-9(-1)}$

$= \dfrac{-6i}{9} = -\dfrac{2}{3}i$ or $0 - \dfrac{2}{3}i$

Note: In the above solution, we multiplied the numerator and denominator by the complex conjugate of $3i$, namely $-3i$. There is a reduction in the end, so the same results can be achieved by multiplying the numerator and denominator by $-i$.

85. $x^2 = -16 \Rightarrow x = \pm\sqrt{-16} = \pm 4i$

Solution set: $\{\pm 4i\}$

87. $x^2 + 12 = 0 \Rightarrow x^2 = -12 \Rightarrow x = \pm\sqrt{-12} \Rightarrow$

$x = \pm 2i\sqrt{3}$

Solution set: $\left\{\pm 2i\sqrt{3}\right\}$

89. $3x^2 + 2 = -4x \Rightarrow 3x^2 + 4x + 2 = 0$

Use the quadratic formula with $a = 3$, $b = 4$, and $c = 2$ to solve for x.

$x = \dfrac{-4 \pm \sqrt{4^2 - 4(3)(2)}}{2(3)} = \dfrac{-4 \pm \sqrt{-8}}{6}$

$= -\dfrac{2}{3} \pm \dfrac{\sqrt{2}}{3}i$

Solution set: $\left\{-\dfrac{2}{3} \pm \dfrac{\sqrt{2}}{3}i\right\}$

91. $x^2 - 6x + 14 = 0$

Use the quadratic formula with $a = 1$, $b = -6$, and $c = 14$ to solve for x:

$x = \dfrac{-(-6) \pm \sqrt{(-6)^2 - 4(1)(14)}}{2(1)} = \dfrac{6 \pm \sqrt{-20}}{2}$

$= 3 \pm i\sqrt{5}$

Solution set: $\left\{3 \pm i\sqrt{5}\right\}$

93. $4\left(x^2 - x\right) = -7 \Rightarrow 4x^2 - 4x + 7 = 0$

Use the quadratic formula with $a = 4$, $b = -4$, and $c = 7$ to solve for x:

$$x = \frac{-(-4) \pm \sqrt{(-4)^2 - 4(4)(7)}}{2(4)}$$

$$= \frac{4 \pm \sqrt{-96}}{8} = \frac{1}{2} \pm \frac{\sqrt{6}}{2} i$$

Solution set: $\left\{ \dfrac{1}{2} \pm \dfrac{\sqrt{6}}{2} i \right\}$

95. $x^2 + 1 = -x \Rightarrow x^2 + x + 1 = 0$

Use the quadratic formula with $a = 1$, $b = 1$, and $c = 1$ to solve for x:

$$x = \frac{-1 \pm \sqrt{1^2 - 4(1)(1)}}{2(1)} = \frac{-1 \pm \sqrt{-3}}{2}$$

$$= -\frac{1}{2} \pm \frac{\sqrt{3}}{2} i$$

Solution set: $\left\{ -\dfrac{1}{2} \pm \dfrac{\sqrt{3}}{2} i \right\}$

97. $i^{25} = i^{24} \cdot i = \left(i^4\right)^6 \cdot i = 1^6 \cdot i = i$

99. $i^{22} = i^{20} \cdot i^2 = \left(i^4\right)^5 \cdot (-1) = 1^5 \cdot (-1) = -1$

101. $i^{23} = i^{20} \cdot i^3 = \left(i^4\right)^5 \cdot i^3 = 1^5 \cdot (-i) = -i$

103. $i^{32} = \left(i^4\right)^8 = 1^8 = 1$

105. $i^{-13} = i^{-16} \cdot i^3 = \left(i^4\right)^{-4} \cdot i^3 = 1^{-4} \cdot (-i) = -i$

107. $\dfrac{1}{i^{-11}} = i^{11} = i^8 \cdot i^3 = \left(i^4\right)^2 \cdot i^3 = 1^2 \cdot (-i) = -i$

109. Answers will vary. Sample answer: Every i^4 factor equals 1, so if the remainder is R, the final product is i^R.

111. The resistive part is $50 + 60 = 110$, and the reactive part is $15i + 17i = 32i$, so the total impedance is $Z = 110 + 32i$.

113. $I = 8 + 6i$, $Z = 6 + 3i$

$E = IZ \Rightarrow$

$E = (8 + 6i)(6 + 3i) = 48 + 24i + 36i + 18i^2$

$= 48 + 60i - 18 = 30 + 60i$

115. $I = 7 + 5i$, $E = 28 + 54i$

$E = IZ \Rightarrow$

$28 + 54i = (7 + 5i)Z$

$$Z = \frac{28 + 54i}{7 + 5i} = \frac{28 + 54i}{7 + 5i} \cdot \frac{7 - 5i}{7 - 5i}$$

$$= \frac{196 - 140i + 378i - 270i^2}{49 - 25i^2}$$

$$= \frac{196 + 238i + 270}{49 + 25} = \frac{466 + 238i}{74}$$

$$= \frac{2(233 + 119i)}{2(37)} = \frac{233}{37} + \frac{119i}{37}$$

117. We need to show that $\left(\dfrac{\sqrt{2}}{2} + \dfrac{\sqrt{2}}{2} i\right)^2 = i$.

$$\left(\frac{\sqrt{2}}{2} + \frac{\sqrt{2}}{2} i\right)^2$$

$$= \left(\frac{\sqrt{2}}{2}\right)^2 + 2 \cdot \frac{\sqrt{2}}{2} \cdot \frac{\sqrt{2}}{2} i + \left(\frac{\sqrt{2}}{2} i\right)^2$$

$$= \frac{2}{4} + 2 \cdot \frac{2}{4} i + \frac{2}{4} i^2 = \frac{1}{2} + i + \frac{1}{2} i^2$$

$$= \frac{1}{2} + i + \frac{1}{2}(-1) = \frac{1}{2} + i - \frac{1}{2} = i$$

119. $x^2 + 4x + 5 = 0$; $x = -2 + i$

$$(-2 + i)^2 + 4(-2 + i) + 5 \overset{?}{=} 0$$

$$\left(4 - 4i + i^2\right) - 8 + 4i + 5 \overset{?}{=} 0$$

$$4 - 4i - 1 - 8 + 4i + 5 \overset{?}{=} 0$$

$$(4 - 1 - 8 + 5) - 4i + 4i \overset{?}{=} 0$$

$$0 = 0$$

Thus, $-2 + i$ is a solution of the equation.

Section 8.2 Trigonometric (Polar) Form of Complex Numbers

1. (a) 2

(b) $2(\cos 0° + i \sin 0°)$

3. (a) $2i$

(b) $2(\cos 90° + i \sin 90°)$

5. (a) $2 + 2i$

(b) $2\sqrt{2}(\cos 45° + i \sin 45°)$

7. The absolute value (or modulus) of a complex number represents the <u>length (or magnitude)</u> of the vector representing it in the complex plane.

9.

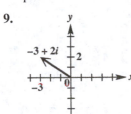

11.

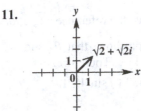

13.

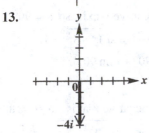

15.

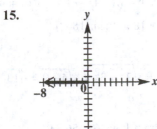

17. $(4-3i)+(-1+2i)=3-i$

19. $(5-6i)+(-5+3i)=-3i$

21. $-3+3i$

23. $(-5-8i)-1=-6-8i$

25. $(7+6i)+3i=7+9i$

27. $\left(\dfrac{1}{2}+\dfrac{2}{3}i\right)+\left(\dfrac{2}{3}+\dfrac{1}{2}i\right)=\dfrac{7}{6}+\dfrac{7}{6}i$

29. $2\left(\cos 45^\circ + i\sin 45^\circ\right)=2\left(\dfrac{\sqrt{2}}{2}+i\dfrac{\sqrt{2}}{2}\right)$

$$=\sqrt{2}+i\sqrt{2}$$

31. $10\left(\cos 90^\circ + i\sin 90^\circ\right)=10\left(0+i\right)$

$$=0+10i=10i$$

33. $4\left(\cos 240^\circ + i\ \sin 240^\circ\right)=4\left(-\dfrac{1}{2}-i\dfrac{\sqrt{3}}{2}\right)$

$$=-2-2i\sqrt{3}$$

35. $3\operatorname{cis}150^\circ = 3\left(\cos\ 150^\circ + i\sin 150^\circ\right)$

$$=3\left(-\dfrac{\sqrt{3}}{2}+\dfrac{1}{2}i\right)=-\dfrac{3\sqrt{3}}{2}+\dfrac{3}{2}i$$

37. $5\operatorname{cis}300^\circ = 5\left(\cos 300^\circ + i\sin 300^\circ\right)$

$$=5\left[\dfrac{1}{2}+\left(-\dfrac{\sqrt{3}}{2}\right)i\right]=\dfrac{5}{2}-\dfrac{5\sqrt{3}}{2}i$$

39. $\sqrt{2}\operatorname{cis}225^\circ = \sqrt{2}\left(\cos 225^\circ + i\ \sin 225^\circ\right)$

$$=\sqrt{2}\left[-\dfrac{\sqrt{2}}{2}+\left(-\dfrac{\sqrt{2}}{2}i\right)\right]$$

$$=-1-i$$

41. $4\left(\cos\left(-30^\circ\right)+i\sin\left(-30^\circ\right)\right)$

$$=4\left(\cos 30^\circ - i\sin 30^\circ\right)=4\left(\dfrac{\sqrt{3}}{2}-\dfrac{1}{2}i\right)$$

$$=2\sqrt{3}-2i$$

43. $-3-3i\sqrt{3}$

Sketch a graph of $-3-3i\sqrt{3}$ in the complex plane.

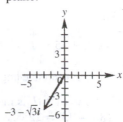

(continued on next page)

(*continued*)

$x = -3$ and $y = -3\sqrt{3}$,

$$r = \sqrt{(-3)^2 + \left(-3\sqrt{3}\right)^2} = \sqrt{9 + 27} = \sqrt{36} = 6$$

and $\tan\theta = \dfrac{-3\sqrt{3}}{-3} = \sqrt{3}$. Thus, the reference

angle for θ is 60°. The graph shows that θ is in quadrant III, so $\theta = 180° + 60° = 240°$. Therefore,

$$-3 - 3i\sqrt{3} = 6\left(\cos 240° + i\sin 240°\right)$$

45. $\sqrt{3} - i$

Sketch a graph of $\sqrt{3} - i$ in the complex plane.

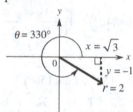

$x = \sqrt{3}$ and $y = -1$, so

$$r = \sqrt{\left(\sqrt{3}\right)^2 + (-1)^2} = \sqrt{3 + 1} = \sqrt{4} = 2.$$

$\tan\theta = \dfrac{-1}{\sqrt{3}} = -\dfrac{\sqrt{3}}{3}$, so the reference angle

for θ is 30°. The graph shows that θ is in quadrant IV, so $\theta = 360° - 30° = 330°$.

Therefore, $\sqrt{3} - i = 2\left(\cos 330° + i\sin 330°\right)$.

47. $-5 - 5i$

Sketch a graph of $-5 - 5i$ in the complex plane.

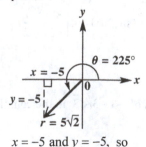

$x = -5$ and $y = -5$, so

$$r = \sqrt{(-5)^2 + (-5)^2} = \sqrt{25 + 25} = \sqrt{50} = 5\sqrt{2}.$$

$\tan\theta = \dfrac{y}{x} = \dfrac{-5}{-5} = 1$, so the reference angle for

θ is 45°. The graph shows that θ is in quadrant III, so $\theta = 180° + 45° = 225°$.

Therefore,

$$-5 - 5i = 5\sqrt{2}\left(\cos 225° + i\sin 225°\right).$$

49. $2 + 2i$

Sketch a graph of $2 + 2i$ in the complex plane.

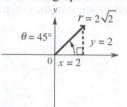

$x = 2, y = 2$, so

$$r = \sqrt{2^2 + 2^2} = \sqrt{4 + 4} = \sqrt{8} = 2\sqrt{2} \text{ and}$$

$\tan\theta = \dfrac{2}{2} = 1$. Thus, the reference angle for

θ is 45°. The graph shows that θ is in quadrant I, so $\theta = 45°$. Therefore,

$$2 + 2i = 2\sqrt{2}\left(\cos 45° + i\sin 45°\right).$$

51. $5i = 0 + 5i$

$0 + 5i$ is on the positive y-axis, so $\theta = 90°$

and $x = 0, y = 5 \Rightarrow r = \sqrt{0^2 + 5^2} = 5$

Thus, $5i = 5\left(\cos 90° + i\sin 90°\right)$.

53. $-4 = -4 + 0i$

$-4 + 0i$ is on the negative x-axis, so $\theta = 180°$

and $x = -4, y = 0$. $r = \sqrt{(-4)^2 + 0^2} = \sqrt{16} = 4$

Thus, $-4 = 4\left(\cos 180° + i\sin 180°\right)$.

55. $2 + 3i$

$$x = 2, y = 3 \Rightarrow r = \sqrt{2^2 + 3^2} = \sqrt{4 + 9} = \sqrt{13}$$

$\tan\theta = \dfrac{3}{2}$

$2 + 3i$ is in quadrant I, so $\theta = 56.31°$.

$$2 + 3i = \sqrt{13}\left(\cos 56.31° + i\sin 56.31°\right)$$

57. $3\left(\cos 250° + i\sin 250°\right) = -1.0261 - 2.8191i$

59. $12i = 0 + 12i$

$$x = 0, y = 12 \Rightarrow r = \sqrt{0^2 + 12^2} = 12$$

$0 + 12i$ is on the positive y-axis, so $0 = 90°$.

$12i = 12(\cos 90° + i\sin 90°)$

61. $3 + 5i$

$$x = 3, y = 5 \Rightarrow r = \sqrt{3^2 + 5^2} = \sqrt{9 + 25} = \sqrt{34}$$

$\tan\theta = \dfrac{5}{3} \Rightarrow \theta = \tan^{-1}\left(\dfrac{5}{3}\right)$

$3 + 5i$ is in quadrant I, so $\theta = 59.04°$.

$$3 + 5i = \sqrt{34}\left(\cos 59.04° + i\sin 59.04°\right)$$

63. The modulus represents the magnitude of the vector in the complex plane, so $z = 1$ represents a circle of radius one centered at the origin.

65. The real part of $z = x + yi$ is 1, the graph of $1 + yi$ is the vertical line $x = 1$.

67. $z = -0.2i$

$$z^2 - 1 = (-0.2i)^2 - 1 = 0.04i^2 - 1 = 0.04(-1) - 1$$
$$= -0.04 - 1 = -1.04$$

The modulus is 1.04.

$$(z^2 - 1)^2 - 1 = (-1.04)^2 - 1 = 0.0816$$

The modulus is 0.0816.

$$\left[(z^2 - 1)^2 - 1\right]^2 - 1 = (0.0816)^2 - 1$$
$$= -0.99334144$$

The modulus is 0.99334144. The moduli do not exceed 2. Therefore, z is in the Julia set.

69. B. $(a + bi) - (c + di) = e + 0i$, so

$b - d = 0 \Rightarrow b = d$. Therefore, the terminal points of the vectors corresponding to $a + bi$ and $c + di$ lie on a horizontal line.

71. A. The difference of the vectors is equal to the sum of one of the vectors plus the opposite of the other vector. If this sum is represented by two parallel vectors, the sum of their absolute values will equal the absolute value of their sum. In order for this sum to be represented by two parallel vectors, their difference must be represented by two vectors with opposite directions.

Section 8.3   The Product and Quotient Theorems

1. When multiplying two complex numbers in trigonometric form, we <u>multiply</u> their absolute values and <u>add</u> their arguments.

3. $\left[5(\cos 150° + i\sin 150°)\right] \cdot \left[2(\cos 30° + i\sin 30°)\right] = \underline{10}\left[\cos \underline{180°} + i\sin \underline{180°}\right] = \underline{-10} + \underline{0}i$

5. $\text{cis}(-1000°) \cdot \text{cis } 1000° = \text{cis } \underline{0°} = \underline{1} + \underline{0}i$

7. $\left[3(\cos 60° + i\sin 60°)\right]\left[2(\cos 90° + i\sin 90°)\right] = 3 \cdot 2\left[\cos(60° + 90°) + i\sin(60° + 90°)\right]$

$$= 6(\cos 150° + i\sin 150°) = 6\left(-\frac{\sqrt{3}}{2} + \frac{1}{2}i\right) = -3\sqrt{3} + 3i$$

9. $\left[4(\cos 60° + i\sin 60°)\right] \cdot \left[6(\cos 330° + i\sin 330°)\right] = 4 \cdot 6\left[(\cos(60° + 330°) + i\sin(60° + 330°)\right]$

$$= 24(\cos 390° + i\sin 390°) = 24(\cos 30° + i\sin 30°)$$

$$= 24\left(\frac{\sqrt{3}}{2} + i\frac{1}{2}\right) = 12\sqrt{3} + 12i$$

11. $\left[2(\cos 135° + i\sin 135°)\right] \cdot \left[2(\cos 225° + i\sin 225°)\right] = 2 \cdot 2\left[\cos(135° + 225°) + i\sin(135° + 225°)\right]$

$$= 4(\cos 360° + i\sin 360°) = 4(1 - 0i) = 4$$

13. $\left[\sqrt{3} \text{ cis } 45°\right]\left[\sqrt{3} \text{ cis } 225°\right] = \sqrt{3} \cdot \sqrt{3}\left[\text{cis}(45° + 225°)\right]$

$$= 3 \text{ cis } 270° = 3(\cos 270° + i\sin 270°) = 3(0 - i) = 0 - 3i \text{ or } -3i$$

15. $[5 \text{ cis } 90°][3 \text{ cis } 45°] = 5 \cdot 3\left[\text{cis}(90° + 45°)\right] = 15 \text{ cis } 135°$

$$= 15(\cos 135° + i\sin 135°) = 15\left(-\frac{\sqrt{2}}{2} + \frac{\sqrt{2}}{2}i\right) = -\frac{15\sqrt{2}}{2} + \frac{15\sqrt{2}}{2}i$$

17. $\dfrac{4(\cos 150° + i\sin 150°)}{2(\cos 120° + i\sin 120°)} = \dfrac{4}{2}\left[\cos(150° - 120°) + i\sin(150° - 120°)\right] = 2(\cos 30° + i\sin 30°) = 2\left(\frac{\sqrt{3}}{2} + \frac{1}{2}i\right)$

$$= \sqrt{3} + i$$

19. $\dfrac{10\left(\cos 50° + i\sin 50°\right)}{5\left(\cos 230° + i\sin 230°\right)} = \dfrac{10}{5}\left[\cos\left(50° - 230°\right) + i\sin\left(50° - 230°\right)\right]$

$= 2\left(\cos\left(-180°\right) + i\sin\left(-180°\right)\right) = 2\left(-1 - 0 \cdot i\right) = -2 + 0i$ or -2

21. $\dfrac{3\,\text{cis}\,305°}{9\,\text{cis}\,65°} = \dfrac{1}{3}\text{cis}\left(305° - 65°\right) = \dfrac{1}{3}\left(\text{cis}\,240°\right) = \dfrac{1}{3}\left(\cos 240° + i\sin 240°\right) = \dfrac{1}{3}\left(-\dfrac{1}{2} - \dfrac{\sqrt{3}}{2}i\right) = -\dfrac{1}{6} - \dfrac{\sqrt{3}}{6}i$

23. $\dfrac{8}{\sqrt{3}+i}$

numerator: $8 = 8 + 0i$ and $r = \sqrt{8^2 + 0^2} = 8$

$\theta = 0°$ because $\cos 0° = 1$ and $\sin 0° = 0$, so $8 = 8\,\text{cis}\,0°$.

denominator: $\sqrt{3}+i$ and $r = \sqrt{\left(\sqrt{3}\right)^2 + 1^2} = \sqrt{3+1} = \sqrt{4} = 2$; $\tan\theta = \dfrac{1}{\sqrt{3}} = \dfrac{\sqrt{3}}{3}$

Because x and y are both positive, θ is in quadrant I, so $\theta = 30°$.

Thus $\sqrt{3}+i = 2\,\text{cis}\,30°$.

$\dfrac{8}{\sqrt{3}+i} = \dfrac{8\,\text{cis}\,0°}{2\,\text{cis}\,30°} = \dfrac{8}{2}\text{cis}\left(0 - 30°\right) = 4\left[\cos\left(-30°\right) + i\sin\left(-30°\right)\right] = 4\left(\dfrac{\sqrt{3}}{2} - \dfrac{1}{2}i\right) = 2\sqrt{3} - 2i$

25. $\dfrac{-i}{1+i}$

numerator: $-i = 0 - i$ and

$r = \sqrt{0^2 + \left(-1\right)^2} = \sqrt{0+1} = \sqrt{1} = 1$

$\theta = 270°$ because $\cos 270° = 0$ and $\sin 270° = -1$, so $-i = 1\,\text{cis}\,270°$.

denominator: $1 + i$

$r = \sqrt{1^2 + 1^2} = \sqrt{1+1} = \sqrt{2}$ and

$\tan\theta = \dfrac{y}{x} = \dfrac{1}{1} = 1$

Because x and y are both positive, θ is in quadrant I, so $\theta = 45°$. Thus,

$1 + i = \sqrt{2}\,\text{cis}\,45°$

$\dfrac{-i}{1+i} = \dfrac{\text{cis}\,270°}{\sqrt{2}\,\text{cis}\,45°} = \dfrac{1}{\sqrt{2}}\text{cis}\left(270° - 45°\right)$

$= \dfrac{\sqrt{2}}{2}\,\text{cis}\,225°$

$= \dfrac{\sqrt{2}}{2}\left(\cos 225° + i\sin 225°\right)$

$= \dfrac{\sqrt{2}}{2}\left(-\dfrac{\sqrt{2}}{2} - i\cdot\dfrac{\sqrt{2}}{2}\right) = -\dfrac{1}{2} - \dfrac{1}{2}i$

27. $\dfrac{2\sqrt{6} - 2i\sqrt{2}}{\sqrt{2} - i\sqrt{6}}$

numerator: $2\sqrt{6} - 2i\sqrt{2}$ and

$r = \sqrt{\left(2\sqrt{6}\right)^2 + \left(-2\sqrt{2}\right)^2} = \sqrt{24+8}$

$= \sqrt{32} = 4\sqrt{2}$

$\tan\theta = \dfrac{-2\sqrt{2}}{2\sqrt{6}} = -\dfrac{1}{\sqrt{3}} = -\dfrac{\sqrt{3}}{3}$

Because x is positive and y is negative, θ is in quadrant IV, so $\theta = -30°$. Thus,

$2\sqrt{6} - 2i\sqrt{2} = 4\sqrt{2}\,\text{cis}\left(-30°\right)$.

denominator: $\sqrt{2} - i\sqrt{6}$ and

$r = \sqrt{\left(\sqrt{2}\right)^2 + \left(-\sqrt{6}\right)^2} = \sqrt{2+6} = \sqrt{8} = 2\sqrt{2}$

$\tan\theta = \dfrac{-\sqrt{6}}{\sqrt{2}} = -\sqrt{3}$

Because x is positive and y is negative, θ is in quadrant IV, so $\theta = -60°$. Thus,

$\sqrt{2} - i\sqrt{6} = 2\sqrt{2}\,\text{cis}\left(-30°\right)$

$\dfrac{2\sqrt{6} - 2i\sqrt{2}}{\sqrt{2} - i\sqrt{6}} = \dfrac{4\sqrt{2}\,\text{cis}\left(-30°\right)}{2\sqrt{2}\,\text{cis}\left(-60°\right)}$

$= \dfrac{4\sqrt{2}}{2\sqrt{2}}\text{cis}\left[-30° - \left(-60°\right)\right]$

$= 2\,\text{cis}\,30° = 2\left(\cos 30° + i\sin 30°\right)$

$= 2\left(\dfrac{\sqrt{3}}{2} + i\dfrac{1}{2}\right) = \sqrt{3} + i$

29. $\left[2.5\left(\cos 35° + i\sin 35°\right)\right]\left[3.0\left(\cos 50° + i\sin 50°\right)\right]$

$= 2.5 \cdot 3.0\left[\cos\left(35° + 50°\right) + i\sin\left(35° + 50°\right)\right]$

$= 7.5\left(\cos 85° + i\sin 85°\right) \approx 0.6537 + 7.4715i$

31. $\left(12\operatorname{cis} 18.5°\right)\left(3\operatorname{cis} 12.5°\right)$

$= 12 \cdot 3\operatorname{cis}\left(18.5° + 12.5°\right)$

$= 36\operatorname{cis} 31° = 36\left(\cos 31° + i\sin 31°\right)$

$\approx 30.8580 + 18.5414i$

33. $\dfrac{45\left(\cos\dfrac{2\pi}{3} + i\sin\dfrac{2\pi}{3}\right)}{22.5\left(\cos\dfrac{3\pi}{5} + i\sin\dfrac{3\pi}{5}\right)}$

$= \dfrac{45}{22.5}\left[\cos\left(\dfrac{2\pi}{3} - \dfrac{3\pi}{5}\right) + i\sin\left(\dfrac{2\pi}{3} - \dfrac{3\pi}{5}\right)\right]$

$= 2\left(\cos\dfrac{\pi}{15} + i\sin\dfrac{\pi}{15}\right) \approx 1.9563 + 0.4158i$

35. $\left[2\operatorname{cis}\dfrac{5\pi}{9}\right]^2 = \left[2\operatorname{cis}\dfrac{5\pi}{9}\right]\left[2\operatorname{cis}\dfrac{5\pi}{9}\right]$

$= 2 \cdot 2\operatorname{cis}\left(\dfrac{5\pi}{9} + \dfrac{5\pi}{9}\right)$

$= 4\left(\cos\dfrac{10\pi}{9} + i\sin\dfrac{10\pi}{9}\right)$

$\approx -3.7588 - 1.3681i$

37. To square a complex number in trigonometric form, square its absolute value and double its argument.

39. $z = r\left(\cos\theta + i\sin\theta\right)$

Because $1 = 1 + 0i = 1\left(\cos 0° + i\sin 0°\right)$,

$\dfrac{1}{z} = \dfrac{1\left(\cos 0° + i\sin 0°\right)}{r\left(\cos\theta + i\sin\theta\right)}$

$= \dfrac{1}{r}\left[\cos\left(0° - \theta\right) + i\sin\left(0° - \theta\right)\right]$

$= \dfrac{1}{r}\left[\cos\left(-\theta\right) + i\sin\left(-\theta\right)\right]$

$= \dfrac{1}{r}\left[\cos\theta - i\sin\theta\right]$

41. $E = 8\left(\cos 20° + i\sin 20°\right)$, $R = 6$, $X_L = 3$,

$I = \dfrac{E}{Z}$, $Z = R + X_L i$

Write $Z = 6 + 3i$ in trigonometric form.

$x = 6$, and $y = 3 \Rightarrow r = \sqrt{6^2 + 3^2} = \sqrt{36 + 9}$

$= \sqrt{45} = 3\sqrt{5}$.

$\tan\theta = \dfrac{3}{6} = \dfrac{1}{2}$, so $\theta \approx 26.6°$.

Thus, $Z = 3\sqrt{5}\operatorname{cis} 26.6°$.

$I = \dfrac{8\operatorname{cis} 20°}{3\sqrt{5}\operatorname{cis} 26.6°} = \dfrac{8}{3\sqrt{5}}\operatorname{cis}\left(20° - 26.6°\right)$

$= \dfrac{8\sqrt{5}}{15}\operatorname{cis}\left(-6.6°\right)$

$= \dfrac{8\sqrt{5}}{15}\left[\cos\left(-6.6°\right) + i\sin\left(-6.6°\right)\right]$

$\approx 1.18 - 0.14i$

43. $Z_1 = 50 + 25i$ and $Z_2 = 60 + 20i$, so

$\dfrac{1}{Z_1} = \dfrac{1}{50 + 25i} \cdot \dfrac{50 - 25i}{50 - 25i} = \dfrac{50 - 25i}{50^2 - 25^2 i^2}$

$= \dfrac{50 - 25i}{2500 - 625\left(-1\right)} = \dfrac{50 - 25i}{2500 + 625}$

$= \dfrac{50 - 25i}{3125} = \dfrac{2}{125} - \dfrac{1}{125}i$

$\dfrac{1}{Z_2} = \dfrac{1}{60 + 20i} \cdot \dfrac{60 - 20i}{60 - 20i} = \dfrac{60 - 20i}{60^2 - 20^2 i^2}$

$= \dfrac{60 - 20i}{3600 - 400\left(-1\right)} = \dfrac{60 - 20i}{3600 + 400}$

$= \dfrac{60 - 20i}{4000} = \dfrac{3}{200} - \dfrac{1}{200}i$

$\dfrac{1}{Z_1} + \dfrac{1}{Z_2} = \left(\dfrac{2}{125} - \dfrac{1}{125}i\right) + \left(\dfrac{3}{200} - \dfrac{1}{200}i\right)$

$= \left(\dfrac{2}{125} + \dfrac{3}{200}\right) - \left(\dfrac{1}{125} + \dfrac{1}{200}\right)i$

$= \left(\dfrac{16}{1000} + \dfrac{15}{1000}\right) - \left(\dfrac{8}{1000} + \dfrac{5}{1000}\right)i$

$= \dfrac{31}{1000} - \dfrac{13}{1000}i$

$Z = \dfrac{1}{\dfrac{1}{Z_1} + \dfrac{1}{Z_2}} = \dfrac{1}{\dfrac{31}{1000} - \dfrac{13}{1000}i}$

$= \dfrac{1000}{31 - 13i} \cdot \dfrac{31 + 13i}{31 + 13i} = \dfrac{1000\left(31 + 13i\right)}{31^2 - 13^2 i^2}$

$= \dfrac{31,000 + 13,000i}{961 - 169\left(-1\right)} = \dfrac{31,000 + 13,000i}{961 + 169}$

$= \dfrac{31,000 + 13,000i}{1130} = \dfrac{3100}{113} + \dfrac{1300}{113}i$

$\approx 27.43 + 11.50i$

In Exercises 45–51, $w = -1 + i$ and $z = -1 - i$.

45. $w \cdot z = \left(-1 + i\right)\left(-1 - i\right)$

$= -1\left(-1\right) + \left(-1\right)\left(-i\right) + i\left(-1\right) + i\left(-i\right)$

$= 1 + i - i - i^2 = 1 - \left(-1\right) = 2$

47. $w \cdot z = \left(\sqrt{2} \text{ cis } 135°\right)\left(\sqrt{2} \text{ cis } 225°\right)$

$\qquad = \sqrt{2} \cdot \sqrt{2}\left[\text{cis } \left(135° + 225°\right)\right]$

$\qquad = 2 \text{ cis } 360° = 2 \text{ cis } 0°$

49. $\dfrac{w}{z} = \dfrac{-1+i}{-1-i} = \dfrac{-1+i}{-1-i} \cdot \dfrac{-1+i}{-1+i} = \dfrac{1-i-i+i^2}{1-i^2}$

$\qquad = \dfrac{1-2i+(-1)}{1-(-1)} = \dfrac{-2i}{2} = -i$

51. $\text{cis } \left(-90°\right) = \cos\left(-90°\right) + i\sin\left(-90°\right)$

$\qquad = 0 + i(-1) = 0 - i = -i$

Section 8.4 DeMoivre's Theorem; Powers and Roots of Complex Numbers

1. Given that $z = 3\left(\cos 30° + i\sin 30°\right)$, it follows that

$z^3 = 27\left(\cos 90° + i\sin 90°\right)$

$\quad = 27\left(0 + i \cdot 1\right)$

$\quad = 0 + 27i$

$\quad = 27i$

3. $\left[\cos 6° + i\sin 6°\right]^{30} = \cos 180° + i\sin 180°$

$\qquad\qquad\qquad\qquad\quad = -1 + 0i$

5. Two. 1 and -1

7. $\left[3\left(\cos 30° + i\sin 30°\right)\right]^3$

$\quad = 3^3\left[\cos\left(3 \cdot 30°\right) + i\sin\left(3 \cdot 30°\right)\right]$

$\quad = 27\left(\cos 90° + i\sin 90°\right)$

$\quad = 27\left(0 + 1 \cdot i\right) = 0 + 27i \text{ or } 27i$

9. $\left(\cos 45° + i\sin 45°\right)^8$

$\quad = \left[\cos\left(8 \cdot 45°\right) + i\sin\left(8 \cdot 45°\right)\right]$

$\quad = \cos 360° + i\sin 360° = 1 + 0 \cdot i \text{ or } 1$

11. $\left[3 \text{ cis } 100°\right]^3$

$\quad = 3^3 \text{ cis } \left(3 \cdot 100°\right)$

$\quad = 27 \text{ cis } 300° = 27\left(\cos 300° + i\sin 300°\right)$

$\quad = 27\left(\dfrac{1}{2} - \dfrac{\sqrt{3}}{2}i\right) = \dfrac{27}{2} - \dfrac{27\sqrt{3}}{2}i$

13. $\left(\sqrt{3} + i\right)^5$

First write $\sqrt{3} + i$ in trigonometric form.

$r = \sqrt{\left(\sqrt{3}\right)^2 + 1^2} = \sqrt{3+1} = \sqrt{4} = 2$ and

$\tan \theta = \dfrac{1}{\sqrt{3}} = \dfrac{\sqrt{3}}{3}$

Because x and y are both positive, θ is in quadrant I, so $\theta = 30°$.

$\sqrt{3} + i = 2\left(\cos 30° + i\sin 30°\right)$

$\left(\sqrt{3} + i\right)^5 = \left[2\left(\cos 30° + i\sin 30°\right)\right]^5$

$\qquad\qquad = 2^5\left[\cos\left(5 \cdot 30°\right) + i\sin\left(5 \cdot 30°\right)\right]$

$\qquad\qquad = 32\left(\cos 150° + i\sin 150°\right)$

$\qquad\qquad = 32\left(-\dfrac{\sqrt{3}}{2} + i\dfrac{1}{2}\right) = -16\sqrt{3} + 16i$

15. $\left(2\sqrt{2} - 2i\sqrt{2}\right)^6$

First write $2\sqrt{2} - 2i\sqrt{2}$ in trigonometric form.

$r = \sqrt{\left(2\sqrt{2}\right)^2 + \left(-2\sqrt{2}\right)^2} = \sqrt{8+8} = \sqrt{16} = 4$

and $\tan \theta = \dfrac{-2\sqrt{2}}{2\sqrt{2}} = -1$

Because x is positive and y is negative, θ is in quadrant IV, so $\theta = 315°$.

$2\sqrt{2} - 2i\sqrt{2} = 4\left(\cos 315° + i\sin 315°\right)^6$

$\left(2\sqrt{2} - 2i\sqrt{2}\right)^6$

$\quad = \left[4\left(\cos 315° + i\sin 315°\right)\right]^6$

$\quad = 4^6\left[\cos\left(6 \cdot 315°\right) + i\sin\left(6 \cdot 315°\right)\right]$

$\quad = 4096\left[\cos 1890° + i\sin 1890°\right]$

$\quad = 4096\left(\cos 90° + i\sin 90°\right)$

$\quad = 4096\left(0 + 1 \cdot i\right) = 0 + 4096i \text{ or } 4096i$

17. $\left(-2 - 2i\right)^5$

First write $-2 - 2i$ in trigonometric form.

$r = \sqrt{\left(-2\right)^2 + \left(-2\right)^2} = \sqrt{4+4} = \sqrt{8} = 2\sqrt{2}$ and

$\tan \theta = \dfrac{-2}{-2} = 1$

Because x and y are both negative, θ is in quadrant III, so $\theta = 225°$.

$-2 - 2i = 2\sqrt{2}\left(\cos 225° + i\sin 225°\right)$

(continued on next page)

(continued)

$(-2-2i)^5$

$= \left[2\sqrt{2} \left(\cos 225° + i\sin 225° \right) \right]^5$

$= \left(2\sqrt{2} \right)^5 \left[\cos \left(5 \cdot 225° \right) + i\sin \left(5 \cdot 225° \right) \right]$

$= 32\sqrt{32} \left(\cos 1125° + i\sin 1125° \right)$

$= 128\sqrt{2} \left(\cos 45° + i\sin 45° \right)$

$= 128\sqrt{2} \left(\dfrac{\sqrt{2}}{2} + \dfrac{\sqrt{2}}{2}i \right) = 128 + 128i$

19. (a) $\cos 0° + i\sin 0° = 1\left(\cos 0° + i\sin 0° \right)$

We have $r = 1$ and $\theta = 0°$. Because

$r^3 \left(\cos 3\alpha + i\sin 3\alpha \right) = 1\left(\cos 0° + i\sin 0° \right)$,

then we have $r^3 = 1 \Rightarrow r = 1$ and

$3\alpha = 0° + 360° \cdot k \Rightarrow \alpha = \dfrac{0° + 360° \cdot k}{3}$

$= 0° + 120° \cdot k = 120° \cdot k$, k any integer.

If $k = 0$, then $\alpha = 0°$.

If $k = 1$, then $\alpha = 120°$.

If $k = 2$, then $\alpha = 240°$.

So, the cube roots are

$\cos 0° + i\sin 0°$, $\cos 120° + i\sin 120°$,

and $\cos 240° + i\sin 240°$.

(b)

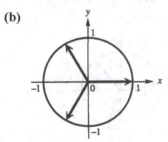

21. (a) Find the cube roots of $8 \operatorname{cis} 60°$

We have $r = 8$ and $\theta = 60°$.

Because $r^3 \left(\cos 3\alpha + i\sin 3\alpha \right)$

$= 8\left(\cos 60° + i\sin 60° \right)$, we have

$r^3 = 8 \Rightarrow r = 2$ and

$3\alpha = 60° + 360° \cdot k \Rightarrow$

$\alpha = \dfrac{60° + 360° \cdot k}{3} = 20° + 120° \cdot k$, k any

integer.

If $k = 0$, then $\alpha = 20° + 0° = 20°$.

If $k = 1$, then $\alpha = 20° + 120° = 140°$.

If $k = 2$, then $\alpha = 20° + 240° = 260°$.

So, the cube roots are

$2 \operatorname{cis} 20°$, $2 \operatorname{cis} 140°$, and $2 \operatorname{cis} 260°$.

(b)

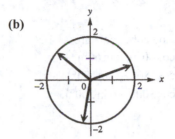

23. (a) Find the cube roots of

$-8i = 8(\cos 270° + i\sin 270°)$

We have $r = 8$ and $\theta = 270°$.

Because $r^3 \left(\cos 3\alpha + i\sin 3\alpha \right)$

$= 8\left(\cos 270° + i\sin 270° \right)$, then we have

$r^3 = 8 \Rightarrow r = 2$ and

$3\alpha = 270° + 360° \cdot k \Rightarrow$

$\alpha = \dfrac{270° + 360° \cdot k}{3} = 90° + 120° \cdot k$, k

any integer.

If $k = 0$, then $\alpha = 90° + 0° = 90°$.

If $k = 1$, then $\alpha = 90° + 120° = 210°$.

If $k = 2$, then $\alpha = 90° + 240° = 330°$.

So, the cube roots are

$2\left(\cos 90° + i\sin 90° \right)$,

$2\left(\cos 210° + i\sin 210° \right)$, and

$2\left(\cos 330° + i\sin 330° \right)$.

(b)

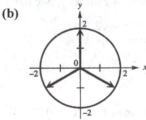

25. (a) Find the cube roots of

$-64 = 64(\cos 180° + i\sin 180°)$

We have $r = 64$ and $\theta = 180°$.

Because $r^3 \left(\cos 3\alpha + i\sin 3\alpha \right)$

$= 64\left(\cos 180° + i\sin 180° \right)$, then we have

$r^3 = 64 \Rightarrow r = 4$ and

$3\alpha = 180° + 360° \cdot k \Rightarrow$

$\alpha = \dfrac{180° + 360° \cdot k}{3} = 60° + 120° \cdot k$, k any

integer.

If $k = 0$, then $\alpha = 60° + 0° = 60°$.

If $k = 1$, then $\alpha = 60° + 120° = 180°$.

If $k = 2$, then $\alpha = 60° + 240° = 300°$.

(continued on next page)

(continued)

So, the cube roots are
$4\left(\cos 60° + i\sin 60°\right),$
$4\left(\cos 180° + i\sin 180°\right),$ and
$4\left(\cos 300° + i\sin 300°\right).$

(b)

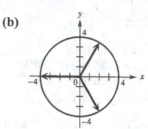

27. (a) Find the cube roots of $1 + i\sqrt{3}$.
We have
$$r = \sqrt{1^2 + \left(\sqrt{3}\right)^2} = \sqrt{1+3} = \sqrt{4} = 2 \text{ and}$$

$\tan\theta = \dfrac{\sqrt{3}}{1} = \sqrt{3}.$ Because θ is in

quadrant I, $\theta = 60°$. Thus,

$$1 + i\sqrt{3} = 2\left(\frac{1}{2} + i\frac{\sqrt{3}}{2}\right)$$
$$= 2\left(\cos 60° + i\sin 60°\right).$$

Because $r^3\left(\cos 3\alpha + i\sin 3\alpha\right)$
$= 2\left(\cos 60° + i\sin 60°\right),$ we have

$r^3 = 2 \Rightarrow r = \sqrt[3]{2}$ and
$3\alpha = 60° + 360° \cdot k \Rightarrow$
$$\alpha = \frac{60° + 360° \cdot k}{3} = 20° + 120° \cdot k, \; k \text{ any}$$
integer.
If $k = 0$, then $\alpha = 20° + 0° = 20°$.
If $k = 1$, then $\alpha = 20° + 120° = 140°$.
If $k = 2$, then $\alpha = 20° + 240° = 260°$.
So, the cube roots are
$\sqrt[3]{2}\left(\cos 20° + i\sin 20°\right),$
$\sqrt[3]{2}\left(\cos 140° + i\sin 140°\right),$ and
$\sqrt[3]{2}\left(\cos 260° + i\sin 260°\right).$

(b)

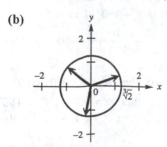

29. (a) Find the cube roots of $-2\sqrt{3} + 2i$.

We have $r = \sqrt{\left(-2\sqrt{3}\right)^2 + 2^2} = \sqrt{12 + 4}$

$$= \sqrt{16} = 4 \text{ and } \tan\theta = \frac{2}{-2\sqrt{3}} = -\frac{\sqrt{3}}{3}.$$

θ is in quadrant II, so $\theta = 150°$. Thus,

$$-2\sqrt{3} + 2i = 4\left(-\frac{\sqrt{3}}{2} + \frac{1}{2}i\right)$$
$$= 4\left(\cos 150° + i\sin 150°\right).$$

Because $r^3\left(\cos 3\alpha + i\sin 3\alpha\right)$
$= 4\left(\cos 150° + i\sin 150°\right),$ we have

$r^3 = 4 \Rightarrow r = \sqrt[3]{4}$ and
$3\alpha = 150° + 360° \cdot k \Rightarrow$
$$\alpha = \frac{150° + 360° \cdot k}{3} = 50° + 120° \cdot k, \; k \text{ any}$$
integer.
If $k = 0$, then $\alpha = 50° + 0° = 50°$.
If $k = 1$, then $\alpha = 50° + 120° = 170°$.
If $k = 2$, then $\alpha = 50° + 240° = 290°$.
So, the cube roots are
$\sqrt[3]{4}\left(\cos 50° + i\sin 50°\right),$
$\sqrt[3]{4}\left(\cos 170° + i\sin 170°\right),$ and
$\sqrt[3]{4}\left(\cos 290° + i\sin 290°\right).$

(b)

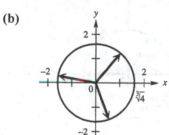

31. Find all the second (or square) roots of
$1 = 1\left(\cos 0° + i\sin 0°\right).$

Because $r^2\left(\cos 2\alpha + i\sin 2\alpha\right)$
$= 1\left(\cos 0° + i\sin 0°\right),$ we have
$2\alpha = 0° + 360° \cdot k \Rightarrow$
$$\alpha = \frac{0° + 360° \cdot k}{2} = 0° + 180° \cdot k = 180° \cdot k, \; k$$

any integer. If $k = 0$, then $\alpha = 0°$.
If $k = 1$, then $\alpha = 180°$. So, the second roots
of 1 are
$\cos 0° + i\sin 0°,$ and $\cos 180° + i\sin 180°$ (or 1
and -1)

(continued on next page)

(continued)

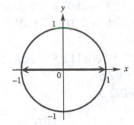

33. Find all the sixth roots of
$$1 = 1(\cos 0° + i \sin 0°).$$

Because $r^6 (\cos 6\alpha + i \sin 6\alpha)$

$= 1(\cos 0° + i \sin 0°)$, we have $r^6 = 1 \Rightarrow r = 1$

and $6\alpha = 0° + 360° \cdot k \Rightarrow$

$$\alpha = \frac{0° + 360° \cdot k}{6} = 0° + 60° \cdot k = 60° \cdot k, \ k \text{ any}$$

integer. If $k = 0$, then $\alpha = 0°$.

If $k = 1$, then $\alpha = 60°$.

If $k = 2$, then $\alpha = 120°$.

If $k = 3$, then $\alpha = 180°$.

If $k = 4$, then $\alpha = 240°$.

If $k = 5$, then $\alpha = 300°$.

So, the sixth roots of 1 are

$\cos 0° + i \sin 0°, \ \cos 60° + i \sin 60°,$

$\cos 120° + i \sin 120°, \ \cos 180° + i \sin 180°,$

$\cos 240° + i \sin 240°,$ and $\cos 300° + i \sin 300°.$

$$\left(\text{or } 1, \ \frac{1}{2} + \frac{\sqrt{3}}{2}i, \ -\frac{1}{2} + \frac{\sqrt{3}}{2}i, \ -1, \right.$$

$$\left. -\frac{1}{2} - \frac{\sqrt{3}}{2}i, \ \text{and } \frac{1}{2} - \frac{\sqrt{3}}{2}i \right)$$

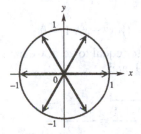

35. Find all the third (cube) roots of
$$i = 1(\cos 90° + i \sin 90°).$$

Because

$r^3 (\cos 3\alpha + i \sin 3\alpha) = 1(\cos 90° + i \sin 90°),$

we have $r^3 = 1 \Rightarrow r = 1$ and

$3\alpha = 90° + 360° \cdot k \Rightarrow$

$$\alpha = \frac{90° + 360° \cdot k}{3} = 30° + 120° \cdot k, \ k \text{ any}$$

integer.

If $k = 0$, then $\alpha = 30° + 0° = 30°$.

If $k = 1$, then $\alpha = 30° + 120° = 150°$.

If $k = 2$, then $\alpha = 30° + 240° = 270°$. So, the

third roots of i are $\cos 30° + i \sin 30°,$

$\cos 150° + i \sin 150°,$ and $\cos 270° + i \sin 270°.$

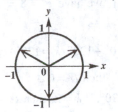

37. $x^3 - 1 = 0 \Rightarrow x^3 = 1$

We have $r = 1$ and $\theta = 0°$.

$x^3 = 1 = 1 + 0i = 1(\cos 0° + i \sin 0°)$

Because $r^3 (\cos 3\alpha + i \sin 3\alpha)$

$= 1(\cos 0° + i \sin 0°)$, we have $r^3 = 1 \Rightarrow r = 1$

and $3\alpha = 0° + 360° \cdot k \Rightarrow$

$$\alpha = \frac{0° + 360° \cdot k}{3} = 0° + 120° \cdot k = 120° \cdot k, \ k$$

any integer. If $k = 0$, then $\alpha = 0°$.

If $k = 1$, then $\alpha = 120°$.

If $k = 2$, then $\alpha = 240°$.

Solution set:

$\{\cos 0° + i \sin 0°, \ \cos 120° + i \sin 120°,$

$\cos 240° + i \sin 240°\}$

or $\left\{ 1, \ -\frac{1}{2} + \frac{\sqrt{3}}{2}i, \ -\frac{1}{2} - \frac{\sqrt{3}}{2}i \right\}$

39. $x^3 + i = 0 \Rightarrow x^3 = -i$

We have $r = 1$ and $\theta = 270°$.

$x^3 = -i = 0 - i = 1(\cos 270° + i \sin 270°)$

Because $r^3 (\cos 3\alpha + i \sin 3\alpha)$

$= 1(\cos 270° + i \sin 270°)$, we have

$r^3 = 1 \Rightarrow r = 1$ and $3\alpha = 270° + 360° \cdot k \Rightarrow$

$$\alpha = \frac{270° + 360° \cdot k}{3} = 90° + 120° \cdot k, \ k \text{ any}$$

integer. If $k = 0$, then $\alpha = 90° + 0° = 90°$.

If $k = 1$, then $\alpha = 90° + 120° = 210°$.

If $k = 2$, then $\alpha = 90° + 240° = 330°$.

Solution set:

$\{\cos 90° + i \sin 90°, \ \cos 210° + i \sin 210°,$

$\cos 330° + i \sin 330°\}$ or

$\left\{ i, \ -\frac{\sqrt{3}}{2} - \frac{1}{2}i, \ \frac{\sqrt{3}}{2} - \frac{1}{2}i \right\}$

41. $x^3 - 8 = 0 \Rightarrow x^3 = 8$

We have $r = 8$ and $\theta = 0°$.

$x^3 = 8 = 8 + 0i = 8(\cos 0° + i \sin 0°)$

Because $r^3(\cos 3\alpha + i \sin 3\alpha)$

$= 8(\cos 0° + i \sin 0°)$, we have $r^3 = 8 \Rightarrow r = 2$

and $3\alpha = 0° + 360° \cdot k \Rightarrow$

$\alpha = \dfrac{0° + 360° \cdot k}{3} = 0° + 120° \cdot k = 120° \cdot k$, k

any integer. If $k = 0$, then $\alpha = 0°$.

If $k = 1$, then $\alpha = 120°$. If $k = 2$, then $\alpha = 240°$.

Solution set:

$\left\{2(\cos 0° + i \sin 0°), 2(\cos 120° + i \sin 120°),\right.$

$\left. 2(\cos 240° + i \sin 240°)\right\}$ or

$\left\{2, -1 + \sqrt{3}i, -1 - \sqrt{3}i\right\}$

43. $x^4 + 1 = 0 \Rightarrow x^4 = -1$

We have $r = 1$ and $\theta = 180°$.

$x^4 = -1 = -1 + 0i = 1(\cos 180° + i \sin 180°)$

Because $r^4(\cos 4\alpha + i \sin 4\alpha)$

$= 1(\cos 180° + i \sin 180°)$, we have

$r^4 = 1 \Rightarrow r = 1$ and $4\alpha = 180° + 360° \cdot k \Rightarrow$

$\alpha = \dfrac{180° + 360° \cdot k}{4} = 45° + 90° \cdot k$, k any

integer. If $k = 0$, then $\alpha = 45° + 0° = 45°$.

If $k = 1$, then $\alpha = 45° + 90° = 135°$.

If $k = 2$, then $\alpha = 45° + 180° = 225°$.

If $k = 3$, then $\alpha = 45° + 270° = 315°$.

Solution set:

$\left\{\cos 45° + i \sin 45°, \cos 135° + i \sin 135°,\right.$

$\left. \cos 225° + i \sin 225°, \cos 315° + i \sin 315°\right\}$ or

$\left\{\frac{\sqrt{2}}{2} + \frac{\sqrt{2}}{2}i, -\frac{\sqrt{2}}{2} + \frac{\sqrt{2}}{2}i, -\frac{\sqrt{2}}{2} - \frac{\sqrt{2}}{2}i, \frac{\sqrt{2}}{2} - \frac{\sqrt{2}}{2}i\right\}$

45. $x^4 - i = 0 \Rightarrow x^4 = i$

We have $r = 1$ and $\theta = 90°$.

$x^4 = i = 0 + i = 1(\cos 90° + i \sin 90°)$

Because $r^4(\cos 4\alpha + i \sin 4\alpha)$

$= 1(\cos 90° + i \sin 90°)$, we have

$r^4 = 1 \Rightarrow r = 1$ and $4\alpha = 90° + 360° \cdot k \Rightarrow$

$\alpha = \dfrac{90° + 360° \cdot k}{4} = 22.5° + 90° \cdot k$, k any

integer. If $k = 0$, then $\alpha = 22.5° + 0° = 22.5°$.

If $k = 1$, then $\alpha = 22.5° + 90° = 112.5°$.

If $k = 2$, then $\alpha = 22.5° + 180° = 202.5°$.

If $k = 3$, then $\alpha = 22.5° + 270° = 292.5°$.

Solution set:

$\left\{\cos 22.5° + i \sin 22.5°, \cos 112.5° + i \sin 112.5°,\right.$

$\cos 202.5° + i \sin 202.5°,$

$\left. \cos 292.5° + i \sin 292.5°\right\}$

47. $x^3 - \left(4 + 4i\sqrt{3}\right) = 0 \Rightarrow x^3 = 4 + 4i\sqrt{3}$

We have

$r = \sqrt{4^2 + \left(4\sqrt{3}\right)^2} = \sqrt{16 + 48} = \sqrt{64} = 8$ and

$\tan \theta = \dfrac{4\sqrt{3}}{4} = \sqrt{3}$. Because θ is in quadrant I,

$\theta = 60°$.

$x^3 = 4 + 4i\sqrt{3} = 8\left(\dfrac{1}{2} + i\dfrac{\sqrt{3}}{2}\right)$

$= 8(\cos 60° + i \sin 60°)$

Because $r^3(\cos 3\alpha + i \sin 3\alpha)$

$= 8(\cos 60° + i \sin 60°)$, we have

$r^3 = 8 \Rightarrow r = 2$ and $3\alpha = 60° + 360° \cdot k \Rightarrow$

$\alpha = \dfrac{60° + 360° \cdot k}{3} = 20° + 120° \cdot k$, k any

integer. If $k = 0$, then $\alpha = 20° + 0° = 20°$.

If $k = 1$, then $\alpha = 20° + 120° = 140°$.

If $k = 2$, then $\alpha = 20° + 240° = 260°$.

Solution set:

$\left\{2(\cos 20° + i \sin 20°), 2(\cos 140° + i \sin 140°),\right.$

$\left. 2(\cos 260° + i \sin 260°)\right\}$

49. $x^3 - 1 = 0 \Rightarrow (x - 1)(x^2 + x + 1) = 0$

Setting each factor equal to zero, we have

$x - 1 = 0 \Rightarrow x = 1$ and

$x^2 + x + 1 = 0 \Rightarrow$

$x = \dfrac{-1 \pm \sqrt{1^2 - 4 \cdot 1 \cdot 1}}{2 \cdot 1} = \dfrac{-1 \pm \sqrt{-3}}{2}$

$= \dfrac{-1 \pm i\sqrt{3}}{2} = -\dfrac{1}{2} \pm \dfrac{\sqrt{3}}{2}i$

Thus, $x = 1, -\frac{1}{2} + \frac{\sqrt{3}}{2}i, -\frac{1}{2} - \frac{\sqrt{3}}{2}i$. We see that

the solutions are the same as Exercise 37.

51. (a) If $z = 0 + 0i$, then $z = 0$, $0^2 + 0 = 0$,

$0^2 + 0 = 0$, and so on. The calculations

repeat as 0, 0, 0, . . ., and will never

exceed a modulus of 2. The point (0, 0) is

part of the Mandelbrot set. The pixel at

the origin should be turned on.

(b) If $z = 1 - 1i$, then $(1-i)^2 + (1-i) = 1 - 3i$.
The modulus of $1 - 3i$ is

$\sqrt{1^2 + (-3)^2} = \sqrt{1+9} = \sqrt{10}$, which is

greater than 2. Therefore, $1 - 1i$ is not
part of the Mandelbrot set, and the pixel
at $(1, -1)$ should be left off.

(c) If $z = -0.5i$, then

$(-0.5i)^2 - 0.5i = -0.25 - 0.5i;$

$(-0.25 - 0.5i)^2 + (-0.25 - 0.5i)$
$= -0.4375 - 0.25i;$

$(-0.4375 - 0.25i)^2 + (-0.4375 - 0.25i)$
$= -0.308593 - 0.03125i;$

$(-0.308593 - 0.03125i)^2$
$\qquad + (-0.308593 - 0.03125i)$
$= -0.214339 - 0.0119629i;$

$(-0.214339 - 0.0119629i)^2$
$\qquad + (-0.214339 - 0.0119629i)$
$= -0.16854 - 0.00683466i$

This sequence appears to be approaching
the origin, and no number has a modulus
greater than 2. Thus, $-0.5i$ is part of the
Mandelbrot set, and the pixel at $(0, -0.5i)$
should be turned on.

53. Using the trace function, we find that the other
four fifth roots of 1 are:
$0.30901699 + 0.95105652i,$
$-0.809017 + 0.58778525i,$
$-0.809017 - 0.5877853i,$
$0.30901699 - 0.9510565i.$

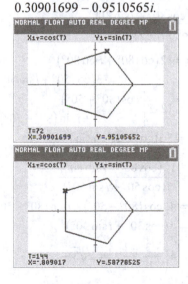

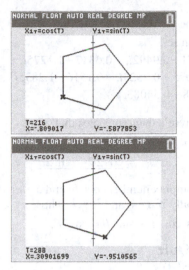

55. $x^2 - 3 + 2i = 0 \Rightarrow x^2 = 3 - 2i$

$r = \sqrt{3^2 + (-2)^2} = \sqrt{9+4} = \sqrt{13} \Rightarrow$

$r^{1/n} = r^{1/2} = \left(\sqrt{13}\right)^{1/2} \approx 1.89883$ and

$\tan\theta = -\dfrac{2}{3}$. Because θ is in quadrant IV,

$\theta \approx 326.3099°$ and

$\alpha = \dfrac{326.3099° + 360° \cdot k}{2} \approx 163.155° + 180° \cdot k,$

where k is an integer.

$x \approx 1.89833 \left(\cos 163.155° + i\sin 163.155°\right),$
$\qquad 1.89833 \left(\cos 343.155° + i\sin 343.155°\right)$

Solution set:
$\left\{-1.8174 + 0.5503i, 1.8174 - 0.5503i\right\}$

57. $x^5 + 2 + 3i = 0 \Rightarrow x^5 = -2 - 3i$

$r = \sqrt{(-2)^2 + (-3)^2} = \sqrt{4+9} = \sqrt{13} \Rightarrow$

$r^{1/n} = r^{1/5} = \left(\sqrt{13}\right)^{1/5} = 13^{1/10} \approx 1.2924$

and $\tan\theta = \dfrac{-3}{-2} = 1.5$

Because θ is in quadrant III, $\theta \approx 236.310°$
and

$\alpha = \dfrac{236.310° + 360° \cdot k}{5} = 47.262° + 72° \cdot k,$

where k is an integer.

$x \approx 1.29239 \left(\cos 47.262° + i\sin 47.262°\right),$
$\qquad 1.29239 \left(\cos 119.262° + i\sin 119.2622°\right),$
$\qquad 1.29239 \left(\cos 191.262° + i\sin 191.2622°\right),$
$\qquad 1.29239 \left(\cos 263.262° + i\sin 263.2622°\right),$
$\qquad 1.29239 \left(\cos 335.262° + i\sin 335.2622°\right)$

(continued on next page)

(continued)

Solution set:

$$\{0.8771+0.9492i, -0.6317+1.1275i,$$
$$-1.2675-0.2524i, -0.1516-1.2835i,$$
$$1.1738-0.54083i\}$$

59. De Moivre's theorem states that

$$(\cos\theta + i\sin\theta)^2 = 1^2(\cos 2\theta + i\sin 2\theta)$$
$$= \underline{\cos 2\theta + i\sin 2\theta}$$

61. Two complex numbers $a + bi$ and $c + di$ are equal only if $a = c$ and $b = d$. Thus, $a = c$ implies $\cos^2\theta - \sin^2\theta = \cos 2\theta$.

Chapter 8 Quiz
(Sections 8.1–8.4)

1. (a) $\sqrt{-24}\cdot\sqrt{-3} = i\sqrt{24}\cdot i\sqrt{3} = i^2\sqrt{72} = -6\sqrt{2}$

(b) $\dfrac{\sqrt{-8}}{\sqrt{72}} = \dfrac{i\sqrt{8}}{\sqrt{72}} = \dfrac{i}{\sqrt{9}} = \dfrac{i}{3} = \dfrac{1}{3}i$

3. (a) $(1-i)^3 = 1^3 - 3(1)^2 i + 3(1)i^2 - i^3$
$$= 1 - 3i + 3(-1) - (-i)$$
$$= 1 - 3i - 3 + i = -2 - 2i$$

(b) $i^{33} = i^{32}\cdot i = \left(i^4\right)^8\cdot i = 1^8\cdot i = i$, or $0 + i$

5. (a) Sketch a graph of $-4i$ in the complex plane.

Because $-4i = 0 - 4i$, we have $x = 0$ and $y = -4$, so $r = \sqrt{0^2 + (-4)^2} = 4$. We cannot find θ by using $\tan\theta = \dfrac{y}{x}$ because $x = 0$. From the graph, we see that $-4i$ is on the negative y-axis, so $\theta = 270°$. Thus,
$$-4i = 4(\cos 270° + i\sin 270°)$$

(b) $1-i\sqrt{3}$

Sketch a graph of $1-i\sqrt{3}$ in the complex plane.

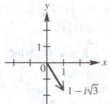

Because $x = 1$ and $y = -\sqrt{3}$,
$$r = \sqrt{1^2 + \left(-\sqrt{3}\right)^2} = \sqrt{1+3} = \sqrt{4} = 2 \text{ and}$$
$$\tan\theta = \dfrac{-\sqrt{3}}{1} = -\sqrt{3}. \text{ The graph shows}$$
that θ is in quadrant IV, so $\theta = 300°$. Therefore,
$$1-i\sqrt{3} = 2(\cos 300° + i\sin 300°)$$

(c) $-3-i$

Sketch a graph of $-3-i$ in the complex plane.

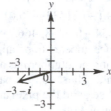

Using a calculator, we find that the reference angle is $18.4°$. The graph shows that θ is in quadrant III, so
$\theta = 180° + 18.4° = 198.4°$. Therefore,
$$-3-i = \sqrt{10}(\cos 198.4° + i\sin 198.4°).$$

7. $w = 12(\cos 80° + i\sin 80°),$
$z = 3(\cos 50° + i\sin 50°)$

(a) $wz = 12(\cos 80° + i\sin 80°)$
$$\cdot 3(\cos 50° + i\sin 50°)$$
$$= 12\cdot 3\left[\begin{array}{l}\cos(80° + 50°)\\ + i\sin(80° + 50°)\end{array}\right]$$
$$= 36(\cos 130° + i\sin 130°)$$

(b) $\dfrac{w}{z} = \dfrac{12(\cos 80° + i\sin 80°)}{3(\cos 50° + i\sin 50°)}$
$$= 4\left[\cos(80° - 50°) + i\sin(80° - 50°)\right]$$
$$= 4(\cos 30° + i\sin 30°)$$
$$= 4\left(\dfrac{\sqrt{3}}{2} + \dfrac{1}{2}i\right) = 2\sqrt{3} + 2i$$

(c) $z^3 = \left[3\left(\cos 50° + i \sin 50°\right)\right]^3$

$= 3^3\left[\cos\left(3 \cdot 50°\right) + i \sin\left(3 \cdot 50°\right)\right]$

$= 27\left(\cos 150° + i \sin 150°\right)$

$= 27\left(-\dfrac{\sqrt{3}}{2} + \dfrac{1}{2}i\right) = -\dfrac{27\sqrt{3}}{2} + \dfrac{27}{2}i$

(d) $w^3 = \left[12\left(\cos 80° + i \sin 80°\right)\right]^3$

$= 12^3\left[\cos\left(12 \cdot 80°\right) + i \sin\left(12 \cdot 80°\right)\right]$

$= 1728\left(\cos 960° + i \sin 960°\right)$

$= 1728\left(-\dfrac{1}{2} - \dfrac{\sqrt{3}}{2}i\right)$

$= -\dfrac{1728}{2} - \dfrac{1728\sqrt{3}}{2}i = -864 - 864i\sqrt{3}$

Section 8.5 Polar Equations and Graphs

1. For the polar equation $r = 3\cos\theta$, if $\theta = 60°$, then $r = \frac{3}{2}$.

3. For the polar equation $r^2 = 4\sin 2\theta$, if $\theta = 15°$, then $r = \pm\sqrt{2}$.

5. II (because $r > 0$ and $90° < \theta < 180°$)

7. IV (because $r > 0$ and $-90° < \theta < 0°$)

9. positive x-axis

11. negative y-axis

For Exercises 13(b)–23(b), answers may vary.

13. (a)

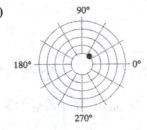

(b) Two other pairs of polar coordinates for $(1, 45°)$ are $(1, 405°)$ and $(-1, 225°)$.

(c) $x = r\cos\theta \Rightarrow x = 1 \cdot \cos 45° = \dfrac{\sqrt{2}}{2}$ and

$y = r\sin\theta \Rightarrow y = 1 \cdot \sin 45° = \dfrac{\sqrt{2}}{2}$, so the

point is $\left(\dfrac{\sqrt{2}}{2}, \dfrac{\sqrt{2}}{2}\right)$.

15. (a)

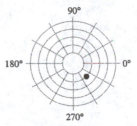

(b) Two other pairs of polar coordinates for $(-2, 135°)$ are $(-2, 495°)$ and $(2, 315°)$.

(c) $x = r\cos\theta \Rightarrow x = (-2)\cos 135° = \sqrt{2}$ and

$y = r\sin\theta \Rightarrow y = (-2)\sin 135° = \sqrt{2}$, so

the point is $\left(\sqrt{2}, -\sqrt{2}\right)$.

17. (a)

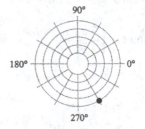

(b) Two other pairs of polar coordinates for $(5, -60°)$ are $(5, 300°)$ and $(-5, 120°)$.

(c) $x = r\cos\theta \Rightarrow x = 5\cos\left(-60°\right) = \dfrac{5}{2}$ and

$y = r\sin\theta \Rightarrow y = 5\sin\left(-60°\right) = -\dfrac{5\sqrt{3}}{2}$,

so the point is $\left(\dfrac{5}{2}, -\dfrac{5\sqrt{3}}{2}\right)$.

19. (a)

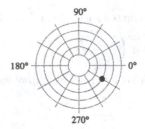

(b) Two other pairs of polar coordinates for $(-3, -210°)$ are $(-3, 150°)$ and $(3, -30°)$.

(c) $x = r\cos\theta \Rightarrow x = (-3)\cos\left(-210°\right) = \dfrac{3\sqrt{3}}{2}$

and

$y = r\sin\theta \Rightarrow y = (-3)\sin\left(-210°\right) = -\dfrac{3}{2}$,

so the point is $\left(\dfrac{3\sqrt{3}}{2}, -\dfrac{3}{2}\right)$.

21. (a)

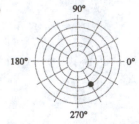

(b) Two other pairs of polar coordinates for

$\left(3, \dfrac{5\pi}{3}\right)$ are $\left(3, \dfrac{11\pi}{3}\right)$ and $\left(-3, \dfrac{2\pi}{3}\right)$.

(c) $x = r\cos\theta \Rightarrow x = 3\cos\dfrac{5\pi}{3} = \dfrac{3}{2}$ and

$y = r\sin\theta \Rightarrow y = 3\sin\dfrac{5\pi}{3} = -\dfrac{3\sqrt{3}}{2}$, so

the point is $\left(\dfrac{3}{2}, -\dfrac{3\sqrt{3}}{2}\right)$.

23. (a)

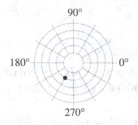

(b) Two other pairs of polar coordinates for

$\left(-2, \dfrac{\pi}{3}\right)$ are $\left(-2, \dfrac{7\pi}{3}\right)$ and $\left(2, \dfrac{4\pi}{3}\right)$.

(c) $x = r\cos\theta \Rightarrow x = -2\cos\dfrac{\pi}{3} = -1$ and

$y = r\sin\theta \Rightarrow y = -2\sin\dfrac{\pi}{3} = -\sqrt{3}$, so the

point is $\left(-1, -\sqrt{3}\right)$.

For Exercises 25(b)–35(b), answers may vary.

25. (a)

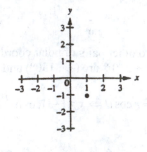

(b) $r = \sqrt{1^2 + (-1)^2} = \sqrt{1+1} = \sqrt{2}$ and

$\theta = \tan^{-1}\left(\dfrac{-1}{1}\right) = \tan^{-1}(-1) = -45°$,

because θ is in quadrant IV.
$360° - 45° = 315°$, so one possibility is

$\left(\sqrt{2}, 315°\right)$. Alternatively, if $r = -\sqrt{2}$,

then $\theta = 315° - 180° = 135°$. Thus, a

second possibility is $\left(-\sqrt{2}, 135°\right)$.

27. (a)

(b) $r = \sqrt{0^2 + 3^2} = \sqrt{0+9} = \sqrt{9} = 3$ and

$\theta = 90°$, because $(0,3)$ is on the positive

y-axis. So, one possibility is $(3, 90°)$.

Alternatively, if $r = -3$, then

$\theta = 90° + 180° = 270°$. Thus, a second

possibility is $(-3, 270°)$.

29. (a)

(b) $r = \sqrt{\left(\sqrt{2}\right)^2 + \left(\sqrt{2}\right)^2} = \sqrt{2+2} = \sqrt{4} = 2$

and $\theta = \tan^{-1}\left(\dfrac{\sqrt{2}}{\sqrt{2}}\right) = \tan^{-1}1 = 45°$,

because θ is in quadrant I. So, one

possibility is $(2, 45°)$. Alternatively, if

$r = -2$, then $\theta = 45° + 180° = 225°$.

Thus, a second possibility is $(-2, 225°)$.

31. (a)

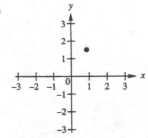

(b) $r = \sqrt{\left(\dfrac{\sqrt{3}}{2}\right)^2 + \left(\dfrac{3}{2}\right)^2} = \sqrt{\dfrac{3}{4} + \dfrac{9}{4}}$

$= \sqrt{\dfrac{12}{4}} = \sqrt{3}$ and $\theta = \arctan\left(\dfrac{3}{2} \cdot \dfrac{2}{\sqrt{3}}\right)$

$= \tan^{-1}\left(\sqrt{3}\right) = 60°$, because θ is in

quadrant I. So, one possibility is

$\left(\sqrt{3}, 60°\right)$. Alternatively, if $r = -\sqrt{3}$,

then $\theta = 60° + 180° = 240°$. Thus, a

second possibility is $\left(-\sqrt{3}, 240°\right)$.

33. (a)

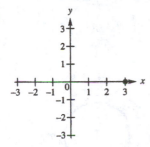

(b) $r = \sqrt{3^2 + 0^2} = \sqrt{9 + 0} = \sqrt{9} = 3$ and

$\theta = 0°$, because $(3, 0)$ is on the positive

x-axis. So, one possibility is $(3, 0°)$.

Alternatively, if $r = -3$, then

$\theta = 0° + 180° = 180°$. Thus, a second

possibility is $(-3, 180°)$.

35. (a)

(b) $r = \sqrt{\left(-\dfrac{3}{2}\right)^2 + \left(-\dfrac{3\sqrt{3}}{2}\right)^2} = \sqrt{\dfrac{9}{4} + \dfrac{27}{4}}$

$= \sqrt{\dfrac{36}{4}} = 3$

$\theta = \tan^{-1}\left(\dfrac{-\dfrac{3\sqrt{3}}{2}}{-\dfrac{3}{2}}\right) = \tan^{-1}\sqrt{3}$. Because

θ is in quadrant III, we have $\theta = 240°$.

So, one possibility is $(3, 240°)$.

Alternatively, if $r = -3$, then

$\theta = 240° - 180° = 60°$. Thus, a second

possibility is $(-3, 60°)$.

37. $x - y = 4$

Using the general form for the polar equation

of a line, $r = \dfrac{c}{a\cos\theta + b\sin\theta}$, with

$a = 1, b = -1,$ and $c = 4,$ the polar equation is

$r = \dfrac{4}{\cos\theta - \sin\theta}.$

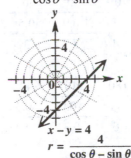

$x - y = 4$

$r = \dfrac{4}{\cos\theta - \sin\theta}$

39. $x^2 + y^2 = 16 \Rightarrow r^2 = 16 \Rightarrow r = \pm 4$

The equation of the circle in polar form is

$r = 4$ or $r = -4.$

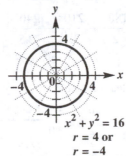

$x^2 + y^2 = 16$

$r = 4$ or

$r = -4$

41. $2x + y = 5$

Using the general form for the polar equation

of a line, $r = \dfrac{c}{a\cos\theta + b\sin\theta}$, with

$a = 2, b = 1,$ and $c = 5,$ the polar equation is

$r = \dfrac{5}{2\cos\theta + \sin\theta}.$

(*continued on next page*)

(continued)

$2x + y = 5$

$r = \dfrac{5}{2\cos\theta + \sin\theta}$

43. C. $r = 3$ represents the set of all points 3 units from the pole.

45. A. $r = \cos 2\theta$ is a rose curve with $2 \cdot 2 = 4$ petals.

47. $r = 2 + 2\cos\theta$ (cardioid)

θ	0°	30°	60°	90°	120°	150°
$\cos\theta$	1	0.9	0.5	0	−0.5	−0.9
$r = 2 + 2\cos\theta$	4	3.7	3	2	1	0.3

θ	180°	210°	240°	270°	300°	330°
$\cos\theta$	−1	−0.9	−0.5	0	0.5	0.9
$r = 2 + 2\cos\theta$	0	0.3	1	2	3	3.7

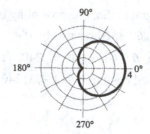

49. $r = 3 + \cos\theta$ (limaçon)

θ	0°	30°	60°	90°	120°	150°
$r = 3 + \cos\theta$	4	3.9	3.5	3	2.5	2.1

θ	180°	210°	240°	270°	300°	330°
$r = 3 + \cos\theta$	2	2.1	2.5	3	3.5	3.9

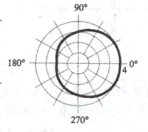

51. $r = 4\cos 2\theta$ (four-leaved rose)

θ	0°	30°	45°	60°	90°	120°	135°	150°
$r = 4\cos 2\theta$	4	2	0	−2	−4	−2	0	2

θ	180°	210°	225°	240°	270°	300°	315°	330°
$r = 4\cos 2\theta$	4	2	0	−2	−4	−2	0	2

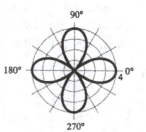

53. $r^2 = 4\cos 2\theta \Rightarrow r = \pm 2\sqrt{\cos 2\theta}$ (lemniscate)

Graph only exists for $[0°, 45°]$, $[135°, 225°]$, and $[315°, 360°]$ because $\cos 2\theta$ must be positive.

θ	0°	30°	45°	135°	150°
$r = \pm 2\sqrt{\cos 2\theta}$	±2	±1.4	0	0	±1.4

θ	180°	210°	225°	315°	330°
$r = \pm 2\sqrt{\cos 2\theta}$	±2	±1.4	0	0	±1.4

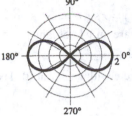

55. $r = 4 - 4\cos\theta$ (cardioid)

θ	0°	30°	60°	90°	120°	150°
$r = 4 - 4\cos\theta$	0	0.5	2	4	6	7.5

θ	180°	210°	240°	270°	300°	330°
$r = 4 - 4\cos\theta$	8	7.5	6	4	2	0.5

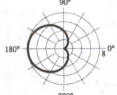

57. $r = 2\sin\theta\tan\theta$ (cissoid)

r is undefined at $\theta = 90°$ and $\theta = 270°$.

θ	0°	30°	45°	60°	90°	120°	135°	150°	180°
$r = 2\sin\theta\tan\theta$	0	0.6	1.4	3	undefined	−3	−1.4	−0.6	0

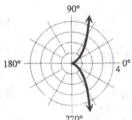

Notice that for $[180°, 360°)$, the graph retraces the path traced for $[0°, 180°)$.

59. Graph $r = \theta$

θ	−360°	−270°	−180°	−90°	0°	90°	180°	270°	360°
θ (radians)	−6.3	−4.7	−3.1	−1.6	0	1.6	3.1	4.7	6.3
$r = \theta$	−6.3	−4.7	−3.1	−1.6	0	1.6	3.1	4.7	6.3

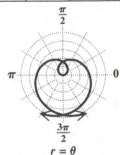

$r = \theta$

61. $r = 2\sin\theta$

Multiply both sides by r to obtain

$r^2 = 2r\sin\theta$.

$r^2 = x^2 + y^2$ and $y = r\sin\theta \Rightarrow x^2 + y^2 = 2y$.

Complete the square on y to obtain

$x^2 + y^2 - 2y + 1 = 1 \Rightarrow x^2 + (y-1)^2 = 1$.

The graph is a circle with center at $(0, 1)$ and radius 1.

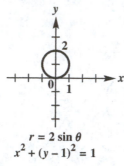

$r = 2\sin\theta$
$x^2 + (y-1)^2 = 1$

63. $r = \dfrac{2}{1-\cos\theta}$

Multiply both sides by $1 - \cos\theta$ to obtain

$r - r\cos\theta = 2$. Substitute $r = \sqrt{x^2+y^2}$ to obtain

$\sqrt{x^2+y^2} - x = 2 \Rightarrow \sqrt{x^2+y^2} = 2 + x \Rightarrow$

$x^2 + y^2 = (2+x)^2 \Rightarrow x^2 + y^2 = 4 + 4x + x^2 \Rightarrow$

$y^2 = 4(1+x)$

The graph is a parabola with vertex at $(-1, 0)$ and axis $y = 0$.

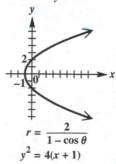

$r = \dfrac{2}{1-\cos\theta}$
$y^2 = 4(x+1)$

65. $r + 2\cos\theta = -2\sin\theta$

$r + 2\cos\theta = -2\sin\theta \Rightarrow$

$r^2 = -2r\sin\theta - 2r\cos\theta \Rightarrow x^2 + y^2 = -2y - 2x$

$x^2 + 2x + y^2 + 2y = 0 \Rightarrow$

$x^2 + 2x + 1 + y^2 + 2y + 1 = 2 \Rightarrow$

$(x+1)^2 + (y+1)^2 = 2$

The graph is a circle with center $(-1, -1)$ and radius $\sqrt{2}$.

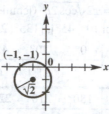

$r = -2\cos\theta - 2\sin\theta$
$(x+1)^2 + (y+1)^2 = 2$

67. $r = 2\sec\theta$

$r = 2\sec\theta \Rightarrow r = \dfrac{2}{\cos\theta} \Rightarrow r\cos\theta = 2 \Rightarrow x = 2$

The graph is a vertical line, intercepting the x-axis at 2.

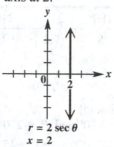

$r = 2\sec\theta$
$x = 2$

69. $r = \dfrac{2}{\cos\theta + \sin\theta}$

Using the general form for the polar equation

of a line, $r = \dfrac{c}{a\cos\theta + b\sin\theta}$, with

$a = 1$, $b = 1$, and $c = 2$, we have $x + y = 2$.

The graph is a line with intercepts $(0, 2)$ and $(2, 0)$.

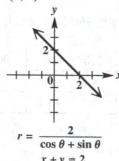

$r = \dfrac{2}{\cos\theta + \sin\theta}$
$x + y = 2$

71. In rectangular coordinates, the line passes through $(1,0)$ and $(0,2)$. So

$$m = \frac{2-0}{0-1} = \frac{2}{-1} = -2 \text{ and}$$

$$(y-0) = -2(x-1) \Rightarrow y = -2x+2 \Rightarrow$$

$2x + y = 2$. Converting to polar form

$$r = \frac{c}{a\cos\theta + b\sin\theta}, \text{ we have:}$$

$$r = \frac{2}{2\cos\theta + \sin\theta}.$$

73. (a) $(r, -\theta)$

(b) $(r, \pi - \theta)$ or $(-r, -\theta)$

(c) $(r, \pi + \theta)$ or $(-r, \theta)$

75.

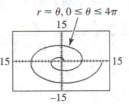

$r = \theta, 0 \le \theta \le 4\pi$

77.

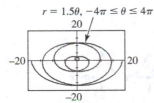

$r = 1.5\theta, -4\pi \le \theta \le 4\pi$

79. $r = 4\sin\theta$, $r = 1 + 2\sin\theta$, $0 \le \theta < 2\pi$

$$4\sin\theta = 1 + 2\sin\theta \Rightarrow 2\sin\theta = 1 \Rightarrow$$

$$\sin\theta = \frac{1}{2} \Rightarrow \theta = \frac{\pi}{6} \text{ or } \frac{5\pi}{6}$$

The points of intersection are

$$\left(4\sin\frac{\pi}{6}, \frac{\pi}{6}\right) = \left(2, \frac{\pi}{6}\right) \text{ and}$$

$$\left(4\sin\frac{5\pi}{6}, \frac{5\pi}{6}\right) = \left(2, \frac{5\pi}{6}\right).$$

Using a graphing calculator, we see that the pole, (0, 0) is also an intersection. However, it is not reached at the same value of theta for each equation, which is why it does not appear as a solution of the equation $4\sin\theta = 1 + 2\sin\theta$.

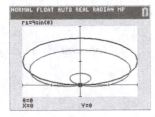

81. $r = 2 + \sin\theta$, $r = 2 + \cos\theta$, $0 \le \theta < 2\pi$

$$2 + \sin\theta = 2 + \cos\theta \Rightarrow \sin\theta = \cos\theta \Rightarrow$$

$$\theta = \frac{\pi}{4} \text{ or } \frac{5\pi}{4}$$

$$r = 2 + \sin\frac{\pi}{4} = 2 + \frac{\sqrt{2}}{2} = \frac{4+\sqrt{2}}{2} \text{ and}$$

$$r = 2 + \sin\frac{5\pi}{4} = 2 - \frac{\sqrt{2}}{2} = \frac{4-\sqrt{2}}{2}$$

The points of intersection are

$$\left(\frac{4+\sqrt{2}}{2}, \frac{\pi}{4}\right) \text{ and } \left(\frac{4-\sqrt{2}}{2}, \frac{5\pi}{4}\right).$$

83. (a) Plot the following polar equations on the same polar axis in radian mode:

$$\text{Mercury: } r = \frac{0.39(1-0.206^2)}{1+0.206\cos\theta}$$

$$\text{Venus: } r = \frac{0.78(1-0.007^2)}{1+0.007\cos\theta}$$

$$\text{Earth: } r = \frac{1(1-0.017^2)}{1+0.017\cos\theta}$$

$$\text{Mars: } r = \frac{1.52(1-0.093^2)}{1+0.093\cos\theta}$$

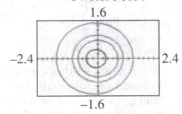

(b) Plot the following polar equations on the same polar axis:

$$\text{Earth: } r = \frac{1(1-0.017^2)}{1+0.017\cos\theta}$$

$$\text{Jupiter: } r = \frac{5.2(1-0.048^2)}{1+0.048\cos\theta}$$

$$\text{Uranus: } r = \frac{19.2(1-0.047^2)}{1+0.047\cos\theta}$$

$$\text{Pluto: } r = \frac{39.4(1-0.249^2)}{1+0.249\cos\theta}$$

(continued on next page)

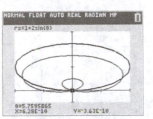

(*continued*)

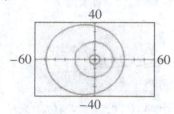

(c) We must determine if the orbit of Pluto is always outside the orbits of the other satellites. Neptune is closest to Pluto, so plot the orbits of Neptune and Pluto on the same polar axes.

Neptune: $r = \dfrac{30.1(1 - 0.009^2)}{1 + 0.009\cos\theta}$

Pluto: $r = \dfrac{39.4(1 - 0.249^2)}{1 + 0.249\cos\theta}$

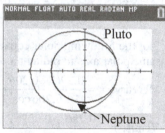

$[-60, 60] \times [-60, 60]$

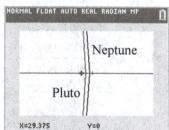

The graph shows that their orbits are very close near the polar axis. Use ZOOM or change your window to see that the orbit of Pluto does indeed pass inside the orbit of Neptune. Therefore, there are times when Neptune, not Pluto, is the farthest planet from the sun. (However, Pluto's average distance from the sun is considerably greater than Neptune's average distance.)

85. $r\sin\theta = k$

87. $r = \dfrac{k}{\sin\theta} \Rightarrow r = k\csc\theta$

89. $r\cos\theta = k$

91. $r = \dfrac{k}{\cos\theta} \Rightarrow r = k\sec\theta$

Section 8.6 Parametric Equations, Graphs, and Applications

1. For the plane curve defined by $x = t^2 + 1$, $y = 2t + 3$, for t in $[-4, 4]$, the ordered pair that corresponds to $t = -3$ is $\underline{(10, -3)}$.

3. For the plane curve defined by $x = \cos t$, $y = 2\sin t$, for t in $[0, 2\pi]$, the ordered pair that corresponds to $t = \frac{\pi}{3}$ is $\underline{\left(\frac{1}{2}, \sqrt{3}\right)}$.

5. C. At $t = 2$, $x = 3(2) + 6 = 12$ and $y = -2(2) + 4 = 0$.

7. A. At $t = 5$, $x = 5$ and $y = 5^2 = 25$.

9. **(a)** $x = t + 2$, $y = t^2$, for t in $[-1, 1]$

t	$x = t + 2$	$y = t^2$
-1	$-1 + 2 = 1$	$(-1)^2 = 1$
0	$0 + 2 = 2$	$0^2 = 0$
1	$1 + 2 = 3$	$1^2 = 1$

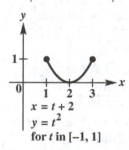

$x = t + 2$
$y = t^2$
for t in $[-1, 1]$

(b) $x - 2 = t$, therefore $y = (x - 2)^2$ or $y = x^2 - 4x + 4$. Because t is in $[-1, 1]$, x is in $[-1 + 2, 1 + 2]$ or $[1, 3]$.

11. **(a)** $x = \sqrt{t}$, $y = 3t - 4$, for t in $[0, 4]$.

t	$x = \sqrt{t}$	$y = 3t - 4$
0	$\sqrt{0} = 0$	$3(0) - 4 = -4$
1	$\sqrt{1} = 1$	$3(1) - 4 = -1$
2	$\sqrt{2} = 1.4$	$3(2) - 4 = 2$
3	$\sqrt{3} = 1.7$	$3(3) - 4 = 5$
4	$\sqrt{4} = 2$	$3(4) - 4 = 8$

(*continued on next page*)

(continued)

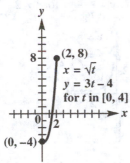

(b) $x = \sqrt{t}$, $y = 3t - 4$

$x = \sqrt{t} \Rightarrow x^2 = t$, so $y = 3x^2 - 4$.
Because t is in $[0, 4]$, x is in
$[\sqrt{0}, \sqrt{4}]$ or $[0, 2]$.

13. (a) $x = t^3 + 1$, $y = t^3 - 1$, for t in $(-\infty, \infty)$

t	$x = t^3 + 1$	$y = t^3 - 1$
-2	$(-2)^3 + 1 = -7$	$(-2)^3 - 1 = -9$
-1	$(-1)^3 + 1 = 0$	$(-1)^3 - 1 = -2$
0	$0^3 + 1 = 1$	$0^3 - 1 = -1$
1	$1^3 + 1 = 2$	$1^3 - 1 = 0$
2	$2^3 + 1 = 9$	$2^3 - 1 = 7$
3	$3^3 + 1 = 28$	$3^3 - 1 = 26$

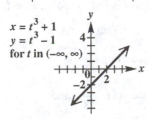

(b) $x = t^3 + 1$, so $x - 1 = t^3$.

$y = t^3 - 1$, so $y = (x - 1) - 1 = x - 2$.
Because t is in $(-\infty, \infty)$, x is in $(-\infty, \infty)$.

15. (a) $x = 2\sin t$, $y = 2\cos t$, for t in $[0, 2\pi]$

t	$x = 2\sin t$	$y = 2\cos t$
0	$2\sin 0 = 0$	$2\cos \theta = 2$
$\frac{\pi}{6}$	$2\sin\frac{\pi}{6} = 1$	$2\cos\frac{\pi}{6} = \sqrt{3}$
$\frac{\pi}{4}$	$2\sin\frac{\pi}{4} = \sqrt{2}$	$2\cos\frac{\pi}{4} = \sqrt{2}$
$\frac{\pi}{3}$	$2\sin\frac{\pi}{3} = \sqrt{3}$	$2\cos\frac{\pi}{3} = 1$
$\frac{\pi}{2}$	$2\sin\frac{\pi}{2} = 2$	$2\cos\frac{\pi}{2} = 0$

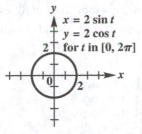

(b) $x = 2\sin t$ and $y = 2\cos t$, so

$\dfrac{x}{2} = \sin t$ and $\dfrac{y}{2} = \cos t$.

$\sin^2 t + \cos^2 t = 1 \Rightarrow$

$\left(\dfrac{x}{2}\right)^2 + \left(\dfrac{y}{2}\right)^2 = 1 \Rightarrow \dfrac{x^2}{4} + \dfrac{y^2}{4} = 1 \Rightarrow$

$x^2 + y^2 = 4$. Because t is in $[0, 2\pi]$, x is
in $[-2, 2]$ because the graph is a circle,
centered at the origin, with radius 2.

17. (a) $x = 3\tan t$, $y = 2\sec t$, for t in $\left(-\frac{\pi}{2}, \frac{\pi}{2}\right)$

t	$x = 3\tan t$	$y = 2\sec t$
$-\frac{\pi}{3}$	$3\tan\left(-\frac{\pi}{3}\right) = -3\sqrt{3}$	$2\sec\left(-\frac{\pi}{3}\right) = 4$
$-\frac{\pi}{6}$	$3\tan\left(-\frac{\pi}{6}\right) = -\sqrt{3}$	$2\sec\left(-\frac{\pi}{6}\right) = \frac{4\sqrt{3}}{3}$
0	$3\tan 0 = 0$	$2\sec 0 = 2$
$\frac{\pi}{6}$	$3\tan\frac{\pi}{6} = \sqrt{3}$	$2\sec\frac{\pi}{6} = \frac{4\sqrt{3}}{3}$
$\frac{\pi}{3}$	$3\tan\frac{\pi}{3} = 3\sqrt{3}$	$2\sec\frac{\pi}{3} = 4$

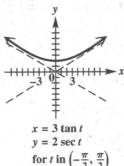

(b) $\dfrac{x}{3} = \tan t$, $\dfrac{y}{2} = \sec t$, and

$1 + \tan^2 t = \left(\dfrac{y}{2}\right)^2 = \sec^2 t$, so

$1 + \left(\dfrac{x}{3}\right)^2 = \left(\dfrac{y}{2}\right)^2 \Rightarrow 1 + \dfrac{x^2}{9} = \dfrac{y^2}{4} \Rightarrow$

$y^2 = 4\left(1 + \dfrac{x^2}{9}\right) \Rightarrow y = 2\sqrt{1 + \dfrac{x^2}{9}}$

This graph is the top half of a hyperbola,
so x is in $(-\infty, \infty)$.

19. (a) $x = \sin t$, $y = \csc t$ for t in $(0, \pi)$

Because t is in $(0, \pi)$ and $x = \sin t$, x is in $(0, 1]$.

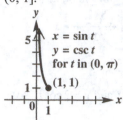

(b) Because $x = \sin t$ and $y = \csc t = \dfrac{1}{\sin t}$, we

have $y = \dfrac{1}{x}$, where x is in $(0, 1]$.

21. (a) $x = t$, $y = \sqrt{t^2 + 2}$, for t in $(-\infty, \infty)$

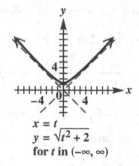

(b) $x = t$ and $y = \sqrt{t^2 + 2}$, so $y = \sqrt{x^2 + 2}$.

Because t is in $(-\infty \ \infty)$ and $x = t$, x is in $(-\infty, \infty)$.

23. (a) $x = 2 + \sin t$, $y = 1 + \cos t$, for t in $[0, 2\pi]$

This is a circle centered at $(2, 1)$ with radius 1 and t is in $[0, 2\pi]$, so x is in $[1, 3]$.

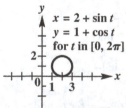

(b) $x = 2 + \sin t \Rightarrow x - 2 = \sin t$

$y = 1 + \cos t \Rightarrow y - 1 = \cos t$

$\sin^2 t + \cos^2 t = 1$, so

$(x - 2)^2 + (y - 1)^2 = 1$, for x in $[1, 3]$.

25. (a) $x = t + 2$, $y = \dfrac{1}{t + 2}$, for $t \neq 2$

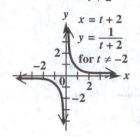

(b) $x = t + 2$ and $y = \dfrac{1}{t + 2} \Rightarrow y = \dfrac{1}{x}$.

Because $t \neq -2$, $x \neq -2 + 2$, $x \neq 0$.

Therefore, x is in $(-\infty, 0) \cup (0, \infty)$.

27. (a) $x = t + 2$, $y = t - 4$, for t in $(-\infty, \infty)$

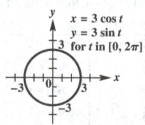

(b) $x = t + 2 \Rightarrow t = x - 2$.

$y = t - 4 \Rightarrow y = (x - 2) - 4 = x - 6$.

Because t is in $(-\infty, \infty)$, x is in $(-\infty, \infty)$.

29. $x = 3\cos t$, $y = 3\sin t$

$x = 3\cos t \Rightarrow \cos t = \dfrac{x}{3}$,

$y = 3\sin t \Rightarrow \sin t = \dfrac{y}{3}$, and

$\sin^2 t + \cos^2 t = 1$. Thus,

$\left(\dfrac{y}{3}\right)^2 + \left(\dfrac{x}{3}\right)^2 = 1 \Rightarrow \dfrac{y^2}{9} + \dfrac{x^2}{9} = 1 \Rightarrow$

$x^2 + y^2 = 9$.

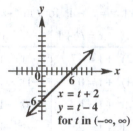

This is a circle centered at the origin with radius 3.

31. $x = 3\sin t, \ y = 2\cos t$

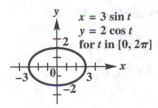

$x = 3\sin t \Rightarrow \sin t = \dfrac{x}{3}, \ \ y = 2\cos t \Rightarrow \cos t = \dfrac{y}{2}$, and $\sin^2 t + \cos^2 t = 1$.

Thus, $\left(\dfrac{x}{3}\right)^2 + \left(\dfrac{y}{2}\right)^2 = 1 \Rightarrow \dfrac{x^2}{9} + \dfrac{y^2}{4} = 1$.

This is an ellipse centered at the origin with axes endpoints $(-3, 0), (3, 0), (0, -2), (0, 2)$.

In Exercises 33–35, answers may vary.

33. $y = (x+3)^2 - 1$

$x = t, \ y = (t+3)^2 - 1$ for t in $(-\infty, \infty)$; $x = t - 3, \ y = t^2 - 1$ for t in $(-\infty, \infty)$

35. $y = x^2 - 2x + 3 = (x-1)^2 + 2$

$x = t, \ y = (t-1)^2 + 2 = t^2 - 2t + 3$ for t in $(-\infty, \infty)$; $x = t + 1, \ y = t^2 + 2$ for t in $(-\infty, \infty)$

37. $x = 2t - 2\sin t, \ y = 2 - 2\cos t$, for t in $[0, 4\pi]$

t	0	$\dfrac{\pi}{2}$	π	$\dfrac{3\pi}{2}$	2π	3π	4π
$x = 2t - 2\sin t$	0	1.14	2π	11.4	4π	6π	8π
$y = 2 - 2\cos t$	0	2	4	2	0	4	0

39. $x = 2\cos t, \ y = 3\sin 2t$

x = 2 cos t, y = 3 sin 2t,
for t in [0, 6.5]

41. $x = 3\sin 4t, \ y = 3\cos 3t$

x = 3 sin 4t, y = 3 cos 3t,
for t in [0, 6.5]

For Exercises 43–45, recall that the motion of a projectile (neglecting air resistance) can be modeled by: $x = (v_0 \cos\theta)t, \ y = (v_0 \sin\theta)t - 16t^2$ for t in $[0, k]$.

43. (a) $x = (v\cos\theta)t \Rightarrow$

$x = (48\cos 60°)t = 48\left(\dfrac{1}{2}\right)t = 24t$

$y = (v\sin\theta)t - 16t^2 \Rightarrow$

$y = (48\sin 60°)t - 16t^2 = 48 \cdot \dfrac{\sqrt{3}}{2}t - 16t^2$

$= -16t^2 + 24\sqrt{3}t$

(b) $t = \dfrac{x}{24}$, so $y = -16\left(\dfrac{x}{24}\right)^2 + 24\sqrt{3}\left(\dfrac{x}{24}\right)$

$= -\dfrac{x^2}{36} + \sqrt{3}x$

(c) $y = -16t^2 + 24\sqrt{3}t$

When the rocket is no longer in flight, $y = 0$. Solve $0 = -16t^2 + 24\sqrt{3}t \Rightarrow$

$0 = t\left(-16t + 24\sqrt{3}\right)$.

$t = 0$ or $-16t + 24\sqrt{3} = 0 \Rightarrow$

$-16t = -24\sqrt{3} \Rightarrow t = \dfrac{24\sqrt{3}}{16} \Rightarrow$

$t = \dfrac{3\sqrt{3}}{2} \approx 2.6$

The flight time is about 2.6 seconds. The horizontal distance at $t = \dfrac{3\sqrt{3}}{2}$ is

$x = 24t = 24\left(\dfrac{3\sqrt{3}}{2}\right) \approx 62$ ft

45. (a) $x = (v\cos\theta)t \Rightarrow x = (88\cos 20°)t$

$y = (v\sin\theta)t - 16t^2 + 2 \Rightarrow$
$y = (88\sin 20°)t - 16t^2 + 2$

(b) $t = \dfrac{x}{88\cos 20°}$, so

$y = 88\sin 20°\left(\dfrac{x}{88\cos 20°}\right)$

$\quad - 16\left(\dfrac{x}{88\cos 20°}\right)^2 + 2$

$\quad = (\tan 20°)x - \dfrac{x^2}{484\cos^2 20°} + 2$

(c) Solving $0 = -16t^2 + (88\sin 20°)t + 2$ by the quadratic formula, we have

$t = \dfrac{-88\sin 20° \pm \sqrt{(88\sin 20°)^2 - 4(-16)(2)}}{(2)(-16)}$

$\quad = \dfrac{-30.098 \pm \sqrt{905.8759 + 128}}{-32} \Rightarrow$

$t \approx -0.064$ or 1.9

Discard $t = -0.064$ because it is an unacceptable answer.
At $t \approx 1.9$ sec, $x = (88\cos 20°)t \approx 161$ ft .
The softball traveled 1.9 sec and 161 feet.

47. (a) $x = (v\cos\theta)t \Rightarrow$

$x = (128\cos 60°)t = 128\left(\dfrac{1}{2}\right)t = 64t$

$y = (v\sin\theta)t - 16t^2 + 8 \Rightarrow$
$y = (128\sin 60°)t - 16t^2 + 8$

$\quad = 128\left(\dfrac{\sqrt{3}}{2}\right)t - 16t^2 + 8$

$\quad = 64\sqrt{3}t - 16t^2 + 8$

$t = \dfrac{x}{64}$, so

$y = 64\sqrt{3}\left(\dfrac{x}{64}\right) - 16\left(\dfrac{x}{64}\right)^2 + 8 \Rightarrow$

$y = -\dfrac{1}{256}x^2 + \sqrt{3}x + 8$.

This is a parabolic path.

(b) Solving $0 = -16t^2 + 64\sqrt{3}t + 8$ by the quadratic formula, we have

$t = \dfrac{-64\sqrt{3} \pm \sqrt{(64\sqrt{3})^2 - 4(-16)(8)}}{2(-16)}$

$\quad = \dfrac{-64\sqrt{3} \pm \sqrt{12,800}}{-32}$

$\quad = \dfrac{-64\sqrt{3} \pm 80\sqrt{2}}{-32} \Rightarrow t \approx -0.07,\ 7.0$

Discard $t = -0.07$ because it gives an unacceptable answer. At $t \approx 7.0$ sec, $x = 64t = 448$ ft. The rocket traveled approximately 7 sec and 448 feet.

49. (a) $x = (v\cos\theta)t \Rightarrow$

$x = (64\cos 60°)t = 64\left(\dfrac{1}{2}\right)t = 32t$

$y = (v\sin\theta)t - 16t^2 + 3 \Rightarrow$
$y = (64\sin 60°)t - 16t^2 + 3$

$\quad = 64\left(\dfrac{\sqrt{3}}{2}\right)t - 16t^2 + 3$

$\quad = 32\sqrt{3}t - 16t^2 + 3$

(b) Solving $0 = -16t^2 + 32\sqrt{3}t + 3$ by the quadratic formula, we have

$t = \dfrac{-32\sqrt{3} \pm \sqrt{(32\sqrt{3})^2 - 4(-16)(3)}}{2(-16)}$

$\quad = \dfrac{-32\sqrt{3} \pm \sqrt{3264}}{-32} = \dfrac{-32\sqrt{3} \pm 8\sqrt{51}}{-32} \Rightarrow$

$t \approx -0.05,\ 3.52$

Discard $t = -0.05$ because it gives an unacceptable answer. At $t \approx 3.52$ sec, $x = 32t \approx 112.6$ ft. The ball traveled approximately 112.6 feet.

(c) To find the maximum height, find the vertex of $y = -16t^2 + 32\sqrt{3}t + 3$.

$$y = -16t^2 + 32\sqrt{3}t + 3$$
$$= -16\left(t^2 - 2\sqrt{3}t\right) + 3$$
$$= -16\left(t^2 - 2\sqrt{3}t + 3 - 3\right) + 3$$
$$y = -16\left(t - \sqrt{3}\right)^2 + 48 + 3$$
$$= -16\left(t - \sqrt{3}\right)^2 + 51$$

The maximum height of 51 ft is reached at $\sqrt{3} \approx 1.73$ sec. because $x = 32t$, the ball has traveled horizontally $32\sqrt{3} \approx 55.4$ ft.

(d) To determine if the ball would clear a 5-ft-high fence that is 100 ft from the batter, we need to first determine at what time is the ball 100 ft from the batter. Because $x = 32t$, the time the ball is 100 ft from the batter is $t = \dfrac{100}{32} = 3.125$ sec. We next need to determine how high off the ground the ball is at this time. Because $y = 32\sqrt{3}t - 16t^2 + 3$, the height of the ball is $y = 32\sqrt{3}(3.125) - 16(3.125)^2 + 3$
≈ 20.0 ft. This height exceeds 5 ft, so the ball will clear the fence.

For Exercises 51–53, other answers are possible.

51. $y = a(x - h)^2 + k$

To find one parametric representation, let $x = t$. We therefore have, $y = a(t - h)^2 + k$. For another representation, let $x = t + h$. We therefore have $y = a(t + h - h)^2 + k = at^2 + k$.

53. $\dfrac{x^2}{a^2} + \dfrac{y^2}{b^2} = 1$

To find a parametric representation, let $x = a \sin t$. We therefore have

$$\frac{(a\sin t)^2}{a^2} + \frac{y^2}{b^2} = 1 \Rightarrow \frac{a^2 \sin^2 t}{a^2} + \frac{y^2}{b^2} = 1 \Rightarrow$$
$$\sin^2 t + \frac{y^2}{b^2} = 1 \Rightarrow \frac{y^2}{b^2} = 1 - \sin^2 t$$
$$y^2 = b^2\left(1 - \sin^2 t\right) \Rightarrow y^2 = b^2 \cos^2 t \Rightarrow$$
$$y = b\cos t$$

55. To show that $r\theta = a$ is given parametrically by $x = \dfrac{a\cos\theta}{\theta}, \ y = \dfrac{a\sin\theta}{\theta}$,

for θ in $(-\infty, 0) \cup (0, \infty)$, we must show that the parametric equations yield $r\theta = a$, where $r^2 = x^2 + y^2$.

$$r^2 = x^2 + y^2 \Rightarrow$$
$$r^2 = \left(\frac{a\cos\theta}{\theta}\right)^2 + \left(\frac{a\sin\theta}{\theta}\right)^2 \Rightarrow$$
$$r^2 = \frac{a^2 \cos^2\theta}{\theta^2} + \frac{a^2 \sin^2\theta}{\theta^2}$$
$$r^2 = \frac{a^2}{\theta^2}\cos^2\theta + \frac{a^2}{\theta^2}\sin^2\theta \Rightarrow$$
$$r^2 = \frac{a^2}{\theta^2}\left(\cos^2\theta + \sin^2\theta\right) \Rightarrow r^2 = \frac{a^2}{\theta^2} \Rightarrow$$
$$r = \pm\frac{a}{\theta} \text{ or just } r = \frac{a}{\theta}$$

This implies that the parametric equations satisfy $r\theta = a$.

57. If $x = f(t)$ is replaced by $x = c + f(t)$, the graph will be translated c units to the right.

Chapter 8   Review Exercises

1. $\sqrt{-9} = i\sqrt{9} = 3i$

3. $x^2 = -81 \Rightarrow x = \pm\sqrt{-81} = \pm 9i$
Solution set: $\{\pm 9i\}$

5. $(1 - i) - (3 + 4i) + 2i = -2 - 3i$

7. $(6 - 5i) + (2 + 7i) - (3 - 2i) = 5 + 4i$

9. $(3 + 5i)(8 - i) = 24 - 3i + 40i - 5i^2$
$= 24 + 37i - 5(-1) = 29 + 37i$

11. $(2 + 6i)^2 = 4 + 2(2)(6i) + (6i)^2$
$= 4 + 24i - 36 = -32 + 24i$

13. $(1 - i)^3 = 1^3 - 3 \cdot 1^2 \cdot i + 3 \cdot 1 \cdot i^2 - i^3$
$= 1 - 3i - 3 - (-i) = -2 - 2i$

15. $\dfrac{25 - 19i}{5 + 3i} = \dfrac{25 - 19i}{5 + 3i} \cdot \dfrac{5 - 3i}{5 - 3i}$

$= \dfrac{125 - 75i - 95i + 57i^2}{5^2 - (3i)^2}$

$= \dfrac{125 - 160i - 57}{25 - (-9)} = \dfrac{68 - 170i}{34}$

$= 2 - 5i$

17. $\dfrac{2+i}{1-5i} = \dfrac{2+i}{1-5i} \cdot \dfrac{1+5i}{1+5i} = \dfrac{2+10i+i+5i^2}{1^2-25i^2}$

$= \dfrac{-3+11i}{26} = -\dfrac{3}{26} + \dfrac{11}{26}i$

19. $i^{53} = i^{52} \cdot i = \left(i^4\right)^{13} \cdot i = i$

21. $\left[5\left(\cos 90° + i\sin 90°\right)\right]$

$\qquad \cdot \left[6\left(\cos 180° + i\sin 180°\right)\right]$

$= 5 \cdot 6\left[\cos\left(90° + 180°\right) + i\sin\left(90° + 180°\right)\right]$

$= 30\left(\cos 270° + i\sin 270°\right)$

$= 30\left(0 - i\right)$

$= 0 - 30i \text{ or } -30i$

23. $\dfrac{2\left(\cos 60° + i\sin 60°\right)}{8\left(\cos 300° + \sin 300°\right)}$

$= \dfrac{2}{8}\left[\cos\left(60° - 300°\right) + i\sin\left(60° - 300°\right)\right]$

$= \dfrac{1}{4}\left[\cos\left(-240°\right) + i\sin\left(-240°\right)\right]$

$= \dfrac{1}{4}\left[\cos\left(240°\right) - i\sin\left(240°\right)\right]$

$= \dfrac{1}{4}\left[-\cos 60° + i\sin 60°\right]$

$= \dfrac{1}{4}\left(-\dfrac{1}{2} + \dfrac{\sqrt{3}}{2}\right) = -\dfrac{1}{8} + \dfrac{\sqrt{3}}{8}i$

25. $\left(\sqrt{3} + i\right)^3$

$r = \sqrt{\left(\sqrt{3}\right)^2 + 1^2} = \sqrt{3+1} = \sqrt{4} = 2$ and

because θ is in quadrant I,

$\tan\theta = \dfrac{1}{\sqrt{3}} = \dfrac{\sqrt{3}}{3} \Rightarrow \theta = 30°$.

$\left(\sqrt{3} + i\right)^3 = \left[2\left(\cos 30° + i\sin 30°\right)\right]^3$

$= 2^3\left[\cos\left(3 \cdot 30°\right) + i\sin\left(3 \cdot 30°\right)\right]$

$= 8\left[\cos 90° + i\sin 90°\right]$

$= 8\left(0 + i\right) = 0 + 8i = 8i$

27. $\left(\cos 100° + i\sin 100°\right)^6$

$= \cos\left(6 \cdot 100°\right) + i\sin\left(6 \cdot 100°\right)$

$= \cos 600° + i\sin 600° = \cos 240° + i\sin 240°$

$= -\cos 60° - i\sin 60° = -\dfrac{1}{2} - \dfrac{\sqrt{3}}{2}i$

29.

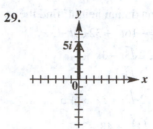

31.

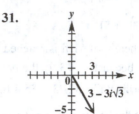

33. $-2 + 2i$

$r = \sqrt{\left(-2\right)^2 + 2^2} = \sqrt{4+4} = \sqrt{8} = 2\sqrt{2}$

Because θ is in quadrant II,

$\tan\theta = \dfrac{2}{-2} = -1 \Rightarrow \theta = 135°$. Thus,

$-2 + 2i = 2\sqrt{2}\left(\cos 135° + i\sin 135°\right)$.

35. $2\left(\cos 225° + i\sin 225°\right)$

$= 2\left(-\cos 45° - i\sin 45°\right)$

$= 2\left(-\dfrac{\sqrt{2}}{2} - \dfrac{i\sqrt{2}}{2}\right) = -\sqrt{2} - i\sqrt{2}$

37. $1 - i$

$r = \sqrt{1^2 + \left(-1\right)^2} = \sqrt{1+1} = \sqrt{2}$ and

$\tan\theta = \dfrac{-1}{1} = -1 \Rightarrow \theta = 315°$, because θ is in

quadrant IV. Thus,

$1 - i = \sqrt{2}\left(\cos 315° + i\sin 315°\right)$.

39. $-4i$

Because $r = 4$ and the point $(0, -4)$ intersects

the negative y-axis, $\theta = 270°$ and

$-4i = 4(\cos 270° + i\sin 270°)$.

41. $z = x + yi$

Because the imaginary part of z is the negative

of the real part of z, we are saying $y = -x$. This

is a line.

43. Convert $1 - i$ to polar form

$r = \sqrt{1^2 + (-1)^2} = \sqrt{1+1} = \sqrt{2}$ and

$\tan\theta = \dfrac{-1}{1} = -1 \Rightarrow \theta = 315°$, because θ is in quadrant IV. Thus,

$1 - i = \sqrt{2}\left(\cos 315° + i\sin 315°\right)$.

$r^3\left(\cos 3\alpha + i\sin 3\alpha\right)$

$= \sqrt{2}\left(\cos 315° + i\sin 315°\right)$, so we have

$r^3 = \sqrt{2} \Rightarrow r = \sqrt[6]{2}$ and

$3\alpha = 315° + 360° \cdot k \Rightarrow$

$\alpha = \dfrac{315° + 360° \cdot k}{3} = 105° + 120° \cdot k,\ k$ any

integer.

If $k = 0$, then $\alpha = 105° + 0° = 105°$.

If $k = 1$, then $\alpha = 105° + 120° = 225°$.

If $k = 2$, then $\alpha = 105° + 240° = 345°$.

So, the cube roots of $1 - i$ are

$\sqrt[6]{2}\left(\cos 105° + i\sin 105°\right)$,

$\sqrt[6]{2}\left(\cos 225° + i\sin 225°\right)$, and

$\sqrt[6]{2}\left(\cos 345° + i\sin 345°\right)$.

45. The number –64 has no real sixth roots because a real number raised to the sixth power will never be negative.

47. $x^4 + 16 = 0 \Rightarrow x^4 = -16$

We have, $r = 16$ and $\theta = 180°$.

$x^4 = -16 = -16 + 0i = 16\left(\cos 180° + i\sin 180°\right)$

$r^4\left(\cos 4\alpha + i\sin 4\alpha\right)$

$= 16\left(\cos 180° + i\sin 180°\right)$, so we have

$r^4 = 16 \Rightarrow r = 2$ and $4\alpha = 180° + 360° \cdot k \Rightarrow$

$\alpha = \dfrac{180° + 360° \cdot k}{4} = 45° + 90° \cdot k,\ k$ any

integer.

If $k = 0$, then $\alpha = 45° + 0° = 45°$.

If $k = 1$, then $\alpha = 45° + 90° = 135°$.

If $k = 2$, then $\alpha = 45° + 180° = 225°$.

If $k = 3$, then $\alpha = 45° + 270° = 315°$.

Solution set:

$\{2\left(\cos 45° + i\sin 45°\right),\ 2\left(\cos 135° + i\sin 135°\right),$

$2\left(\cos 225° + i\sin 225°\right),$

$2\left(\cos 315° + i\sin 315°\right)\}$

49. $x^2 + i = 0 \Rightarrow x^2 = -i$

We have $r = 1$ and $\theta = 270°$.

$x^2 = -i = 0 - i = 1\left(\cos 270° + i\sin 270°\right)$

$r^2\left(\cos 2\alpha + i\sin 2\alpha\right)$

$= 1\left(\cos 270° + i\sin 270°\right)$, so we have

$r^2 = 1 \Rightarrow r = 1$ and $2\alpha = 270° + 360° \cdot k \Rightarrow$

$\alpha = \dfrac{270° + 360° \cdot k}{2} = 135° + 180° \cdot k,\ k$ any

integer. If $k = 0$, then $\alpha = 135° + 0° = 135°$.

If $k = 1$, then $\alpha = 135° + 180° = 315°$.

Solution set: $\{\cos 135° + i\sin 135°,$

$\cos 315° + i\sin 315°\}$

51. $\left(-1, \sqrt{3}\right)$

$r = \sqrt{(-1)^2 + \left(\sqrt{3}\right)^2} = \sqrt{1+3} = \sqrt{4} = 2$ and

$\theta = \tan^{-1}\left(-\dfrac{\sqrt{3}}{1}\right) = \tan^{-1}\left(-\sqrt{3}\right) = 120°$,

because θ is in quadrant II. Thus, the polar coordinates are (2, 120°).

53. $r = 4\cos\theta$ is a circle.

θ	0°	30°	45°	60°	90°	120°	135°	150°	180°
$r = 4\cos\theta$	4	3.5	2.8	2	0	−2	−2.8	−3.5	−4

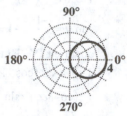

$r = 4\cos\theta$
Graph is retraced in the
interval (180°, 360°).

55. $r = 2\sin 4\theta$ is an eight-leaved rose.

θ	0°	7.5°	15°	22.5°	30°	37.5°	45°
$r = 2\sin 4\theta$	0	1	$\sqrt{3}$	2	$\sqrt{3}$	1	0

θ	52.5°	60°	67.5°	75°	82.5°	90°	52.5°
$r = 2\sin 4\theta$	−1	$-\sqrt{3}$	−2	$-\sqrt{3}$	−1	0	−1

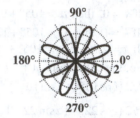

$r = 2\sin 4\theta$
The graph continues to form
eight petals for the interval
[0°, 360°).

57. $r = \dfrac{3}{1+\cos\theta}$

$r = \dfrac{3}{1+\cos\theta} \Rightarrow r(1+\cos\theta) = 3 \Rightarrow$

$r + r\cos\theta = 3 \Rightarrow \sqrt{x^2+y^2} + x = 3 \Rightarrow$

$\sqrt{x^2+y^2} = 3 - x$

$x^2 + y^2 = (3-x)^2 \Rightarrow x^2 + y^2 = 9 - 6x + x^2 \Rightarrow$

$y^2 = 9 - 6x \Rightarrow y^2 + 6x - 9 = 0 \Rightarrow y^2 = -6x + 9$

$y^2 = -6\left(x - \dfrac{3}{2}\right)$ or $y^2 + 6x - 9 = 0$

59. $r = 2 \Rightarrow \sqrt{x^2+y^2} = 2 \Rightarrow x^2 + y^2 = 4$

61. $y = x^2 \Rightarrow r\sin\theta = r^2\cos^2\theta \Rightarrow$

$\sin\theta = r\cos^2\theta \Rightarrow r = \dfrac{\sin\theta}{\cos^2\theta} \Rightarrow$

$r = \dfrac{\sin\theta}{\cos\theta} \cdot \dfrac{1}{\cos\theta} = \tan\theta\sec\theta$

$r = \tan\theta\sec\theta$ or $r = \dfrac{\tan\theta}{\cos\theta}$

63. Suppose (r,θ) is on the graph, $-\theta$ reflects this point with respect to the x-axis, and $-r$ reflects the resulting point with respect to the origin. The net result is that the original point is reflected with respect to the y-axis. The correct choice is B.

65. If (r,θ) lies on the graph, $(r,-\theta)$ would reflect that point across the x-axis. Therefore, there is symmetry about the x-axis. The correct choice is C.

67. $x = 2$

$x = r\cos\theta \Rightarrow r\cos\theta = 2 \Rightarrow$

$r = \dfrac{2}{\cos\theta}$ or $r = 2\sec\theta$

69. $x + 2y = 4$

$x = r\cos\theta$ and $y = r\sin\theta$, so we have

$r\cos\theta + 2r\sin\theta = 4 \Rightarrow$

$r(\cos\theta + 2\sin\theta) = 4 \Rightarrow r = \dfrac{4}{\cos\theta + 2\sin\theta}$

71. $x = t + \cos t$, $y = \sin t$ for t in $[0, 2\pi]$

t	0	$\frac{\pi}{6}$	$\frac{\pi}{3}$	$\frac{\pi}{2}$	$\frac{3\pi}{4}$	π
$x = t + \cos t$	1	$\frac{\pi}{6} + \frac{\sqrt{3}}{2} \approx 1.4$	$\frac{\pi}{3} + \frac{1}{2} \approx 1.5$	$\frac{\pi}{2} \approx 1.6$	$\frac{3\pi}{4} - \frac{\sqrt{2}}{2} \approx 1.6$	$\pi - 1 \approx 2.1$
$y = \sin t$	0	$\frac{1}{2} = 0.5$	$\frac{\sqrt{3}}{2} \approx 1.7$	1	$\frac{\sqrt{2}}{2} \approx 0.7$	0

t	$\frac{7\pi}{6}$	$\frac{5\pi}{4}$	$\frac{4\pi}{3}$	$\frac{3\pi}{2}$	$\frac{7\pi}{4}$	2π
$x = t + \cos t$	$\frac{7\pi}{6} - \frac{\sqrt{3}}{2} \approx 2.8$	$\frac{5\pi}{4} - \frac{\sqrt{2}}{2} \approx 3.2$	$\frac{4\pi}{3} - \frac{1}{2} \approx 3.7$	$\frac{3\pi}{2} \approx 4.7$	$\frac{7\pi}{4} + \frac{\sqrt{2}}{2} \approx 6.2$	$2\pi + 1 \approx 7.3$
$y = \sin t$	$-\frac{1}{2} = -0.5$	$-\frac{\sqrt{2}}{2} \approx -0.7$	$-\frac{\sqrt{3}}{2} \approx -0.9$	-1	$-\frac{\sqrt{2}}{2} \approx -0.7$	0

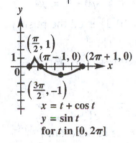

$x = t + \cos t$
$y = \sin t$
for t in $[0, 2\pi]$

73. $x = \sqrt{t-1}$, $y = \sqrt{t}$, for t in $[1, \infty)$

$x = \sqrt{t-1} \Rightarrow x^2 = t-1 \Rightarrow t = x^2 + 1$, so substitute $x^2 + 1$ for t in the equation for y to obtain $y = \sqrt{x^2 + 1}$. Because t is in $[1, \infty)$, x is in $[\sqrt{1-1}, \infty)$ or $[0, \infty)$.

75. $x = 5 \tan t$, $y = 3 \sec t$, for t in $\left(-\frac{\pi}{2}, \frac{\pi}{2} \right)$

$\frac{x}{5} = \tan t$, $\frac{y}{3} = \sec t$, and $1 + \tan^2 t = \sec^2 t$, so

we have $1 + \left(\frac{x}{5} \right)^2 = \left(\frac{y}{3} \right)^2 \Rightarrow 1 + \frac{x^2}{25} = \frac{y^2}{9} \Rightarrow$

$9\left(1 + \frac{x^2}{25} \right) = y^2 \Rightarrow y = \sqrt{9\left(1 + \frac{x^2}{25} \right)} \Rightarrow$

$y = 3\sqrt{1 + \frac{x^2}{25}}$.

y is positive because $y = 3 \sec t > 0$ for t in $\left(-\frac{\pi}{2}, \frac{\pi}{2} \right)$. Because t is in $\left(-\frac{\pi}{2}, \frac{\pi}{2} \right)$ and

$x = 5 \tan t$ is undefined at $-\frac{\pi}{2}$ and $\frac{\pi}{2}$, x is in $(-\infty, \infty)$.

77. $x = \cos 2t$, $y = \sin t$ for t in $(-\pi, \pi)$

$\cos 2t = \cos^2 t - \sin^2 t$ (double angle formula)

$\cos^2 t + \sin^2 t = 1$, so we have

$\cos^2 t + \sin^2 t = 1 \Rightarrow$

$\left(\cos^2 t - \sin^2 t \right) + 2\sin^2 t = 1 \Rightarrow$

$x + 2y^2 = 1 \Rightarrow 2y^2 = -x + 1 \Rightarrow 2y^2 = -(x-1)$

$y^2 = -\frac{1}{2}(x-1)$ or $2y^2 + x - 1 = 0$

$x \geq 0 + 5 = 5$ t is in $(-\pi, \pi)$ and $\cos 2t$ is in $[-1, 1]$, x is in $[-1, 1]$.

79. (a) $x = (v \cos \theta)t \Rightarrow x = (118 \cos 27°)t$ and

$y = (v \sin \theta)t - 16t^2 + h \Rightarrow$

$y = (118 \sin 27°)t - 16t^2 + 3.2$

(b) $t = \dfrac{x}{118 \cos 27°}$, so we have

$y = 118 \sin 27° \cdot \dfrac{x}{118 \cos 27°}$

$\qquad - 16\left(\dfrac{x}{118 \cos 27°} \right)^2 + 3.2$

$= 3.2 - \dfrac{4}{3481 \cos^2 27°} x^2 + (\tan 27°)x$

(c) Solving $0 = -16t^2 + (118\sin 27°)t + 3.2$ by the quadratic formula, we have

$$t = \frac{-118\sin 27° \pm \sqrt{(118\sin 27°)^2 - 4(-16)(3.2)}}{2(-16)} \Rightarrow$$

$t \approx -0.06, \; 3.406$

Discard $t = -0.06$ sec because it is an unacceptable answer. At $t = 3.4$ sec, the baseball traveled

$x = (118\cos 27°)(3.406) \approx 358$ ft.

Chapter 8   Test

1. (a) $\sqrt{-8} \cdot \sqrt{-6} = i\sqrt{8} \cdot i\sqrt{6} = i^2\sqrt{48} = -4\sqrt{3}$

(b) $\dfrac{\sqrt{-2}}{\sqrt{8}} = \dfrac{i\sqrt{2}}{\sqrt{8}} = \dfrac{i}{\sqrt{4}} = \dfrac{1}{2}i$

(c) $\dfrac{\sqrt{-20}}{\sqrt{-180}} = \dfrac{i\sqrt{20}}{i\sqrt{180}} = \dfrac{1}{\sqrt{9}} = \dfrac{1}{3}$

2. $w = 2 - 4i, \; z = 5 + i$

(a) $w + z = (2 - 4i) + (5 + i) = 7 - 3i$

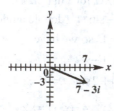

(b) $w - z = (2 - 4i) - (5 + i) = -3 - 5i$

(c) $wz = (2 - 4i)(5 + i) = 10 + 2i - 20i - 4i^2$
$= 10 - 18i - 4(-1) = 14 - 18i$

(d) $\dfrac{w}{z} = \dfrac{2 - 4i}{5 + i} = \dfrac{2 - 4i}{5 + i} \cdot \dfrac{5 - i}{5 - i}$

$= \dfrac{10 - 2i - 20i + 4i^2}{5^2 - i^2} = \dfrac{10 - 22i + 4(-1)}{5^2 - (-1)}$

$= \dfrac{6 - 22i}{26} = \dfrac{3}{13} - \dfrac{11}{13}i$

3. (a) $i^{15} = i^{12+3} = i^{12} \cdot i^3 = (i^4)^3 \cdot i^3 = 1(-i) = -i$

(b) $(1 + i)^2 = (1 + i)(1 + i) = 1 + i + i + i^2$
$= 1 + 2i + (-1) = 2i$

4. $2x^2 - x + 4 = 0$
Use the quadratic formula with $a = 2$, $b = -1$, and $c = 4$:

$x = \dfrac{-(-1) \pm \sqrt{(-1)^2 - 4(2)(4)}}{2(2)} = \dfrac{1 \pm \sqrt{1 - 32}}{4}$

$= \dfrac{1 \pm \sqrt{-31}}{4} = \dfrac{1 \pm i\sqrt{31}}{4} = \dfrac{1}{4} \pm \dfrac{\sqrt{31}}{4}i$

Solution set: $\left\{\dfrac{1}{4} \pm \dfrac{\sqrt{31}}{4}i\right\}$

5. (a) $3i$

$r = \sqrt{0^2 + 3^2} = \sqrt{0 + 9} = \sqrt{9} = 3$

The point $(0, 3)$ is on the positive y-axis, so, $\theta = 90°$. Thus,

$3i = 3(\cos 90° + i\sin 90°)$.

(b) $1 + 2i$

$r = \sqrt{1^2 + 2^2} = \sqrt{1 + 4} = \sqrt{5}$

Because θ is in quadrant I,

$\theta = \tan^{-1}\left(\dfrac{2}{1}\right) = \tan^{-1} 2 \approx 63.43°$. Thus,

$1 + 2i = \sqrt{5}(\cos 63.43° + i\sin 63.43°)$.

(c) $-1 - \sqrt{3}i$

$r = \sqrt{(-1)^2 + (-\sqrt{3})^2} = \sqrt{1 + 3} = \sqrt{4} = 2$

Because θ is in quadrant III,

$\theta = \tan^{-1}\left(\dfrac{-\sqrt{3}}{-1}\right) = \tan^{-1}\sqrt{3} = 240°$.

Thus, $-1 - \sqrt{3}i = 2(\cos 240° + i\sin 240°)$

6. (a) $3(\cos 30° + i\sin 30°) = 3\left(\dfrac{\sqrt{3}}{2} + \dfrac{1}{2}i\right)$

$= \dfrac{3\sqrt{3}}{2} + \dfrac{3}{2}i$

(b) $4 \operatorname{cis} 40° = 3.06 + 2.57i$

(c) $3(\cos 90° + i\sin 90°) = 3(0 + 1 \cdot i)$
$= 0 + 3i = 3i$

7. $w = 8(\cos 40° + i\sin 40°)$,
$z = 2(\cos 10° + i\sin 10°)$

(a) wz

$= 8 \cdot 2\left[\cos(40° + 10°) + i\sin(40° + 10°)\right]$

$= 16(\cos 50° + i\sin 50°)$

(b) $\dfrac{w}{z} = \dfrac{8}{2}\left[\cos\left(40° - 10°\right) + i\sin\left(40° - 10°\right)\right]$

$= 4\left(\cos 30° + i\sin 30°\right) = 4\left(\dfrac{\sqrt{3}}{2} + \dfrac{1}{2}i\right)$

$= 2\sqrt{3} + 2i$

(c) $z^3 = \left[2\left(\cos 10° + i\sin 10°\right)\right]^3$

$= 2^3\left(\cos 3 \cdot 10° + i\sin 3 \cdot 10°\right)$

$= 8\left(\cos 30° + i\sin 30°\right)$

$= 8\left(\dfrac{\sqrt{3}}{2} + \dfrac{1}{2}i\right) = 4\sqrt{3} + 4i$

8. Find all the fourth roots of
$-16i = 16\left(\cos 270° + i\sin 270°\right)$.

$r^4\left(\cos 4\alpha + i\sin 4\alpha\right)$

$= 16\left(\cos 270° + i\sin 270°\right)$, so we have

$r^4 = 16 \Rightarrow r = 2$ and $4\alpha = 270° + 360° \cdot k \Rightarrow$

$\alpha = \dfrac{270° + 360° \cdot k}{4} = 67.5° + 90° \cdot k$, k any

integer. If $k = 0$, then $\alpha = 67.5°$.

If $k = 1$, then $\alpha = 157.5°$.

If $k = 2$, then $\alpha = 247.5°$. If $k = 3$, then $\alpha = 337.5°$.

The fourth roots of $-16i$ are 2(cos 67.5° + i sin 67.5°), 2(cos 157.5° + i sin 157.5°), 2(cos 247.5° + i sin 247.5°), and 2(cos 337.5° + i sin 337.5°).

9. Answers may vary.

(a) (0, 5)

$r = \sqrt{0^2 + 5^2} = \sqrt{0 + 25} = \sqrt{25} = 5$

The point (0, 5) is on the positive y-axis. Thus, $\theta = 90°$. One possibility is (5, 90°). Alternatively, if $\theta = 90° - 360° = -270°$, a second possibility is (5, −270°).

(b) (−2, −2)

$r = \sqrt{\left(-2\right)^2 + \left(-2\right)^2} = \sqrt{4 + 4} = \sqrt{8} = 2\sqrt{2}$

Because θ is in quadrant III,

$\theta = \tan^{-1}\left(\dfrac{-2}{-2}\right) = \tan^{-1} 1 = 225°$. One

possibility is $\left(2\sqrt{2}, 225°\right)$. Alternatively, if $\theta = 225° - 360° = -135°$, a second possibility is $\left(2\sqrt{2}, -135°\right)$.

10. (a) (3, 315°)

$x = r\cos\theta \Rightarrow$

$x = 3\cos 315° = 3 \cdot \dfrac{\sqrt{2}}{2} = \dfrac{3\sqrt{2}}{2}$ and

$y = r\sin\theta \Rightarrow$

$y = 3\sin 315° = 3\left(-\dfrac{\sqrt{2}}{2}\right) = \dfrac{-3\sqrt{2}}{2}$

The rectangular coordinates are

$\left(\dfrac{3\sqrt{2}}{2}, \dfrac{-3\sqrt{2}}{2}\right)$.

(b) (−4, 90°)

$x = r\cos\theta \Rightarrow x = -4\cos 90° = 0$ and
$y = r\sin\theta \Rightarrow y = -4\sin 90° = -4$

The rectangular coordinates are $\left(0, -4\right)$.

11. $r = 1 - \cos\theta$ is a cardioid.

θ	0°	30°	45°	60°
$r = 1 - \cos\theta$	0	0.1	0.3	0.5

θ	90°	135°	180°	225°
$r = 1 - \cos\theta$	1	1.7	2	1.7

θ	270°	315°	360°
$r = 1 - \cos\theta$	1	0.3	0

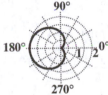

$r = 1 - \cos\theta$

12. $r = 3\cos 3\theta$ is a three-leaved rose,

θ	0°	30°	45°	60°	90°
$r = 3\cos 3\theta$	3	0	−2.1	−3	0

θ	120°	135°	150°	180°
$r = 3\cos 3\theta$	3	2.1	0	−3

$r = 3\cos 3\theta$

Graph is retraced in the interval (180°, 360°).

13. (a) $r = \dfrac{4}{2\sin\theta - \cos\theta} = \dfrac{4}{-1\cdot\cos\theta + 2\sin\theta}$, so we can use the general form for the polar equation of a line,

$r = \dfrac{c}{a\cos\theta + b\sin\theta}$, with

$a = -1$, $b = 2$, and $c = 4$, we have $-x + 2y = 4$ or $x - 2y = -4$. The graph is a line with intercepts $(-4, 0)$ and $(0, 2)$.

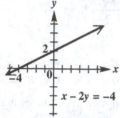

$x - 2y = -4$

(b) $r = 6$ represents the equation of a circle centered at the origin with radius 6, namely $x^2 + y^2 = 36$.

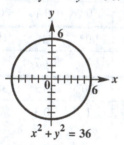

$x^2 + y^2 = 36$

14. $x = 4t - 3$, $y = t^2$ for t in $[-3, 4]$

t	x	y
−3	−15	9
−1	−7	1
0	−3	0
1	1	1
2	5	4
4	13	16

$x = 4t - 3 \Rightarrow t = \dfrac{x+3}{4}$ and $y = t^2$, so we have

$y = \left(\dfrac{x+3}{4}\right)^2 = \dfrac{1}{4}(x+3)^2$, where x is in

$[-15, 13]$

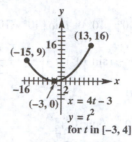

$(13, 16)$
$(-15, 9)$
$(-3, 0)$
$x = 4t - 3$
$y = t^2$
for t in $[-3, 4]$

15. $x = 2\cos 2t$, $y = 2\sin 2t$ for t in $\left[0, 2\pi\right]$

t	0	$\frac{\pi}{8}$	$\frac{\pi}{4}$	$\frac{3\pi}{8}$
x	2	$\sqrt{2}$	0	$-\sqrt{2}$
y	0	$\sqrt{2}$	2	$\sqrt{2}$

t	$\frac{\pi}{2}$	$\frac{5\pi}{8}$	$\frac{3\pi}{4}$	π
x	−2	$-\sqrt{2}$	0	2
y	0	$-\sqrt{2}$	−2	0

t	$\frac{5\pi}{4}$	$\frac{3\pi}{2}$	$\frac{7\pi}{4}$	2π
x	0	−2	0	2
y	2	0	−2	0

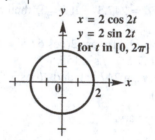

$x = 2\cos 2t$
$y = 2\sin 2t$
for t in $[0, 2\pi]$

$x = 2\cos 2t \Rightarrow \cos 2t = \dfrac{x}{2}$,

$y = 2\sin 2t \Rightarrow \sin 2t = \dfrac{y}{2}$, and

$\cos^2(2t) + \sin^2(2t) = 1$, so we have

$\left(\dfrac{x}{2}\right)^2 + \left(\dfrac{y}{2}\right)^2 = 1 \Rightarrow \dfrac{x^2}{4} + \dfrac{y^2}{4} = 1 \Rightarrow$

$x^2 + y^2 = 4$, where x is in $[-1, 1]$.

16. $z = -1 + i$

$z^2 - 1 = (-1 + i)^2 - 1 = 1 - i - i + i^2 - 1$
$= -2i - 1 = -1 - 2i$

Because

$r = \sqrt{(-1)^2 + (-2)^2} = \sqrt{1 + 4} = \sqrt{5} > 2$, z is not in the Julia set.

Appendices

Appendix A   Equations and Inequalities

1. An <u>equation</u> is a statement that two expressions are equal.

3. A linear equation is a <u>first-degree equation</u> because the greatest degree of the variable is 1.

5. A <u>contradiction</u> is an equation that has no solution.

7. $5x + 4 = 3x - 4$
$2x + 4 = -4$
$2x = -8 \Rightarrow x = -4$
Solution set: $\{-4\}$

9. $6(3x - 1) = 8 - (10x - 14)$
$18x - 6 = 8 - 10x + 14$
$18x - 6 = 22 - 10x$
$28x - 6 = 22$
$28x = 28 \Rightarrow x = 1$
Solution set: $\{1\}$

11. $\dfrac{5}{6}x - 2x + \dfrac{4}{3} = \dfrac{5}{3}$
$6 \cdot \left[\dfrac{5}{6}x - 2x + \dfrac{4}{3} \right] = 6 \cdot \dfrac{5}{3}$
$5x - 12x + 8 = 10$
$-7x + 8 = 10$
$-7x = 2 \Rightarrow x = -\dfrac{2}{7}$
Solution set: $\left\{ -\dfrac{2}{7} \right\}$

13. $3x + 5 - 5(x + 1) = 6x + 7$
$3x + 5 - 5x - 5 = 6x + 7$
$-2x = 6x + 7$
$-8x = 7 \Rightarrow x = \dfrac{7}{-8} = -\dfrac{7}{8}$
Solution set: $\left\{ -\dfrac{7}{8} \right\}$

15. $2\left[x - (4 + 2x) + 3 \right] = 2x + 2$
$2(x - 4 - 2x + 3) = 2x + 2$
$2(-x - 1) = 2x + 2$
$-2x - 2 = 2x + 2$
$-2 = 4x + 2$
$-4 = 4x$
$-1 = x$
Solution set: $\{-1\}$

17. $0.2x - 0.5 = 0.1x + 7$
$10(0.2x - 0.5) = 10(0.1x + 7)$
$2x - 5 = x + 70$
$x - 5 = 70$
$x = 75$
Solution set: $\{75\}$

19. $-4(2x - 6) + 8x = 5x + 24 + x$
$-8x + 24 + 8x = 6x + 24$
$24 = 6x + 24$
$0 = 6x \Rightarrow 0 = x$
Solution set: $\{0\}$

21. $4(2x + 7) = 2x + 22 + 3(2x + 2)$
$8x + 28 = 2x + 22 + 6x + 6$
$8x + 28 = 8x + 28$
$28 = 28 \Rightarrow 0 = 0$
identity; $\{\text{all real numbers}\}$

23. $2(x - 8) = 3x - 16$
$2x - 16 = 3x - 16$
$-16 = x - 16 \Rightarrow 0 = x$
conditional equation; $\{0\}$

25. $4(x + 7) = 2(x + 12) + 2(x + 1)$
$4x + 28 = 2x + 24 + 2x + 2$
$4x + 28 = 4x + 26$
$28 = 26$
contradiction; $\varnothing$

27. D is the only one set up for direct use of the zero-factor property.
$(3x - 1)(x - 7) = 0$
$3x - 1 = 0$ or $x - 7 = 0$
$x = \dfrac{1}{3}$ or $\quad x = 7$
Solution set: $\left\{ \dfrac{1}{3}, 7 \right\}$

29. All of the equations can be solved using the quadratic formula.

31. $x^2 - 5x + 6 = 0$
$(x - 2)(x - 3) = 0$
$x - 2 = 0 \Rightarrow x = 2$ or $x - 3 = 0 \Rightarrow x = 3$
Solution set: $\{2, 3\}$

33. $5x^2 - 3x - 2 = 0$
$(5x + 2)(x - 1) = 0$
$5x + 2 = 0 \Rightarrow x = -\frac{2}{5}$ or $x - 1 = 0 \Rightarrow x = 1$
Solution set: $\left\{-\frac{2}{5}, 1\right\}$

35. $-4x^2 + x = -3$
$0 = 4x^2 - x - 3$
$0 = (4x + 3)(x - 1)$
$4x + 3 = 0 \Rightarrow x = -\frac{3}{4}$ or $x - 1 = 0 \Rightarrow x = 1$
Solution set: $\left\{-\frac{3}{4}, 1\right\}$

37. $x^2 - 100 = 0$
$(x + 10)(x - 10) = 0$
$x + 10 = 0 \Rightarrow x = -10$ or $x - 10 = 0 \Rightarrow x = 10$
Solution set: $\{-10, 10\}$

39. $4x^2 - 4x + 1 = 0$
$(2x - 1)^2 = 0$
$2x - 1 = 0 \Rightarrow 2x = 1 \Rightarrow x = \frac{1}{2}$
Solution set: $\left\{\frac{1}{2}\right\}$

41. $25x^2 + 30x + 9 = 0$
$(5x + 3)^2 = 0$
$5x + 3 = 0 \Rightarrow 5x = -3 \Rightarrow x = -\frac{3}{5}$
Solution set: $\left\{-\frac{3}{5}\right\}$

43. $x^2 = 16$
$x = \pm\sqrt{16} = \pm 4$
Solution set: $\{\pm 4\}$

45. $x^2 - 27 = 0$
$x^2 = 27$
$x = \pm\sqrt{27} = \pm 3\sqrt{3}$
Solution set: $\left\{\pm 3\sqrt{3}\right\}$

47. $(3x - 1)^2 = 12$
$3x - 1 = \pm\sqrt{12}$
$3x = 1 \pm 2\sqrt{3} \Rightarrow x = \frac{1 \pm 2\sqrt{3}}{3}$
Solution set: $\left\{\frac{1 \pm 2\sqrt{3}}{3}\right\}$

49. $x^2 - 4x + 3 = 0$
$a = 1, b = -4, c = 3$
$x = \dfrac{-b \pm \sqrt{b^2 - 4ac}}{2a}$
$= \dfrac{-(-4) \pm \sqrt{(-4)^2 - 4(1)(3)}}{2(1)} = \dfrac{4 \pm \sqrt{16 - 12}}{2}$
$= \dfrac{4 \pm \sqrt{4}}{2} = \dfrac{4 \pm 2}{2} = \dfrac{6}{2} = 3$ or $\dfrac{2}{2} = 1$
Solution set: $\{1, 3\}$

51. $2x^2 - x - 28 = 0$
$a = 2, b = -1, c = -28$
$x = \dfrac{-b \pm \sqrt{b^2 - 4ac}}{2a}$
$= \dfrac{-(-1) \pm \sqrt{(-1)^2 - 4(2)(-28)}}{2(2)}$
$= \dfrac{1 \pm \sqrt{1 + 224}}{4} = \dfrac{1 \pm \sqrt{225}}{4} = \dfrac{1 \pm 15}{4}$
$= \dfrac{16}{4} = 4$ or $-\dfrac{14}{4} = -\dfrac{7}{2}$
Solution set: $\left\{-\frac{7}{2}, 4\right\}$

53. $x^2 - 2x - 2 = 0$
$a = 1, b = -2, c = -2$
$x = \dfrac{-b \pm \sqrt{b^2 - 4ac}}{2a}$
$= \dfrac{-(-2) \pm \sqrt{(-2)^2 - 4(1)(-2)}}{2(1)} = \dfrac{2 \pm \sqrt{4 + 8}}{2}$
$= \dfrac{2 \pm \sqrt{12}}{2} = \dfrac{2 \pm 2\sqrt{3}}{2} = 1 \pm \sqrt{3}$
Solution set: $\left\{1 \pm \sqrt{3}\right\}$

55. $x^2 - 6x = -7$
$x^2 - 6x + 7 = 0$
Let $a = 1, b = -6,$ and $c = 7$.
$x = \dfrac{-b \pm \sqrt{b^2 - 4ac}}{2a}$
$= \dfrac{-(-6) \pm \sqrt{(-6)^2 - 4(1)(7)}}{2(1)} = \dfrac{6 \pm \sqrt{36 - 28}}{2}$
$= \dfrac{6 \pm \sqrt{8}}{2} = \dfrac{6 \pm 2\sqrt{2}}{2} = 3 \pm \sqrt{2}$
Solution set: $\left\{3 \pm \sqrt{2}\right\}$

57. $x^2 - x - 1 = 0$

Let $a = 1, b = -1,$ and $c = -1.$

$$x = \frac{-b \pm \sqrt{b^2 - 4ac}}{2a}$$

$$= \frac{-(-1) \pm \sqrt{(-1)^2 - 4(1)(-1)}}{2(1)}$$

$$= \frac{1 \pm \sqrt{1 + 4}}{2} = \frac{1 \pm \sqrt{5}}{2}$$

Solution set: $\left\{ \frac{1 \pm \sqrt{5}}{2} \right\}$

59. $-2x^2 + 4x + 3 = 0$

$a = -2, b = 4, c = 3$

$$x = \frac{-b \pm \sqrt{b^2 - 4ac}}{2a}$$

$$= \frac{-4 \pm \sqrt{4^2 - 4(-2)(3)}}{2(-2)} = \frac{-4 \pm \sqrt{16 + 24}}{-4}$$

$$= \frac{-4 \pm \sqrt{40}}{-4} = \frac{-4 \pm 2\sqrt{10}}{-4} = \frac{2 \pm \sqrt{10}}{2}$$

Solution set: $\left\{ \frac{2 \pm \sqrt{10}}{2} \right\}$

61. F. The inequality $x < -6$ includes all real numbers less than -6 not including -6. The correct interval notation is $(-\infty, -6)$.

63. A. The inequality $-2 < x \le 6$ includes all real numbers from -2 to 6, not including -2, but including 6. The correct interval notation is $(-2, 6]$.

65. I. The inequality $x \ge -6$ includes all real numbers greater than or equal to -6, so it includes -6. The correct interval notation is $[-6, \infty)$.

67. B. The interval shown on the number line includes all real numbers between -2 and 6, including -2, but not including 6. The correct interval notation is $[-2, 6)$.

69. E. The interval shown on the number line includes all real numbers less than -3, not including -3, and greater than 3, not including 3. The correct interval notation is $(-\infty, -3) \cup (3, \infty)$,

71. Answers will vary. Sample answer: A square bracket is used to show that a number is part of the solution set, while a parenthesis is used to indicate that a number is not part of the solution set.

73. $-2x + 8 \le 16 \Rightarrow -2x + 8 - 8 \le 16 - 8 \Rightarrow$

$$-2x \le 8 \Rightarrow \frac{-2x}{-2} \ge \frac{8}{-2} \Rightarrow x \ge -4$$

Solution set: $[-4, \infty)$

75. $-2x - 2 \le 1 + x$

$-2x - 2 + 2 \le 1 + x + 2$

$-2x \le x + 3 \Rightarrow -2x - x \le 3 \Rightarrow$

$$-3x \le 3 \Rightarrow \frac{-3x}{-3} \ge \frac{3}{-3} \Rightarrow x \ge -1$$

Solution set: $[-1, \infty)$

77. $3(x + 5) + 1 \ge 5 + 3x$

$3x + 15 + 1 \ge 5 + 3x \Rightarrow 16 \ge 5$

The inequality is true when x is any real number.

Solution set: $(-\infty, \infty)$

79. $8x - 3x + 2 < 2(x + 7)$

$5x + 2 < 2x + 14$

$5x + 2 - 2x < 2x + 14 - 2x$

$3x + 2 < 14$

$3x + 2 - 2 < 14 - 2$

$3x < 12$

$$\frac{3x}{3} < \frac{12}{3} \Rightarrow x < 4$$

Solution set: $(-\infty, 4)$

81. $\frac{4x + 7}{-3} \le 2x + 5$

$$(-3)\left(\frac{4x + 7}{-3} \right) \ge (-3)(2x + 5)$$

$4x + 7 \ge -6x - 15$

$4x + 7 + 6x \ge -6x - 15 + 6x$

$10x + 7 \ge -15$

$10x + 7 - 7 \ge -15 - 7$

$10x \ge -22$

$$\frac{10x}{10} \ge \frac{-22}{10} \Rightarrow x \ge -\frac{11}{5}$$

Solution set: $\left[-\frac{11}{5}, \infty \right)$

83. $-5 < 5 + 2x < 11$

$-5 - 5 < 5 + 2x - 5 < 11 - 5$

$-10 < 2x < 6$

$$\frac{-10}{2} < \frac{2x}{2} < \frac{6}{2}$$

$-5 < x < 3$

Solution set: $(-5, 3)$

85.
$$10 \le 2x + 4 \le 16$$
$$10 - 4 \le 2x + 4 - 4 \le 16 - 4$$
$$6 \le 2x \le 12$$
$$\frac{6}{2} \le \frac{2x}{2} \le \frac{12}{2}$$
$$3 \le x \le 6$$
Solution set: [3, 6]

87.
$$-11 > -3x + 1 > -17$$
$$-11 - 1 > -3x + 1 - 1 > -17 - 1$$
$$-12 > -3x > -18$$
$$\frac{-12}{-3} < \frac{-3x}{-3} < \frac{-18}{-3}$$
$$4 < x < 6$$
Solution set: (4, 6)

89.
$$-4 \le \frac{x+1}{2} \le 5$$
$$2(-4) \le 2\left(\frac{x+1}{2}\right) \le 2(5)$$
$$-8 \le x + 1 \le 10$$
$$-8 - 1 \le x + 1 - 1 \le 10 - 1$$
$$-9 \le x \le 9$$
Solution set: [−9, 9]

Appendix B Graphs of Equations

1. The point $(-1, 3)$ lies in quadrant <u>II</u> in the rectangular coordinate system.

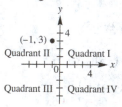

3. Any point that lies on the x-axis has y-coordinate equal to <u>0</u>.

5. The x-intercept of the graph of $2x + 5y = 10$ is <u>(5, 0)</u>. Find the x-intercept by letting $y = 0$ and solving for x.
$$2x + 5(0) = 10 \Rightarrow 2x = 10 \Rightarrow x = 5$$

7.–13.

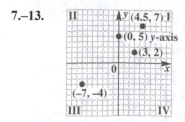

15. (a)

x	y

x	y
0	−2
4	0
2	−1

y-intercept:
$x = 0 \Rightarrow$
$y = \frac{1}{2}(0) - 2 = -2$

x-intercept:
$y = 0 \Rightarrow$
$0 = \frac{1}{2}x - 2 \Rightarrow$
$2 = \frac{1}{2}x \Rightarrow 4 = x$

additional point

(b)

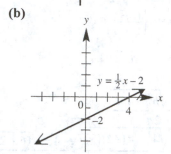

17. (a)

x	y
0	$\frac{5}{3}$
$\frac{5}{2}$	0
4	−1

y-intercept:
$x = 0 \Rightarrow$
$2(0) + 3y = 5 \Rightarrow$
$3y = 5 \Rightarrow y = \frac{5}{3}$

x-intercept:
$y = 0 \Rightarrow$
$2x + 3(0) = 5 \Rightarrow$
$2x = 5 \Rightarrow x = \frac{5}{2}$

additional point

(b)

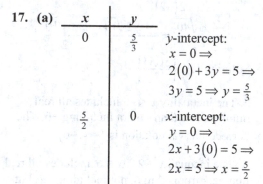

19. (a)

x	y
0	0
1	1
−2	4

x- and y-intercept:
$0 = 0^2$

additional point

additional point

(b)

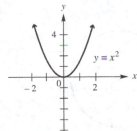

21. (a)

x	y	
3	0	x-intercept:
		$y = 0 \Rightarrow$
		$0 = \sqrt{x-3} \Rightarrow$
		$0 = x - 3 \Rightarrow 3 = x$
4	1	additional point
7	2	additional point

no y-intercept:

$$x = 0 \Rightarrow y = \sqrt{0-3} \Rightarrow y = \sqrt{-3}$$

(b)

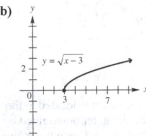

23. (a)

x	y	
0	2	y-intercept:
		$x = 0 \Rightarrow$
		$y = |0-2| \Rightarrow$
		$y = |-2| \Rightarrow y = 2$
2	0	x-intercept:
		$y = 0 \Rightarrow$
		$0 = |x-2| \Rightarrow$
		$0 = x-2 \Rightarrow 2 = x$
-2	4	additional point
4	2	additional point

(b)

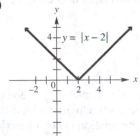

25. (a)

x	y	
0	0	x- and y-intercept:
		$0 = 0^3$
-1	-1	additional point
2	8	additional point

(b)

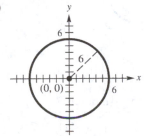

27. The circle with equation $x^2 + y^2 = 49$ has center with coordinates (0, 0) and radius equal to 7.

29. The graph of $(x-4)^2 + (y+7)^2 = 9$ has center with coordinates (4, –7).

31. This circle has center (3, 2) and radius 5. This is graph B.

33. This circle has center (–3, 2) and radius 5. This is graph D.

35. (a) Center (0, 0), radius 6

$$\sqrt{(x-0)^2 + (y-0)^2} = 6$$
$$(x-0)^2 + (y-0)^2 = 6^2 \Rightarrow x^2 + y^2 = 36$$

(b)

y

6

6

(0, 0) x

6

$x^2 + y^2 = 36$

37. (a) Center (2, 0), radius 6

$$\sqrt{(x-2)^2 + (y-0)^2} = 6$$
$$(x-2)^2 + (y-0)^2 = 6^2$$
$$(x-2)^2 + y^2 = 36$$

(b)

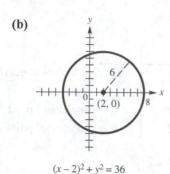

$(x-2)^2 + y^2 = 36$

39. (a) Center $(0, 4)$, radius 4

$$\sqrt{\left(x-0\right)^2 + \left(y-4\right)^2} = 4$$
$$x^2 + \left(y-4\right)^2 = 16$$

(b)

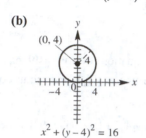

$x^2 + (y-4)^2 = 16$

41. (a) Center $(-2, 5)$, radius 4

$$\sqrt{\left[x-(-2)\right]^2 + \left(y-5\right)^2} = 4$$
$$\left[x-(-2)\right]^2 + \left(y-5\right)^2 = 4^2$$
$$\left(x+2\right)^2 + \left(y-5\right)^2 = 16$$

(b)

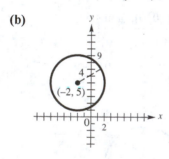

$(x+2)^2 + (y-5)^2 = 16$

43. (a) Center $(5, -4)$, radius 7

$$\sqrt{\left(x-5\right)^2 + \left[y-(-4)\right]^2} = 7$$
$$\left(x-5\right)^2 + \left[y-(-4)\right]^2 = 7^2$$
$$\left(x-5\right)^2 + \left(y+4\right)^2 = 49$$

(b)

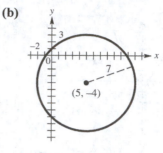

$(x-5)^2 + (y+4)^2 = 49$

45. (a) Center $\left(\sqrt{2}, \sqrt{2}\right)$, radius $\sqrt{2}$

$$\sqrt{\left(x-\sqrt{2}\right)^2 + \left(y-\sqrt{2}\right)^2} = \sqrt{2}$$
$$\left(x-\sqrt{2}\right)^2 + \left(y-\sqrt{2}\right)^2 = 2$$

(b)

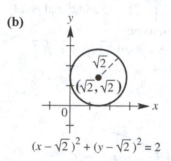

$(x-\sqrt{2})^2 + (y-\sqrt{2})^2 = 2$

47. The center of the circle is located at the midpoint of the diameter determined by the points $(1, 1)$ and $(5, 1)$. Using the midpoint formula, we have $C = \left(\dfrac{1+5}{2}, \dfrac{1+1}{2}\right) = (3,1)$.

The radius is one-half the length of the diameter: $r = \dfrac{1}{2}\sqrt{\left(5-1\right)^2 + \left(1-1\right)^2} = 2$

The equation of the circle is

$$\left(x-3\right)^2 + \left(y-1\right)^2 = 4$$

Appendix C   Functions

1. The relation is a function because for each different x-value there is exactly one y-value. This correspondence can be shown as follows.

$\{5, 3, 4, 7\}$ x-values

$\downarrow \quad \downarrow \quad \downarrow \quad \downarrow$

$\{1, 2, 9, 8\}$ y-values

3. Two ordered pairs, namely $(2, 4)$ and $(2, 6)$, have the same x-value paired with different y-values, so the relation is not a function.

5. The relation is a function because for each different x-value there is exactly one y-value. This correspondence can be shown as follows.

{−3, 4, −2} x-values

{1, 7} y-values

7. The relation is a function because for each different x-value there is exactly one y-value. This correspondence can be shown as follows.

{3, 7, 10} x-values

{−4} y-values

9. Two sets of ordered pairs, namely (1, 1) and (1, −1) as well as (2, 4) and (2, −4), have the same x-value paired with different y-values, so the relation is not a function.
domain: {0, 1, 2}; range: {−4, −1, 0, 1, 4}

11. The relation is a function because for each different x-value there is exactly one y-value.
domain: {2, 3, 5, 11, 17}; range: {1, 7, 20}

13. The relation is a function because for each different x-value there is exactly one y-value. This correspondence can be shown as follows.

{0, −1, −2} x-values

{0, 1, 2} y-values
Domain: {0, −1, −2}; range: {0, 1, 2}

15. This graph represents a function. If you pass a vertical line through the graph, one x-value corresponds to only one y-value.
domain: $(-\infty, \infty)$; range: $(-\infty, \infty)$

17. This graph does not represent a function. If you pass a vertical line through the graph, there are places where one value of x corresponds to two values of y.
domain: $[3, \infty)$; range: $(-\infty, \infty)$

19. This graph represents a function. If you pass a vertical line through the graph, one x-value corresponds to only one y-value.
domain: $(-\infty, \infty)$; range: $(-\infty, \infty)$

21. $y = x^2$ represents a function because y is always found by squaring x. Thus, each value of x corresponds to just one value of y. x can be any real number. Because the square of any real number is not negative, the range would be zero or greater.

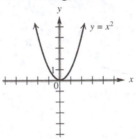

domain: $(-\infty, \infty)$; range: $[0, \infty)$

23. The ordered pairs (1, 1) and (1, −1) both satisfy $x = y^6$. This equation does not represent a function. Because x is equal to the sixth power of y, the values of x are nonnegative. Any real number can be raised to the sixth power, so the range of the relation is all real numbers.

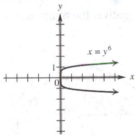

domain: $[0, \infty)$ range: $(-\infty, \infty)$

25. $y = 2x - 5$ represents a function because y is found by multiplying x by 2 and subtracting 5. Each value of x corresponds to just one value of y. x can be any real number, so the domain is all real numbers. Because y is twice x, less 5, y also may be any real number, and so the range is also all real numbers.

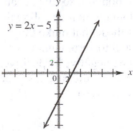

domain: $(-\infty, \infty)$; range: $(-\infty, \infty)$

27. For any choice of x in the domain of $y = \sqrt{x}$, there is exactly one corresponding value of y, so this equation defines a function. Because the quantity under the square root cannot be negative, we have $x \geq 0$. Because the radical is nonnegative, the range is also zero or greater.

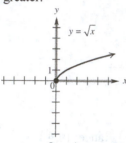

domain: $[0, \infty)$; range: $[0, \infty)$

29. For any choice of x in the domain of $y = \sqrt{4x + 1}$ there is exactly one corresponding value of y, so this equation defines a function. Because the quantity under the square root cannot be negative, we have $4x + 1 \geq 0 \Rightarrow 4x \geq -1 \Rightarrow x \geq -\frac{1}{4}$. Because the radical is nonnegative, the range is also zero or greater.

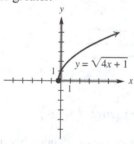

domain: $\left[-\frac{1}{4}, \infty\right)$; range: $[0, \infty)$

31. Given any value in the domain of $y = \frac{2}{x-3}$, we find y by subtracting 3, then dividing into 2. This process produces one value of y for each value of x in the domain, so this equation is a function. The domain includes all real numbers except those that make the denominator equal to zero, namely $x = 3$. Values of y can be negative or positive, but never zero. Therefore, the range will be all real numbers except zero.

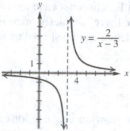

domain: $(-\infty, 3) \cup (3, \infty)$;

range: $(-\infty, 0) \cup (0, \infty)$

33. B. The notation $f(3)$ means the value of the dependent variable when the independent variable is 3.

35. $f(x) = -3x + 4$
$f(0) = -3 \cdot 0 + 4 = 0 + 4 = 4$

37. $g(x) = -x^2 + 4x + 1$
$g(-2) = -(-2)^2 + 4(-2) + 1$
$ = -4 + (-8) + 1 = -11$

39. $f(x) = -3x + 4$
$f\left(\frac{1}{3}\right) = -3\left(\frac{1}{3}\right) + 4 = -1 + 4 = 3$

41. $g(x) = -x^2 + 4x + 1$
$g\left(\frac{1}{2}\right) = -\left(\frac{1}{2}\right)^2 + 4\left(\frac{1}{2}\right) + 1 = -\frac{1}{4} + 2 + 1 = \frac{11}{4}$

43. $f(x) = -3x + 4$
$f(p) = -3p + 4$

45. $f(x) = -3x + 4$
$f(-x) = -3(-x) + 4 = 3x + 4$

47. $f(x) = -3x + 4$
$f(x + 2) = -3(x + 2) + 4$
$ = -3x - 6 + 4 = -3x - 2$

49. $f(x) = -3x + 4$
$f(2m - 3) = -3(2m - 3) + 4$
$ = -6m + 9 + 4 = -6m + 13$

51. **(a)** $f(2) = 2$ **(b)** $f(-1) = 3$

53. **(a)** $f(2) = 15$ **(b)** $f(-1) = 10$

55. **(a)** $f(2) = 3$ **(b)** $f(-1) = -3$

57. **(a)** $f(-2) = 0$ **(b)** $f(0) = 4$

    (c) $f(1) = 2$ **(d)** $f(4) = 4$

59. (a) $f(-2) = -3$ (b) $f(0) = -2$

(c) $f(1) = 0$ (d) $f(4) = 2$

61. (a) $(-2, 0)$ (b) $(-\infty, -2)$

(c) $(0, \infty)$

63. (a) $(-\infty, -2); (2, \infty)$

(b) $(-2, -2)$ (c) none

65. (a) $(-1, 0); (1, \infty)$

(b) $(-\infty, -1); (0, 1)$

(c) none

Appendix D   Graphing Techniques

1. To graph the function $f(x) = x^2 - 3$, shift the graph of $y = x^2$ down $\underline{3}$ units.

3. The graph of $f(x) = (x + 4)^2$ is obtained by shifting the graph of $y = x^2$ to the <u>left</u> 4 units.

5. The graph of $f(x) = -\sqrt{x}$ is a reflection of the graph of $f(x) = \sqrt{x}$ across the <u>x</u>-axis.

7. To obtain the graph of $f(x) = (x + 2)^3 - 3$, shift the graph of $y = x^3$ to the left $\underline{2}$ units and down $\underline{3}$ units.

9. (a) B; $y = (x - 7)^2$ is a shift of $y = x^2$, 7 units to the right.

(b) D; $y = x^2 - 7$ is a shift of $y = x^2$, 7 units down.

(c) E; $y = 7x^2$ is a vertical stretch of $y = x^2$, by a factor of 7.

(d) A; $y = (x + 7)^2$ is a shift of $y = x^2$, 7 units to the left.

(e) C; $y = x^2 + 7$ is a shift of $y = x^2$, 7 units up.

11. (a) B; $y = x^2 + 2$ is a shift of $y = x^2$, 2 units upward.

(b) A; $y = x^2 - 2$ is a shift of $y = x^2$, 2 units downward.

(c) G; $y = (x + 2)^2$ is a shift of $y = x^2$, 2 units to the left.

(d) C; $y = (x - 2)^2$ is a shift of $y = x^2$, 2 units to the right.

(e) F; $y = 2x^2$ is a vertical stretch of $y = x^2$, by a factor of 2.

(f) D; $y = -x^2$ is a reflection of $y = x^2$, across the x-axis.

(g) H; $y = (x - 2)^2 + 1$ is a shift of $y = x^2$, 2 units to the right and 1 unit upward.

(h) E; $y = (x + 2)^2 + 1$ is a shift of $y = x^2$, 2 units to the left and 1 unit upward.

(i) I; $y = (x + 2)^2 - 1$ is a shift of $y = x^2$, 2 units to the left and 1 unit down.

13. $f(x) = 3|x|$

x	$|x|$	$f(x) = 3|x|$
-2	2	6
-1	1	3
0	0	0

x	$|x|$	$f(x) = 3|x|$
1	1	3
2	2	6

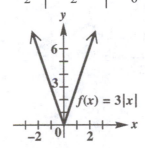

15. $f(x) = \frac{2}{3}|x|$

x	$h(x) = |x|$	$f(x) = \frac{2}{3}|x|$
-3	3	2
-2	2	$\frac{4}{3}$
-1	1	$\frac{2}{3}$
0	0	0
1	1	$\frac{2}{3}$
2	2	$\frac{4}{3}$
3	3	2

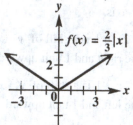

17. $f(x) = 2x^2$

x	$h(x) = x^2$	$f(x) = 2x^2$
-2	4	8
-1	1	2
0	0	0
1	1	2
2	4	8

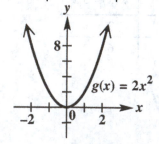

19. $f(x) = \frac{1}{2}x^2$

x	$h(x) = x^2$	$f(x) = \frac{1}{2}x^2$
-2	4	2
-1	1	$\frac{1}{2}$
0	0	0
1	1	$\frac{1}{2}$
2	4	2

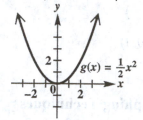

21. $f(x) = -\frac{1}{2}x^2$

x	$h(x) = x^2$	$f(x) = -\frac{1}{2}x^2$
-3	9	$-\frac{9}{2}$
-2	4	-2
-1	1	$-\frac{1}{2}$
0	0	0
1	1	$-\frac{1}{2}$
2	4	-2
3	9	$-\frac{9}{2}$

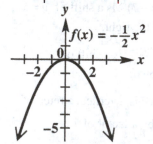

23. $f(x) = -3|x|$

x	$h(x) = |x|$	$f(x) = -3|x|$
−2	2	−6
−1	1	−3
0	0	0
1	1	−3
2	2	−6

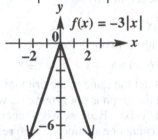

25. $h(x) = \left|-\frac{1}{2}x\right|$

x	$f(x) = |x|$	$h(x) = \left|-\frac{1}{2}x\right|$ $= \left|-\frac{1}{2}\right||x| = \frac{1}{2}|x|$
−4	4	2
−3	3	$\frac{3}{2}$
−2	2	1
−1	1	$\frac{1}{2}$
0	0	0
1	1	$\frac{1}{2}$
2	2	1
3	3	$\frac{3}{2}$
4	4	2

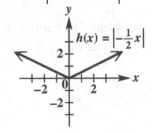

27. $h(x) = \sqrt{4x}$

x	$f(x) = \sqrt{x}$	$h(x) = \sqrt{4x} = 2\sqrt{x}$
0	0	0
1	1	2
2	$\sqrt{2}$	$2\sqrt{2}$
3	$\sqrt{3}$	$2\sqrt{3}$
4	2	4

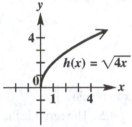

29. $f(x) = -\sqrt{-x}$

x	$h(x) = \sqrt{-x}$	$f(x) = -\sqrt{-x}$
−4	2	−2
−3	$\sqrt{3}$	$-\sqrt{3}$
−2	$\sqrt{2}$	$-\sqrt{2}$
−1	1	−1
0	0	0

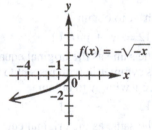

31. (a) The point that is symmetric to (5, −3) with respect to the x-axis is (5, 3).

(b) The point that is symmetric to (5, −3) with respect to the y-axis is (−5, −3).

(c) The point that is symmetric to (5, −3) with respect to the origin is (−5, 3).

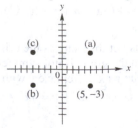

33. (a) The point that is symmetric to $(-4, -2)$ with respect to the x-axis is $(-4, 2)$.

(b) The point that is symmetric to $(-4, -2)$ with respect to the y-axis is $(4, -2)$.

(c) The point that is symmetric to $(-4, -2)$ with respect to the origin is $(4, 2)$.

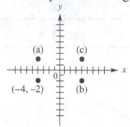

35. $y = x^2 + 5$

Replace x with $-x$ to obtain

$y = (-x)^2 + 5 = x^2 + 5$. The result is the same as the original equation, so the graph is symmetric with respect to the y-axis. Because y is a function of x, the graph cannot be symmetric with respect to the x-axis. Replace x with $-x$ and y with $-y$ to obtain

$-y = (-x)^2 + 2 \Rightarrow -y = x^2 + 2 \Rightarrow y = -x^2 - 2$.

The result is not the same as the original equation, so the graph is not symmetric with respect to the origin. Therefore, the graph is symmetric with respect to the y-axis only.

37. $x^2 + y^2 = 12$

Replace x with $-x$ to obtain

$(-x)^2 + y^2 = 12 \Rightarrow x^2 + y^2 = 12$.

The result is the same as the original equation, so the graph is symmetric with respect to the y-axis. Replace y with $-y$ to obtain

$x^2 + (-y)^2 = 12 \Rightarrow x^2 + y^2 = 12$

The result is the same as the original equation, so the graph is symmetric with respect to the x-axis. Because the graph is symmetric with respect to the x-axis and y-axis, it is also symmetric with respect to the origin.

39. $y = -4x^3 + x$

Replace x with $-x$ to obtain

$y = -4(-x)^3 + (-x) \Rightarrow y = -4(-x^3) - x \Rightarrow$

$y = 4x^3 - x$.

The result is not the same as the original equation, so the graph is not symmetric with respect to the y-axis. Replace y with $-y$ to obtain $-y = -4x^3 + x \Rightarrow y = 4x^3 - x$.

The result is not the same as the original equation, so the graph is not symmetric with respect to the x-axis. Replace x with $-x$ and y with $-y$ to obtain

$-y = -4(-x)^3 + (-x) \Rightarrow -y = -4(-x^3) - x \Rightarrow$

$-y = 4x^3 - x \Rightarrow y = -4x^3 + x$.

The result is the same as the original equation, so the graph is symmetric with respect to the origin. Therefore, the graph is symmetric with respect to the origin only.

41. $y = x^2 - x + 8$

Replace x with $-x$ to obtain

$y = (-x)^2 - (-x) + 8 \Rightarrow y = x^2 + x + 8$.

The result is not the same as the original equation, so the graph is not symmetric with respect to the y-axis. Because y is a function of x, the graph cannot be symmetric with respect to the x-axis. Replace x with $-x$ and y with $-y$ to obtain $-y = (-x)^2 - (-x) + 8 \Rightarrow$

$-y = x^2 + x + 8 \Rightarrow y = -x^2 - x - 8$.

The result is not the same as the original equation, so the graph is not symmetric with respect to the origin. Therefore, the graph has none of the listed symmetries.

43. $f(x) = x^2 - 1$

This graph may be obtained by translating the graph of $y = x^2$ 1 unit downward.

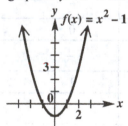

45. $f(x) = x^2 + 2$

This graph may be obtained by translating the graph of $y = x^2$ 2 units upward.

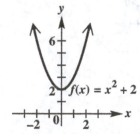

47. $g(x) = (x-4)^2$

This graph may be obtained by translating the graph of $y = x^2$ 4 units to the right.

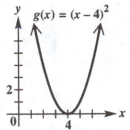

49. $g(x) = (x+2)^2$

This graph may be obtained by translating the graph of $y = x^2$ 2 units to the left.

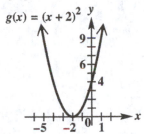

51. $g(x) = |x| - 1$

The graph is obtained by translating the graph of $y = |x|$ 1 unit downward.

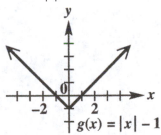

53. $h(x) = -(x+1)^3$

This graph may be obtained by translating the graph of $y = x^3$ 1 unit to the left. It is then reflected across the x-axis.

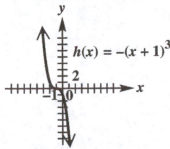

55. $h(x) = 2x^2 - 1$

This graph may be obtained by translating the graph of $y = x^2$ 1 unit down. It is then stretched vertically by a factor of 2.

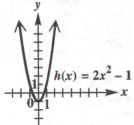

57. $f(x) = 2(x-2)^2 - 4$

This graph may be obtained by translating the graph of $y = x^2$, 2 units to the right and 4 units down. It is then stretched vertically by a factor of 2.

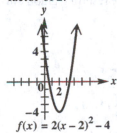

59. $f(x) = \sqrt{x+2}$

This graph may be obtained by translating the graph of $y = \sqrt{x}$ two units to the left.

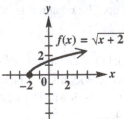

61. $f(x) = -\sqrt{x}$

This graph may be obtained by reflecting the graph of $y = \sqrt{x}$ across the x-axis.

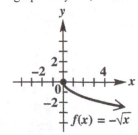

63. $f(x) = 2\sqrt{x} + 1$

This graph may be obtained by stretching the graph of $y = \sqrt{x}$ vertically by a factor of two and then translating the resulting graph one unit up.

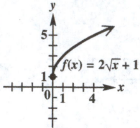

65. $g(x) = \frac{1}{2}x^3 - 4$

This graph may be obtained by stretching the graph of $g(x) = x^3$ vertically by a factor of $\frac{1}{2}$, then shifting the resulting graph down four units.

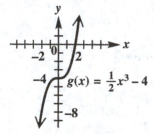

67. It is the graph of $f(x) = |x|$ translated 1 unit to the left, reflected across the x-axis, and translated 3 units up. The equation is $y = -|x + 1| + 3$.

69. It is the graph of $g(x) = \sqrt{x}$ translated 4 units to the left, stretched vertically by a factor of 2, and translated four units down. The equation is $y = 2\sqrt{x + 4} - 4$.